PRINCIPLES
OF
MINERALOGY

PRINCIPLES
OF
MINERALOGY

SECOND EDITION

WILLIAM H. BLACKBURN

UNIVERSITY OF WINDSOR

WILLIAM H. DENNEN

PROFESSOR EMERITUS,
UNIVERSITY OF KENTUCKY

WCB **Wm. C. Brown Publishers**

Dubuque, Iowa · Melbourne, Australia · Oxford, England

Book Team

Executive Editor *Jeffrey L. Hahn*
Developmental Editor *Robert Fenchel*
Production Editor *Sue Dillon*
Designer *Elise A. Lansdon/Anna Manhart*
Art Editor *Jodi Wagner*
Photo Editor *Carrie Burger*
Art Processor *Brenda A. Ernzen*
Visuals/Design Developmental Consultant *Donna Slade*

Wm. C. Brown Publishers
A Division of Wm. C. Brown Communications, Inc.

Vice President and General Manager *Beverly Kolz*
Vice President, Publisher *Earl McPeek*
Vice President, Director of Sales and Marketing *Virginia S. Moffat*
Marketing Manager *Christopher T. Johnson*
Advertising Manager *Janelle Keeffer*
Director of Production *Colleen A. Yonda*
Publishing Services Manager *Karen J. Slaght*
Permissions/Records Manager *Connie Allendorf*

Wm. C. Brown Communications, Inc.

President and Chief Executive Officer *G. Franklin Lewis*
Corporate Senior Vice President, President of WCB Manufacturing *Roger Meyer*
Corporate Senior Vice President and Chief Financial Officer *Robert Chesterman*

Cover image: Gold—Paymaster Mine, South Porcupine, Ontario.
Courtesy of Royal Ontario Museum. © Calvin Nicholls/Nicholls Design Inc.

Copyedited by Martha Morss

The credits section for this book begins on page 399 and is considered an extension of the copyright page.

A Times Mirror Company

Library of Congress Catalog Card Number: 92-75621

ISBN 0-697-15078-X

Printed in the United States of America by Wm. C. Brown Communications, Inc., 2460 Kerper Boulevard, Dubuque, IA 52001

10 9 8 7 6 5 4 3

To Charlotte and Ramona

Contents

Preface

This book is intended for use in an introductory college course in mineralogy. It treats the subject as a study of natural phenomena that are common to all minerals. Emphasizing principles, mineralogy is presented in terms of the fundamental geometrical, chemical, and physical relationships of matter rather than simply as a course in mineral recognition. The book has a threefold goal: first, to provide students in geology and allied fields with a mineralogical background that will aid them in studies where the nature of the solid state is an important consideration; second, to provide a text that is broad enough to serve as a base for continuing mineralogical studies and detailed enough to bridge the gap between introductory mineralogical work and advanced studies in crystallography, crystal chemistry, petrology, and geochemistry; and third, to provide a convenient source of descriptive material for the more common minerals and mineral groups.

The book is divided into three parts that separate theoretical, practical, and descriptive matter. This provides an organization similar to that of most mineralogy courses composed of lecture and laboratory components. The eight chapters of Part One cover the theoretical aspects of the solid state that are of direct and general concern in mineralogy. Topics include the nature of matter, symmetry, crystallography, crystal forms, the nature of the atom, bonding, chemical and geometric variation in minerals, the physical chemistry of mineral stability, the crystallization and habits of minerals, plus a review of the relationships between mineralogy and the broader geologic processes. These topics require a knowledge of high school physics and elementary college chemistry and geology. They provide, however, the conceptual background necessary for understanding the geometrical and chemical parameters that determine the kind, number, variations, and stability of atomic arrangements that minerals assume. The level of mathematics used is generally limited to college algebra and trigonometry, and some knowledge of differential calculus is required in the discussion of mineral stability.

Part Two is devoted to the practical aspects of mineralogic study. Included are chapters on the physical properties of minerals, the interaction of light and matter, instrumental methods of mineralogic analysis, and the use of x rays in the study of minerals. Part Three includes the descriptions of individual mineral species or series with information applicable to appropriate mineral groups. Some 178 minerals are described in detail and over 740 related species are mentioned. In addition, more than 450 varieties and synonyms are given. The four chapters of Part Two, coupled with the mineral descriptions and the determinative tables of Appendix B, are designed to form the core of the laboratory portion of an introductory course in mineralogy.

The "Suggested Readings" given at the end of each chapter include works that afford the student a different perspective on or more detailed coverage of the subject and, where relevant, classical references to the mineralogic literature. Further references are given following the appendices.

Teaching/learning aids include highlighted notes and boxes. The notes provide historic or scientific background information or present a geologic example. The numbered boxes present specific aspects of mineralogy in more detail than is taught in many mineralogy courses. Their use is optional and is left to the discretion of the instructor.

Mineralogy embraces so many aspects of geology, physics, and chemistry that, in a textbook of this size, the choices of subjects and the order in which they are presented must be somewhat arbitrary. The chapters of Part Two are self-contained and may be studied in any order.

No order is prescribed for the chapters of Part One. The authors have found, however, that an understanding of the nature of crystal growth and habit as well as many aspects of chemical and geometrical variations in crystals demands a background in crystal chemistry. In turn, the concepts and nomenclature of crystal chemistry often require a previous study of crystallography. These two subjects are, however, sometimes best treated concurrently. In some universities, the mineralogy course may comprise a full academic year of study. Such a course can include all the chapters of this book plus supplementary material in optical mineralogy, x-ray mineralogy, or advanced study of the rock-forming minerals. In situations where mineralogy is taught in half an academic year, a shortened, less quantitative course results from excluding chapters 6, 8, 10, 11, 12, and parts of chapters 4 and 7.

In the years that have passed since the publication of the first edition we have received many helpful and constructive comments from professionals and students alike. Many of their suggestions have been incorporated into the second edition. Chapters 2 and 3 on symmetry and crystallography have been reorganized and, in part, rewritten. The principal changes were to introduce the concepts of indexing and stereographic projection earlier in the treatment. Methods of crystallographic and crystal chemical calculations, formerly collected into a single chapter, have been dispersed throughout the book to accompany appropriate theory. The descriptive matter of Part Three has been thoroughly revised to include recent mineral data and to update nomenclature according to the Commission on New Minerals and Mineral Names of the International Mineralogical Association. Accompanying this revision was the elimination of many discredited species and variants. Complete descriptions of alunite, turquoise, crocoite,

and chloritoid have been added. A complete review of the whole of the first edition has resulted in a simpler syntax and, hopefully, the elimination of typographic glitches.

Helpful suggestions were received from Peter Bayliss, University of Calgary, Rodney Metcalf, University of Nevada–Las Vegas, Herb Yates and Jeff Steiner, City College of New York, and Paul Holm, University of Windsor. Iain Samson and Terry Smith, University of Windsor, read and class-tested portions of the manuscript. Ihsan Al-Aasm supplied photographs and consulted on mineral names of Arabic origin. Tom Weiland, Georgia Southwestern College, Charles Ritter, University of Dayton, and Paul Farnham, University of St. Thomas, provided excellent and constructive reviews of the first edition.

We extend a special thanks to those who provided reviews for the second edition. They are: Tom Weiland, Georgia Southwestern College; Charles J. Ritter, University of Dayton; Paul R. Farnham, Chair., Dept. of Geology, University of St. Thomas; Karl A. Riggs, Mississippi State University; Andrew Campbell, New Mexico Tech.; H. Richard Naslund, State University of New York at Binghamton; Ellen P. Metzger, San Jose State University; and Robert B. Furlong, Wayne State University. Their thoughtful comments and suggestions are greatly appreciated.

The authors also thank Deborah MacDonald, University of Windsor, for her photography and Bob Gait, Royal Ontario Museum, for his help in obtaining photographs. Elizabeth Chandler and Susan Temporal, University of Windsor, stepped into the breach when necessary. To these and all the staff at Wm. C. Brown Publishers who have helped in the preparation and production of the second edition, we are greatly indebted.

Part One

Theoretical Considerations

Introduction to Mineralogy

Matter and Its States
The Crystalline State
Minerals
A Historical Perspective
Concluding Remarks
Suggested Readings

We, as the thinking residents of our planet, are vitally concerned with the materials that make up the earth. We ultimately depend on terrestrial raw materials for essentially everything that makes our life more than mere subsistence. Raw materials, of course, can be used by unthinking plants and animals, but the refinement, combination, and development of natural substances into the products that support today's technological society require a keen understanding of energy and matter in its varied forms. Geologists and mineralogists contribute to society their understanding of the earth's natural materials. They have learned to find and interpret the probable meaning and history of particular mineral associations, and they know about their properties and uses.

Rocks and their constituent minerals have been used throughout history as tools, decorations, and raw materials for manufacturing. Natural human curiosity has led to the continual study of minerals throughout recorded history, and a great deal was learned about their shapes and macroscopic properties. Little was known about the true nature of minerals, however, until the twentieth century.

Early mineralogy dealt almost exclusively with the external forms and properties of minerals. It was recognized that crystals had faces, though these were thought to be generally few in number. It was also known that the angles between adjacent faces were constant. Further, properties such as hardness, electrical and thermal conductivity, and magnetic susceptibility could be related to crystal geometry. Although it was suspected that external properties were controlled by something within the crystal, it was not until the early part of this century that the internal arrangement of atoms could be determined using x-ray techniques. This was indeed a great advance in the understanding of minerals. The realization that the type and internal arrangement of particles on the atomic scale control the macroscopic properties of a mineral formed the principal basis of modern mineralogic studies.

Matter and Its States

Matter, the substance of the physical universe, can occur in various physical forms and can change from one form to another by such processes as crystallization, sublimation, solution, vaporization, and melting. Matter can be grouped chemically into those substances that make up or resemble the chemistry of living organisms and those that do not. The former are combinations of carbon and chiefly such elements as hydrogen, nitrogen, and oxygen. Such *organic* substances include all compounds of carbon except the very simplest, such as C, CO, CO_2, or simple compounds containing the radicals $(CO_3)^{2-}$ or CN^-. All other substances are considered to be *inorganic*. Matter may also be subdivided according to whether or not it is formed into compounds by artificial or natural means.

All matter is made from elementary particles—neutrons, protons, and electrons—that in various numbers and combinations compose the more than ninety naturally occurring chemical elements and their numerous isotopes. Each atom is composed of a central nucleus made up of protons and neutrons in roughly equal numbers and surrounded by an exterior array of electrons equal to the number of nuclear protons. The exact nature of these subatomic particles is not known, but certain characteristics of their mass, charge, and motion are understood. Protons and neutrons are of about equal mass and are about 1,800 times heavier than electrons. Neutrons carry no electrical charge. Protons carry one unit of positive charge and electrons one unit of negative charge.

Electrons are affected by electrical fields other than the field associated with the nucleus, and they can be displaced by external phenomena such as the passage of light, electrical current, or the near approach of another atom. To form some compounds, electrons are transferred from one atom to another. This results in positively and negatively charged atoms, respectively **cations** and **anions**, which then mutually attract. In other instances, electrons may alternate rapidly among a few or many nuclei. In essence, both cases result in a kind of electronic glue that binds atoms together.

Solids, liquids, and gases are distinguished by the degree to which their constituent atoms or molecules stay in continuous contact. Substances at 0° Kelvin (−273° C) contain no thermal energy and their constituent atoms or molecules are at rest. At temperatures above 0° K, the added thermal energy is converted to kinetic energy, resulting in random motion of the atoms (translational, vibrational, and rotational) within the constraints imposed by chemical linkages or collisions with other atoms. The higher the temperature, the greater the heat content and the greater the kinetic energy. For a gas, the distance between collisions depends on the distance between atoms, and at low pressure the mean free path may be relatively long. If pressure is increased, however, the separation of the atoms is reduced until they are in virtually continuous contact but still unbound, not having reached the *critical point* after which the gas becomes indistinguishable from a liquid. Thermal motion may still effect interchanges of adjacent atoms or molecules and thus cause diffusive mixing of the supercritical fluid.

When adjacent atoms are fixed in regular positions by interatomic bonding, the diffusion of atoms through the mass is greatly reduced although not completely eliminated. Regularity of position and strong interatomic bonding distinguish the crystalline, or truly solid, state from the glassy, or pseudosolid, state. The truly solid state is the crystalline state in which atoms are all bonded to their neighbors in some regular way. In glasses, the bonding lacks regularity. Indeed, solids can be arranged along a spectrum based on the amount of long-range ordering of atoms from glassy to crystalline solids.

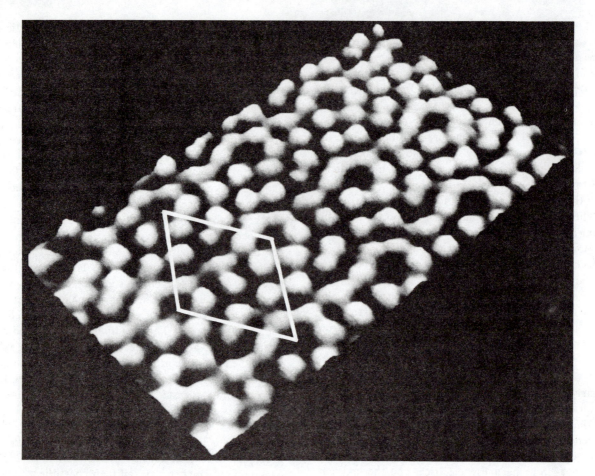

Figure 1.1
The surface atoms of silicon metal are shown magnified about ten million times by a scanning tunneling microscope (STM). A unit cell is outlined. Note how the unit cell can be repeated to describe the structure. (Courtesy of International Business Machines Corporation, Research Division, Yorktown Heights, NY.)

Distinguishing between solids, liquids, and gases is not always easy. Nonperfect gases and liquids are the rule rather than the exception, and intermediate physical states between liquids and crystals, such as glasses and colloids, are common.

The Crystalline State

The atoms of a crystal are fixed in position by chemical bonds, and the disposition of the atoms in space is periodically repeated along any straight line through the structure. Substances are considered to be crystalline if a regular atomic arrangement persists over distances that are very great compared with the distances between the individual atoms.

An assemblage of atoms arranged in a regular manner in three dimensions together with their bonding arrangement constitutes the crystalline **structure,** and the imaginary geometrical framework to which the structure is related is the **lattice.**

For the mineralogist, the terms lattice and structure are distinct. Misuse of these terms, however, is widespread in scientific writing and the distinction has become blurred. Properly, a structure refers to the actual matter and its arrangement in a mineral. In contrast, a lattice is a regular geometric array of imaginary points that can serve as a reference system for the actual positions of atoms. Because the lattice is an array of imaginary points, it says nothing about the atoms themselves. Thus, it is inappropriate to speak of "lattice energy," "lattice defects," "lattice vibration," or "lattice strain." One could, however, speak of lattice coordinates or lattice directions because such terms are mathematical and intangible.

A fundamental characteristic of a crystal that results from the repetitive nature of its components is that it can be subdivided into **unit cells.** These are small volumes of identical size, orientation, and atomic constitution that, taken together, fill all space within the crystal. The unit cell is like a small template, a plan for constructing a crystal of any size, as shown in figure 1.1. Crystal structures are aggregates of atoms so combined as to equal or

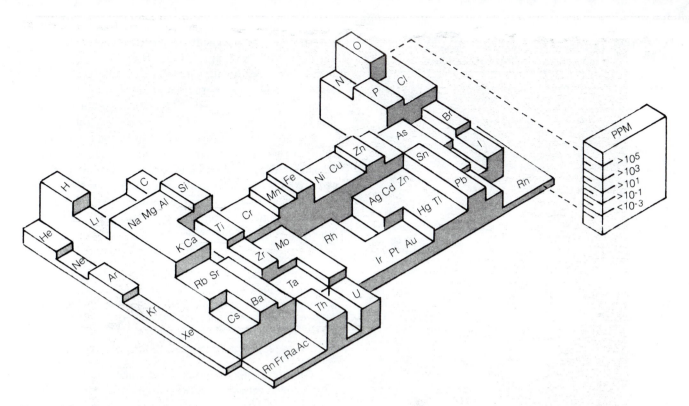

Figure 1.2
The crustal abundance of the elements.

approach the lowest potential energy arrangement possible with the available constituents at the time and conditions of the crystallization. In other words, during crystallization atoms will arrange themselves as snugly as possible to use the maximum available bonding energy.

Minerals

In many nonscientific contexts, the word mineral is bandied about with considerable breadth of meaning. News reporters, politicians, and economists have used the word to refer to naturally occurring compounds as varied as quartz sand, table salt, diamonds, organic-rich rocks such as coal and oil shale, water, and petroleum. In fact, in a scientific sense, **minerals** are a particular subdivision of matter defined as *naturally occurring, homogeneous, solid, inorganic substances, usually crystalline, that have fixed limits of chemical composition and physical properties.* Minerals are thus by definition a relatively restricted grouping of the possible forms in which matter may be found. For example, the term mineral does not include water (a liquid) but does include ice (a solid). Coal (an assemblage of carbon-rich organic compounds) is not a mineral, but the term mineral does include graphite. Minerals do not include natural glasses, which are not crystalline and have wide compositional limits, but they do include glassy appearing but crystalline quartz and feldspar. They do not strictly include artificial gems, although these may be otherwise identical to natural gems. The term **mineraloid** is often used for natural materials that lack definite structure or chemical composition.

The number of naturally occurring, inorganic, crystalline substances (minerals) that may be found is sharply limited by several factors—requirements of spatial geometry, terrestrial abundances of the elements, levels of naturally available energy, and atomic parameters. In all, only about 3,200 mineral species are known, about 100 are fairly common, and only a few dozen are truly abundant. The paucity of mineral species is underscored when compared with the extraordinary variety in the biological world; for example, there are more than 10,000 species of inchworms alone!

Geometrical analysis of periodically repeating points in space has revealed only fourteen distinct regular arrays or lattices. These arrays of imaginary points are based on only six parallelepipeds, or primitive cells, and only thirty-two patterns of points about any given point are possible. The possible arrangements of atoms cannot exceed these geometrical limits. Indeed, crystal structures may be grouped into six great crystal systems, consistent with the six primitive cells, and into thirty-two crystal classes that correspond to the thirty-two possible point groups. Further, the symmetrically disposed atoms must be placed so that they can form chemical bonds with their neighbors. The dual restrictions of atomic arrangement and bond formation sharply limit the possible numbers of different crystalline substances.

The abundance of elements in the earth's crust is our inheritance from the formation of the elements and the evolution of the earth. The measured values are shown graphically in figure 1.2 and tabulated in table 1.1, both of which show that oxygen and silicon predominate and

Table 1.1 Abundances of the More Common Elements in the Crust of the Earth

Element	Symbol	Percent by Volume	Percent by Weight
Oxygen	O	93.8	46.60
Silicon	Si	0.9	27.72
Aluminum	Al	0.5	8.13
Iron	Fe	0.4	5.00
Calcium	Ca	1.0	3.63
Sodium	Na	1.3	2.83
Potassium	K	1.8	2.59
Magnesium	Mg	0.3	2.09
		100.0	98.59
Titanium	Ti		0.440
Hydrogen	H		0.140
Phosphorus	P		0.118
Manganese	Mn		0.100
Sulfur	S		0.0520
Carbon	C		0.0320
Chlorine	Cl		0.0314
Rubidium	Rb		0.0310
Fluorine	F		0.0300
Strontium	Sr		0.0300
Barium	Ba		0.0250
Zirconium	Zr		0.0220
Chromium	Cr		0.0200
Vanadium	V		0.0150
Zinc	Zn		0.0132
All other elements			less than 0.0100

the familiar cubic rock-salt structure. Galena, PbS, is another mineral whose atoms have this same arrangement although its interatomic bonds are of a different kind. In calcite, $CaCO_3$, the alternation of calcium ions and carbonate radicals (Ca^{2+} and $(CO_3)^{2-}$) may occur in much the same manner, but the large size and triangular shape of the carbonate radical requires a distortion from simple cubic to rhombic geometry; that is, the cube is slightly squashed along a body diagonal. Numerous packing geometries are known, but the tendency of ions having the same radius ratios to aggregate with like geometries allows the same arrangements to occur in many different minerals.

Ionic radii for the elements are given in Appendix A. Note, however, that the ionic radius of an atom varies depending on the other ion involved. Most tables of ionic radii, including the one in this book, are for metal-oxygen bonds. Ionic radii are expressed in Angstroms (Å) where $1Å = 10^{-10}$m.

The prevalence of oxygen and silicon in the earth's crust coupled with their ability to join into a stable silicate radical control the nature of the most abundant minerals—the silicates. On an atomic scale, the charges and sizes of these two ions are such that a group of four tightly packed oxygen ions leaves just enough space for a silicon ion to be nestled in the central hole of the oxygen group. This arrangement may be described as an $(SiO_4)^{4-}$ tetrahedron with oxygen ions located on tetrahedral corners and a silicon ion at the center. Each oxygen ion carries a charge of $2-$; half of this charge goes to neutralize the $4+$ charge on the central silicon ion and the remainder to give the tetrahedral radical a net charge of $4-$.

In some silicate structures, the $(SiO_4)^{4-}$ building block alternates with appropriately placed and charged cations for neutralization, as in zircon, $Zr^{4+}(SiO_4)^{4-}$. In others, one or more of the oxygen ions is common between adjacent tetrahedra which then become complex radicals articulated into pairs, chains, sheets, or frameworks of silica tetrahedra. These complex radicals, in turn, alternate with various cations to form other silicate structures.

Carbonates, sulfates, phosphates, and borates are examples of mineral classes in which a strongly bonded radical made up of a positive ion other than silicon is surrounded by oxygen ions to form a negatively charged structural unit. These radicals alternate with cations, as in the tetrahedral unit structures of the silicates, but do not, in general, form complex articulations. The interaction of positive and negative ions often results in a mutual deformation from their normally spherical shapes. Small, highly charged positive ions are usually the most effective in deforming or polarizing their neighbors. Thus, the size and polarizing capability of the central cation in the radical determine both the number of oxygen ions that participate and the shape of the grouping. Table 1.2 lists some of the more common radicals.

that only eight elements by themselves make up over 98% of the crust. By weight the crust is about 50% oxygen, but the atmosphere is only about 20% oxygen. Further, in terms of volume, oxygen composes over 90% of the crust! The eight dominant elements—oxygen, silicon, aluminum, iron, calcium, sodium, potassium, and magnesium—are not only abundant but also fairly uniformly dispersed and so are everywhere available for mineral formation. Of these elements, only oxygen forms negative ions (O^{2-}), and most of the common minerals may be anticipated to be oxygen bearing. Further, since O^{2-} is larger than most common cations, minerals in which it is a significant component may be described as being constructed of more or less closely packed oxygen ions with cations regularly distributed throughout the interstices. It should be no surprise that most minerals are various combinations of these eight elements and that oxygen and silicon-oxygen compounds are met everywhere in the study of rocks and minerals.

The more common mineral structures are composed of alternating positively and negatively charged units that may be either single atoms or ions or small ionic groupings called **radicals.** These units are packed together in different ways according to their relative sizes. For example, sodium and chlorine ions, whose ionic radii are 1.10 and 1.72 Å respectively, pack together in a cubic array of alternating Na^+ and Cl^-, creating the mineral halite with

Table 1.2 **Examples of Nonsilicate Radicals**

Mineral Class	Radical	Shape	Example
Hydroxides	$(OH)^-$	Oxygen ion size (H^+ is a proton.)	Gibbsite, $Al(OH)_3$
Carbonates	$(CO_3)^{2-}$	Plane triangle	Calcite, $Ca(CO_3)$
Borates	Complex radicals	Triangles and tetrahedra	Borax, $Na_2(B_4O_7) \cdot 10H_2O$
Sulfates	$(SO_4)^{2-}$	Tetrahedron	Anhydrite, $Ca(SO_4)$
Phosphates	$(PO_4)^{3-}$	Tetrahedron	Apatite, $Ca_5(PO_4)_3F$
Arsenates	$(AsO_4)^{3-}$	Tetrahedron	Mimetite, $Pb_3(AsO_4)Cl$
Tungstates	$(WO_4)^{2-}$	Distorted tetrahedron	Scheelite, $Ca(WO_4)$
Molybdates	$(MoO_4)^{2-}$	Distorted tetrahedron	Wulfenite, $Pb(MoO_4)$

Another large and important group of oxygen-bearing minerals is that in which O^{2-} alone alternates with one or more cations. These minerals, the oxides, are represented by such species as ice, H_2O; periclase, MgO; corundum, Al_2O_3; rutile, TiO_2; and spinel, $MgAl_2O_4$.

Minerals that do not use oxygen in their makeup, although subordinate in abundance, are fairly common and often economically important. Their classes include the halides, in which chlorine and fluorine play the anionic role of oxygen, and sulfides, in which sulfur acts as the anion. Other oxygen-free mineral families are the native elements and the sulfosalts. The latter have arsenic, antimony, or bismuth and sulfur in a complex radical reminiscent of the silicates. Some examples of these oxygen-free minerals are listed below.

Halides: halite, $NaCl$; fluorite, CaF_2

Sulfides: chalcocite, Cu_2S; bornite, Cu_5FeS_4; galena, PbS; covellite, CuS; pyrite, FeS_2

Sulfosalts: pyrargyrite, Ag_3SbS_3; enargite, Cu_3AsS_4; bournonite, $PbCuSbS_3$

Native elements: sulfur, S; graphite, C; gold, Au

A Historical Perspective

Minerals have been used as tools, decorations, and trade commodities since the earliest times of human existence. Prehistoric cave painters used black manganese oxides and red iron-oxide pigments. Ornamentations of turquoise, garnet, and amethyst were known to the early peoples, and tools of nephrite (jade) were used as trade goods. Metals have been in use for some 4,000 years, requiring an early knowledge of sources, extraction, and refining methods to recover iron, copper, zinc, lead, and silver. Mercury was in use in Egypt as early as the fifteenth or sixteenth century B.C., and the Old Testament of the Bible refers to the use of gold, silver, copper, tin, lead, and iron. However, essentially no written records remain attesting to mineralogic studies related to metal production.

Many ancient Greek scholars refer to minerals, rocks, and ores. Aristotle (384–322 B.C.), for example, mentions sulfur, realgar, and ochre but gives no descriptions or proposed origins. His student and successor as head of the Lyceum at Athens was Theophrastus (372–287 B.C.), who among his many writings produced a volume titled *Concerning Stones*. Although it is quite short, the work appears to be a fragment of a much larger manuscript. In his treatise, Theophrastus mentions only a few minerals but uses these to develop an empirical classification based on the effects of heating. The work, perhaps the first mineralogy text, remained a reference for 1,800 years and demonstrates that there was a sizable body of knowledge about minerals, especially concerning practical uses, available to the Greek philosophers. From the surviving works, however, it is difficult to tell how many minerals were known at that time.

In an early work, Lucretius (99–55 B.C.) discusses the chemical elements as determined by the Greeks (earth, air, fire, water) in terms of their properties and combinations. He proceeds to a discussion of the constitution of matter, including minerals, which he says is composed of atoms of these elements, in his *Atomic Theory*. Pliny (A.D. 23–79) presents the most complete review of mineralogic knowledge for Roman times in his work *Natural History*. Included are rather detailed descriptions of minerals, their properties, uses, and methods of extraction. The state of knowledge with regard to mineral origins was, however, apparently still quite primitive. For example, Pliny describes quartz ("rock crystal") in considerable detail, giving information on luster, inclusions, stains, and surficial features. He wonders that quartz crystals always have six faces and that, although the crystal terminations might not always have the same form, their faces are recurring. Pliny, however, ascribes the origin of quartz to a peculiar, very hard form of ice! Pliny might also be described as the first mineral conservationist. He complains that, although minerals are a very important and nonrenewable resource, their extraction is not primarily for the common good but rather for simple profit.

Little literature is available on mineralogy (or on any other science, for that matter) from the medieval period and, except for some works by Arab authors, nothing of real importance appeared until the mid-sixteenth century. A series of books by the German physician Georgius Agricola—*Bernannus* (1540), *De Natura Fossilium* (1546), and *De Re Metallica* (1556)—established mineralogy as a science. The last, Agricola's best-known work, was principally a review of mining and metallurgical practices of the time. *De Natura Fossilium*, however, has been described as the first textbook of mineralogy. In it Agricola presents a classification of minerals based on physical properties such as color, density, luster, taste, shape, hardness, and solubility, all still used today for mineral identification. Also of significance is the metallurgical work

Pirotecnia, by Vannoccio Biringuccio of Siena, Italy, published in 1540, which includes a lengthy discussion of the metallic ores and some other minerals of commercial interest. Although the Renaissance period ushered in a whole new interest in scientific thought, it was not until 1669 that the next big advance was made in mineralogy. This was the observation by the Dane Niels Stensen (commonly latinized to Nicolaus Steno) that the angles between the adjacent faces of quartz crystals were constant no matter what habit the crystal assumed. An instrument to measure these interfacial angles, a goniometer, was developed by Carangeot in 1780, and soon after (1783) Romé de l'Isle confirmed Steno's hypothesis and proposed a law of constancy of interfacial angles.

About the same time, an important school was established at Freiberg under the direction of A. G. Werner (1750–1817). Werner and his students, by studying the raw material of the mining industry of their day, found many new minerals and several new elements, including cobalt, manganese, molybdenum, nickel, and tungsten. Werner also proposed a chemical classification of minerals not too much different from that in use today.

The concept of the unit cell, described earlier, is an outgrowth of the work of R. J. Haüy, who is often described as the father of mathematical crystallography. He proposed in 1784 that minerals were formed by the stacking of tiny individual blocks, or "integral molecules." In 1801 Haüy also stated the theory of rational indices that, *for all crystals of the same species, a simple mathematical relation exists between all planes that are possible crystal faces.* This is a fundamental law of crystallography.

Significant advances in mineralogy were made in the early part of the nineteenth century. Atomic theory was more fully developed, and mineral chemistry had advanced to the point where scientists realized that each mineral species had a recognizable composition. W. H. Wollaston developed the reflecting goniometer in 1809, allowing interfacial angles to be measured with greater accuracy and on smaller crystals. This led to a classification of crystals and crystal forms. The Swedish chemist Jons Jakob Berzelius (1779–1848) and his student Eilhard Mitscherlich (1794–1863) made considerable advances in the chemistry of minerals and devised the basic chemical classification of minerals in use today.

The colonial expansion of the nineteenth century was accompanied by much exploration for mineral resources, resulting in the discovery of many new minerals. In 1815, minerals were first examined under the microscope by the French naturalist P. L. A. Cordier, and in 1828 William Nicol invented the polarizing microscope, a device still important to mineralogists and geologists for observing the behavior of light in crystals.

The next great advance in mineralogy was made following the demonstration of the diffraction of x rays in crystals by W. Friedrich and P. Knipping under the direction of Max von Laue, in 1912 at the University of Munich. This was immediately followed by the first determination of a crystal structure (sphalerite, ZnS) by the English physicists W. H. Bragg, at Cambridge, and his son W. L. Bragg, at Leeds. Since their time, many mineral structures have been determined, and many new minerals have been found. The relation between atomic structure and mineral properties is more thoroughly understood today, owing in large part to specialized scientific instrumentation. The scanning electron microscope (SEM) affords visual observation of mineral grains at magnifications of up to 2×10^6. The transmission electron microscope (TEM) magnifies the internal structure of minerals millions of times to produce facsimile representations of the atomic structure. The electron microprobe (EMP) allows chemical analysis of a mineral at a spot as small as 1 micrometer in diameter. Electron microprobes can generate chemical data on very small grains of minerals in situ and allow examination of chemical variation within a single grain.

Earlier we said that the number of mineral species is controlled by available chemicals, geometry, and energy. It is interesting to consider the number of species in a historical perspective. Figure 1.3 presents a chronology of mineralogy in two ways. The total number of recognized mineral species is plotted against time. On the same time scale, the number of new mineral species found for each twenty-year period since 1800 is shown. Clearly, the number of known species is, to a large extent, determined by technological advances. For example, a significant and continuous increase in number of new species followed the invention of the polarizing microscope, but its effect seems to wane around the turn of the century. The innovation of x-ray diffraction techniques, however, brought about a new resurgence in mineral identification that has been supported and spurred by the application of the electron microprobe. Will such rapid expansion continue? There must certainly be geometric and chemical constraints on the possible number of species, but there are many new environments that are just now becoming available to us. It is to be expected, for example, that with planetary exploration will come the discovery of many new minerals that have developed under different chemical, atmospheric, gravimetric, temperature, and pressure conditions.

Concluding Remarks

Minerals are natural examples of the crystalline state. All of their properties—chemical, physical, and geometric—as well as their formation, distribution, analysis, and uses are appropriate topics for mineral study.

Mineralogy today specifically encompasses the disciplines of crystallography, crystal chemistry, and crystal physics, but it is closely associated as well with essentially all fields of geologic study. It is indeed difficult to find a geologic problem that does not require in some way an application of mineralogic principles. Thus, the identifi-

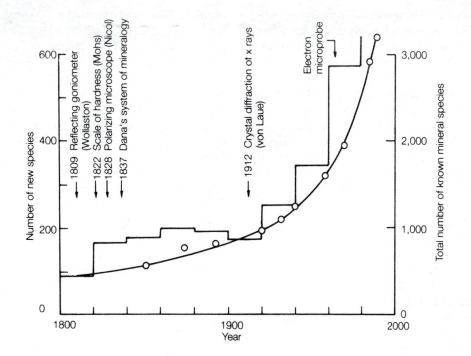

Figure 1.3

Mineral species. Each bar represents the number of minerals discovered for twenty-year periods from 1800. The curve represents the total number of known mineral species. (Data from Mandarino, 1977; Fleischer, 1991.)

cation of minerals and an understanding of mineralogic relations and stability is fundamental to our understanding of the physical universe.

Minerals are studied on the basis of the constraints of their origin: geometric, chemical, and energetic. We consider the geometric constraints through a study of symmetry and crystallography. The chemical controls on mineral formation are determined by investigation of crystal chemistry and the compositional variations of species. The study of mineralogic stability and reactions reveals the energetic constraints on minerals.

As with other sciences, mineralogy has developed, loaned, borrowed, and discarded many concepts and techniques over its long history. Among the more important scientific developments rooted in mineralogy are the concepts of symmetry and the probing of crystalline matter with x rays. Borrowings from chemistry and physics include the understanding of the nature and interactions of atoms as described by quantum mechanics, physical chemistry, and thermodynamics. Mineralogic discards of the recent past include qualitative systems of analysis and unwieldy systems of crystallographic and mineralogic nomenclature. The next victim will probably be morphological crystallography whose utility is limited to crystals exhibiting external faces whereas x rays examine the internal arrangement of atoms.

Studying the theoretical aspects of mineralogy helps us understand mineral formation, variation, and stability, but the practical topics of mineralogic study are also important. Minerals are the stuff of rocks, and the identification of minerals and mineralogic associations is essential in the investigation of geologic processes. The student is

urged to practice the skills of mineral identification while keeping in mind the theoretical principles of mineralogy.

Suggested Readings

Adams, F. D. [1938] 1950. *The Birth and Development of the Geological Sciences.* New York: Dover.

Agricola, G. [1530] 1806. *Bernannus.* Translated from the Latin by F. A. Schmid, Freiberg.

Agricola, G. [1546] 1810. *De Natura Fossilium.* Translated from the Latin by E. Lehmann, Freiberg.

Agricola, G. [1556] 1950. *De Re Metallica.* Translated from the Latin in 1912 by H. C. Hoover and L. H. Hoover. New York: Dover.

Berry, L. G., and Mason, B. H. 1983. *Mineralogy.* 2d. ed., revised by R. V. Dietrich. New York: W. H. Freeman.

Biringuccio, V. [1540] 1942. *The Pirotecnia.* Translated from the Latin by C. S. Smith and M. T. Gnudi. New York: American Institute of Mining and Metallurgical Engineers.

Caley, E. R., and Richards, J. F. C. 1956. *Theophrastus on Stones.* Columbus: Ohio State Univ. Press.

Fleischer, M. 1991. *Glossary of Mineral Species.* 6th ed. Tucson, AZ: Mineralogical Record.

Haüy, R. J. [1801] 1822. *Traite de Mineralogie.* 2d. ed. Paris: Louis.

Klein, C., and Hurlbut, C. S. 1985. *Manual of Mineralogy (after J. D. Dana).* 20th ed. New York: Wiley.

Werner, A. G. [1774] 1962. *On the External Characters of Minerals.* Translated by A. V. Carozzi. Champaign: Univ. of Illinois Press.

Zoltai, T., and Stout, J. H. 1984. *Mineralogy: Concepts and Principles.* Minneapolis, MN: Burgess.

Symmetry

On the large scale, nature does not seem to be particularly ordered in terms of lines, or planes, or the regular positioning of its elements. Humans, on the other hand, find intellectual satisfaction in the orderly arrangement of their works. It is thus understandable that we are attracted to those natural materials that show a regularity of design, such as the petals of a flower, the shape of a seashell, or the polyhedral form of a crystal.

Crystals are indeed a remarkable natural phenomenon, and in the past they were thought to have a variety of magical properties. The old magic is gone in the present world, but the real properties of crystals are as strange and fascinating as anything dreamed of in earlier times. Consider, for example, that quartz changes an oscillating electrical signal into an ultrasonic pulse in a quartz-tuned radio, that scheelite glows when bathed in "black light," that galena acts as an electrical rectifier in a crystal radio set, that ruby generates a homogeneous light beam in a laser, and that calcite splits light into two parts, to name but a very few properties.

Structure and Lattice

Nearly all minerals are crystalline solids composed of atoms or ions held in an orderly, three-dimensional array by interatomic forces. Such arrays of atoms are called **crystal structures** and are characterized by the periodic duplication of any grouping of atoms along any line through the structure. This regular and periodic arrangement of constituents is the salient feature of crystals. In addition to controlling many interesting and useful properties, this regularity allows physical examination of the structure by x rays and permits geometrical analysis of possible and particular atomic arrangements. The periodicity also provides a means of identification, because the internal structure of a mineral is reflected in its external shape.

The regular distribution of atoms in space can be most readily studied by relating atom locations to a coordinate system. In crystallography, these reference coordinate systems may be of the usual orthogonal kind but are just as likely to be somewhat more generalized. Several possible generalizations, all showing the location of point $x = 3$, $y = 5$, are given in figure 2.1.

A three-dimensional network of regularly arranged points to which the atoms in a crystal are related is called a lattice. Although lattice points may represent atom locations, it is not necessary that lattice points and atom centers be coincident. The atoms in a structure must be symmetrically disposed with respect to the lattice, but the lattice itself is an imaginary geometrical framework. The actual array of atoms is properly termed the structure. Figure 2.2 shows the relationship between a lattice and

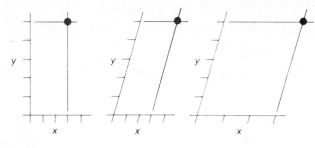

Figure 2.1
Coordinate systems.

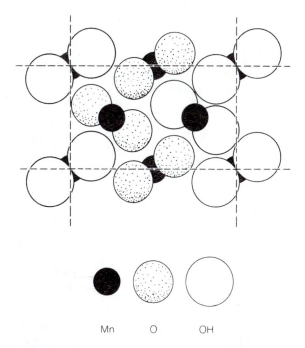

Mn   O   OH

Figure 2.2
Manganite, MnO(OH), showing its structure (atomic locations) and lattice. The lattice points are at the intersections of the dashed lines. Note that no atom is centered on a lattice point in this structure.

the structure of the mineral manganite, MnO(OH). Points of the lattice lie at the intersections of the dashed lines; in this structure no atom lies on a lattice point.

Lattices

A space lattice to which a three-dimensional structure may be referred can be generated by successive **translations** of an initial point. Repeated application of a translation of given length and direction, t_1, to an initial point generates a row of points or a linear lattice:

The repeated application of some other translation not along the same line (noncolinear), t_2, moves the entire row parallel to itself, thus generating a planar array of lattice points:

A third translation not in the same plane (noncoplanar), t_3, when applied to such a plane lattice, generates a three-dimensional array of points called a *space lattice,* or simply lattice.

In considerations of symmetry, the effect of a translation is to move not only a lattice point but all the surrounding space and its contained matter. The effect of a translation in crystals, then, is to duplicate the entire pattern of atoms at some distance and direction from the origin.

The **rational features** of a lattice and its related structure are determined by the lattice points. A rational point is a lattice point itself, a rational line connects two lattice points, and a rational plane passes through at least three lattice points not in a straight line. All other points, lines, or planes are considered irrational.

Tiling a Plane

Because a plane lattice is generated by two noncolinear translations, it must consist of points lying on the corners of identical parallelograms. In turn, each parallelogram may be divided by a diagonal into two congruent triangles. If the geometrically distinct kinds of triangles are tabulated, their number must equal the number of distinctively different kinds of plane lattices that can exist.

The geometrically different kinds of plane triangles, first recognized by the Greek geometer, Euclid, who flourished about 300 B.C., are given in table 2.1. When combined into parallelograms, these five different triangles yield the five distinct mesh shapes shown in figure 2.3: a general parallelogram, a diamond, a rectangle, a square, and a 120° rhombus. No other geometrically distinct plane lattice shapes are possible.

If the regularly spaced imaginary points that constitute a lattice are connected by imaginary lines, a plane lattice is seen to be made up of small identical units, or cells, much as a tiled floor is composed of individual tiles. The cells of a plane lattice, like the tiles of a floor, must fill all space or "tile the plane." One cannot, for example, tile a floor entirely with pentagonal or octagonal tiles.

A plane is tiled with any of the five different parallelograms by moving a cell by unit translations, and if the

Table 2.1	Five Distinct Kinds of Triangles and Mesh Shapes	
Triangle	**Sides**	**Angles**
Scalene	Unequal	Unequal
Isosceles	Two equal	Two equal
Right	Unequal	Unequal; one is 90°
Equilateral	Three equal	Three equal
Right isosceles	Two equal	Two equal; third is 90°

Triangles

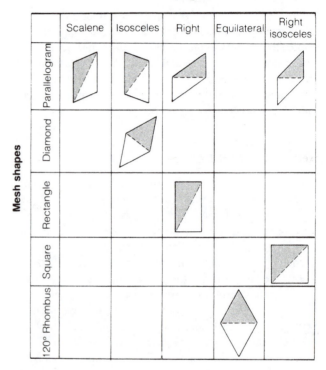

Figure 2.3
The five distinct kinds of triangles and the five distinct mesh shapes.

Figure 2.4
Choice of cells. Cells *A, B, C,* and *D* are reduced primitive cells; *E* is a multiple cell.

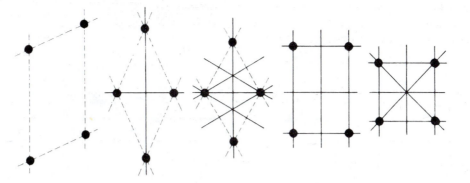

Figure 2.5
Planes of reflection symmetry (mirrors, *m*) consistent with the five plane lattice meshes are shown as solid lines.

cell represents a region containing space and matter, these also are translated and reproduced at the new site. The units, or **motifs,** of any pattern, including atomic arrays, may thus be related to a lattice and reproduced by translation.

In crystallography, it is usually necessary to approach the relationship of the lattice and structure by fitting an appropriate lattice to an existing structure. As a practical exercise, try correlating the pattern of a periodically repeated motif, as in figured wallpaper, a masonry wall, or orchard plantings, with an imaginary array of regularly spaced points in a plane lattice. This exercise will provide both practice in choosing cells and insight into geometrical restrictions that may arise. Several possible choices are shown in figure 2.4. In *A, B,* and *C,* the shortest translations have been chosen to outline the cell. The spatial surroundings of each of the lattice points may be seen to be identical, and it should be noted that only a single motif (or lattice point) is represented by these cells. A lattice point occurs where the corners of four cells meet, and therefore 1/4 point per corner lies in a given cell (1/4 × 4 = 1). Cells that include one lattice point, such as these, are termed **primitive cells.**

Obviously, many different cells containing a single lattice point may be chosen, including cell *D* with longer translations. Cells defined by the shortest translations are distinguished as **reduced cells.** Another choice is represented by *E,* in which shortest translations are not used but the translations are at right angles. More than one motif (and lattice point) is included in this **multiple cell** which, in this instance, contains two lattice points, one internal plus another distributed over the corners. Whichever cell is chosen as the "tile" is termed the unit cell.

The general considerations in the choice of a unit cell to serve as a reference for the structure are (1) to keep the translations short, (2) to provide as highly specialized a lattice geometry as possible, and (3) to have the cell shape comparable with the shape of the crystal.

Periodic repetition of a plane lattice in some noncoplanar direction generates a space lattice. A space lattice is thus a stack of plane lattices in parallel position separated by translation t_3. Another way of visualizing a space lattice is to recognize that it is composed of three intersecting sets of plane lattices, any two of which have a translation in common. Three noncoplanar translations thus serve to define the space lattice. The cells of space lattices provide the fundamental reference coordinates for the division of crystalline matter into **crystal systems.**

Because a space lattice can be visualized as being built up from intersecting plane lattices, the number of parallelepipeds that can be generated and that will "tile space" is again sharply limited. Only six geometrically distinct primitive cells can be formed: a cube (square-square-square), a square prism (square-rectangle-rectangle), a hexagonal prism (120° rhombus-rectangle-rectangle), a brick-shaped cell (rectangle-rectangle-rectangle), a brick-shaped cell deformed to make one face a parallelogram (parallelogram-rectangle-rectangle), and a cell whose faces are all general parallelograms (parallelogram-parallelogram-parallelogram).

A useful exercise at this point is to construct these cells using cardboard, pipe cleaners, toothpicks and gumdrops, or some other versatile construction medium.

Symmetry of a Lattice

Symmetry may be defined as the correspondence in size, form, and arrangement of the parts of a recurring array (for example, patterned wallpaper) with respect to a point, line, or plane. Put another way, it is the regular arrangement of parts with reference to corresponding parts. The **elements of symmetry** are those simple geometric operations that change the position of a distinctive and recurring form, shape, or figure of the array—a motif—but not its size or shape.

Translation, previously discussed, is one element of symmetry, and the others—reflection, rotation, and inversion—are taken up in this section.

Lattice points, including their surrounding space and its contents are motifs that can be related or interchanged by the elements of symmetry. For example, the points of the five plane lattices in figure 2.5 are consistent with **reflection** planes (mirrors) normal to the page as shown by

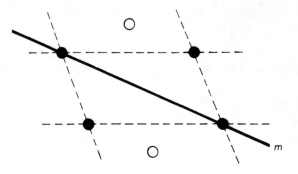

Figure 2.6
Nonconsistent mirror.

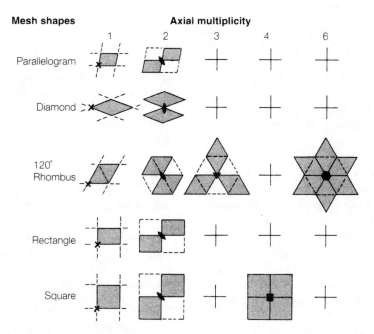

Mesh shapes	Axial multiplicity				
	1	2	3	4	6
Parallelogram					
Diamond					
120° Rhombus					
Rectangle					
Square					

Figure 2.7
Consistency of rotation axes with the five plane mesh shapes. The
axial multiplicity is the plane symmetry of the pattern.

the solid lines. For extended lattices, it is apparent that only a limited number of such mirrors can be consistently located in the mesh. This is because the geometry of reflection requires these mirrors either to pass through a point or to lie midway between two points and perpendicular to the line joining them. A nonconsistent mirror *m* is shown in figure 2.6. Its location would require the presence of points (open circles) that are not part of the mesh.

Another element of symmetry related to lattices is **rotation.** In rotation the operation of an axis rotates points and their surroundings through fixed angles. In figure 2.7, the five plane lattices are shown together with a consistent axis. The axes must be normal to the page (except for the 1-fold case), and their multiplicity is the number of sequential rotary operations required to complete one full rotation. These axes are commonly termed 1-, 2-, 3-, 4-, and 6-fold axes and are symbolized as follows:

In symmetry notation the axes are simply designated as 1, 2, 3, 4, and 6. Their rotary operations are respectively 360°, 180°, 120°, 90°, and 60°.

Figure 2.8 shows the location of all consistent axes and mirrors with respect to the five plane lattices. In figure 2.8, note that some axes of the same multiplicity, such as the 4-fold axes related to a square mesh, are not always separated by unit translations. The location of such **nonequivalent axes** with respect to the mesh may be found by an analysis of the combined operation of a rotation and a translation.

In figure 2.9, the operation A_α is the rotation through the angle α around the axis A so that lines 1 and 2 are symmetrically disposed with respect to the perpendicular of a translation t. Note that the rotation of A_α brings line 1 to line 2. If this rotary operation is followed by a translation t, A is brought to A' and line 2 to line 3. The initial and final positions of the line are 1 and 3, and obviously there has been no motion of the line at B. The net motion of the line as a result of rotation and translation is thus

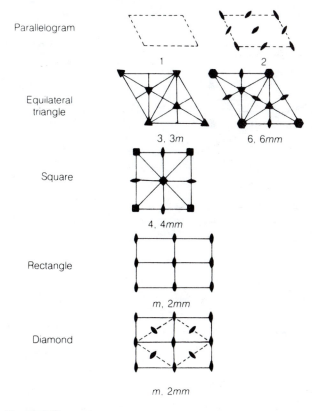

Parallelogram

1   2

Equilateral triangle

3, 3*m*   6, 6*mm*

Square

4, 4*mm*

Rectangle

m, 2*mm*

Diamond

m, 2*mm*

Figure 2.8
Axes and mirrors consistent with the five plane lattices. The symbols are standard crystallographic notation.

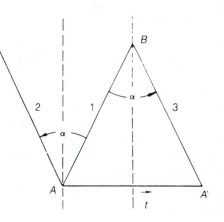

Figure 2.9
Combination of rotation and perpendicular translation.

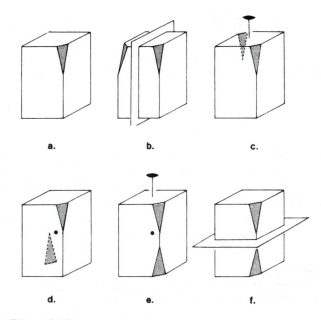

a.   b.   c.

d.   e.   f.

Figure 2.10
Illustration of the results of the operation of the symmetry elements of reflection, rotation, and inversion as they might cause duplication of a crystal face. (a) Initial condition. (b) Reflection. (c) Rotation through 180° (2-fold axis). (d) Inversion. (e) Combined 180° rotation and inversion (= f). (f) Reflection (= e).

The symbols used in figure 2.8 conform to the **International Notation** that will be used throughout the text. The symbolism is derived from the notation developed simultaneously by C. H. Hermann (1898–1961) and C. V. Mauguin (1878–1958) and thus is often referred to as Hermann-Mauguin Notation. The notation was approved by a group of crystallographers meeting in Zurich in 1930 and first appeared in the *International Tables for Symmetry and Crystal Structure* in 1935.

Numbers refer to axes of a given multiplicity; the axis of highest multiplicity is given first. A bar over the number indicates an axis combined with an inversion center, or rotoinversion axis. Mirrors are designated by an *m* and are parallel to an axis if written on its line, as in 3*m*, or perpendicular to an axis if written as a fractional denominator, as in 4/*m*. Three mutually perpendicular 2-fold axes with a mirror perpendicular to each would be written 2/*m* 2/*m* 2/*m*. Usually the principal axis is given first, but some crystallographers give it last, as in *mm*2 rather than 2*mm*, or 2/*m*$\overline{3}$ for $\overline{3}$2/*m*.

equivalent to a rotation of α about *B*. A useful generalization from this construction is that an axis *B* of equal multiplicity to axis *A* lies on the perpendicular bisector of *t*. In order to ensure that all possible axial locations have been found, it is necessary to consider the interaction of

rotation with each of the different translations of the mesh (t_1, t_2) and the mesh diagonals (vector sum of t_1 and t_2).

A final element of symmetry is **inversion.** An inversion center *i* repeats motifs as though they were inverted by a simple lens. Inversion, however, does not have the status of a fundamental element of symmetry because the same result may be obtained by some combination of the first three elements. Figure 2.10 illustrates the results of the operation of the symmetry elements of reflection, rotation, and inversion as they might cause duplication of a crystal face.

Identical Operations

The symmetry elements of translation (t), reflection (m), rotation (n), and inversion (i) act not only on lattice points, space, and matter but also upon themselves. Thus, when they are simultaneously present, they must interact in a consistent manner. Most lattices are consistent with several different symmetry elements. They may be considered to have been generated by an alternation of the several operations. The consistency of such operation sets may be shown by an **identical operation** in which the sequence of operations is tested to see that it returns a motif to some initial position and does not add points to the existing lattice.

For example, a consistent combination of reflection and rotation is illustrated in figure 2.11a. The two mirrors, m_1 and m_2, are normal to the page and at right angles to each other. The mirror m_1 transforms an initial point 1 into point 2, and m_2 transforms point 2 into point 3. A 2-fold axis A, normal to the plane of the page at the intersection of the mirrors, can return point 3 to point 1 by a rotary operation of 180°. The particular sequence of reflection and rotation described is thus an identical operation and may be written analytically as

$$m_1 \cdot \underset{90}{m_2} \cdot A_{180} = 1$$

where the dot is read "followed by" and 1 indicates that the operation sequence returns a motif to its initial position. The order of the operations is not important, and the angular operation of the axis will always be twice the angle between the mirrors.

Another identical operation is illustrated in figure 2.11b. The reflection of the motif at point 1 by mirror m yields point 2. Inversion of point 2 through the inversion center i lying in the mirror plane produces the motif at point 3. A 2-fold axis, normal to the mirror and passing through the inversion center, can transform the motif at point 3 to its original position at point 1. Analytically,

$$m \cdot i \cdot A_{180} = 1$$

The combination of a translation and a rotation was illustrated in figure 2.9, in which $A_\alpha \cdot t_1 = B_\beta$ or $A_\alpha \cdot t_1 \cdot B_\beta = 1$ and A and B have the same multiplicity, that is, $\alpha = \beta$. If the interaction of a 4-fold axis and a translation is examined following the same construction, as in figure 2.12a, then the operation of A'_{90} generates the configuration shown in figure 2.12b, and a translation of the configuration yields figure 2.12c. Continued operations of translation and rotation of this nature are only consistent with a plane lattice having a square mesh, because $t_1 = t_2$ and the angle between these translations is 90°. It is also apparent that 4-fold axes cannot be consistent with any other plane lattice.

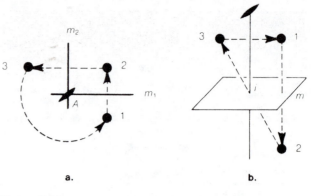

a. **b.**

Figure 2.11
(a) Consistent combination of two mirrors and a rotation axis.
(b) Consistent combination of reflection, inversion, and rotation.

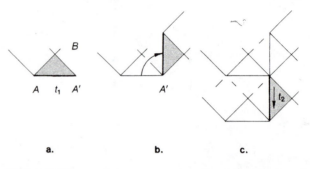

a. **b.** **c.**

Figure 2.12
Combined translation and rotation. In (a) the basic configuration of figure 2.9 is reproduced; in (b) the operation of A'_{90} rotates the configuration 90° clockwise; and in (c) the configuration in (b) is translated by t_2.

Handedness

Lattices are composed of imaginary, dimensionless, and shapeless points. However, the atoms and atomic groupings that make up a structure have both size and shape. Asymmetrical atomic groupings may be present in minerals. Such configurations are asymmetrical motifs that must be taken into account when considering the symmetry of crystal structures. As an example, consider the alternation of asymmetric atomic groupings (motifs) in the structure of the copper carbonate mineral azurite shown in figure 17.11.

The repetition of an asymmetrical motif can be carried out in such a way that the repeated motif is no longer congruent with the original one. Such a relationship is shown in the symmetry between a left and right hand, which are obviously related but just as obviously different. These figures

are all asymmetrical, but they all have the same handedness and would coincide when placed one on top of the

Box 2.1

Defining Cell Symmetry

For practice, assign the appropriate elements of symmetry to a patterned wallpaper or a brick wall. Remember that the key to the presence of a symmetry element is whether it produces the original pattern after successive operations. First find the direction and shortest distance between a repeating motif to define the tile or cell. Then determine whether any mirrors or rotary axes are present. Figure 2.13 shows how to do this for a brick wall. Cell **a** is the smallest possible cell and is termed primitive. It has, however, lower symmetry (fewer elements of symmetry) than other possible choices. Cell **b** is a better choice, having two mirrors at 90° and a 2-fold axis. The best choice is cell **c** which has the same elements of symmetry but is smaller.

Now assign elements of symmetry to such objects as a cube (don't miss the diagonals), a brick, a tent, and a stovepipe elbow.

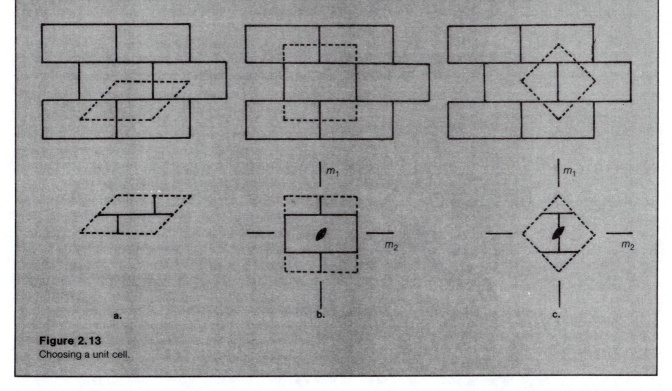

Figure 2.13
Choosing a unit cell.

other by movements in the plane of the paper. Such figures are **congruent.** On the other hand, these figures

cannot be made congruent by any movement in the plane of the paper. Such figures are **enantiomorphous.** An analogous situation exists in three dimensions, and both enantiomorphous groups of atoms within a crystal and enantiomorphous crystals of the same mineral can exist.

The operation of either a mirror or an inversion center generates an enantiomorphous motif that is identical in all respects to the original in size and shape, but the two cannot be made to occupy the same space. Symmetry operations that generate such enantiomorphous motifs are termed **improper** to distinguish them from **proper** operations that generate congruent motifs.

The repetition of an asymmetric motif by translation or proper rotation produces a sequence of congruent motifs (see figure 2.14a), whereas repetition by reflection or inversion produces a series of enantiomorphous pairs (see figure 2.14b).

Subsymmetries

Symmetry sets consistent with the various kinds of plane lattices may contain a number of subsymmetries that are the same as those of the related but more specialized cell. A square lattice, for example, contains all the symmetry elements of a rectangle, which in turn contains those of a parallelogram. Various possible subsymmetries for a square mesh are shown in figure 2.15.

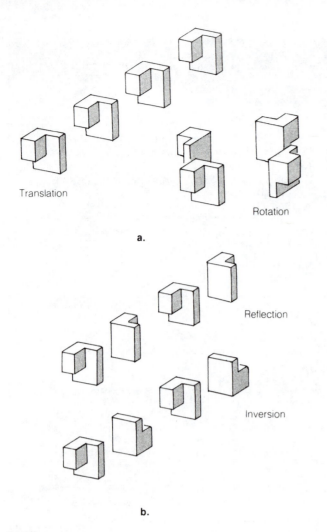

Translation

Rotation

a.

Reflection

Inversion

b.

Figure 2.14

(a) Repetition of congruent motifs. (b) Repetition of enantiomorphous motifs.

Space Lattices

The periodic repetition of a plane lattice in some noncoplanar direction generates a space lattice that is a stack of identical plane lattices in parallel position separated by a translation t_3. The location of the successive plane lattices in the stack must be consistent with the symmetry elements of the zero level. The number of space lattice types is, therefore, sharply limited, because axes and mirrors associated with the plane lattice must pass through an equivalent position in each successive layer.

A stack of plane lattices consistent with an n-fold axis can be devised by starting with a plane lattice having an n-fold axis location and adding successive levels in such a way as to have this axis intersect n-fold axial locations in the upper levels. The relative placement of the zero and first levels is conveniently described by giving the horizontal components of t_3 in terms of the fractions x and y of the translations t_1 and t_2, respectively, plus the perpendicular distance z between the sheets. The location of A' with respect to A in figure 2.16 using this system is $x = \frac{1}{3} t_1$, $y = \frac{1}{2} t_2$, z, or simply $\frac{1}{3}$, $\frac{1}{2}$, z.

The general possibilities for location of successive levels with respect to a given zero level are illustrated in figure 2.17. The zero level (figure 2.17a) has 2-fold axes normal to the page at positions represented by the solid circles. Distinct ways of stacking successive levels are shown by the plan views (figure 2.17b) slightly offset for easier visualization and by three-dimensional representations (figure 2.17c). The x, y components of the t_3 coordinates that may be chosen are the x, y components of one 2-fold axis with respect to another on the zero level. The distinct possibilities in this instance are 0 0, 0 ½, ½ 0, and ½ ½.

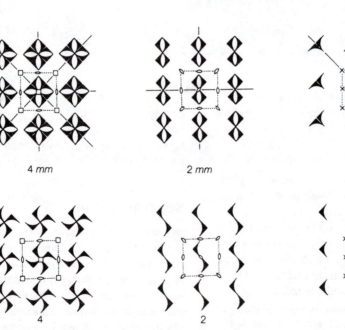

4 mm

2 mm

m

4

2

1

Figure 2.15

Subsymmetries. Different patterns are generated by the replication of the same motif on identical square meshes. The symmetry elements of 1 are included in all others; those of 2 are included in 2mm and 4mm; and the symmetry elements of m, 2mm, and 4 are included in 4mm.

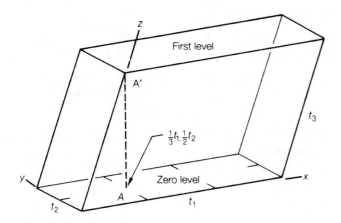

Figure 2.16
Coordinates of t_3.

Table 2.2		General Space Lattice Types	
Name	**Symbol**	**Location of Additional Points**	**Total Number of Lattice Points per Cell**
Primitive	*P*	—	1
Body-centered	*I*	Center of cell	2
Rhombohedral	*R*	Two points along the body diagonal of the cell	3
A-centered	*A*	Center of front and back faces	2
B-centered	*B*	Center of side faces	2
C-centered	*C*	Center of top and bottom faces	2
Face-centered	*F*	Centers of all faces	4

As successive levels are added, obviously some upper plane lattice must fall directly over the plane lattice of the zero level. Comparable points in these two levels can be connected to form an orthogonal cell. Such cells are outlined by dashed lines in figure 2.17c. The orthogonality gained by the use of these multiple cells is of considerable aid in visualizing and describing space lattices.

Multiple cells have lattice points on their faces or in their interiors, in addition to those on their corners. Cells with interior points are called **body-centered** (*I*) for space lattices based on parallelogram, square, rectangle, and diamond plane lattice meshes and **rhombohedral** (*R*) for space lattices based on a 120° rhombus. Cells with lattice points on their faces are termed **A-, B-,** or **C-centered** for point pairs on the front and back, side, or top and bottom faces or **face-centered** (*F*) if points are present on all six faces of the cell. The various cells that may be derived are listed in table 2.2.

Cells Based on a General Parallelogram

A general parallelogram plane lattice mesh is consistent with plane symmetries 1 and 2 (see figure 2.7). In symmetry 1, the operation of a proper 1-fold axis returns a motif to its initial position in a single operation, so such

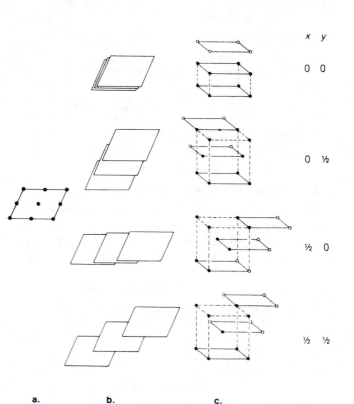

Figure 2.17
General possibilities for stacking plane lattices.

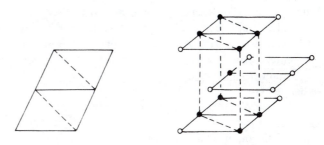

Figure 2.18
Alternate choices of cells.

an axis need not be perpendicular to the plane lattice. In consequence, t_3 coordinates are not restricted to simple fractional parts of t_1 and t_2 but may have any value between 0 and 1. The unit cell usually chosen in this instance is primitive, that is, a parallelepiped outlined by the shortest t_1, t_2, and t_3 translations. This space lattice is designated as 1*P*. The angular difference from orthogonality in real crystals is often slight, and a structure based on this cell may be easily mistaken for one of higher symmetry.

A general parallelogram consistent with plane symmetry 2 is that described in the previous section and illustrated by figure 2.17. Possible distinct t_3 coordinates in this instance are 0 0 *z*, 0 ½ *z*, ½ 0 *z*, and ½ ½ *z*. These give rise to orthogonal cells with parallelogram bases that are called, respectively, *P, B, A,* and *I* (see table 2.2). The choice of an alternate cell, as shown in figure 2.18 by dashed lines, however, can transform a face-centered point to a body-centered point. Consequently, the only distinct lattice types consistent with plane symmetry 2 are primitive and body-centered cells. These are called 2P and 2I, respectively. It should be noted that 2I is sometimes designated 2C, as in the *International Tables for X-ray Crystallography*. This may be confusing since C usually designates a point on a face rather than an interior point.

Cells Based on Rectangles and Diamonds

Rectangular and diamond-shaped plane lattice meshes are consistent with plane symmetries m and 2mm (see figure 2.7). The possible t_3 coordinates for a rectangle are the same as for a parallelogram. The coordinates 0 0 z define a primitive brick-shaped cell. The coordinates 0 ½ z and ½ 0 z define a space lattice from which orthogonal B-centered and A-centered cells may be chosen. These are, however, not distinct lattice types since a rotation of 90° around the z direction will transform one into the other. The coordinates ½ ½ z define a space lattice from which a body-centered orthogonal cell may be chosen. Distinct space lattice types consistent with a rectangular plane lattice are thus composed of primitive, A- or B-centered, and body-centered cells that are designated as 222P, 222A, and 222I, respectively. The designation 222 refers to three intersecting 2-fold axes.

The only possible t_3 coordinates consistent with the 2mm symmetry of a diamond-shaped plane lattice are 0 0 z and ½ ½ z. The first set defines a primitive cell having a diamond-shaped base. A more convenient cell, however, is C-centered, as shown in figure 2.19. This lattice type is not distinct from A- or B-centered cells because one can be derived from the other by a rotation of 90°.

The coordinates ½ ½ z define a body-centered cell having a diamond base. This cell is geometrically difficult to deal with, and it is customary to choose a face-centered cell as shown in figure 2.20. This constitutes the only space lattice different from those derived from a rectangular plane lattice. It is designated as 222F.

Cells Based on a Square

A square plane lattice is consistent with plane symmetries 4 and 4mm. The two permissible locations of 4-fold axes with respect to the plane net are at 0 0 and ½ ½. The two possible values of t_3 are then 0 0 z and ½ ½ z. The first defines an orthogonal primitive cell and the second a space lattice from which a body-centered orthogonal cell may be chosen. These two cells are identified as 4P and 4I respectively.

A cubic space lattice may also be derived from a square plane lattice by setting t_3 equal to t_1 and t_2. The possible coordinates of t_3 in this instance are 0 0 1 and ½ ½ ½, which define primitive and body-centered cubes designated respectively as 23P and 23I. Because consistent 4-fold axis locations will also exist parallel to the plane

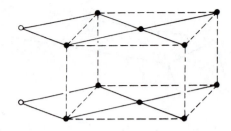

Figure 2.19
Relation of a cell with a diamond base to one with a rectangular base.

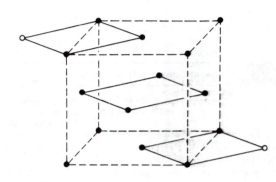

Figure 2.20
Derivation of space lattice type 222F.

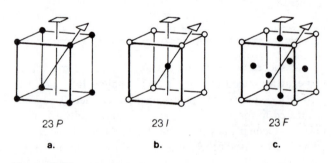

23 P   23 I   23 F

a.   b.   c.

Figure 2.21
Cubic cells. (a) 23P, (b) 23I, (c) 23F. The characteristic lattice points are blackened.

mesh and passing through the centers of the vertical faces, a face-centered cell 23F may also be chosen. Figure 2.21 illustrates the cubic cells.

Cells Based on a 120° Rhombus

Plane symmetries 3, 3m, and 6mm are consistent with a plane mesh composed of 120° rhombi. Permissible positions of 3-fold axes consistent with this mesh are at 0 0, ⅔ ⅓, and ⅓ ⅔ (see figure 2.7). The coordinates of t_3 may, therefore, be 0 0 z, ⅔ ⅓ z, or ⅓ ⅔ z. The first coordinate set defines a primitive cell (see figure 2.22a) that is designated 3P. The other two coordinate sets define cell types

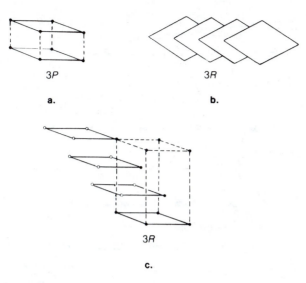

Figure 2.22

Cells based on a 120° rhombus. (a) 3P, (b) and (c) 3R. (After Buerger, 1963.)

Table 2.3	Summary of Specialized Space Lattices		
Plane Lattice Type	**t_3 Coordinates**	**Space Lattice Designation**	**Primitive Cell**
Parallelogram	xyz	1P	1P
Parallelogram	00z	2P	2P
Parallelogram	½ ½ z	2I	
Rectangle	00z	222P	222P
Rectangle	½ ½ z	222I	
Rectangle	0½ z	222A	
Diamond	½ ½ z	222F	
Square	00z	4P	4P
Square	½ ½ z	4I	
Square	001	23P	23P
Square	½ ½ ½	23I	
Square	0½ ½	23F	
120° Rhombus	00z	3P	3P
120° Rhombus	½ ½ z	3R	
Five plane lattice types	Fourteen space lattices		Six primitive cells

that can be transformed into one another by a rotation of 180° about a normal to the plane lattices. The only distinct nonprimitive cell is shown in figures 2.22b and 2.22c. This cell contains two interior lattice points located on the long body diagonal of the cell. It is closely related to the little-used primitive rhombohedral cell and is designated as 3R.

At submultiples of t_1 and t_2, 6-fold axes do not occur (see figure 2.7). The only possible t_3 coordinates are therefore 0 0 z, and the cell is primitive. This cell is identical with 3P and is so designated.

A total of fourteen symmetrically specialized kinds of space lattices have now been described. These are called the **Bravais lattices** after the crystallographer who in 1848 first showed that there are just fourteen. These lattices represent all the possible specialized space lattices. They are listed in table 2.3, which summarizes the symmetrical groupings to this point.

When the lattice points whose position defines a cell are connected by lines for easier visualization, the six primitive cells found are parallelepipeds whose faces are portions of plane lattices defined by t_1 and t_2, t_2 and t_3, and t_1 and t_3. In other words, these cells can be considered to be outlined by the intersection of three pairs of nonparallel plane lattices with the cell corners being the intersections of the set. In a space lattice, the plane symmetry sets associated with each face of the cell must interact. Thus, the only collections of symmetry elements consistent with a space lattice will be combinations of plane symmetries in specialized positions and angular relationships that do not generate irrational lattice points.

The coordinate designations of x, y, and z have been used to this point in their usual geometric sense. When dealing with crystalline matter, however, it is conventional to replace them with a, b, and c. Crystals are oriented for examination or description by placing the a-axis front (+) to back (−), or pointing *at* you, the b-axis right (+) to left (−), or *broadside*, and the c-axis up (+) and down (−).

Indexing

The selection of a unit cell makes it possible to describe the crystallographic position of rational lines and planes consistent with a lattice, and thus the lines and planes of real crystals. For example, cleavage planes and crystal faces and edges can be described in terms of the chosen lattice.

The generally accepted notation for describing the orientation of lines and planes with respect to crystallographic axes, called **indexing,** was first developed in 1825 by the Reverend Dr. W. Whewell, a professor of mineralogy at Cambridge University. His method was popularized by one of his students, Professor W. H. Miller, in his *Treatise on Crystallography* (1839). As a consequence, this system of notation is now popularly, and incorrectly, called the Miller indices.

The basic geometrical assumption of indexing is that any lattice point can serve as the origin of coordinates. Other important concepts are that any two points of a lattice can be connected by a straight line, three noncolinear points define a plane, and lines and planes are motifs that can undergo translation. Conventionally, the crystallographic directions, a, b, and c (i.e., the cell edges) are chosen for the coordinate system, which is not necessarily orthogonal.

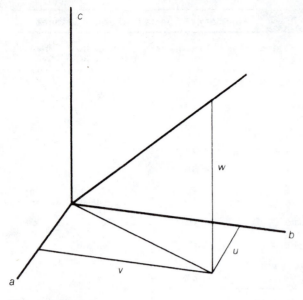

Figure 2.23
Coordinates [*uvw*] of a line.

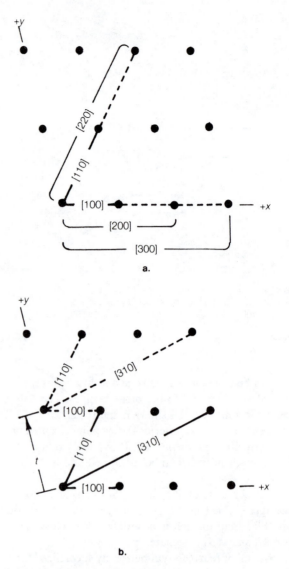

Figure 2.24
Repetition of rational lines by extension and translation. (a) The line [100] may be extended to [200], [300], and so on. Similarly, line [110] may be extended to [220], and so on. The index of lowest common denominator is always used, in this case, [100] and [110]. (b) The solid lines [100], [110], and [310] are repeated by translation *t* to the dashed line positions.

Notation for Lines

Any lattice point can be taken as an origin, and any other lattice point reached from it by a vector that is the linear sum of three vectors defined by the edges of the unit cell:

$$T = ut_1 + vt_2 + wt_3$$

where *u*, *v*, and *w* are integers. The orientation of any line in a lattice or crystal may thus be readily specified by using three integers that describe the number of unit translations in each translation direction. Because a line can be translated parallel to itself so that it passes through an origin, the **line index** is simply the *a*, *b*, and *c* coordinates of a point on the line relative to the chosen origin, or the integer coefficients *u*, *v*, and *w* (see figure 2.23). Conventionally, the line index is given in square brackets as [*uvw*].

Because it is the attitude and not the length of a line that is important, the integers representing *u*, *v*, and *w* are reduced to their least common denominator. In figure 2.24a line [220] is simply an extension of [110], and [200] and [300] are extensions of [100]. Parallel lines have the same index because the origin is arbitrary and a translation will generate one from the other (see figure 2.24b). Therefore, any rational line can be repeated by extension and non-colinear translation into a set of parallel lines having constant direction and separation, as shown for the planar mesh in figure 2.25. Note that the intercepts of these lines with the cell edges divide the edges into an integral number of parts equal to the reciprocal of the translation coordinates, respectively 3, 2, and 0 for the line [230] shown.

The orientation and index for some common lines in a crystal are shown in figure 2.26 with respect to the crystallographic axes *a*, *b*, and *c*. Some particular lines that may be readily visualized are

- lines parallel to edges of cells: [100], [010], and [001], which are respectively parallel to the *a*-, *b*-, and *c*-axes
- face diagonals of cells: [011] and [0$\bar{1}$1] on the front and back faces, [101] and [$\bar{1}$01] on the side faces, and [110] and [$\bar{1}$10] on the top and bottom faces

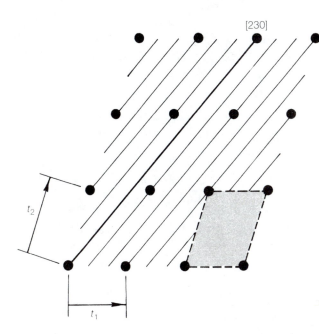

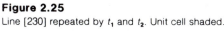

Figure 2.25
Line [230] repeated by t_1 and t_2. Unit cell shaded.

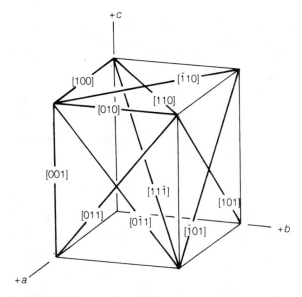

Figure 2.26
Indices of some common lines.

- body diagonals of cells: [111], [11$\bar{1}$], [1$\bar{1}$1], and [$\bar{1}$11] (not shown)

A bar over the index number indicates that it is negative. The line index [0$\bar{1}$1] is read "zero, bar one, one."

Notation for Planes

The attitude of crystallographic planes is of particular mineralogic importance. The **plane indices** are used to distinguish the different faces on a crystal and to characterize crystal forms (chapter 3), identify the surface of

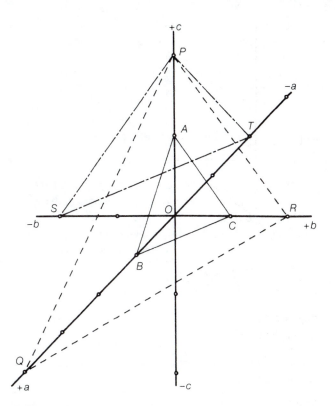

Figure 2.27
Coordinate axes showing unit translations in the *a*, *b*, and *c* directions.

juncture in twinned crystals (chapter 5), and locate the directions of mineral cleavage (chapter 9).

In crystallography, the actual position of a plane in space is not as important as is its position or attitude relative to other planes in the crystal. Figure 2.27 shows a plane *ABC* cutting the *a*, *b*, and *c* axes, respectively. Note that the distances *OA*, *OB*, and *OC* are all unit translations, that is, the intercepts are 1, 1, 1. The plane *ABC*, nearest to the origin, is called the **parametral plane** because it reflects the parameters of the chosen cell. A second plane *PQR* in figure 2.27 has intercepts 4, 2, 2 on *a*, *b*, and *c*, respectively. The position of *PQR* relative to *ABC* is given by *A/P*, *B/Q*, *C/R*, or ¼, ½, ½. The index of the plane *PQR* is given by clearing the fractions, yielding 211 (read "two one one"). The index of the parametral plane is, of course, 111 ("one one one").

The plane *PST* in figure 2.27 has *a*, *b*, and *c* intercepts equal to −2, −2, and 2, respectively. In relation to *ABC*, the position of *PST* is $1/-2$, $1/-2$, $1/2$, and the index of *PST* is $\bar{2}\bar{2}2$. Note that $\bar{2}\bar{2}2$ is parallel to $\bar{1}\bar{1}1$ and may be moved to that position by a simple translation. By convention the indices are reduced to their smallest integer values; thus $\bar{2}\bar{2}2$ is given as $\bar{1}\bar{1}1$ (read "bar one, bar one, one").

A plane parallel to an axis will intercept that axis only at infinity. Consider, for example, a plane intersecting the *a*-axis at t_1, the *b*-axis at $2t_2$, and parallel to *c*. The reciprocals of the intercepts are $1/1$, $1/2$, $1/\infty$ giving the indices 210.

Rational crystallographic planes must parallel or intersect the crystallographic axes at integral numbers representing unit translations from the origin. For example, the shaded plane shown in figure 2.28 has intercepts on the a-, b-, and c-axes of $-t_1$, $3t_2$, and $-2t_3$. Such a plane is a motif that can be moved parallel to itself by consistent translations. The general equation for a plane intersecting axes a, b, and c at distances x, y, and z from the origin is

$$a/x + b/y + c/z = 1$$

For the rational plane nearest to the origin (the parametral plane), which is parallel to all other planes in the stack, intercept $x = t_1/h$, $y = t_2/k$, and $z = t_3/l$, where h, k, and l are integers. These intercepts, although differing in absolute length, are each unit translations. Thus

$$\frac{a}{1/h} + \frac{b}{1/k} + \frac{c}{1/l} = 1$$

and

$$ha + kb + lc = 1$$

The integer variables h, k, and l are usually small and are called the indices of the plane.

For the specific case illustrated in figure 2.28, the values of the translations in the a-, b-, and c- directions are -1, $+3$, and -2, respectively. Taking the reciprocals and clearing fractions gives the indices of the plane hkl as $\overline{6}2\overline{3}$ ("bar six, two, bar three").

In summary, to determine the indices of a particular plane,

1. Choose the reference axes.
2. Choose the parametral plane abc.
3. Express the intercepts of the plane as a/h, b/k, c/l, or the intercepts of the parametral plane divided by the intercepts of the plane in question.
4. Clear the fractions to express the indices as simple rational numbers or zero.

Different styles of brackets indicate the nature of the plane being described. Parentheses () are used for a particular plane such as a crystal face, and braces {} for symmetrically related sets of planes such as a cleavage or a crystallographic form. In x-ray diffraction studies a single unbracketed index designates a family of parallel planes. Thus, planes parallel to the (110) face produce an x-ray response designated 110 for 110, 220, 330, and so on.

Numerically equal indices do not describe planes of identical attitude in different lattices because the indices refer to translation distances that differ from lattice to lattice. The planes shown in figure 2.29 are all (111). To illustrate further the relations of intercepts and indices, consider figure 2.30 which shows the traces of rational planes parallel to the c-axis as they cut the ab plane. The

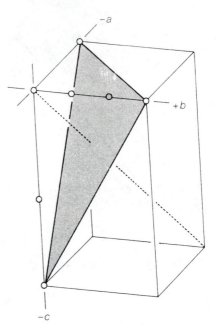

Figure 2.28
Index of a plane.

variation in the spacing d for planes of identical index but based on translations of different length assumes great importance in x-ray diffraction studies. The spacing d is a readily determined parameter that allows different mineral species to be identified.

When the exact indices are unknown, a **type symbol** (hkl) in which h, k, and l represent simple whole numbers serves to locate the plane approximately. For example, an (hkl) plane cuts all three crystallographic axes or would do so if extended. Similarly, $(hk0)$ cuts the a- and b- axes and is parallel to c, $(0kl)$ is parallel to a and cuts b and c, and so on. When a plane is parallel to two axes and intersects the third, $(0k0)$ for example, the plane is parallel to (010) and is so designated. Thus, (100), (010), and (001) intersect the a-, b-, and c-axes, respectively, and are parallel to the plane of the other two axes.

Lattices based on a 120° rhombus ($3P$ or $3R$) are unique in that the cells define a coordinate system of four axes. Three of the axes are of equal length and are designated as a_1, a_2, and a_3. The fourth axis, c, corresponds to the 3- or 6-fold symmetry axis and is customarily oriented vertically, perpendicular to the plane of the a-axes. In this system, the indices take the form $hkil$ or $hk \cdot l$, the appropriate brackets being used for lines or planes.

Except for the (0001) plane perpendicular to the c-axis, the geometry of this lattice requires both positive and negative terms in the index. In figure 2.31 several planes perpendicular to the page and parallel to the crystallographic c-axis ($l = 0$) are indexed to illustrate these

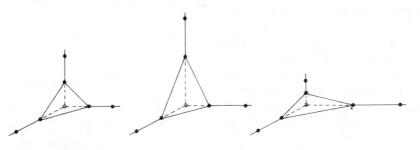

Figure 2.29
Different attitudes of a (111) plane.

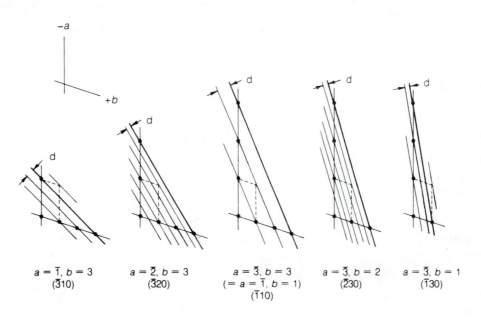

Figure 2.30
Traces of planes parallel to *c*-axis on *ab* plane for different intercepts.

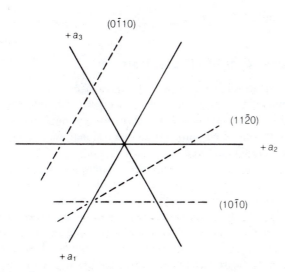

Figure 2.31
Indices in the hexagonal system.

relations. If the planes are inclined to the page, they will intersect the *c*-axis and $(11\bar{2}0)$, for example, will become $(11\bar{2}l)$ or $\{11\bar{2}l\}$. A quick check on the correctness of hexagonal indices is that the sum of the first two digits times -1 should equal the third digit. For example, for $(11\bar{2}0)$, $(1+1) \times (-1) = -2$ and for $(0\bar{1}11)$, $(0-1) \times (-1) = 1$.

Examples of using indices to identify different faces on crystals are given in figure 2.32. The application of indices to crystal forms is covered in chapter 3.

Zones

It is evident from examining crystals that several faces have edges that are mutually parallel. Such faces are referred to as **zones,** and the common direction of their mutually parallel edges is the **zone axis.** Edges, or zones, are rational lines with indices $[uvw]$.

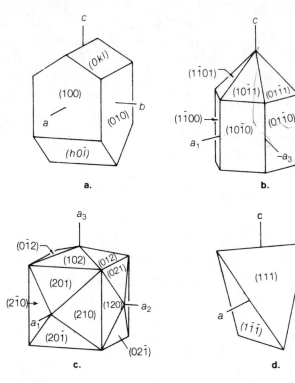

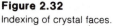

Figure 2.32
Indexing of crystal faces.

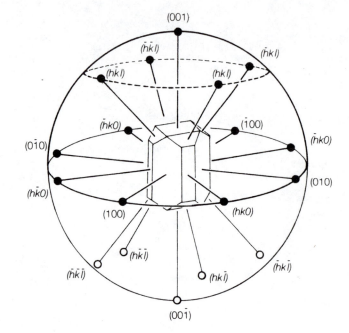

Figure 2.33
Development of a spherical projection on the upper hemisphere of a crystal.

The index of a zone axis or an edge can be calculated if the indices of the two intersecting faces are known. The equation of a plane is

$$ha + kb + lc = 1$$

When the plane is translated parallel to itself so that it passes through the origin, its equation becomes

$$ha + kb + lc = 0$$

If two planes with indices $h_1k_1l_1$ and $h_2k_2l_2$ forming a zone are so translated, then

$$h_1a + k_1b + l_1c = 0$$

$$h_2a + k_2b + l_2c = 0$$

The ratio of coordinates of a point on the line of intersection gives the index of the zone axis:

$$u:v:w = (k_1 \cdot l_2 - k_2 \cdot l_1):(l_1 \cdot h_2 - l_2 \cdot h_1): (h_1 \cdot k_2 - h_2 \cdot k_1)$$

A determinant can be useful for calculating a zone index, as shown below. Remember to pay attention to the sign of the indices.

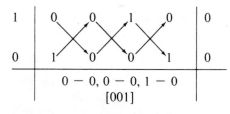

As an example, consider the zone formed by the (100) and (010) faces on the crystal of figure 2.32a. The determinant is constructed as follows:

The index of the zone axis is [001].

Stereographic Projection

A map showing the relative position of crystallographic components and real features of a crystal is highly desirable in mineral study. Such three-dimensional information is converted into a two-dimensional plot using a **stereographic projection.**

The oriented lattice or crystal is imagined to lie at the center of a sphere. Lines perpendicular to the planes or faces are projected from the crystal to intersect the sphere (see figure 2.33), and the point of each intersection is identified by its longitude (ϕ) and colatitude (ρ), as shown in figure 2.34. The positive (east) end of the b-axis is taken as $\phi = 0$; positive angles are measured clockwise and negative angles counterclockwise from this point. Because only one hemisphere, usually the upper, is conventionally used, values of ρ are always positive and are measured from the positive end of the c-axis (north pole) for points on the upper hemisphere and vice versa in the lower hemisphere.

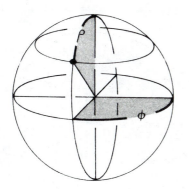

Figure 2.34
Location of a point on a spherical surface using longitude (φ) and colatitude (ρ).

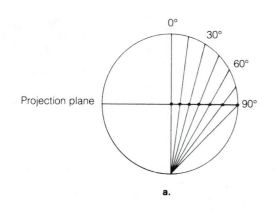

a.

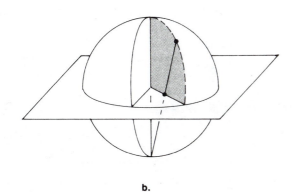

b.

Figure 2.35
Stereographic projection. (a) Vertical section through a sphere; the length of a degree on the plane of the stereographic projection increases from the center outward. (b) The principle of constructing a stereographic point from a spherical projection.

The spherical projection thus obtained must be converted to a flat surface for any practical use. No system of plotting the whole sphere on a flat surface has been found to be satisfactory, and for crystallographic problems only one hemisphere is commonly plotted. The stereographic projection is used in preference to other possible projections because (1) the entire hemisphere can be

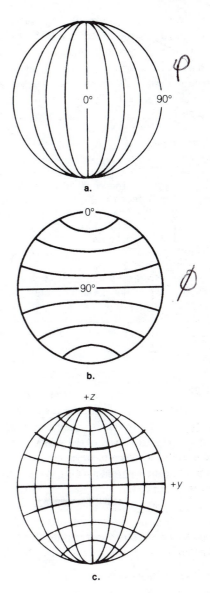

a.

b.

c.

Figure 2.36
Construction of a stereonet. The net (c) combines regularly spaced meridional circles (a) and latitudinal circles (b).

plotted, (2) symmetry about the major axes is generally apparent, (3) symmetry mirrors and axes are readily plotted, and (4) certain x-ray data plot easily and naturally on these projections.

The conversion of points on a sphere to a two-dimensional plot that preserves their angular relationships is done by projecting a point on the upper hemisphere to the equatorial plane along a line connecting the point to the south pole, as shown in figure 2.35 (vice versa for points plotted on the lower hemisphere). The two-dimensional representation of the angular relations of crystal faces and elements of symmetry is facilitated by using a meridian stereonet, or Wulff net, developed by the Russian crystallographer G. V. Wulff (1863–1925). The net is generated by first projecting great circles onto the plane of the prime meridian as shown in figure 2.36a.

In this projection, true angles of longitude (ϕ) are shown on the x-axis, 0° at the center and 90° on the periphery. Next, small circles representing colatitudes (ρ) are projected to the same plane. Figure 2.36b shows the result in which true angles are shown on the z-axis, 90° at the center and 0° at the poles. Finally, the two projections are combined, as in figure 2.36c, into the stereonet.

The Wulff net reproduces angles between the lines on a sphere without distortion. It is thus often referred to as an *equal-angle* projection. The Wulff net is most often used in mineralogic studies because knowing the precise angles between faces and symmetry elements is very important.

Another device, the **Schmidt Net,** preserves the spatial distribution of the original data. It is constructed so that angular distances are not reduced in the central part of the diagram, and thus it is an *equal-area* projection. The Schmidt net is often employed in structural geology, where spatial relationships between planar elements are important for statistical analysis.

Using a Stereonet

In use, the stereonet is usually 20 cm in diameter with 2° spacing of the great and small circles. It is covered with tracing paper that is pinned at the center of the net. Then the outer or primitive circle is traced, and a *reference mark* is placed at the positive end of the y-axis ($\phi = 0$ position). A somewhat more elegant stereographic plotting board consists of a circular raised area carrying a Wulff net surrounded by a rotatable annular ring, as shown in figure 2.37. The ring has corks embedded in it to which a piece of tracing paper may be thumbtacked.

To plot a pole if ϕ and ρ are known, follow three steps:

1. Locate and mark ϕ on the circumference of the primitive circle counting from $\phi = 0$ on the north end of the y-axis, clockwise for positive angles and counterclockwise for negative angles.
2. The ρ value for the face pole must lie along the straight line between the center of the net and the mark made in step 1. Because true angles are shown only on the east-west and north-south axes, rotate the paper overlay until the mark made in step 1 is on such an axis. Measure the value of ρ *outward* from the center and plot the point. This is the pole position of the particular face on the sphere projected to the stereonet. Commonly, face poles projected to the upper hemisphere are indicated on stereographic projections by a small circle and those projected to the lower hemisphere by a cross, but usage varies.
3. Rotate the tracing paper to its original position and repeat the steps above to locate the next pole.

Some general features of the arrangement of points plotted on a stereonet make it easy to interpret the morphology of a crystal. Horizontal faces, (001), have vertical

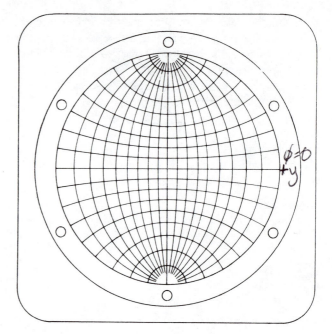

Figure 2.37
Stereographic plotting board with Wulff net.

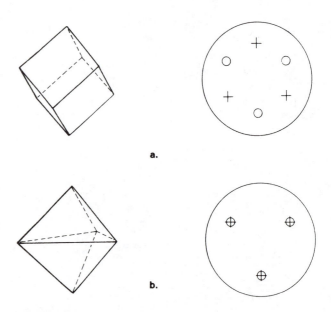

Figure 2.38
Use of a lower-hemisphere projection to distinguish a rhombohedron (a) from a trigonal dipyramid (b). Points projected from the lower hemisphere are plotted as crosses.

face normals and plot in the center of the stereonet. Vertical faces, ($hk0$), have horizontal face normals and plot on the periphery of the net. Zones, collections of symmetrically equivalent faces having parallel edges, plot along a great circle of the projection. The zone axis is perpendicular to this circle.

Occasionally it is useful to combine upper- and lower-hemisphere presentations. For example, the plot of only the top faces of a rhombohedron would not distinguish it from a trigonal dipyramid (see figure 2.38). In such cases, the face poles intersecting the lower hemisphere are projected toward the north pole and recorded at their intersection with the equatorial plane as crosses. Faces intersecting the upper hemisphere are projected toward the south pole and are represented by circles.

A number of aspects of crystal symmetry are readily shown by plotting on a stereonet. The arrangement of the elements of symmetry in the thirty-two crystal classes is represented in this manner in figure 2.48. A stereonet plot showing the location of the axes and planes of symmetry of a cube is shown in figure 2.39. Figures 2.39a and 2.39c are perspective drawings of a cube showing its axes of symmetry and mirror planes, respectively. Figures 2.39b and 2.39d show these symmetry elements in stereographic projection. Figure 2.39e is a stereographic projection of all the symmetry elements of a cube.

An Example of Stereographic Projection

The following outline presents a step-by-step sequence for measuring, plotting, and indexing the faces of the orthorhombic crystal (class $2/m\,2/m\,2/m$) shown in figure 2.40.

1. Orient the crystal by selecting positions for the axes, in this instance choosing the greatest length as the c direction and the shortest as a, thus making the large front face (100).

2. Measure in sequence all the interfacial angles across vertical edges. Starting with the front face (100) and moving around the crystal to the right, the sequence of interfacial angles would be 25°, 65°, 65°, 25°, 25°, 65°, 65°, and 25°. Figure 2.41a shows a top view of the crystal as it would appear looking along the c-axis.

3. Plot the vertical faces at their appropriate angular positions around the perimeter of the projection (see figure 2.41b). Because normals to vertical faces are horizontal, all ($hk0$) faces will plot on the perimeter.

4. The third face is 25° plus 65° or 90° from (100) and is best selected as (010). The face between (100) and (010) is the largest face cutting both the a- and b-axes and by rule should have the simplest index, (110).

5. Next, determine the interfacial angles from (100) to the face above it, and from that to the top face on the crystal; do the same from (010). The normals to these faces all lie on mirror planes and are readily plotted by counting degrees inward from the perimeter. The two interfacial angles between (100) and the top face are 38.5° and 51.5°. The top face is thus 90° from (100) and is most simply considered to be (001). The simplest index for the ($h0l$) intermediate face is (101). Angles between (010) and (001) are 40.5° and 49.5°, and the single intermediate face ($0kl$) may be labeled (011), although this is an erroneous conclusion, as will be

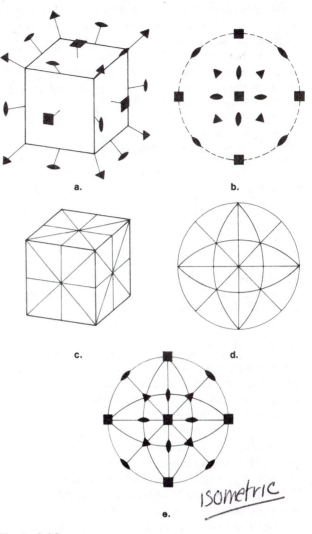

a.   b.

c.   d.

isometric

e.

Figure 2.39
Stereonet plot of the symmetry elements of a cube. (a) Rotational elements. (b) Stereographic plot of rotary elements. (c) Reflection elements. (d) Stereographic plot of reflection elements. (e) Complete stereonet plot.

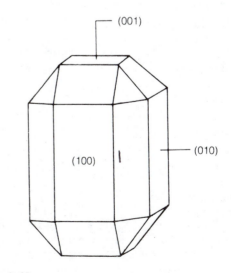

(001)

(100)   (010)

Figure 2.40
Orthorhombic crystal showing indices of some faces.

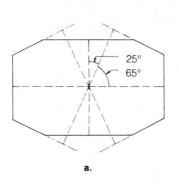

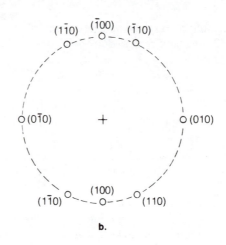

b.

Figure 2.41
Top view of orthorhombic crystal of figure 2.40 (a) and its indexing (b).

shown later. These faces, all lying on mirror traces, are plotted in figure 2.42.

Each of the eight triangular faces of the remaining form cut all three axes and are components of an orthorhombic dipyramid (111). These faces do not lie on a plane of symmetry and must be plotted at the intersection of angular distances from two other faces. The interfacial angles measured are

$$(100) \wedge (111) = 54°,$$

$$(010) \wedge (111) = 79°,$$

$$(110) \wedge (111) = 50°.$$

To locate the position of the face normals, start at (100) on the stereonet projection and count 54° toward the center. Then draw along that line toward the right as in figure 2.43a. Next, rotate (010) to the point originally occupied by (100), count inward 79°, and trace a line left as in figure 2.43b. The point of intersection is the location of (111) at 54° from (100) and 79° from (010). As a check, rotate (110) into the position initially occupied by (100) and count up to the intersection, which should be 50° (see figure 2.43c).

This completes the mechanics of plotting face positions on the projection, but there is an error in indexing. The form indicated as (011) is actually (021). This will be apparent only after a check is made for consistency of indexing with axial ratios.

Because the interfacial angle between (100) and (110) is 25°, and because in axial ratios the b-axis is assigned a relative value of 1, the relative length of the a-axis can be determined as follows.

The axes a and b can be drawn at right angles as in figure 2.44a. The face normal to (110) is

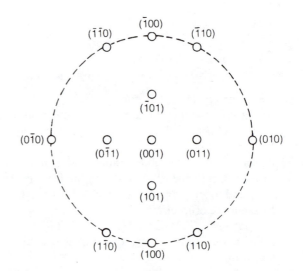

Figure 2.42
Stereogram with tentative indices.

drawn at 25° to the a-axis, and (110) is constructed normal to it. The tangent of angle α is

$$a/b = a/1 = a$$

and

$$\tan 25° = 0.466 = a$$

Therefore, the axial ratios of the mineral, assuming (110) is correctly indexed, are $a:b:c = 0.466:1:c$.

The axial ratio of $a:c$ can be similarly determined using (101), as in figure 2.44b:

$$\cot \beta = c/a$$

$$\cot 38.5° = c/0.466$$

$$c = 0.586$$

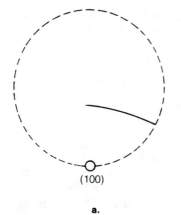

a.

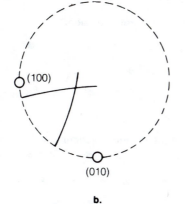

b.

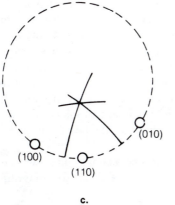

c.

Figure 2.43
Location of face normals.

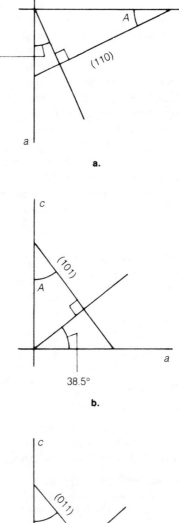

a.

b.

c.

Figure 2.44
Consistency of indexing.

Thus, assuming that the faces are correctly indexed, $a:b:c = 0.466:1:0.586$, and (011) can be checked, as shown in figure 2.44c:

$\tan \gamma = c/b = c/1 = c$, where γ is the complement of 40.5°, which is 49.5°

and

$\tan 49.5° = \tan \gamma = 1.171$

This result is apparently double the previous figure of $c = 0.586$. Thus, either (101) or (011) is an incorrect index. The error can be corrected by assuming that (011) is (021) or that (101) is (102). The latter supposition would also require that (110) be (120). The simpler decision is to assume (011) was erroneously indexed and correct it to (021). The accuracy of the (111) can also be checked using similar procedures, but the method is a bit more laborious.

Point Groups

As we have seen, lattice points and motifs can be reproduced by the operation of a limited number of elements of symmetry that, combined in certain permissible ways, will generate the complete lattice or pattern. All possible combinations of these elements should thus yield all of the possible symmetries that can be assumed by periodically repeated motifs. The motifs may represent the faces on a real crystal or relate its directional physical properties to its atomic arrangement.

In crystallography, the various symmetries can be organized by considering the distribution of points or motifs around a central unmoving point. In this way, combining possible and nonidentical elements of symmetry generates **point groups.** The basic symmetry element used is a rotary axis in consistent combinations with itself and other elements of symmetry.

The possible cases are

n	Single, proper, n-fold axes that are consistent with the five plane lattice meshes when n is 1, 2, 3, 4, or 6. When n is greater than 1, the axis must be perpendicular to the lattice plane.
$\bar{n}$	Single, improper n-fold axes arising from the combination of an n-fold proper axis with an inversion center, $n \cdot i = \bar{n}$. These lattices are centrosymmetrical and must be consistent with inversion centers.
$\dfrac{n}{\bar{n}}$	Coincident proper and improper axes of the same multiplicity. When $\bar{n}$ is even, the set becomes n/m; when $\bar{n}$ is odd, the set is equivalent to $\bar{n}$.
$n_1 n_2 n_3$	Intersecting, consistent sets of proper axes.
$n_1 \bar{n}_2 \bar{n}_3$ $\bar{n}_1 n_2 \bar{n}_3$ $\bar{n}_1 \bar{n}_2 n_3$	Intersecting, consistent sets of one proper and two improper axes. A "double negative" is required for final congruence after sequential operations.
$\dfrac{n_1}{\bar{n}_1} \dfrac{n_2}{\bar{n}_2} \dfrac{n_3}{\bar{n}_3}$	Combined, intersecting, and consistent sets of proper and improper axes.

The n point groups consistent with a lattice may all be found by trial and error along the lines of the example shown in figure 2.12. The point groups are five in number,

are conventionally designated 1, 2, 3, 4, and 6, and may be illustrated diagrammatically:

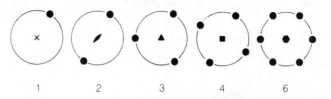

Single improper axes arise from the combination of one of the proper n groups with an inversion center on the axis or with a mirror perpendicular to it. The basic geometry is shown in figure 2.11b. Following the diagrammatic representations, with solid circles for motifs above the plane of the page and open circles for motifs below the plane of the page, the various cases can be illustrated:

Rotation and inversion (rotoinversion)

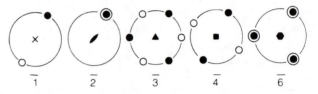

Rotation and reflection (rotoreflection)

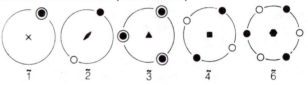

Comparison of the rotoinversion and rotoreflection patterns reveals that only five symmetries are represented, since $\tilde{1} = \bar{2}$, $\tilde{2} = \bar{1}$, $\tilde{3} = \bar{6}$, $\tilde{4} = \bar{4}$, and $\tilde{6} = \bar{3}$. It is conventional to use rotoinversion notation ($\bar{n}$), and five further symmetries designated $\bar{1}$, $\bar{2}$, $\bar{3}$, $\bar{4}$, and $\bar{6}$ are thus identified. Since $\bar{2}$ is geometrically equivalent to a mirror perpendicular to the axis, it is conventionally designated as m. Also, $\bar{6}$ is symmetrically identical to $3/m$ (see following discussion), and the latter notation is conventionally used.

Coincident proper and improper axes having the same multiplicity may also be illustrated diagrammatically:

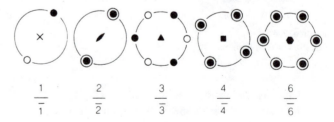

By comparison with the $\bar{n}$ groups it may be seen that $1/\bar{1}$ equals $\bar{1}$ and that $3/\bar{3}$ is the equivalent of $\bar{3}$. Thus, no new point groups arise from these combinations. On the other hand, $2/\bar{2}$, $4/\bar{4}$, and $6/\bar{6}$ are three new symmetries. The operation of such axial combinations, however, is equivalent to that of a proper axis normal to a mirror. In consequence, these groups are conventionally designated $2/m$, $4/m$, and $6/m$.

In general, rotation about two intersecting axes is equivalent to some other single rotation about a third axis passing through the common center. This is demonstrated by the motion of a point on the surface of a sphere resulting from rotation about an axis normal to the plane of the motion. In figure 2.45 rotation around axis A through some angle α causes an initial point to follow path 1. Rotation on axis B through an angle β then causes the point to follow path 2. The same final point may be reached by a single rotary operation of γ around axis C, with the point following path 3. The counterclockwise rotation $\bar{A}_\alpha$ followed by the counterclockwise rotation $\bar{B}_\beta$ is thus equivalent to the single clockwise rotation C_γ. This is an identical operation and may be rewritten in the form $\bar{A}_\alpha \cdot \bar{B}_\beta \cdot C_\gamma = 1$.

Three axes so related are designated by a set of three numbers. Thus, 432 signifies a set of three intersecting axes having multiplicities of 4, 3, and 2. Successive operations of these axes, when combined at appropriate angles, will reproduce an initial point.

In the general case, any pair of axes, A_α and B_β, can be combined at any angle of intersection, and such a combination requires the presence of a third rotation axis, C_γ. In crystals, however, the only permissible values of α, β, and γ are those of 1-, 2-, 3-, 4-, and 6-fold rotary operations. A limited number of combinations can be anticipated that will repeat the initial motif after three successive rotary operations. This is because the axial set must meet fixed requirements of angular displacement and interaxial angles in order for an initial motif to be reproduced.

There are six permissible $n_1 n_2 n_3$ combinations of intersecting axes—222, 322, 422, 622, 332, and 432—whose angular relations are shown in figure 2.46.

Combined axes may be proper or improper providing only that the initial motif is reproduced by successive operation of the three intersecting axes. This condition may be fulfilled by the successive operation of three proper axes or of a proper axis and two improper axes. The combinations of one improper and two proper or of three improper axes will produce an enantiomorph of the initial motif and cannot be combined in a consistent set.

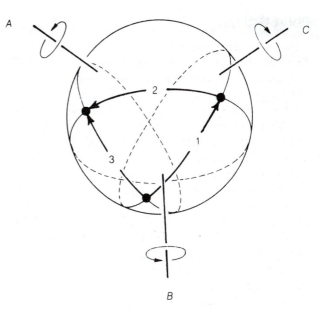

Figure 2.45
Combination of three rotations.

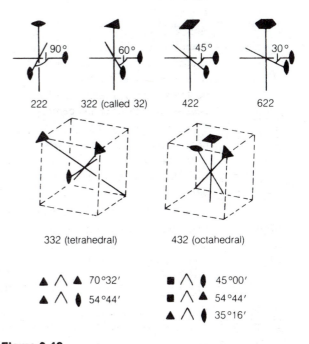

Figure 2.46
Permissible combinations of three intersecting axes. (After Buerger, 1963.)

Table 2.4   The Thirty-Two Crystal Classes (International Notation)

Axial Type	Classes						Number of Classes
n	1	2	3	4	6		5
$\bar{n}$	$\bar{1}$	$\bar{2}=m$	$\bar{3}$	$\bar{4}$	$\bar{6}=\dfrac{3}{m}$		5
$\dfrac{n}{\bar{n}}$	$\dfrac{1}{\bar{1}}=\bar{1}$	$\dfrac{2}{\bar{2}}=\dfrac{2}{m}$	$\dfrac{3}{\bar{3}}=\bar{3}$	$\dfrac{4}{\bar{4}}=\dfrac{4}{m}$	$\dfrac{6}{\bar{6}}=\dfrac{6}{m}$		3
$n_1n_2n_3$	222	322 called 32	422	622	332 called 23	432	6
$n_1\bar{n}_2\bar{n}_3$ $\bar{n}_1n_2\bar{n}_3$ $\bar{n}_1\bar{n}_2n_3$	$2\bar{2}2=2mm$	$3\bar{2}2=3mm$ called 3m; $3\bar{2}2=\bar{3}m2=\bar{3}\dfrac{2}{m}$	$4\bar{2}2=4mm$; $\bar{4}\bar{2}2=\bar{4}2m$	$6\bar{2}2=6mm$; $6\bar{2}2=\bar{6}m2$		$\bar{4}3\bar{2}=\bar{4}3m$	7
$\dfrac{n_1}{\bar{n}_1}\dfrac{n_2}{\bar{n}_2}\dfrac{n_3}{\bar{n}_3}$	$\dfrac{2}{\bar{2}}\dfrac{2}{\bar{2}}\dfrac{2}{\bar{2}}=\dfrac{2}{m}\dfrac{2}{m}\dfrac{2}{m}$	$\dfrac{3}{\bar{3}}\dfrac{2}{\bar{2}}\dfrac{2}{\bar{2}}=\bar{3}\dfrac{2}{m}\dfrac{2}{m}$ called $\bar{3}\dfrac{2}{m}$	$\dfrac{4}{\bar{4}}\dfrac{2}{\bar{2}}\dfrac{2}{\bar{2}}=\dfrac{4}{m}\dfrac{2}{m}\dfrac{2}{m}$	$\dfrac{6}{\bar{6}}\dfrac{2}{\bar{2}}\dfrac{2}{\bar{2}}=\dfrac{6}{m}\dfrac{2}{m}\dfrac{2}{m}$	$\dfrac{3}{\bar{3}}\dfrac{3}{\bar{3}}\dfrac{2}{\bar{2}}=\bar{3}\bar{3}\dfrac{2}{m}$ called $\dfrac{2}{m}\bar{3}$	$\dfrac{4}{\bar{4}}\dfrac{3}{\bar{3}}\dfrac{2}{\bar{2}}=\dfrac{4}{m}\bar{3}\dfrac{2}{m}$	6

Total number of classes = 32

A total of seven distinct point-group symmetries can be generated by the combined operations of proper and improper axes:

$2\bar{2}2$ (which equals $2mm$, see figure 2.47)
$3\bar{2}2$ (which equals $3mm$ and is called $3m$)
$4\bar{2}2$ (or $4mm$)
$\bar{4}\bar{2}2$ (which is equivalent to $\bar{4}2m$)
$6\bar{2}2$ (or $6mm$)
$\bar{6}22$ (which is the same as $\bar{6}m2$)
$\bar{4}3\bar{2}$ (which equals $\bar{4}3m$)

Combined axes of the type $n/\bar{n}$ may intersect in $(n_1/\bar{n}_1)(n_2/\bar{n}_2)(n_3/\bar{n}_3)$ sets resulting in six distinct symmetries:

$2/\bar{2}\ 2/\bar{2}\ 2/\bar{2}$ (which is equivalent to $2/m\ 2/m\ 2/m$)
$3/\bar{3}\ 2/\bar{2}\ 2/\bar{2}$ (which equals $\bar{3}\ 2/m\ 2/m$ and is called $\bar{3}\ 2/m$)
$4/\bar{4}\ 2/\bar{2}\ 2/\bar{2}$ (or $4/m\ 2/m\ 2/m$)
$6/\bar{6}\ 2/\bar{2}\ 2/\bar{2}$ (or $6/m\ 2/m\ 2/m$)
$3/\bar{3}\ 3/\bar{3}\ 2/\bar{2}$ (which equals $\bar{3}\ \bar{3}\ 2/m$ and is conventionally called $2/m\ \bar{3}$)
$4/\bar{4}\ 3/\bar{3}\ 2/\bar{2}$ (which is equivalent to $4/m\ \bar{3}\ 2/m$)

All the symmetry elements or their combinations, thirty-two in number, that are consistent with space lattices and that reproduce an initial motif have now been examined, and no new sets may be found by new combinations. These axial sets were developed by examining the symmetrical distribution of lattice points about some particular lattice point. Because this distribution is the same for all lattice points in the same lattice, they describe the overall symmetry of a particular lattice. In such an arrangement, a single point (at the intersection of axes, in a mirror, or at a center) will be unmoved by the operation of the symmetry elements. Such sets of symmetry elements are called **crystallographic point groups.** The lattice provides a reference framework to which atoms of a structure are related, and crystals must therefore conform to point-group symmetry.

The symmetry of crystals is shown by the internal arrangement of constituent atoms, by the directional control of certain physical and chemical properties, and by the distribution of faces bounding the crystalline solid. A crystal face represents some particular rational plane that is symmetrically repeated by the point group. Other rational planes are also repeated, and the result is a polyhedral solid bounded by collections of symmetrically disposed faces. The mineralogical divisions based on these polyhedra and corresponding to point-group symmetry are the **crystal classes.** Table 2.4 summarizes the crystal classes with respect to consistent axes and axial sets.

The distribution of symmetry elements in the thirty-two crystal classes is shown in figure 2.48. The diagrams represent the top view of a sphere whose outline is dotted unless there is an equatorial (horizontal) mirror, in which case it is shown as a solid line. Other mirrors projected on the spherical surface are shown as solid lines. The symbols for the vertical axes are centered, and the symbols for the horizontal axes are on the periphery of the circular projection. Those for inclined axes have intermediate positions. Improper axes, alone or in combination, are represented by a small open circle within the axis symbol.

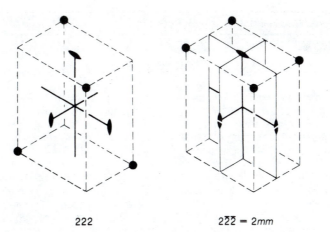

222   2$\overline{2}\overline{2}$ = 2mm

Figure 2.47
Comparison of 222 with 2 $\overline{2}$ $\overline{2}$ and identity of 2 $\overline{2}$ $\overline{2}$ with 2mm.

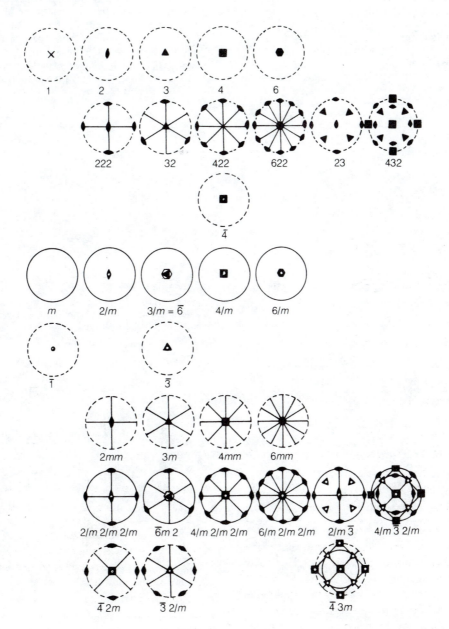

Figure 2.48
Symmetry of the thirty-two crystal classes.

Box 2.2

An Introduction to Space Groups

Our discussion of the geometrical properties of an orderly and periodic lattice has been concerned so far with the symmetrical disposition of points or motifs about a point. In a space lattice, however, the points have neighbors, and all have identical point-group symmetry. The interaction of symmetry elements between point groups generates two further (hybrid) symmetry elements that combine translations between point groups and reflections or rotations within point groups. These new elements of symmetry are a glide reflection and a screw axis.

Figure 2.49 illustrates the interaction of a translation t acting on a mirror m at some arbitrary angle to its plane and reproducing the mirror at m'. If now a motif lying on the mirror is introduced and the elements of figure 2.49a are allowed to act, the product, a **glide reflection,** is as shown in figure 2.49b. When coupled with another translation in the plane of the mirror (see figure 2.49c), the pattern is consistent with a new element of symmetry, a glide plane g, lying parallel to and midway between the mirrors. The operation of a glide plane is thus one of translation parallel to a mirror and a reflection across it as shown in figure 2.50.

Proper rotary motions may be symmetrically combined with lattice translations to generate **screw axes** as shown in figure 2.51. The possible distinct combinations for a 4-fold proper axis are illustrated in figure 2.52. Two-, 3-, and 6-fold screw axes also exist.

When the symmetries of the thirty-two point groups (table 2.4, figure 2.48) are combined with the fourteen space lattices (table 2.3) and glide reflections and screw axes are taken into account, 230 **space groups** are found.

Symbols for space groups are characterized by (1) a prefix giving the general lattice type (P, A, B, C, I, F, R) as shown in table 2.2, (2) the several particular symmetry axes of the space group that have the same rotational component as those of the corresponding point group, and (3) the several glide planes that are parallel to the corresponding mirrors in the point group. Thus, a primitive cell consistent with the screw axis 4_3 of figure 2.52 would be designated $P\,4_3$, and the space group having the same axis in a body-centered cell as $I\,4_3$. Similarly, glide planes in the space groups consistent with the point group $4mm$ are designated $P\,4mm$ and $I\,4mm$.

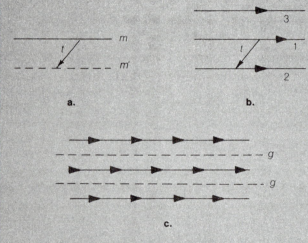

a.

b.

c.

Figure 2.49

Glide reflection. (a) A mirror m interacting with a nonorthogonal, noncolinear translation t, is reproduced at m'. (b) A motif 1, lying on a mirror, is reproduced at 2 by translation and at 3 by reflection. (c) Developed pattern incorporating glide planes g.

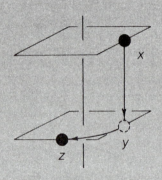

Figure 2.51

Screw axis operation. A point or motif at x is translated to y and rotated to z. The rotary motion must be consistent with the axial multiplicity, 4 in this instance.

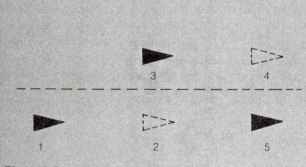

Figure 2.50

Operation of a glide reflection. A motif at 1 is translated to 2 and reflected to 3; translation to 4 and reflection to 5 complete the operation. Note that each glide translation component parallel to a glide plane is ½ t.

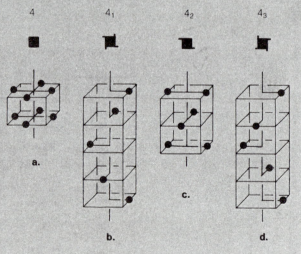

a.

b.

c.

d.

Figure 2.52

4-fold screw axes. A proper 4-fold rotation (a), combined with different translations parallel to the axis, yields three distinct screw axes. Screw axes (b) and (d) are respectively left- and right-handed and have an enantiomorphous relationship.

Concluding Remarks

We stated in chapter 1 that the controls on the number and types of possible minerals are geometric, chemical, and energetic. In this chapter we have begun to investigate the geometric constraints on mineral formation.

The atoms or ions in crystalline materials are arranged in orderly arrays that are repeated along any line through the structure. Geometrical analysis of crystal structures may be made by referring them to an imaginary array of points in space, a lattice. Lattices may be linear, planar, or three-dimensional, and they may be described using standard point-group or space-group notation.

The external symmetry of crystals is sometimes a readily determined feature. Therefore, it is an extremely useful criterion for mineral identification, classification, and understanding of the directional properties that minerals possess. Several different kinds of notation have been used to designate the symmetrically unique point groups or crystal classes. International Notation has been used to this point and will be followed throughout the text.

In the next chapter we will use the concepts of symmetry developed here to establish a geometric classification of crystals. We will also see how the elements of symmetry determine one of the more striking features of a crystal, its external shape.

Suggested Readings

Bloss, F. D. 1971. *Crystallography and Crystal Chemistry.* New York: Holt, Rinehart and Winston. A clearly written review of symmetry and crystallography with excellent illustrations.

Buerger, M. J. [1956] 1960. *Elementary Crystallography.* Rev. printing. New York: Wiley. A detailed and closely reasoned presentation of the geometrical interplay that describes the crystalline state together with morphological results is given in chapters 1 to 9.

———. 1970. *Contemporary Crystallography.* New York: McGraw-Hill. Chapter 2 is particularly germane.

International Tables for X-ray Crystallography, 1969, vol. 1, N. F. M. Henry and K. Lonsdale, eds. International Union of Crystallography, Kynoch Press, Birmingham, England.

McKie, D., and McKie, C. 1986. *Essentials of Crystallography.* Oxford: Blackwell.

Phillips, F. C. 1972. *An Introduction to Crystallography.* 4th ed. New York: Wiley.

Crystallography

Now that we have looked at the conceptual aspects of symmetry, we can proceed to connect this geometrical exercise with real matter. In this chapter, we provide a working terminology for describing crystals.

Crystal Systems and Classes

The translations of a lattice provide a natural coordinate system for crystallographic reference. Further, the distinctively different combinations of lattice transformations give rise to six types of primitive cells. It follows, then, that the natural coordinate axes for crystals will be the edges of the primitive cells. These axes are called the **crystallographic axes.** As indicated in table 2.3, except for the triclinic system, each primitive cell has several related multiple cells. Mineral structures based on such a group of related cells are called a **crystal system.** The parallel sets of terminology for lattices and structures defined to this point are summarized in table 3.1.

Crystallographic axes are labeled and oriented according to accepted conventions, as illustrated in figure 3.1. The translation directions of the unit cell chosen for the axes are labeled a, b, and c except when a particular symmetry causes two or more of them to be symmetrically identical. In this case, the equivalent axes are labeled a_1, a_2, and so on. The interaxial angles are designated α, β, and γ. The angle between axial pairs can readily be remembered because it completes the "abc" group, i.e., αbc, $a\beta c$, and $ab\gamma$.

It is customary to orient crystals and crystal drawings with the c-axis vertical, the b-axis, running left and right (broadside), and the a-axis running front and back.

Table 3.2 summarizes the essential features of the six crystal systems, and the following paragraphs provide detailed information as to the labeling and orientation of axes in each system. A review of axial relationships in the six crystal classes is given in Box 3.1.

Triclinic System (Classes 1 and $\bar{1}$)

This system is based on a perfectly general parallelepiped cell, usually a reduced cell outlined by the three shortest translations of the lattice. The cell edges are conventionally taken as a, b, and c, and the interaxial angles as α, β, and γ.

The coordinate origin of a triclinic cell may be chosen so that α, β, and γ are all acute (Type I) or all obtuse (Type II). The c-axis may be taken as either the longest or shortest translation, but b is always greater than a. Some of these conventions for orientation and labeling are illustrated in figure 3.2.

Table 3.1	Terminology for Lattices and Structures	
Lattice Terms	**Structure Terms**
Conjugate translations of a primitive cell	Crystallographic axes
6 primitive cells	6 crystal systems
14 Bravais lattices	14 unit cell types
32 crystallographic point groups	32 crystal classes

Monoclinic System (Classes 2, m, 2/m)

The axes of this system coincide with the cell edges of a right prism having a parallelogram base; that is, one axis is perpendicular to the plane of the other two, which are inclined to one another.

Two conventional assignments of axes are used, as illustrated in figure 3.3. Either b or c may be selected as the unique 2-fold axis.

Orthorhombic System (Classes 222, 2mm, 2/m 2/m 2/m)

The three orthogonal axes of this system correspond to the edges of a brick-shaped cell. The axes are always labeled according to length so that b is longer than a, but sometimes c has been taken as the shortest and sometimes as the longest axis. Contemporary usage is that $c < a < b$.

Tetragonal System (Classes 4, $\bar{4}$, 4/m, 422, 4mm, 42m, 4/m 2/m 2/m)

The axes of this system correspond of the edges of a right prism having a square base. The 4-fold axis is always chosen as c, and the a- and b-axes are equivalent and so become a_1 and a_2.

Hexagonal System (Classes 3, $\bar{3}$, 32, 3m, $\bar{3}$2/m, 6, $\bar{6}$, 622, 6mm, 6m2, 6/m, 6/m 2/m 2/m)

Hexagonal axes correspond to the edges of a right prism having a 120° rhombus as a base. The unique 3- or 6-fold axis is always chosen as c. The edges of a rhombus are equal and are repeated by the 3-fold axis to three equivalent axes separated by 120°. Extension of these axes through the origin provides the same repetition sequence as required by the 6-fold axis. All axes other than c are equivalent and designated a_1, a_2, and a_3. The hexagonal system is thus referred to a set of four axes in which c is perpendicular to the plane of three equal coplanar axes separated by 120°.

Structures that can be referred to the rhombohedral space lattice 3R are sometimes related to a set of rhombohedral axes. These axes have the advantage of corresponding to a primitive cell, but because this cell is difficult

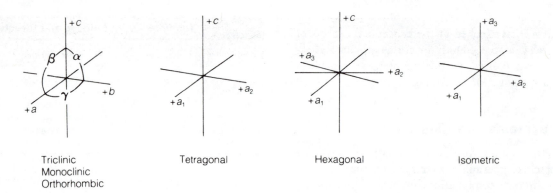

+c β α +b γ +a

Triclinic
Monoclinic
Orthorhombic

Tetragonal

Hexagonal

Isometric

Figure 3.1
Crystallographic axes.

Table 3.2	The Crystal Systems						
Name	**Description**	**Coordinate System**	**Cell Types**				
			P	I(R)		C	F
Triclinic	$a \neq b \neq c$ $\alpha \neq \beta \neq \gamma$						
Monoclinic	$a \neq b \neq c$ $\gamma \neq \alpha = \beta = 90°$						
Orthorhombic	$a \neq b \neq c$ $\alpha = \beta = \gamma = 90°$						
Tetragonal	$a_1 = a_2 \neq c$ $\alpha = \beta = \gamma = 90°$						
Hexagonal	$a_1 = a_2 = a_3 \neq c$ $\beta_1 = \beta_2 = \beta_3 = 90°$ $\gamma_1 = \gamma_2 = \gamma_3 = 120°$						
Isometric	$a_1 = a_2 = a_3$ $\alpha = \beta = \gamma = 90°$						

to work with and the symmetry of a crystal does not distinguish it from a hexagonal cell, it is probably best to use hexagonal axes.

Isometric System (Classes 23, 2/m $\overline{3}$, 432, $\overline{4}3m$, 4/m $\overline{3}2/m$)

The reference cell for this system is a cube that has equal orthogonal edges. Since all of the edges are the same, they are labeled a_1, a_2, and a_3.

Interfacial Angles and Axial Ratios

The science of crystallography was initiated by the Danish scientist Steno (latinized from Niels Stensen), who in 1669 demonstrated that no matter how badly a quartz crystal might be distorted the angles between adjacent faces remain the same. The angle measured from one prism hexagonal face to its neighbor remains 120° regardless of the crystal shape, as shown in figure 3.4. This law of **constancy of interfacial angles** was established by Romé de l'Isle in 1783 and remains the fundamental fact of morphological crystallography.

Box 3.1

Axial Relations in the Six Crystal Systems

Isometric system

Axes a_1, a_2, and a_3 are at right angles and of equal length.

Orient with a_1 and a_2 in the horizontal plane, a_1 front to back, a_2 left to right, and a_3 vertical.

Tetragonal system

Axes a_1, a_2, and c are at right angles. Axes a_1 and a_2 are of equal length; the lengths of a_1 and a_2 are taken as unity and c as longer or shorter.

Orient with a_1 and a_2 in the horizontal plane, a_1 front to back, and c vertical.

Hexagonal system

Axes a_1, a_2, and a_3 lie in the same plane separated by 120° and are of equal length. The c-axis is perpendicular to the plane of the a-axes and may be longer or shorter than they are.

Orient with $+a_1$ to the left, $+a_2$ to the right, and $+a_3$ to the left rear; c is vertical.

Orthorhombic system

Axes a, b, and c are at right angles and of unequal length. The b-axis is taken as unity; the lengths of a and c are not the same and may be greater or less than b. Modern convention is $c < a < b$.

Orient with a and b in the horizontal plane, b left to right, and c vertical.

Monoclinic system

Axes are of unequal length. In the first setting, the unique axis (2 or $\bar{2}$) is taken as c. This axis is perpendicular to the ab plane and is oriented vertically. The a and b axes lie in a horizontal plane with the obtuse angle γ between $+a$ and $-b$. In the second setting of older usage, the unique 2-fold axis is taken as b and set horizontally perpendicular to the ac plane.

Orient with the c-axis vertical and the $+a$-axis inclined downward to the front at an angle of β with $+c$ (see figure 3.3).

Triclinic system

Axes a, b, and c are inclined to each other by the angles α (b to c), β (a to c), and γ (a to b). The axes are of unequal length, b is taken as unity, and the lengths of a and c may be greater or less than unity. Modern convention is that $c < a < b$.

Orient with the c-axis vertical and the a-axis inclined downward and to the front.

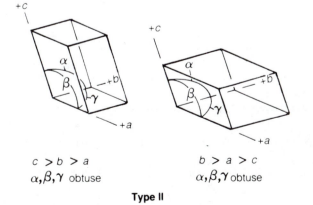

$c > b > a$
α, β, γ acute

$b > a > c$
α, β, γ acute

Type I

$c > b > a$
α, β, γ obtuse

$b > a > c$
α, β, γ obtuse

Type II

Figure 3.2

Some orientations for the triclinic system.

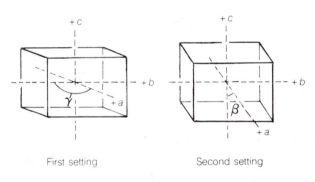

First setting

Second setting

Figure 3.3

Conventional orientations for the monoclinic system.

In 1780 Carangeot developed a device, a **goniometer,** consisting of a pair of movable arms for measuring interfacial angles. This device could measure the dihedral angle between adjacent crystal faces. A simple **contact goniometer** is diagrammed in figure 3.5. Contact goniometers are useful only for measurements of large crystals, but the most perfect and complex crystals are usually small. For these crystals more sophisticated apparatus are needed for highly reliable measurements.

In 1809 Wollaston developed a reflecting goniometer, shown diagrammatically in figure 3.6, in which a beam of light reflected from a crystal face serves as a measuring arm. In use, the crystal under study is mounted at the center of a graduated circle, a reflection is obtained, and

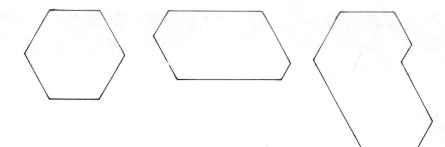

Figure 3.4
Cross sections of regular and distorted quartz prisms. All internal angles between adjacent faces measure 120°.

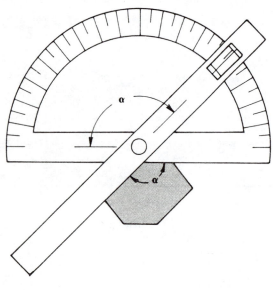

Figure 3.5
Contact goniometer.

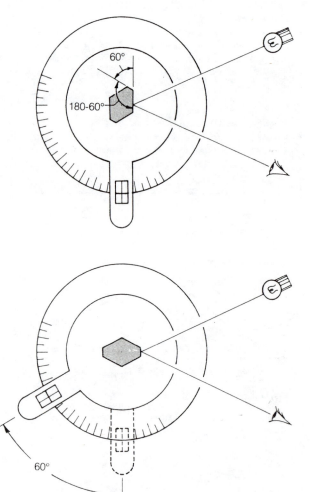

Figure 3.6
Reflecting goniometer.

the angular reading recorded. The crystal is then rotated until an adjacent face reflects into the eyepiece. The angular difference of the two readings is that between the normals of adjacent faces, and since angles with mutually perpendicular sides are equal, this reading is the same as the interfacial angle. The interfacial angle is the external angle between adjacent faces and is the supplement of the internal angle between the same two faces. Modern, two-circle, goniometers are made so that both horizontal and vertical angles may be read with only one mounting of the crystal.

The concept of crystal structures composed of unit cells bounded by translations of various length was well understood before cell sizes could be determined by x-ray techniques. Although the actual cell dimensions were unknown, the relative lengths of the unit translations t_1, t_2, and t_3 were known by calculations from interfacial angles using faces of known index.

To follow an example of these calculations, consider the mineral celestine, $Sr(SO_4)$, of class $2/m\ 2/m\ 2/m$.

Assume the following interfacial angles have been measured, $(110) \wedge (010) = 32.62°$—read "(110) makes an angle with (010) of 32.62°"—and $(001) \wedge (011) = 37.93°$.

The convention for orthorhombic axial ratios is that $a:b:c = a:1:c$. To begin the analysis, we make a c-axis

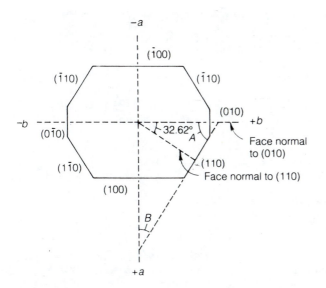

Figure 3.7
A *c*-axis projection of a celestine crystal showing the construction of (110) ∧ (010).

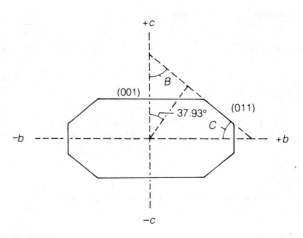

Figure 3.8
An *a*-axis projection of a celestine crystal showing the construction of (010) ∧ (011).

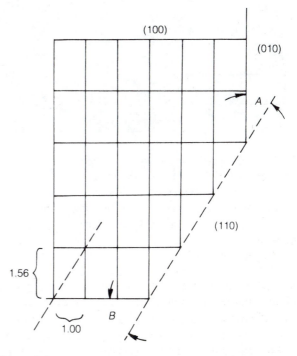

Figure 3.9
Determination of interfacial angles and axial ratios in celestine using the unit cell dimensions.

projection (see figure 3.7), that shows (110) as the hypotenuse of a right triangle whose sides are the positive ends of the *a* and *b* crystallographic axes. Angle *A* opposite side +*a* is the complement of the angle measured between the face normals to (110) and (010) of 32.62°. Thus, $A = 90° - 32.62° = 57.38°$.

The tangent of angle $A = +a/+b = +a$, since *b* is taken as unity and tan 57.38° = 1.562. Therefore, the axial ratio is 1.562 : 1 : *c*.

Next, we make a new projection of the crystal perpendicular to the *a*-axis (see figure 3.8), which shows the (011) face and the +*b*- and +*c*-axes as the sides of a right triangle. The angle *B* measures 52.07° and is the complement of the measured interfacial angle of 37.93°. The tangent of 52.07° = *c* = 1.283. Therefore the axial ratio of celestine is 1.562 : 1 : 1.283.

Axial ratios in the other crystal systems can be determined by the same procedure with the provision that the angles α, β, and γ must be considered in the monoclinic and triclinic systems. In the tetragonal and hexagonal systems $a : c = 1 : c$, and in the monoclinic and triclinic systems, as in the orthorhombic system, $a : b : c = a : 1 : c$. Isometric axes are of the same length, and therefore the axial ratios are always $a_1 : a_2 : a_3 = 1 : 1 : 1$.

The angular relations of crystal faces may also, and more readily, be determined if the dimensions of the unit cell are known. These dimensions (in Ångstrom units, $1 Å = 10^{-10} m$) have been determined by x-ray techniques for most minerals and are given as part of a mineral's description in part 3.

For celestine, the cell dimensions are $a = 8.539$, $b = 5.352$, $c = 6.866$. Taking *b* as unity, these dimensions are in the ratio of $a : b : c = 1.56 : 1 : 1.28$, as previously found.

A plan of the *ab* plane of celestine using these ratios is given in figure 3.9, from which it may be seen that the tangent of the interfacial angle *A* between (010) and (110)

is 1.00/1.56, and the interfacial angle is 32.62°, as previously measured. From the same figure, (100) ∧ (110) = *B*, tan *B* = 1.56/1.00, and this interfacial angle is 57.34°.

Crystal Forms

The symmetrical relations that exist between atoms or molecules of a crystal cannot be observed without the use of x ray or similar techniques, and then only indirectly. One consequence of this symmetry, however, can often be seen and is used by the mineralogist. That is the external shape, or **morphology,** of crystals.

Box 3.2

The Nomenclature of Forms

Greek words are used to name the forms of crystals (and other geometric solids). Note that *-gonal* refers to an angle and *-hedron* to a face.

Numbers

mono	one	penta	five
bi, di	two	hexa	six
tri, tris	three	octa	eight
tetra	four	dodeca	twelve

Face Shapes

scaleno	scalene triangle
rhombo	rhombus-shaped; an equal-sided parallelogram
trapezo	trapezoid-shaped; a four-sided plane figure with unequal sides

Solids

tetrahedron	4-faced solid; each face is an equilateral triangle
hexahedron (cube)	6-faced solid; each face is a square
octahedron	8-faced solid; each face is an equilateral triangle
dodecahedron	12-faced solid; each face is a rhombus
diploid	12-faced solid bounded by 5-sided planes; the most common example is a pyritohedron often exhibited by pyrite, FeS_2
scalenohedron	solid-faceted with scalene triangles
gyroid	24-faced solid; no mineral representative
tetartoid	isometric solid lacking mirrors and inversion center; similar to a diploid

Using this nomenclature, we can describe some very complex forms. For example, a tetragonal scalenohedron is a solid having a 4-fold axis of symmetry and faceted with scalene triangles. A dihexagonal bipyramid is two 12-sided pyramids related by a mirror. A tristetrahedron has each of its four faces modified into three isosceles triangles.

When minerals crystallize in an environment where unhindered growth is possible, the exterior shape of each crystal reflects the symmetrical relations among its constituents. The external surfaces are then plane **faces,** each of which is parallel to a stack of rational planes of the lattice. The faces that actually develop are controlled by growth rate, environment, and the nature of the compound, but such faces always have simple indices.

The symmetry of crystals requires that parallel stacks of rational planes be chemically and physically as well as geometrically equivalent. Equivalent faces must, therefore, behave identically during growth, assuming that the external environment is also symmetrical. Under such conditions, faces corresponding to the symmetries of point groups arise. Consider, for example, the symmetrically equivalent spikes on a snow crystal, the alternation of striated faces on a pyrite cube, the parallel cleavage planes of mica, or the differential hardness of andalusite.

Crystallographic studies emphasize the geometrical equivalency of symmetrically disposed crystal faces rather than the symmetrically controlled distribution of certain physical and chemical properties such as solubility, heat conduction, hardness, or cleavage. The latter properties, however, are often very important. Geometrical equivalency (or nonequivalency) is directly related to physical and chemical properties. Thus, in crystallography, a **form** is a symmetrically repeated set of planar motifs, all of which have the same relation to the elements of symmetry and display the same chemical and physical properties.

For simplicity, a form is designated by its simplest positive indices and enclosed in braces { }. Forms of importance in mineralogy identify the symmetrically disposed planes of parting, cleavage, twin planes, and crystal faces. The basic terminology of forms is described in Box 3.2.

The plane faces in a form have like positions with respect to the symmetry elements. If one face is present, a form includes all the faces required by the symmetry set; that is, the face is a motif that is repeated by the operation of the symmetry elements. A face with positive indices is conventionally used to represent all of the symmetrically equivalent faces of a form.

Forms are divided into closed forms which completely enclose space, as in a cube, and open forms which have open ends or sides. Most forms, and all open forms, occur in combination with other forms. The forms of all crystal systems except the isometric can be grouped into five widely occurring form types—pedions, pinacoids, prisms, pyramids, and dipyramids—and six more restricted types—domes, sphenoids, scalenohedrons, trapezohedrons, disphenoids, and rhombohedrons.

The names of closed forms are usually prefixed by a term that gives the symmetry of the form and provides a unique geometrical description of the solid. Closed forms are described as tetragonal, symmetrical with respect to a 4-fold axis and thus having a square cross section normal to the axis; trigonal, symmetrical with respect to a 3-fold axis and having an equilateral triangular cross section; and hexagonal, symmetrical with respect to a 6-fold axis and having a hexagonal cross section.

In general, two, or even three, examples of the same form may occur on a single crystal. These may be designated as first-, second-, and third-order forms. Apatite, $6/m$, is a prime example exhibiting an $\{h0hl\}$ dipyramid, an $\{hh2hl\}$ dipyramid rotated 30° with respect to the first, and an $\{hkil\}$ dipyramid rotated at an arbitrary angle.

Types of Forms

Pedions (Monohedrons)   A pedion is a form consisting of only one face.

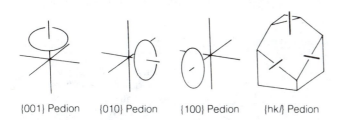

{001} Pedion   {010} Pedion   {100} Pedion   {hkl} Pedion

Pinacoids (Parallelohedrons)   Any form consisting of only two parallel faces is a pinacoid. Three pinacoids—basal, front, and side—are recognized by specific names. The remainder are best prefixed by the indices of one face, as in a {101} pinacoid.

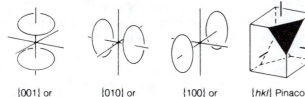

{001} or   {010} or   {100} or   {hkl} Pinacoid
Basal pinacoid Side pinacoid   Front pinacoid

Prisms   A prism is a multifaced open form with 3, 4, 6, 8, or 12 faces whose intersections are parallel. Such sets of planes with one direction in common are called zones. A zone axis is the axis of symmetry parallel to the intersection of the zone faces.

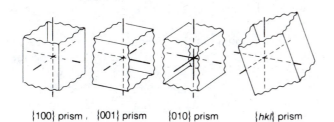

{100} prism   {001} prism   {010} prism   {hkl} prism

Pyramids   A pyramid is an open form consisting of 3, 4, 6, 8, or 12 regularly arrayed faces diverging from a point.

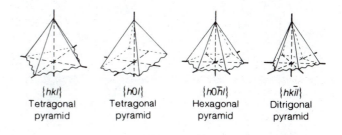

{hkl}   {h0l}   {h0h̄l}   {hkil}
Tetragonal   Tetragonal   Hexagonal   Ditrigonal
pyramid   pyramid   pyramid   pyramid

Dipyramids   A dipyramid is a pyramid repeated as if reflected by a mirror normal to the principal symmetry axis to create a closed form.

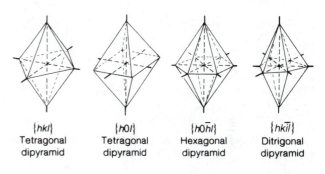

{hkl}   {h0l}   {h0h̄l}   {hkīl}
Tetragonal   Tetragonal   Hexagonal   Ditrigonal
dipyramid   dipyramid   dipyramid   dipyramid

Domes and Sphenoids (Dihedrons)   A dome is an open form composed of two nonparallel faces symmetrical with respect to a mirror. A sphenoid is an open form composed of two faces symmetrical with respect to a 2-fold axis.

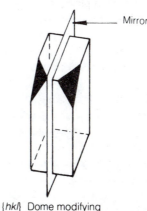 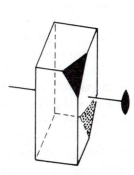

{hkl} Dome modifying   {hkl} Sphenoid modifying
a monoclinic crystal   a monoclinic crystal

Disphenoids   A disphenoid is a 4-faced closed form generally found as a modifying form, although disphenoids of chalcopyrite are common as simple crystals.

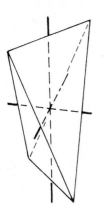

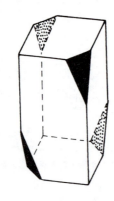

{hkl} + Disphenoid

{hkl} + Disphenoid modifying a tetragonal crystal

Scalenohedrons   A scalenohedron is a closed form consisting of eight scalene triangular faces in the tetragonal system and twelve scalene faces in the hexagonal system. The term is used mainly to describe hexagonal scalenohedral crystals of calcite.

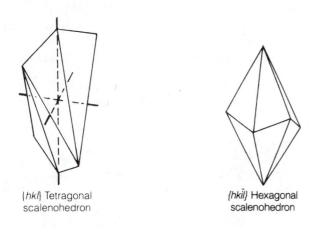

{hkl} Tetragonal scalenohedron

{hkil} Hexagonal scalenohedron

Trapezohedrons   A trapezohedron is a closed form occurring in the hexagonal, tetragonal, and isometric systems characterized by trapezoid-shaped faces. The form never occurs in the hexagonal and tetragonal systems except as a minor modification, where it has historical importance as the very rare form distinguishing the true symmetry of quartz. Isometric trapezohedrons are considered separately.

Tetragonal trapezohedron

Hexagonal trapezohedron

Trapezohedral modification of quartz crystal

Rhombohedrons   The rhombohedron is an important 6-faced closed form in the hexagonal system having the indices $\{h0\bar{h}l\}$, $\{0hhl\}$, $\{hh\overline{2h}l\}$, or $\{h2h\bar{h}l\}$.

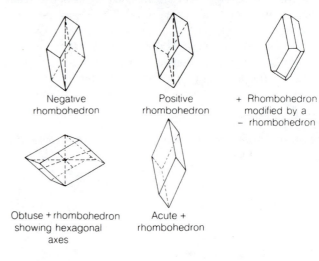

Negative rhombohedron

Positive rhombohedron

+ Rhombohedron modified by a − rhombohedron

Obtuse + rhombohedron showing hexagonal axes

Acute + rhombohedron

Forms in the Isometric System

The isometric system is unique in that all forms are closed. This system is of special interest because a number of important minerals, such as diamond, fluorite, galena, garnet, magnetite, and pyrite, have isometric forms that are easily recognized.

Isometric forms have the following possible indices: $\{100\}$, $\{110\}$, $\{111\}$, $\{hk0\}$, $\{hll\}$, $\{hhl\}$, and $\{hkl\}$. The complex shapes assumed by isometric crystals are mostly derived from the simple forms of a cube, a tetrahedron, or an octahedron. To see how this takes place, imagine adjacent faces of a cube as rubber sheets stretched on rigid frames. If each face is equipped with hooks and eyes at its center, pulling on the square faces first deforms each into four identical isosceles triangles arranged as a short pyramid. A cube with such "deformed" faces is a tetrahexahedron, having four faces on each of six sides. With continued pulling, a point is reached at which adjacent triangular faces make a single plane and the form becomes a dodecahedron (12-sided) whose faces are rhombic. Further stretching would produce a six-pointed, star-shaped solid. Crystals with reentrant angles are not found, however, and such shapes can be left for topologists to consider.

Other examples of the generation of complex shapes from simple forms may be visualized by stretching the rubber sheet at the midpoint of a face edge as well as at the face centers. For example, center-stretched faces on a tetrahedron yield a tristetrahedron and center- plus edge-stretching yields a hextetrahedron.

Some of these so-called rubber-sheet-related forms are suggested by the shading of the isometric forms shown at the top of p. 47.

Positive and Negative Forms

With one face given, the total number of faces in a form depends on the symmetry of the class. For example, a $\{111\}$ modification of a cube requires eight faces in the 432, $2/m\bar{3}$, and $4/m\bar{3}2/m$ classes, but only four in the 23 and

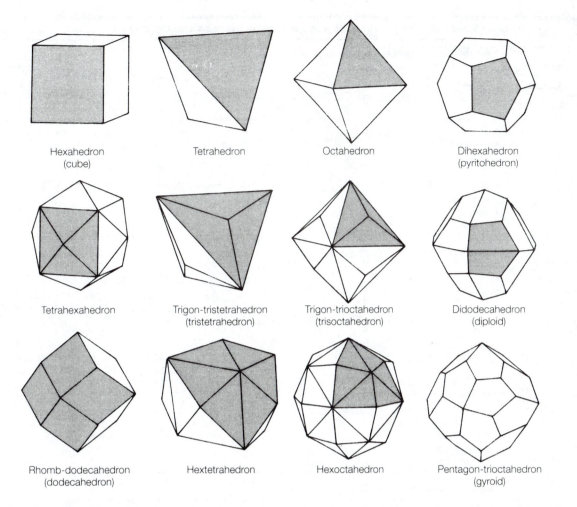

| Hexahedron (cube) | Tetrahedron | Octahedron | Dihexahedron (pyritohedron) |

| Tetrahexahedron | Trigon-tristetrahedron (tristetrahedron) | Trigon-trioctahedron (trisoctahedron) | Didodecahedron (diploid) |

| Rhomb-dodecahedron (dodecahedron) | Hextetrahedron | Hexoctahedron | Pentagon-trioctahedron (gyroid) |

43m classes. The illustration shows examples of 4-faced modifications of a cube {100}.

Cube {100} modified by:

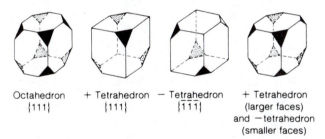

Octahedron {111}   + Tetrahedron {111}   − Tetrahedron {$\bar{1}\bar{1}\bar{1}$}   + Tetrahedron (larger faces) and −tetrahedron (smaller faces)

Tetrahedral forms may be positive (+) or negative (−). Those with a face that intersects the positive ends of all three axes are called positive tetrahedrons, {111}, and those with a face that cuts the negative end of all axes are negative, {$\bar{1}\ \bar{1}\ \bar{1}$}. This is a common rule for the positive-negative nomenclature, but the rules do not always apply so directly.

Right and Left Forms

Enantiomorphous or right and left forms, are recognized in certain crystal classes. These are distinct forms that cannot be converted to one another by proper symmetry operations. Therefore, a given enantiomorphous form (on quartz, for example) can be termed unequivocally as a right- or left-handed form.

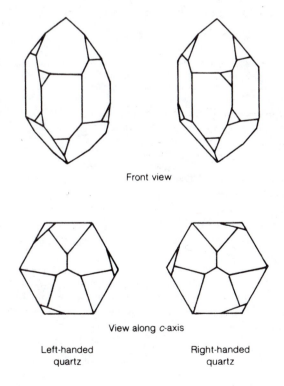

Front view

View along c-axis

Left-handed quartz   Right-handed quartz

Table 3.3   The Distribution of Forms Among the Various Crystal Systems and Classes

Triclinic-Monoclinic-Orthorhombic Systems

Number of Faces	Form	1	$\bar{1}$	2	m	$\frac{2}{m}$	222	2mm	$\frac{2}{m}\frac{2}{m}\frac{2}{m}$
1	Pedion	X		X	X			X	
2	Pinacoid		X	X	X	X	X	X	X
2	Dome			X	X			X	
2	Sphenoid								
4	Prism						X	X	X
4	Disphenoid						X		
4	Pyramid							X	
8	Dipyramid								X

Tetragonal System

Number of Faces	Form	4	$\bar{4}$	$\frac{4}{m}$	422	4mm	$\bar{4}2m$	$\frac{4}{m}\frac{2}{m}\frac{2}{m}$
1	Pedion	X					X	
2	Pinacoid		X	X	X		X	X
4	Tetragonal prism	X	X	X	X	X	X	X
4	Tetragonal pyramid	X				X		
4	Tetragonal disphenoid		X				X	
8	Ditetragonal prism				X	X	X	X
8	Tetragonal dipyramid			X	X	X	X	X
8	Tetragonal trapezohedron				X			
8	Tetragonal scalenohedron						X	
8	Ditetragonal pyramid					X		
16	Ditetragonal dipyramid							X

Hexagonal System

Number of Faces	Form	3	$\bar{3}$	32	3m	$\bar{3}\frac{2}{m}$	6	$\frac{3}{m}$	$\frac{6}{m}$	622	6mm	$\bar{6}m2$	$\frac{6}{m}\frac{2}{m}\frac{2}{m}$
1	Pedion	X			X		X				X		
2	Pinacoid		X	X		X		X	X	X		X	X
3	Trigonal prism	X		X	X			X					
3	Trigonal pyramid	X			X								
6	Ditrigonal prism				X							X	
6	Hexagonal prism		X	X	X	X	X	X	X	X	X	X	X
6	Trigonal dipyramid							X				X	
6	Rhombohedron		X	X		X							
6	Trigonal trapezohedron			X									
6	Ditrigonal pyramid				X								
6	Hexagonal pyramid						X				X		
12	Hexagonal dipyramid								X				X
12	Hexagonal scalenohedron					X							
12	Dihexagonal prism									X	X		X
12	Ditrigonal bipyramid											X	
12	Hexagonal trapezohedron									X			
12	Dihexagonal pyramid										X		
24	Dihexagonal dipyramid												X

Combined Forms

Most crystals exhibit forms in combination, but of course forms occurring together on the same crystal are limited to those consistent with the symmetry of that class. For example, a pinacoid, a tetragonal prism, and a tetragonal disphenoid are possible forms of class $\bar{4}$ (see table 3.3), and these three forms could be found in combination. However, the combination of a tetragonal prism and a tetragonal disphenoid is not consistent with the symmetry of this class. If these two forms were found in combination, the crystal class would have to be **4mm**.

Open forms cannot exist except when combined with other forms in such a way as to enclose space with the combination. Some combinations of open forms are illus-

Table 3.3 (continued)

Isometric System

Number of Faces	Form	Class 23	432	$\frac{2}{m}\bar{3}$	$\bar{4}3m$	$\frac{4}{m}\bar{3}\frac{2}{m}$
4	Tetrahedron	X			X	
6	Cube (hexahedron)	X	X	X	X	X
8	Octahedron		X	X		X
12	Dodecahedron	X	X	X	X	X
12	Pyritohedron	X		X		
12	Tristetrahedron	X			X	
12	Deltohedron	X			X	
12	Tetartoid	X				
24	Tetrahexahedron		X		X	X
24	Trapezohedron		X	X		X
24	Trisoctahedron		X	X		X
24	Hextetrahedron				X	
24	Diploid			X		
24	Gyroid		X			
48	Hexoctahedron					X

trated in figure 3.10. A small face of one form that changes the principal form by beveling is called a **modification.** Forms occurring on opposite ends of a crystal and serving to close major open forms are called **terminations.**

Closed forms and open forms may exist in combination. This feature is illustrated in figure 3.11, which shows the shapes that arise as the result of the combination of various open forms with a tetragonal dipyramid.

Finally, a closed form may be combined with another closed form, as illustrated in figure 3.12, which shows the shapes resulting from the combination of a cube and an octahedron or tetrahedron.

A trivial constraint to the permissible combination of forms, but one that sometimes bothers beginning students, is that only one form of a given index may be present on one crystal. For example, there cannot be two {100} forms on a single crystal. More than one {h0l} may be present, however, since the precise indices might be {101}, {102}, {103}, and so on. Figure 3.13 is a real-life example of a combined form in which an earlier {h0l} face has been overgrown with another.

A crystal may only display forms that are consistent with its symmetry, and the use of table 3.3 will often lead to a unique characterization of the crystal class. The distribution of minor faces is especially useful in this characterization. Figure 3.14 shows various (hkl) modifications that distinguish the classes in the orthorhombic system.

Real versus Ideal Crystals

Perfect crystals that display identical development of equivalent faces are so rare as to be effectively nonexistent. The cause of this deviation from perfection lies in the complexities of crystal growth, which are discussed in chapter 7. The essential features of crystalline matter as reflected in crystal form, however, are *faces with simple rational indices,* identifiable by small integers, and *constant interfacial angles.*

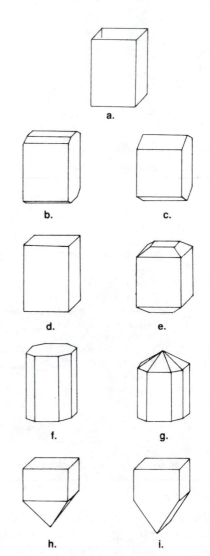

Figure 3.10

Combinations of open forms. (a) Prism. (b) Modified by dome, terminated by pinacoid. (c) Terminated by dome. (d) Terminated by pinacoid. (e) Modified by dipyramid, terminated by pinacoid. (f) Modified by prism, terminated by pinacoid. (g) Modified by prism, terminated by pyramid and pedion. (h) Terminated by pedion and pyramid. (i) Terminated by pedion and dome.

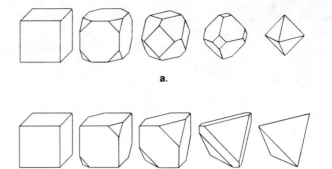

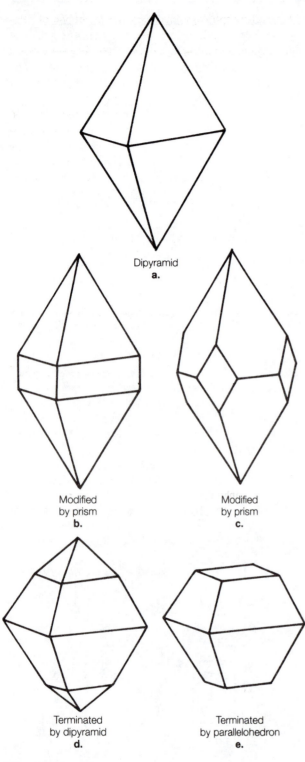

Dipyramid
a.

Modified
by prism
b.

Modified
by prism
c.

Terminated
by dipyramid
d.

Terminated
by parallelohedron
e.

Figure 3.11
Combinations of open and closed forms. (a) {hkl} dipyramid.
(b) Modified by {110} prism. (c) Modified by {100} prism.
(d) Terminated by {hkl} dipyramid. (e) Terminated by {001}
parallelohedron.

Figure 3.12
Combinations of closed forms. (a) Cube and octahedron. (b) Cube
and tetrahedron.

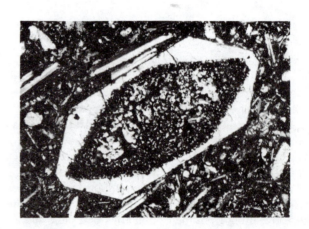

Figure 3.13
Photomicrograph of a zircon crystal, c-axis in plane of page, showing
two stages of growth with different (hkl) faces. (Photograph by W. H.
Dennen.)

show no external crystal form because of mutual interference among growing crystals in an aggregate. Misshapen crystals may develop in a nonuniform environment. The nutrient fluid, for example, may be in motion (see figure 7.24), or impurities in the fluid from which the crystal grows may inhibit the development of certain faces. Whatever their external shape, however, crystals exhibit an orderly and characteristic internal arrangement of their constituent atoms. This orderliness can be observed indirectly with x-ray diffraction apparatus or with a petrographic microscope.

Regardless of how distorted or incomplete a crystal may be, the angles between the faces present remain constant. The ideal form, therefore, may be deduced from measurement of the nonideal form. Precise identification of forms relies on the measurement of interfacial angles by goniometry followed by reference to tables or calculation of face positions and plotting on coordinate paper.

Departures from the ideal are generally seen in the difference in size of symmetrically equivalent faces and in alternations of several faces on a minuscule scale, leading to striations, steps, or lineage structure. Some specimens

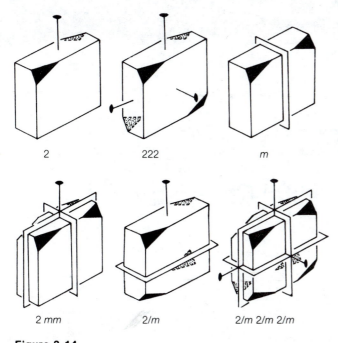

2   222   m

2 mm   2/m   2/m 2/m 2/m

Figure 3.14
Minor faces consistent with classes in the orthorhombic system.

Concluding Remarks

The crystalline state is characterized by the orderly repetition of motifs in space. In the last two chapters we have examined the fundamental geometric aspects of motif symmetries without regard for the constituent particles. We have not yet, however, investigated the particles themselves. In minerals, as in any chemical compound, the particles that make up the motifs are atoms or ions that themselves possess symmetry. The nature of the atoms in a compound, their symmetry, and the symmetry of their bonds play a very significant role in determining the symmetry of the structure.

Crystal morphology has been the backbone of classical mineralogy and, historically, has been the means of gaining insight into the nature of the crystalline state. Most mineral specimens, however, are not found as crystals, and modern study of mineral geometry is carried out by probing mineral structures with x rays.

In the next chapter we will combine aspects of crystallography, physics, and chemistry to present the fundamentals of crystal chemistry. We will examine the architecture of atoms, the mechanisms and characters of various bond types, and the effect of crystallochemical symmetry on the physical properties of a mineral.

Suggested Readings

Bishop, A. C. 1967. *An Outline of Crystal Morphology.* London: Hutchinson. A beginning text treating crystal morphology in a systematic way.

Bloss, F. D. 1971. *Crystallography and Crystal Chemistry, an Introduction.* New York: Holt, Rinehart and Winston. An elegant and detailed treatment of crystallography and crystallographic measurements.

Buerger, M. J. 1971. *Introduction to Crystal Geometry.* New York: McGraw-Hill. A text written as the simplest of a series on crystallography by an acknowledged leader in the field of crystallographic mineralogy.

Miller, W. H. 1839. *Treatise on Crystallography.* Cambridge: Cambridge University Press.

Phillips, F. C. 1971. *An Introduction to Crystallography.* 4th ed. London: Longman Group. A historical approach to crystallography that discusses crystal measurement by goniometers, crystal projection, and drawings as a prelude to internal structure and x-ray diffraction studies.

Sands, D. E. 1975. *Introduction to Crystallography.* New York: W. A. Benjamin.

Whittaker, E. S. W. 1981. *Crystallography: An Introduction for Earth Science Students.* New York: Oxford University Press.

4

Fundamentals of Crystal Chemistry

In the two previous chapters, we have examined some of the geometric aspects of mineralogy through a study of symmetry and crystallography. To attain a full understanding of crystallographic phenomena, we must now consider the chemistry of minerals on the atomic scale. We will first examine the nature of atoms and their properties. This will be followed by a discussion of how atoms combine in various arrangements and how these combinations and arrangements determine many of the physical properties that distinguish one mineral from another.

The Architecture of Atoms

The distinctive properties of the different kinds of atoms and their arrangement in space determine whether a mineral species will form and, once formed, what its chemical and physical properties will be.

Atoms

All atoms appear to be aggregates of three constituents—protons, neutrons, and electrons. The exact nature of these fundamental units of matter is unknown, but following convention we will treat protons and neutrons as very small, very dense spheres and electrons as rather diffuse units of matter. Each of these fundamental particles has associated with it a distinctive mass and electrical charge, as shown in table 4.1. Note the discrepancy in mass between a neutron and a proton. This can be accounted for as energy, according to the relationship $e = mc^2$, after Einstein.

As a first approximation, an atom consists of a nucleus about which a planetary array of electrons revolves. The balance of electrical forces is such that the atom is externally neutral. The nucleus is a compact aggregate composed of protons and neutrons and is held together by powerful binding forces. The electrical charge of a nucleus is neutralized by the presence of an appropriate number of orbital electrons.

Nuclear Progression

Every atom contains an integral number of protons in its nucleus, and the number of protons determines the particular atomic species, or **element.** Naturally occurring elements have nuclei with a regular progression from one proton (in hydrogen) to ninety-two protons (in uranium). The number of protons in a nucleus is given by the atomic number Z of the element. The nucleus of every element except hydrogen also contains neutrons in numbers about equal to the protons.

Most hydrogen atoms have only a proton as their nucleus, but small amounts of heavy hydrogen (deuterium), with a proton and a neutron in the nucleus, do exist.

Table 4.1	Properties of Atomic Particles	
	Mass Unit	Electronic Unit of Charge
Proton	1.00758	+ 1
Electron	0.00055	− 1
Neutron	1.00896	0

The number of neutrons plus protons largely determines the mass of the atom because the mass of the electron is relatively small. The atomic weight of an element is ordinarily given in terms of a unit approximated by the mass of a hydrogen atom. Modern usage defines the atomic weight in terms of the carbon nuclide ^{12}C taken as 12.000.

The number of neutrons, unlike the number of protons, is not rigidly fixed for each element, and two atoms of a given element may have different numbers of neutrons in their respective nuclei. The different, although related, atoms are called **isotopes,** and may be described in terms of their mass number A, the sum of the protons and neutrons. As an example, the nucleus of lithium contains three protons ($Z = 3$) and either three or four neutrons, and two isotopes of lithium exist ($A = 6$ or $A = 7$). Naturally occurring lithium is made up of 92.5% ^7Li and 7.5% ^6Li, and the atomic weight of lithium may be calculated as a weighted average to be 6.9.

The Electronic Cloud

The conventional laws of mechanics are concerned with the interaction of mass and energy at the macroscopic level and do not completely hold at the submicroscopic levels that must be considered in atomic studies. Several conceptual adjustments must be made.

First, the wave properties associated with all matter are very important at the atomic scale. Indeed, the wave properties of electrons are as important as their mass, and only by considering electrons as waves may some phenomena be understood. Electrons may be described only in very general terms as elementary units of matter with both particulate and wave properties.

Relativity also assumes an important role at the atomic level, as does indeterminacy of electron speed for a given position and of position for a given speed. Electron space-time coordinates can, therefore, be given only in terms of probability.

At the atomic level, energy is not continuous but is composed of discrete and indivisible packets called **quanta.** One quantum is the smallest energy change that can occur. The energy changes within and between atoms are whole-number multiples of quanta and prescribe the manner by which electrons may neutralize the electrical field surrounding a nucleus. This neutralization is a simple electrostatic balance. The location and motions of the electrons within the field are, however, governed by quantum considerations.

The system used to specify the particular energy status of electrons is known as **quantum mechanics.** The quantum state of an electron may be specified by considering the kinds of energy it possesses, both kinetic and potential. This energy is conveniently described in terms of (1) the size of an electron's orbit, (2) its total angular momentum, (3) the magnetic contribution due to the motion of a charged electron in an electric field, and (4) the spin of an electron on its own axis.

The energy of an electron in an orbit depends on the *size* of the orbit. Electrons can occupy only those orbits (which are generally elliptical) whose energy differences are regulated by quantum changes proportional to $1/n^2$, where n is called the **principal** or **total quantum number** and may have values of 1, 2, 3, and so on.

The angular momentum of an electron depends on the *shape* of its orbit. Changes in the eccentricity of the orbit may take place only in whole quantum steps designated by the **serial** or **azimuthal quantum number** l, which takes the values of 0, 1, 2, 3, . . . , $n - 1$, there being n values of l possible. For example, if n is 3, l may be 0, 1, or 2. These quantum values of l are usually expressed in a different way because spectral line production related to changes in l was used early to investigate the electron envelope of atoms. Spectral lines arising from electrons in $l = 0, 1, 2,$ and 3 were termed *s, p, d,* and *f,* and this nomenclature is in use today. Thus, if $l = 0, 1, 2,$ or 3, atoms are said to be in an *s, p, d,* or *f* state.

Following conventional usage, *s, p, d,* and *f* refer to spectral line character described by early workers as sharp, principal, diffuse, and fundamental.

The **magnetic quantum number** (m_l) describes the energy levels into which the preceding states are split under the influence of a magnetic field. Moving electrons have a magnetic field associated with them just as electrical current flowing in a wire does. The interaction of these fields is responsible for small changes in electron energy.

The values of m_l are dependent on the azimuthal quantum number l, which in turn is dependent on the principal quantum number n. A total of $2l + 1$ values of m_l are possible for each value of l, and these m_l quantum numbers may be obtained from the series

$$l, l - 1, l - 2, . . . , 0, . . . , -l,$$

which reduces to the numerical sequence

$$. . . , + 5, + 4, + 3, + 2, + 1, 0, - 1, - 2, - 3, - 4, - 5, . . .$$

If $l = 2$, there will be $2(2) + 1 = 5$ possible values of m_l, and these are $+ 2, + 1, 0, - 1, - 2$. If $l = 4$, the nine possible m_l values are $+ 4, + 3, + 2, + 1, 0, - 1, - 2, - 3, - 4$. If $l = 0$, the only possible value of m_l is 0.

The **spin quantum number** (m_s) is required because an electron may be considered to have extension in space and to be spinning with a constant angular momentum. This spin may be in either of two directions with reference to an arbitrary coordinate system, and m_s is assigned values of $+ \frac{1}{2}$ or $- \frac{1}{2}$. The spin quantum number is independent of the preceding quantum numbers.

Quantum mechanical studies have provided a powerful means for investigating and understanding atomic systems. However, certain phenomena cannot be clearly described in this manner, and a further development known as **wave mechanics** is also used to describe atomic systems. The two viewpoints of quantum and wave mechanics are analogous to the treatment of light as either waves or particles. The two approaches are not mutually exclusive but are only different treatments of the same fundamental conditions. A useful analogy is to picture a moving boat with its wake and consider the mathematical and mensurative techniques that could best describe this phenomenon.

The wave mechanical approach has been especially fruitful in studies of the more complex atomic systems and of the quantum conditions themselves. In essence, the motion of any particle, such as an electron, has an associated wave motion and wavelength. The only wave paths that occur are those that provide standing waves with vibration modes that are not destroyed by interference. The frequencies that do occur are those whose energy values are separated by integral quantum differences. The electron orbits of the planetary model are thus replaced by vibrating surfaces, or shells, having an integral number of nodes.

The actual position of an electron on this surface is not strictly definable. However, the distance of an electron from the nucleus and the spacing of electrons in a shell may be given in terms of probability. Figure 4.1 shows the radial electron density in a rubidium atom $(Z = 37)$ in which the highest densities (most probable radial locations of electrons) define the electron shells. The spatial distribution of electrons in a shell is such that the greatest interelectron distance is attained in order to separate the like charges. Two electrons occupy antipodal positions, three occupy the corners of an equilateral triangle, and four, six, and eight, respectively, occupy the corners of a tetrahedron, octahedron, and cube.

The vibratory surfaces have the nucleus as a center only if an electron is in the *s* state $(l = 0)$. For all other values of l, the nodal surfaces are tangent to the center of the atom and may be conveniently thought of as probability volumes occupying different positions with respect to a set of orthogonal axes. Thus, *s* electrons occupy a probability volume whose center is the origin of coordinates, *p* electrons occupy volumes on the orthogonal axes (see figure 4.2), and *d* electrons are along axial and diagonal directions (see Box 5.2).

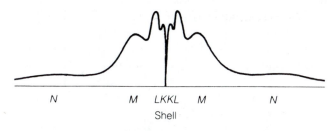

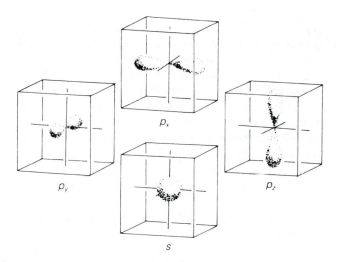

N   M LKKL M   N
Shell

Figure 4.1
Electron distribution in a rubidium atom. (Calculated using the method of Hartree, 1947.)

Table 4.2	Quantum Number Combinations for 3p					
n	3	3	3	3	3	3
l	1	1	1	1	1	1
m_l	1	1	0	0	-1	-1
m_s	$+\frac{1}{2}$	$-\frac{1}{2}$	$+\frac{1}{2}$	$-\frac{1}{2}$	$+\frac{1}{2}$	$+\frac{1}{2}$

Figure 4.2
Spatial arrangement of s and p orbitals.

Table 4.3	Distribution of Electrons According to Quantum States				
n (number of orbits or shells)	1	2	3	4	etc.
l (eccentricity, subshells, orbitals)	0	0 1	0 1 2	0 1 2 3	
Symbolic representation of n and l combinations	1s	2s 2p	3s 3p 3d	4s 4p 4d 4f	
Number of m_l quanta ($= 2l + 1$)	1	1 3	1 3 5	1 3 5 7	
Number of spin quanta ($+\frac{1}{2}, -\frac{1}{2}$)	2	2	2	2	
Number of electrons in subgroup	2	2 6	2 6 10	2 6 10 14	
Number of electrons in principal group ($= 2n^2$)	2	8	18	32	
Shell designation	K	L	M	N	etc.

Pauli Exclusion Principle

The amount of energy available per atom at ordinary temperatures is small compared with the difference in energy between two quantum states, and it would be natural to expect that all the electrons in an atom would go to the quantum state of lowest energy. This is not the case, however, and the situation is summed up by the Pauli exclusion principle, which seems to have universal validity: *In any system, no two electrons can have all four quantum numbers the same.* This principle applies not only to single atoms but also to systems in which two or more atoms share the same electron(s).

Periodic Classification of the Elements

The principles in the preceding sections can be used to arrange the different elements in an orderly manner according to their electronic configurations. Each proton in the nucleus must be balanced by an extranuclear electron of equal and opposite charge in order to maintain the neutrality of the atom. The number of protons, given by the atomic number, thus establishes the number of electrons that are required.

Principles

The arrangement of the electrons around the nucleus is that which has the lowest energy and is also consistent with the Pauli principle. The maximum number of electrons for any orbit may be determined, therefore, by finding all possible combinations of quantum numbers without violating Pauli's exclusion principle (table 4.2). As an example, in the 3p state ($n = 3$, $l = 1$, $m_l = 1$, 0, -1, and $m_s = +\frac{1}{2}$ and $-\frac{1}{2}$) the possible combinations of quantum numbers, as shown in table 4.2, are 6. Therefore, six electrons completely fill the 3p state.

An extension of this simple analysis to other possible values of n and l gives the maximum number of electrons in any group or subgroup, as shown in table 4.3.

Each s orbital can accommodate two electrons, the three p orbitals can contain a total of six electrons, the five d orbitals a total of ten electrons, and so on. The filling of orbitals up to boron ($Z = 5$) presents no problem because it will have two electrons in 1s, two electrons in 2s, and one in 2p, that is, $1s^2 2s^2 2p^1$. Carbon ($Z = 6$), however, has six electrons, and its configuration can be described as $1s^2 2s^2 2p^2$, but this tells us nothing about the distribution of electrons among the three energetically equivalent p orbitals.

The term *orbital* conveniently sidesteps the issue of whether the electron is thought of as a wave or as a particle.

Hund's rule states that the electrons will distribute themselves in such a way as to minimize the interelectron repulsion. They will spread out and occupy energetically equivalent orbitals singly with their spins in the same direction. Pairing of electrons in each of the *p* orbitals is not required since their m_l values are different, although their *n, l,* and m_s quantum numbers are the same. For example, the configuration of carbon might be better described as $1s^22s^22p_x^12p_y^1$. Actually, there is no way of knowing that the second electron enters $2p_y$, but it is convenient to visualize it that way. The configuration $1s^22s^22p_y^12p_z^1$ would be indistinguishable from the previous one.

Atoms have a paramagnetism corresponding to the number of unpaired electrons they possess. This relationship becomes important for certain elements in a high-spin state—those with many unpaired electrons—and can lead to high magnetic susceptibility of their compounds. The natural high magnetic susceptibility of almandine, $Fe_3Al_2(SiO_4)_3$, is a good example.

The principal groups enumerated in table 4.3 are related to groups of orbitals usually referred to as shells, and from the nucleus outward they are designated *K, L, M, N,* and so on. The subgroups, which are related to orbital eccentricity, are usually referred to as subshells.

The addition of electrons around a nucleus in order to neutralize its field begins with those quantum states for which the energy is at a minimum and proceeds to higher states as the lower ones become filled. The *K* shell is filled first with two electrons in the 1*s* state; then the *L* shell is filled by two electrons in 2*s* followed by six electrons in 2*p*.

The 3*s* and 3*p* states of the *M* shell are filled in sequence. They are, however, not followed directly by the 3*d* electrons but rather by the 4*s* electrons. The reason for this is that the energy of the 4*s* state is actually lower than that of 3*d* owing to the high eccentricity of the latter's orbit (see figure 4.2 and Box 5.2). This same situation arises in all succeeding shells; the *d* state always follows the next larger *s* state. A similar situation obtains for the 4*f* state, which is not filled until after 5*p* and 6*s*. The sequence in which the shells are filled is shown diagrammatically in figure 4.3.

The outermost shell of any atom can never contain more than eight electrons (two in the *s* state and six in the *p* state). The *s* and *p* subshells of a given shell are always filled before any addition to shells of higher *n* is made.

Such an eight-electron shell is developed periodically as electrons are added to balance the progressive increase of protons in the nucleus. This electron configuration is energetically very stable and will be used as a reference

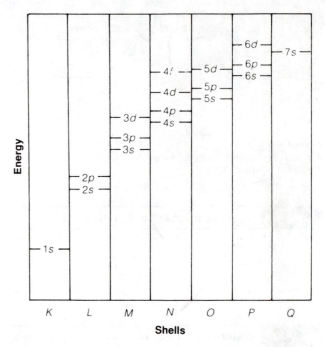

Figure 4.3
Shell-filling sequence.

shell in future discussions. When an eight-electron shell is completed, the next two electrons always go into the *s* subshell of the next larger shell. Following the filling of this *s* subshell, electrons may enter the higher energy subshells of either the same or smaller shells.

The orderly addition of electrons to satisfy the energy requirements of the atomic system is presented in a simplified tabulation in table 4.4.

The Periodic Table

The ninety-two naturally occurring elements have nuclei that contain from one to ninety-two protons. Each proton must be electrically balanced by an extranuclear electron, and the arrangement of these electrons, especially in the outermost (valence) shells, largely establishes the chemical properties of the elements.

The elements were first arranged by Mendeleev (1869) in approximate order of increasing atomic weights in rows, or **periods,** such as elements appearing in columns, or **groups,** displayed similar chemical properties. We now know that the basis for this arrangement is the electronic configuration of the atoms; the similarity among elements in a group results from the similarity in their electronic structure. Chemical similarity within a group varies, but from it, in Mendeleev's day, the chemistry of new elements was predicted prior to their discovery, and today new synthetic elements are identified by their properties and assumed positions on the periodic table.

Additions of electrons in successive shells repeat configurations that were present in previous shells, and a periodicity of chemical behavior results. The periodic classification of the elements (figure 4.4) epitomizes the

Table 4.4 Electronic Buildup of the Elements

Atomic Number	Number of Electrons in Shell						
	K	L	M	N	O	P	Q
1–2	1–2						
3–10	2	1–8					
11–18	2	8	1–8				
19–28	2	8	8–16	1–2			
29–36	2	8	18	1–8			
37–46	2	8	18	8–18	1–2		
47–54	2	8	18	18	1–8		
55–57	2	8	18	18	8–9	1–2	
58–71	2	8	18	19–32	9	2	
72–78	2	8	18	32	10–17	1–2	
79–86	2	8	18	32	18	1–8	
87–92	2	8	18	32	18	8–12	1–2

Completed shells   *Shells being filled*

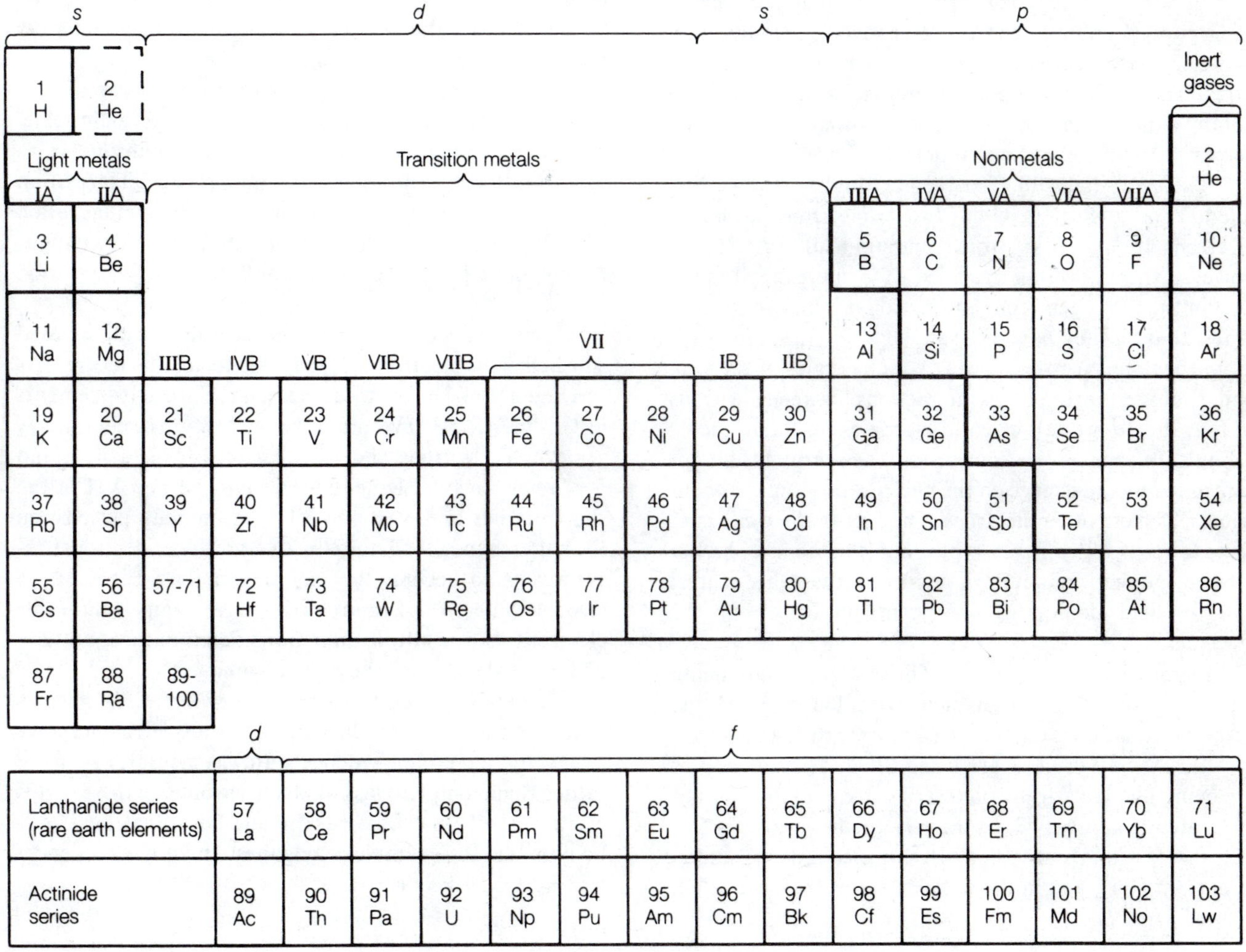

Figure 4.4
Periodic table of the elements showing orbitals being completed. Helium is shown twice to indicate the presence of *s* electrons and its inclusion in the Noble Gases. Atomic weights and ionic radii for common valences are given in Appendix A.

chemical relationships that arise. Elements in the same group (column) have the same number of outer electrons and have similar chemical properties. A sequence of elements in a period (row) will exhibit a continuous change of properties.

Helium (He), with two $1s$ electrons, has a complete outer shell (K). Similarly, eight-electron outer shells are complete at atomic numbers 10, 18, 36, 54, and 86 (Group VIII). These elements, along with He, have lowest-energy outer-shell configurations $ns^2\,np^6$ and have no significant tendency to participate in chemical reactions. Thus they are the most stable chemically and are referred to as the **noble gases.** Group VII elements, referred to as the **halogens,** have outer-shell configurations $ns^2\,np^5$. They have a strong tendency to add an electron to complete a noble gas configuration and form singly charged anions. On the other hand, the **alkali metals** (Group I of the short periods and IA of the long periods), with $ns^2 np^6 (n+1)s$ configurations, tend to lose an outer s electron to form monovalent cations with the noble gas structure. Similarly, elements of Groups II and VI tend to, respectively, lose and gain two electrons to form divalent ions with noble gas structures. The tendency to gain or lose electrons in order to obtain a noble gas configuration continues toward the center of the table, although more weakly.

The d subshells fill after the next higher s subshell is filled. Thus calcium (Ca) has two $4s$ electrons but no $3d$ electrons. In proceeding from scandium (Sc) to zinc (Zn), however, the $3d$ orbitals are filled. It is interesting to note that in the hydrogen atom all electron orbitals are degenerate, that is, they have the same energy. In atoms with more than one electron, however, the energies of the s, p, and d subshells are split by interactions between electrons so that the $3d$ orbitals are at a higher energy than the $4s$ orbitals. In a multiple-electron atom the energy field is not controlled by the charge on the nucleus alone but has contributions from other electrons. The fact that we can continue to speak of hydrogen-type orbitals in atoms of higher atomic number, however, is evidence that the central nuclear field dominates over contributions from the electrons.

Elements in which the d or f orbital groups are being filled are referred to as **transition series.** For example, the first transition series contains elements 21 to 30 (Sc to Zn) in which the $3d$ orbitals are being filled. The $4d$ orbitals are being filled in a transitions series from atomic number 39 to atomic number 48. In the **rare earths** (elements 57–71), the $4f$ orbitals are filled after $6s$ electrons are added in barium (Ba), and in the **actinide** series the $5f$ subshell is filled after $7s$.

Some Intrinsic Properties of Atoms

Valence

The state of lowest free energy for an atom is attained when it has a completed outer shell rather than when it is electrically neutral. As a consequence, atoms tend to exchange or share electrons in such a way as to gain completed outer shells even at the expense of electrical neutrality. Whenever several kinds of atoms are brought together, brisk trading in electrons and development of new atomic groupings occur if such activity favors the decreased free-energy status of the various atoms. Such activities are conveniently designated as chemical reactions.

The loss or gain of electrons by an atom is known as **ionization.** Electron loss results in a net gain of positive charge on the atom, which is numerically equal to the number of electrons lost, whereas the opposite is true for a gain in electrons. Such electrically charged atoms are called ions or, more specifically, cations (attracted to the cathode) if the net charge is positive and anions (attracted to the anode) if the net charge is negative. The numerical value of the residual charge is the **valence.**

The valences of elements that usually occur in natural materials can be found readily in the periodic table. The general rule is that the atom loses or gains electrons to attain the nearest noble-gas configuration. Thus Li, in losing one electron, takes on the electronic configuration of He; Sc loses three electrons to attain the Ar configuration; and I gains one electron to have the configuration of Xe.

Elements in Group I have one electron outside a closed subshell. This electron is loosely bound, and its loss results in a valence of $1+$ for these elements. Similarly, elements in Groups II, III, IV, and V have two to five chemically removable electrons and valences of $2+$, $3+$, $4+$, and $5+$, respectively. Elements in Groups VI and VII in the short periods and VIB and VIIB in the long periods, on the other hand, tend to gain two and one electrons, respectively, to become $2-$ and $1-$ anions. Elements in Groups VIA, VIIA, and VIIIA have many removable electrons as a result of their transitional nature, but $2+$ and $3+$ valences are the most common.

The above generalizations for the valences of various elements cannot be applied rigidly. Many elements have two or more common valence states as a result of partial rather than complete loss of electrons outside their eight-electron shell. In such a case, a number of valences may be possible; for example, manganese, may have valences of $2+$, $3+$, $4+$, and $7+$.

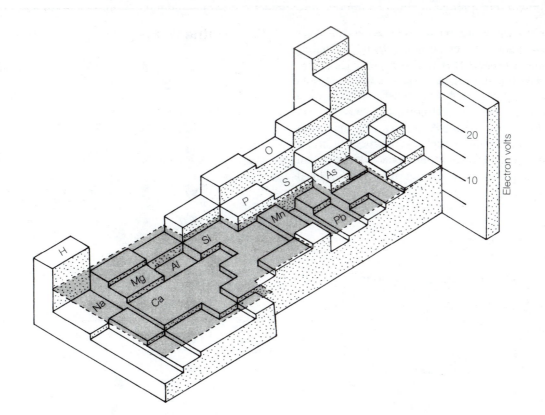

Figure 4.5
Relation of the first ionization potential of the elements to the periodic table. Elements present in 10% or more of mineral species identified; shaded plane at ten electron volts. For other elements, see figure 4.4.

Changes in valence are usually described by the terms oxidation and reduction. **Oxidation** is the process whereby electrons are lost, as in the change from Na° to Na^+ or from Fe^{2+} to Fe^{3+}. **Reduction** is the process whereby electrons are gained, as in the change from Cl° to Cl^- or Mn^{4+} to Mn^{2+}.

Shielding

The presence in an atom of a number of moving electrons results in a complex and continuously varying electrical field intensity in and beyond the outer shells. The electrons are attracted to the nucleus by coulombic forces and repel each other with similar forces.

The force of coulombic attraction is directly proportional to the product of the charges q and q' and inversely proportional to the square of their separation, r; that is, $F = qq'/r^2$.

A solution for the energy distribution in such a complex system cannot be achieved without drastic approximations, but a measure of the magnitude of the field affecting an outer electron may be obtained experimentally. The **ionization potential** of an atom is a measure of the work required to remove an electron from the atom; it is equal to the forces that hold it to the atomic system. Elements with high ionization potentials must have a large proportion of nuclear charge "leaking through" to and beyond the outer electrons to ensure they are held together strongly. The situation can be described as one of poor **shielding** of the nuclear charge from outer electrons and extra-atomic points by intermediate electronic shells. The opposite situation of weakly bound electrons and consequent low ionization potential is found when shielding is efficient. Ionization potential is thus a direct measure of the tightness with which an outer electron is bound to the atom. Figure 4.5 illustrates the manner in which the potential required to remove the first electron from a neutral atom varies from element to element across the periodic table.

The different shielding efficiencies of the elements affect the magnitude of the electrical field in their immediate vicinity with important consequences for their neighbors. Each atom will prey on its neighbor's electrons to the extent it is capable. Atoms with the strongest external fields (poorest shielding) will be the most successful hunters. As an atom is exposed to a stronger and stronger pull from a neighboring atom, it first is distorted from its normally spherical form; that is, it is polarized. Eventually it may lose one or more of its valence electrons to its neighbor and thus is ionized. Remember that changes in the energy status of the electrons concerned occur in quantized increments. Both polarization, with the consequent possibility that outer shells of the reacting atoms will overlap, and the transfer of electrons from one atom to another have important consequences for chemical bonding.

Bonds and Bonding

The electrical fields associated with atoms are capable of interacting with those of other atoms at distances of many atom diameters. The existence of these long-range forces is the cause of adsorption and friction between smooth surfaces. Such forces are important in bringing atoms close enough together that their respective outer shells interact and result in a chemical bond. Many factors contribute to this close interatomic linkage, and the usual bond combines several mechanisms in its formation. In the following sections, the principal mechanisms contributing to a chemical bond are discussed as pure cases, but keep in mind that a real bond is a mixture of types.

When atoms or ions are brought closely together, they gain stability by mutually attaining noble-gas configurations (eight-electron outer shells). For ions, these shells may be formed in the process of ionization, and the ions bond by the coulombic attraction of the spherical ions. This is called **ionic bonding.** For neutral atoms, bonding may be accomplished by a sharing of electrons in their respective incomplete shells to form a single eight-electron shell surrounding the reacting atoms. This is called **covalent,** or homopolar, bonding. Bonding of atoms can also be accomplished by a sharing of electrons from incomplete shells among a number of atoms. In this case, the electrons could be compared to a gas filling the interstices in a pile of marbles. This is called **metallic bonding.** The three bonding mechanisms are contrasted in figure 4.6, which shows diagrammatically the different paths of valence electrons in the different bond types. In ionic bonding, the valence electrons have paths around only a single nucleus. The extension of this path to include adjacent nuclei that mutually share the electrons results in a covalent bond. Complex and nonrepeating paths characterize the metallic bond.

These three bonding mechanisms are not mutually exclusive, and most solids are bonded by some combination of the pure types. Because bonds are dynamic systems, any particular electron can be envisaged as changing from one path to another. If the change from path to path is accomplished in a regular manner, the bond resonates between two pure types. For example, an electron may generally follow a covalent path but occasionally circulate

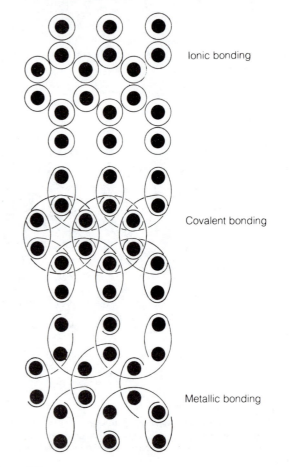

Figure 4.6
Electron paths for different kinds of bonds.

among a number of atoms or orbit around a single nucleus. The covalent bond in this instance would then exhibit some metallic or ionic characteristics. A sharp distinction is not implied when bonds are described as ionic, covalent, or metallic, although it is convenient to describe mineral bonding in terms of a predominant bond type. The gradational nature of chemical bonds is illustrated in figure 4.7. Corners of the triangle represent pure bond types, and typical minerals or mineral groups are located on the edges in positions representing the mixture of bond types present in them.

Ionic Bonding

The ionic bonding is usually developed between elements in the flank groups on the periodic table. Elements in Groups I to IV in the short periods and IA to IVA in the long periods have shielding efficiencies such that electrons may be readily removed from the atom, causing it to become a positive ion (cation). On the other hand, elements of Groups VI and VII of the short periods and VIB and VIIB of the long periods have a high affinity for electrons that produce a negative ion (anion) when added to a neutral atom. In effect, electrons are transferred from well-shielded to poorly shielded atoms in the process of ionization.

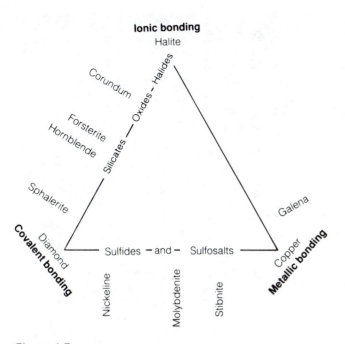

Ionic bonding

Halite

Corundum

Forsterite
Hornblende

Silicates — Oxides – Halides

Sphalerite

Diamond

Covalent bonding

Sulfides — and — Sulfosalts

Copper

Metallic bonding

Galena

Nickeline

Molybdenite

Stibnite

Figure 4.7
The gradational nature of the chemical bond in minerals.

Ions of opposite sign, once formed, are mutually attracted by coulombic forces until their respective noble-gas configurations are in contact. The ions in the aggregates thus formed are so disposed that the ionic charges of the individual ions are neutralized and the whole aggregate is electrically neutral. The proportion of + and − ions present in such an aggregate is controlled by the valences of the participating ions. There will be equal numbers of cationic and anionic elements if their + and − valences are numerically equal; (for example, Na^+ + $Cl^- \rightarrow NaCl$, and Ca^{2+} + $O^{2-} \rightarrow CaO$). Also, the proportions of elements of different valences will be reciprocal to their neighbors' valences; for example $2Fe^{3+}$ + $3O^{2-} \rightarrow Fe_2O_3$; Ca^{2+} + $2F^- \rightarrow CaF_2$; Fe^{2+} + $2Cr^{3+}$ + $4O^{2-} \rightarrow FeCr_2O_4$; and Ca^{2+} + $(CO_3)^{2-} \rightarrow CaCO_3$.

Ionic bonds result from the simple electrostatic attraction between ions of unlike charge. The electrostatically neutral solids thus tend to have low electrical and thermal conductivity. Melted ionic substances or solutions of such compounds, however, have free ions which afford electrical conductivity by ion transfer. Ionic minerals are soluble in polar solvents such as water and inorganic acids. The bond strength depends in general on the sizes and valences of the participating ions. Thus, ionic minerals melt at moderate to high temperatures and have low to moderate hardness. They are usually brittle and often display distinct cleavage. Ions participating in ionic bonding surround themselves with as many like ions of opposite charge as possible. Such an arrangement leads to a nondirectionality of bond strengths resulting in crystals of high symmetry without directional differences in their physical properties. The relatively open structure of ionically

bonded minerals results in their transparency or translucency. Examples of ionically bonded minerals include halite, NaCl, and fluorite, CaF_2.

Covalent Bonding

The covalent bond predominates in the formation of chemical compounds from elements of Groups IV to VII in the short periods and from *B* subgroup in the long periods. This bond forms when poorly shielded atoms react; each such atom has a strong affinity for electrons, but neither will relinquish its own. A compromise to this stalemate is effected by the mutual deformation of outer electron shells and the interpenetration of these shells in such a way as to make the desired electrons common to the reacting atoms. A single eight-electron outer shell is developed by this pooling or sharing of electrons between the reacting atoms, and the participation of the various atoms in this mutually shared, stable shell gives rise to a very strong bond.

Participation of atoms in a common stable shell implies, of course, that the shared electrons do not violate the Pauli exclusion principle. This is ordinarily accomplished by restricting the bond formed to that in which electrons of opposed spin participate. Thus, two hydrogen atoms, each having a single $1s$ electron, may bond covalently into a diatomic hydrogen molecule H_2 if the electrons have opposed spins.

The requirements of the Pauli principle may also be met by transferring a few electrons to orbits of different eccentricity only slightly higher energetically than their normal orbits. Thus, in carbon, which has two electrons with unpaired spins (in $2p$), a very small energy increment will transfer one of the $2s$ electrons to $2p$, and four electrons with unpaired spins (one in $2s$ and three in $2p$) become available for bonding. The small increase in energy required for the formation of covalent bonds by this mechanism may be likened to making a small investment for a large guaranteed return.

The strength of a covalent bond is a direct function of the energy that is localized in the bond. In certain situations resonance of the electron between two or more bonding combinations can increase the energy of the bond. For reacting atoms with unpaired spins of both s and p electrons, s-s and p-p bonds could form, but a stronger bond is produced by resonance of electrons between s and p states. Diagrammatically, this may be shown as follows:

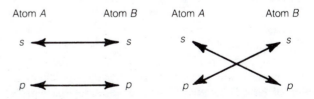

Atom *A*　　　Atom *B*　　　Atom *A*　　　Atom *B*

s ⟷ s　　　s　　s

p ⟷ p　　　p　　p

The two s-p bonds formed by resonating electrons are stronger than the sum of the s-s and the p-p bonds formed by nonresonating electrons.

Box 4.1

Another View of Bonding

An alternate means of viewing the amount of nuclear charge that leaks through the electronic cloud may be gained through the concept of **electronegativity.** This is the tendency of an atom to acquire a negative charge as measured by the electrostatic force at its periphery. Although no exact theoretical or experimental means has been devised to determine this affinity of an atom for electrons, relative negativity (electronegativity) values are popular and widely used.

The original work on electronegativity was done by the American chemist Linus Pauling, who in 1948 provided a scale using the usual valence state for ions. The scale, shown in table 4.5, is relative to a value of 4.0 taken for fluorine. The greater the numerical difference between an element pair, the more ionic is their bond. Thus, the Ti-O bonds in rutile, TiO_2, $(3.5 - 1.5 = 2.0)$ are more ionic than the Si-O bonds in quartz $(3.5 - 1.8 = 1.7)$. Likewise, CuO, with an electronegativity of 1.6, is more ionic than CuS, with a value of 0.6. The use of electronegativities to estimate bond types becomes quite important when elemental substitutions in minerals must be considered.

Table 4.5	The Pauling Electronegativity Scale															
Li	Be											B	C	N	O	F
1.0	1.5											2.0	2.5	3.0	3.5	4.0
Na	Mg											Al	Si	P	S	Cl
0.9	1.2											1.5	1.8	2.1	2.5	3.0
K	Ca	Sc	Ti	V	Cr	Mn	Fe	Co	Ni	Cu	Zn	Ga	Ge	As	Se	Br
0.8	1.0	1.3	1.5	1.6	1.6	1.5	1.8	1.8	1.8	1.9	1.6	1.6	1.8	2.0	2.4	2.8
Rb	Sr	Y	Zr	Nb	Mo	Te	Ru	Rh	Pd	Ag	Cd	In	Sn	Sb	Te	I
0.8	1.0	1.2	1.4	1.6	1.8	1.9	2.2	2.2	2.2	1.9	1.7	1.7	1.8	1.9	2.1	2.5
Cs	Ba	La-Lu	Hf	Ta	W	Re	Os	Ir	Pt	Au	Hg	Tl	Pb	Bi	Po	At
0.7	0.9	1.1–1.2	1.3	1.5	1.7	1.9	2.2	2.2	2.2	2.4	1.9	1.8	1.8	1.9	2.0	2.2
Fr	Ra	Ac	Th	Pa	U	Np-No										
0.7	0.9	1.1	1.3	1.5	1.7	1.3										

Data from Pauling, L. 1960. *The Nature of the Chemical Bond.* 3d ed. Cornell University Press, Ithaca.

The approximately linear relationship between the Pauling electronegativity values and the first ionization potentials of the elements is shown in figure 4.8a. Remember that the first ionization potential is inversely proportional to the work function involved in the concept of electronic shielding. The difference in electronegativity of any pair of atoms determines the percent ionic character of their bond. The conversion is shown in figure 4.8b. The original Pauling scale electronegativities have been largely supplanted in modern work by values adjusted for the relative compactness of the electronic clouds of ions of different size and charge, i.e., by values that involve a measure of the surface charge density of an ion.

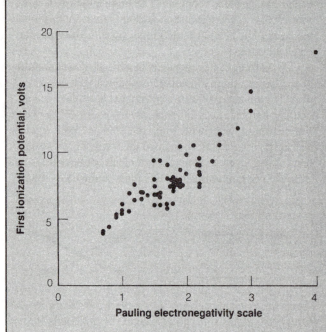

Figure 4.8a
The relation between the electronegativities and the first ionization potentials of the elements.

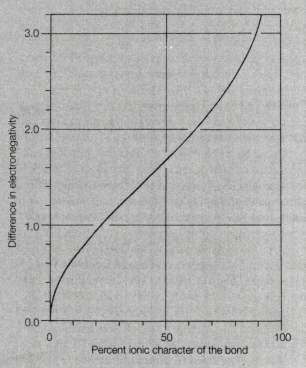

Figure 4.8b
Conversion of electronegativity values to the percent ionic character of a bond.

Covalent bonds occur by the sharing of electrons between ions to produce stable and complete noble-gas configurations for both. The participating ions are very reluctant to give up electrons and the resultant strong bonds are reflected in the physical properties of covalently bonded minerals. Covalently bonded materials are generally insoluble except in very strong acids, and solution does not produce free ions. Thus, covalently bonded substances are poor electrical conductors both in the solid and liquid states. They have high melting points and hardnesses and are generally brittle. Covalent bonds tend to be highly directional. Thus, the crystals of covalent minerals are usually of lower symmetry, and their physical properties can vary with crystallographic direction. Covalent minerals are generally translucent to opaque. Minerals exhibiting pure covalent bonding are rare. Diamond, however, is an excellent example; each carbon atom is coordinated with four adjacent carbons, each of which shares an electron with the central carbon (see figure 13.8).

Metallic Bonding

The external fields of very poorly shielded atoms affect not only the outer electrons of their neighbors but also the electrons in deeper shells. The interaction of such atoms raises the electrons of neighboring atoms to levels of higher energy, and bonding reactions following the principles of covalent sharing occur. This interaction is very complex, because each atom in common metal structures is surrounded by a number of similar neighbors. An individual electron participating in the bond will circulate through the structure much like a dancer in the Virginia reel and cannot be considered as having a fixed relation to any given nucleus. The numerous reacting atoms must satisfy the Pauli restrictions, however, and the discrete quantum states that characterize the interaction in covalent bonding must be split up into a number of slightly differing energy levels so that all the reacting atoms can participate in the bonding. The energy zones so developed may vary from those of negligible width (covalent bonds) through discontinuous zones (metallocovalent or metalloidal bonds) to continuous zones (metallic bonds).

The individual electron in an aggregate of atoms so bonded is in continuous transition from one quantum state to another and as such is essentially free of alliances to any one atom. The application of an electrical potential to the aggregate causes such free electrons to migrate readily through the structure, thus producing the electron flow familiar to us as electric current. Metallic bonding is found principally among elements of the long periods in the transition metals of Group VIIIA and Groups IB and IIB.

Metallically bonded substances are made up of close-packed arrays of positively charged metal ions, each of which has given up an electron. The electrons are rela-

Figure 4.9
Change in coordination.

tively free to move throughout the structure, contributing to the electronic configuration of all ions and the overall electrostatic neutrality. This free movement of electrons results in the high electrical and thermal conductivity observed in metals. Bond strength is low to moderate and depends on the size of the metal ion concerned and the closeness of ion packing. Bond strengths are approximately equal in all directions. Thus metallically bonded minerals usually have variable melting points, low to moderate hardnesses, and high symmetry. They also tend to be sectile, ductile, and malleable. The close-packed metallic structures make these minerals opaque. Native gold, silver, and copper are excellent examples.

Spatial Distribution of Bonds

An ionic bond is one that has no directional properties. The spatial distribution of the bonds formed around a given ion is only a function of the location of neighboring ions, and bond direction in ionically bonded structures is based on the sizes of the ions involved.

Many more anions of a given size can pack at uniform spacing around a large cation than one of smaller radius. The geometry of this packing is fixed in terms of the ratio of the ionic radii (see table 4.6). The minimum radius-ratio values in the table represent arrangements in which all ions are in contact. The same arrangement is retained with increasing size of the central ion until the radius ratio for the next larger number is reached. Figure 4.9 illustrates this situation for a hypothetical change from a triangular to a square arrangement.

The number of nearest neighbors (ions in contact) to any ion is called its **coordination.** Each ion is bonded to or coordinated by its nearest neighbors, and the coordination polyhedra are stacked to make up the crystal structure. Usually these coordination polyhedra share their corner ions because this arrangement allows the greatest distance between ions of like sign, but sometimes edges or even faces are shared.

The relative strength of the ionic bond linking one ion to another is a function of both the valence of the ion and the number of neighboring ions to which it is bonded. If,

| Table 4.6 | Packing of Ions |

Minimum Radius Ratio Cation/Anion	Number of Nearest Anion Neighbors	Geometrical Arrangement of Nearest Neighbors		
0.155	3	Corners of an equilateral triangle		
0.255	4	Corners of a tetrahedron		
0.414	6	Corners of an octahedron		
0.732	8	Corners of a cube		
1.0	12	Corners of a cuboctahedron		

for example, Na^+ is surrounded by six Cl^- neighbors, as in the mineral halite, the Na^+-Cl^- relative bond strength is

valence of Na/number of neighbors $= 1/6$

In this structure, the number of Na^+ ions around each Cl^- ion must also be 6, and the Cl^--Na^+ bond strength is also $1/6$. Similarly, for SiF_4, the Si^{4+}-F^- bond is $4/4$ and the F^--Si^{4+} bond is $1/1$.

For a silicate such as muscovite, $KAl(AlSi_3O_{10})(OH)_2$, the relative bond strengths (RBS) are as follows. The superscripts IV and VI refer to coordination.

Element	Valence	O and OH neighbors	RBS
K	1 +	12	0.08
^{VI}Al	3 +	6	0.50
^{IV}Al	3 +	4	0.75
Si	4 +	4	1.00

Obviously, the K-O bonds are weakest and rupture (cleavage) may be anticipated on planes of potassium ions.

In ionic compounds, which comprise the majority of minerals, those structures in which the bond strengths throughout the crystal are essentially equal are termed **isodesmic,** whereas those in which some bond strengths are markedly different are called **anisodesmic.** A necessary part of an anisodesmic structure is the presence of tightly bonded groups that are linked together by means of bonds of lesser strength. For example, in $Ca(CO_3)$ the C-O bonds are stronger than the Ca-O bonds, and the discrete $(CO_3)^{2-}$ groups present in the structure retain their identity even when the mineral is dissolved. These tightly bound groups, called **radicals** or **complex ions,** are always composed of a central cation coordinated by anions. Typical radicals are $(CO_3)^{2-}$, $(SiO_4)^{4-}$, $(SO_4)^{2-}$, and $(PO_4)^{3-}$.

The valences of C and Ca are 4+ and 2+, and they can surround themselves with three and six oxygen neighbors, respectively. The relative strength of the C-O bond is thus 4/3 and that of the Ca-O bond is 2/6.

The shapes of these radicals reflect the shielding efficiency of the positive element. Radicals containing a cation with efficient shielding have regular geometric shapes. On the other hand, radicals containing cations with poor shielding efficiency may be deformed in the presence of external electrical fields. For example, in the sequence $(SiO_4)^{4-}$, $(MoO_4)^{4-}$, $(PtO_4)^{2-}$, the shielding efficiency decreases from silicon to platinum and the radicals are, respectively, tetrahedral, distorted tetrahedral, and square in shape. This sequence is illustrated in figure 4.10.

The spatial distribution of covalent bonds is related to the spatial distribution of the probability volumes of electrons in the various states within an atom. The probability of finding an electron at some particular distance from the nucleus is illustrated in figure 4.1. The separation of electrons in any particular shell is governed by probability, with the electrons distributed in such a way as to be at maximum separation. For the two electrons in the s state, such positions can be on opposite sides of the nucleus, whereas the six electrons in the p state would be spaced at equal distances and their statistical positions would correspond to the corners of an octahedron. Bonds arising because of the overlap of shells and sharing of electrons might thus be expected to be strongest in the direction of the concentration of these probability volumes (see figure 4.11).

The s state is spherically symmetrical, and atoms bonded by paired electron spins between s electrons have no directional preference for the bond. Because only a single s state is present in the valence shell of a given atom, however, the s-s bond is restricted to the formation of diatomic molecules.

The probability volumes of p electrons lie along mutually perpendicular axes (p_x, p_y, p_z), each of which may accommodate two electrons. The p-p bonds tend, therefore, to be at right angles to one another. For example, the water molecule H_2O is held together by p bonds (the $1s$ electron of the hydrogen atom being elevated to the p state), and the angle H-O-H is 104° 31′. The deviation of this angle from 90° is accounted for by the partial ionic character of the bond and a consequent repulsion between the hydrogen atoms.

Only half the electrons needed to fill a quantum state are available for bonding because of the necessity of having opposed spins.

The combination of s and p states provides the possibility of four bonds, because one s electron and three p electrons are available for bonding. These bonds are oriented in such a way as to attain the maximum interbond angle and are thus directed toward the corners of a tetrahedron. As an example, the free energy of a system of carbon atoms is minimized by making the bonding energies as large as possible, which in turn is accomplished by

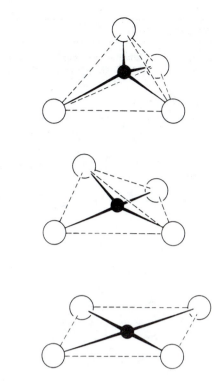

Figure 4.10
Distortion of a coordination polyhedron by polarization.

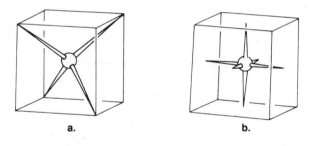

Figure 4.11
Spatial distribution of bonds. (a) Tetrahedral sp^3. (b) Octahedral d^2sp^3 and sp^3d^2.

s-p hybridization. The resulting structure is a network of carbon atoms on tetrahedral corners with bonds directed along tetrahedral edges, as in diamond.

Hybridization of d, s, and p states occurs when the energies of the d state are close to that of the s and p states, as in the transition elements. Theoretically, nine bonds could be formed by such hybridization—one from s, three from p, and five from d—but in reality the d state is only partially filled, and a maximum of two d electrons are available for bonding. The maximum number of d-s-p bonds is thus six, and such bonds are found to be directed toward the corners of an octahedron.

The spatial distribution of metallic bonds cannot be readily determined by examining the bonding mechanism, but the nature of the bond implies that it is nondirectional. Only three common arrangements of atoms in

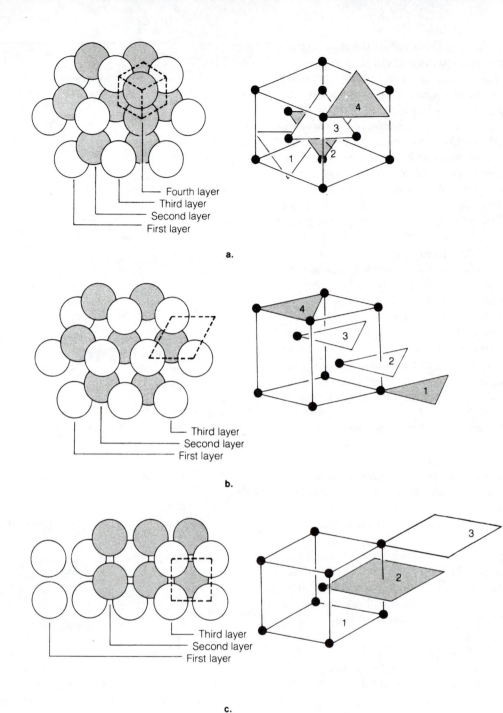

First layer
Second layer
Third layer
Fourth layer

a.

First layer
Second layer
Third layer

b.

First layer
Second layer
Third layer

c.

Figure 4.12
Closely packed arrangements of atoms in metallic structures. The
atoms are shown undersized for clarity and alternate layers are
shaded. (a) Cubic close-packed (*ccp*). (b) Hexagonal close-packed
(*hcp*). (c) Body-centered cubic (*bcc*).

metallic structures are known: cubic close-packed (ccp),
hexagonal close-packed (hcp), and body-centered cubic
(bcc) (see figure 4.12). In the first two arrangements, a
given atom is surrounded by twelve neighbors and may be
thought of as having twelve symmetrically disposed bonds.
In the third arrangement, a given atom has eight neighbors located on the corners of a cube, and the bonds thus
lie along cube diagonals.

The Size and Shape of Atoms and Ions

The actual size of an atom or ion depends not only on itself
but also on its environment. Specifically, the size of an atom
depends on (1) its degree of ionization, (2) the kinds of
neighbors it has and the number and arrangement of those
neighbors, and (3) the manner in which it is bonded.

The size of an atom or ion is thus variable and is controlled by factors intrinsic to the element in balance with

external factors. Values of ionic radii have been calculated from wave functions or determined from the atomic spacing of various salts (see chapter 12) and differ depending on the method of measurement and kind of salt used. Because the radius of ion A in situation X is not the same as its radius in situation Y, a given radius can be assumed only when the atoms in a specific situation exactly correspond to the situation in which the radius was determined. However, the factors that affect the radius difference of an ion in different situations are reasonably well known, and thus it is possible to estimate an atomic or ionic radius in a new situation with good accuracy by appropriate corrections to the measured value. The principal factors that affect the size of atoms and ions are given in table 4.7 and described in the following paragraphs.

An atom can be thought of as a positive nucleus embedded in a ball of negative electricity that represents the distribution of electrons through space. The electron density of this ball is greatest near the nucleus and decreases rapidly as distance from the nucleus increases (see figure 4.1), so that the greater part of the atom is within a few Ångstroms of the nucleus.

The density of electrons around the nucleus is not uniform, however, but appears as a series of shells of decreasing density. Also, the distance of the outer shell from the nucleus does not vary linearly as new shells are added, since the inner shells are compressed to such an extent that the overall atomic radii are not greatly different. This drawing in of electronic shells is due to increasing strength of the nuclear field caused by increasing numbers of nuclear protons. In crystals of the elements, the range of atomic radii is only about 0.8 to 2.6 Å, and lithium, atomic number 3, is about the same size as uranium, atomic number 92.

The equilibrium position of the outer electronic shell controls the size of the atom. This equilibrium distance is related to the strength of attraction of the nucleus for outer electrons less the shielding effects of intermediate electron shells.

Ionization of atoms markedly changes the effective radius of the atom. Should an electron be lost, the radius of the resulting cation is less than that of the atom from which it is formed because of the proportionally greater attractive power of the nuclear field per electron.The opposite is true when anions are formed by addition of electrons to the electronic envelope of an atom. The changes in atomic radius as a result of ionization are illustrated in table 4.8.

If a radius change occurs when a neutral atom is ionized, it follows that further valence changes will result in different ionic radii and that the smaller radii will be associated with higher positive valences. The change of ionic radius with valence is especially well illustrated by the various ions of manganese, whose radii for $2+$, $3+$, and $4+$ cations are respectively, 0.75, 0.66, and 0.62 Å.

The mechanics of chemical bonding imply that the size of a given atom must be different when it bonds in

Table 4.7	Effects of Various Factors on Atomic or Ionic Size	
Factor	Change	Effect on Radius
Nuclear charge	Increase	Decrease
Ionization	Electron loss	Decrease
Bond type	Metallic to covalent to ionic	Decrease
Coordination	Decrease	Decrease
Temperature	Decrease	Decrease
Polarization	Sphericity	Two or three radii

Table 4.8	Radius Changes with Ionization					
Atom or Ion	Nuclear Charge	Electrons in Shell				Radius, Å
		K	L	M	N	
K	19	2	8	8	1	2.31
K+	19	2	8	8		1.59
Cl	17	2	8	7		0.97
Cl-	17	2	8	8		1.72

different ways because of electron transfer and electronic shell overlap. A cation participating in an ionic bond loses electrons and its effective size is diminished. On the other hand, when the cation is covalently bonded it retains its electrons, at least temporarily, and its effective size is diminished relatively less. The opposite is true for an anion. The effect of bond type on the radius of an atom or ion is evident in aluminum, whose radius in ionically bound structures is 0.61 Å, in covalently bound structures is 1.30 Å, and in metallic aluminum is 1.37 Å.

The radius of an atom or ion is affected slightly by the number of neighbors that surrounds it as well as by the manner in which it is bonded to them. The effect is a small but systematic decrease of radius with decrease in coordination on the order of 2–3% for the usual changes in the number of nearest neighbors. Aluminum, for example, has radii of 1.37 and 1.43 Å, respectively, with six and twelve neighbors in the metallic structures. In ionic structures, the radius of Al^{3+} is 0.61 Å in 6-fold coordination and 0.47 Å in 4-fold coordination.

One of the more important parameters in mineralogical studies is the size of the atoms in their more usual valence states. Figure 4.13 presents this basic information in relation to the periodic table. Ionic and covalent radii are listed in Appendix A.

In a general way, two ions undergoing reaction are attracted by their respective electrostatic charges inversely as the square of the distance separating them. This is true up to some finite distance at which the like charges of their respective nuclei limit their approach. The force of repulsion at close approach appears to follow an inverse 6- to 8-power law, and in consequence reacting atoms to the first approximation are rigid spheres. The effect of one

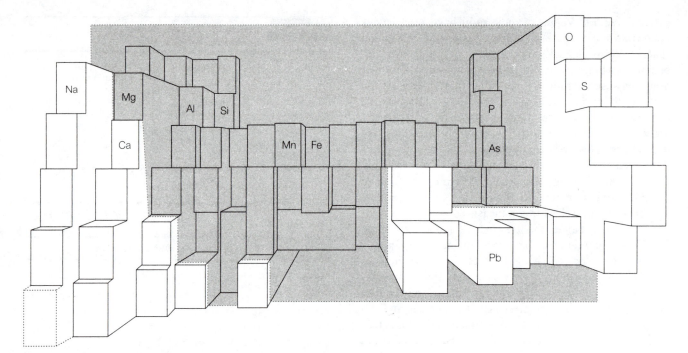

Figure 4.13

Relation of the radii of ions to the periodic table. Elements present in 10% or more of mineral species identified; shaded plane at one Ångstrom. For other elements, see figure 4.4.

ion on another, however, is such that deforming forces are at work. Should an ion be small, highly charged, and poorly shielded, it will suffer very little distortion if it is brought into the electrical influence of another ion; but if such an ion is juxtaposed to a large ion of low charge, the smaller ion will tend to deform the larger ion (and incidently change its radius) in such a way as to neutralize its own charge most effectively.

The ability of one ion to deform another is called its **polarizing power,** and the reaction of an ion to such deforming forces is called **polarizability.** Most cations have low polarizability, and most anions, with the exception of fluorine, are polarizable. The amount of polarization present in a structure is essentially the result of the polarizing power of the cations and the polarizability of the anions. As the degree of polarization increases, the tendency to form self-limiting units increases because the anion field is distorted in such a way that its attraction for another cation is lessened. Eventually, layer structures form and finally discrete molecules.

Polarizability and polarizing power show fairly regular periodic changes from one ion to another, as illustrated in figure 4.14. Polarizability decreases from left to right and increases from top to bottom in the periodic table; the reverse is true for polarizing power.

A final factor affecting atomic or ionic radii is thermal motion. At all temperatures above absolute zero, ions in crystals are vibrating rapidly (on the order of 10^{13} vibrations per second) around some neutral position. Naturally,

Table 4.9	Root-Mean-Square Amplitudes of Vibration in NaCl (Å)	
T° C	Na+	Cl−
−190	0.15	0.13
20	0.24	0.23
225	0.32	0.28
625	0.58	0.58

the effective radius of the ion must be larger than its true or rest radius, and careful measurements in halite, NaCl, show the difference to be quite significant. Root-mean-square amplitudes at different temperatures are given in table 4.9. For example, if the radii of Na+ and Cl− in 6-fold coordination are 1.10 and 1.72 Å, respectively, the vibrational amplitudes at room temperature are on the order of ± 10–20% of the rest radius.

Thus, thermal vibration, in which an ion reigns over a volume larger than itself, contributes to such phenomena as order-disorder (discussed in chapter 5) and diffusion and establishes their dependence on temperature. To visualize the effect, consider figure 4.15, in which thermal domains are dashed, and note the increased size of the gateway available to the central ion if it moves at the correct time. As coordination increases, the *relative* increase in gateway size decreases. This fact, coupled with the decreasing probability of appropriate movements of the coordinating ions, suggests that interchange will become increasingly difficult.

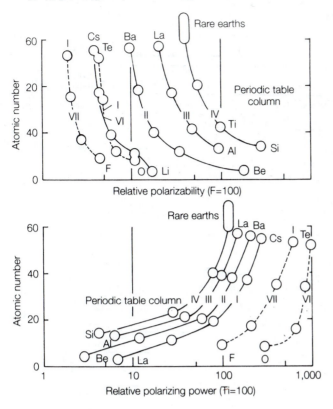

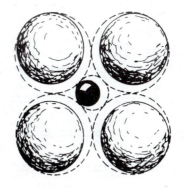

Figure 4.14
Relative polarizing power and polarizability of some ions as a periodic function.

Figure 4.15
Thermal domains. The dashed lines represent the volumes swept out by rapidly vibrating atoms.

Mineral Classification

Crystal chemical laws can be used as a basis for mineral classification, but an arbitrary application of these laws could lead to an artificial system which bears little resemblance to commonly accepted mineral groupings. The bases used for the classical subdivisions are, however, ultimately dependent on crystal chemical phenomena. This concept controls the arrangement of mineral descriptions in part 3.

Minerals, of course, constitute only a small fraction of the number of possible compounds that can be classified according to crystal chemical principles. Many crystal chemical divisions do not apply to the classification of minerals. The scheme used in this book provides, as far as possible, a classification that (1) can be used with a minimum of information about a given mineral, (2) divides minerals into groups of roughly equivalent size, and (3) presents minerals in an order compatible with other schemes of classification.

Minerals are usually subdivided on the basis of the anion or radical present and are further described in terms of the symmetry of their crystalline form and overall chemistry. The terms commonly used to describe minerals and mineralogical groupings are as follows:

Mineral class—chemical grouping based on anion or anionic group, usually comprising:

Native elements

Sulfides (including selenides and tellurides)

Sulfosalts

Oxides (distinguished as simple oxides, hydrated oxides and hydroxides, and multiple oxides)

Halides

Carbonates

Nitrates

Borates

Phosphates, vanadates, arsenates, tungstates, molybdates, uranates, chromates

Silicates

Mineral type or family—minerals showing a similarity of chemical type, such as the zeolite group
Mineral group—minerals showing a similarity of crystallography and structure, such as the rhombohedral carbonates
Mineral species—distinct individual minerals
Mineral series—minerals related by solid solution, such as the plagioclase series
Mineral variety—variant of a mineral species, usually a distinctive habit or color resulting from small amounts of chemical substitution

The approximate distribution of the known minerals among the various classes is shown in figure 4.16.

The crystal chemical classification of minerals used in this text is based on predominant bond type, the presence or absence of radicals, which indicate the relative strengths of cation-anion bonds (ionic structures only), the general chemical content, and the general structural geometry. A scheme based on these principles is shown in table 4.10.

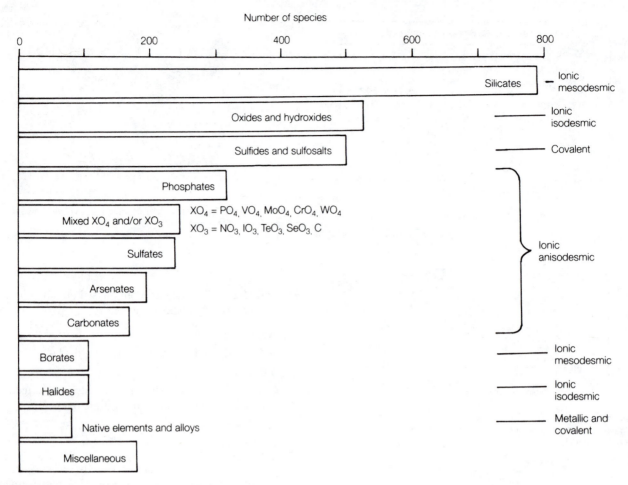

Figure 4.16
Approximate distribution of mineral species among the chemical classes.

In this scheme, bond type is the primary classifier. While this parameter has some failings because of the gradational nature of the chemical bond, the major mineralogical groupings can be readily and naturally made on the basis of bond type.

In the ionic group of minerals the relative strengths of the cation-anion bonds are of considerable importance in differentiating the wide variety of substances in this group. The relative bond strength of a cation is defined as its valence divided by the number of anions to which it is bonded. The strength of all the cation-anion bonds may be qualitatively equal. The direct test of this feature is to determine if the strength of each bond reaching an anion is numerically less than one-half the anion valence. If there is no bond reaching the anion whose strength is more than one-half the anion charge, then the anion is not linked more strongly to one cation than another. Thus, such structures are isodesmic.

Other ionic structures that contain cation-anion bonds of different strengths are called anisodesmic structures. In such structures the strength of some bond reaching an anion is numerically greater than one-half the anion charge. No anion can be linked to two of these more powerful cations, and therefore this cation and its coordinated

anions form a discrete group in the crystal. Such anionic groups, or radicals, are bound together by forces that are stronger than those binding it to the rest of the structure, and the group may retain its identity even when the mineral structure is broken down.

A special case of bond distribution arises when the cation-anion bond is equal to exactly one-half of the charge of the anion. Because of the importance of the minerals in which this circumstance is present—the silicates—these structures make up a separate subdivision in table 4.10. Anionic groups occur, and intergroup sharing leads to interlocked radicals or complex ions. Such structures are called **mesodesmic.**

Ionic structures that have all of their cation-anion bonds of approximately equal strength have formulas of the type $A_m X_p$ or $A_m B_n X_p$, where A and B are cations, X an anion, and m, n, and p are integers. AX structures are *simple structures* because they contain one cation type, as in $NaCl$, CaF_2, or Al_2O_3. ABX structures, which contain two or more kinds of cations, are *multiple structures,* as represented by such compounds as $CaTiO_4$ and $MgAl_2O_4$.

Ionic structures in which the cation-anion bonds are not all of equal strength always consist of an anionic group

Table 4.10   A Scheme of Mineral Classification

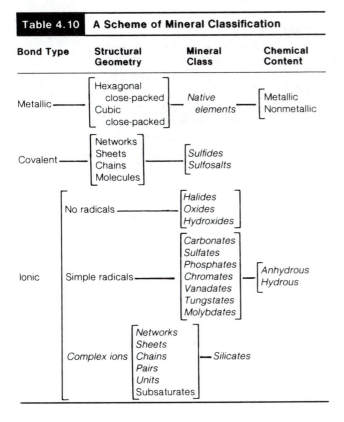

Bond Type	Structural Geometry	Mineral Class	Chemical Content
Metallic	Hexagonal close-packed / Cubic close-packed	*Native elements*	Metallic / Nonmetallic
Covalent	Networks / Sheets / Chains / Molecules	*Sulfides / Sulfosalts*	
Ionic — No radicals		*Halides / Oxides / Hydroxides*	
Ionic — Simple radicals		*Carbonates / Sulfates / Phosphates / Chromates / Vanadates / Tungstates / Molybdates*	Anhydrous / Hydrous
Ionic — *Complex ions*	Networks / Sheets / Chains / Pairs / Units / Subsaturates	*Silicates*	

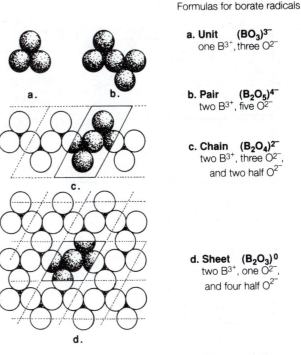

a. Unit $(BO_3)^{3-}$
one B^{3+}, three O^{2-}

b. Pair $(B_2O_5)^{4-}$
two B^{3+}, five O^{2-}

c. Chain $(B_2O_4)^{2-}$
two B^{3+}, three O^{2-},
and two half O^{2-}

d. Sheet $(B_2O_3)^0$
two B^{3+}, one O^{2-},
and four half O^{2-}

Figure 4.17
Oxygen sharing in borates. Boron, small black dots; oxygen large circles. Oxygen ions in sampling motif stippled.

---

or radical coordinated by one cation with the groups being linked by the action of a different cation. In minerals, the anion is almost always oxygen. These structures might be subdivided by the shape of the anion group, but as a practical matter it is preferable to subdivide on the basis of whether or not the mineral contains water or hydroxyl groups, that is, whether it is hydrous or anhydrous. The role of water in the structure is not relevant in the classification scheme, and hydrous minerals may contain either structural water or coordinated water.

Structural water, such as that found in zeolites, merely occupies interstices in the structure and is essentially passive, contributing nothing of importance to the electrical picture. In some minerals, and especially in minerals undergoing weathering reactions, water may coordinate with a cation in a manner analogous to an ordinary anion.

In the special case of mineral structures in which the cation-anion bond strength equals one-half the anion charge, the anionic groups link or polymerize by sharing anions to form complex ions. The resultant pairs, chains, sheets and networks of complex ions are in turn bound together by other cations. Consider, for example, borates. The actual structures of borates are more complex than shown here because of the ability of boron to coordinate four as well as three oxygen ions and the tendency of $(OH)^-$ to substitute for O^{2-}.

Boron, valence $3+$, coordinates three oxygens, valence $2-$, in a plane triangular radical. The boron-oxygen bond strength, determined by the coordination of boron and its valence, is unity and is numerically equal to one-half the anion charge. The borate group, $(BO_3)^{3-}$, which shares no oxygens with adjacent groups in the structure, is shown in figure 4.17a. If one oxygen ion is shared between two of these triangular radicals, the pair of $(BO_3)^{3-}$ units has the group formula $(B_2O_3)^{4-}$ as shown in figure 4.17b. For two shared oxygens per group, a single chain with the group formula $(BO_2)^-$ is formed, as shown in figure 4.17c. A sheet of borate groups results when all three of the oxygen ions are shared between adjacent triangular groups. The group formula is $(BO_{1.5})^0$ or B_2O_3 and the sharing is shown in figure 4.17d. Further sharing is possible among complex ions containing greater numbers of anions per cation. In the silicates such sharing, as we will see later, is the foundation for classifying this extensive mineral class.

Covalently bonded structures are found in only a limited number of minerals. Simple covalent minerals usually have the formula type A_2X, AX, AX_2, or A_2X_3 and multiple covalent compounds have formulas of the type $A_mB_nX_p$. (Compare with simple and multiple ionic structures.) Only a very few native metals are known as representatives of the group of metallically bonded minerals.

Suggested Readings

Bloss, F. D. 1971. *Crystallography and Crystal Chemistry: An Introduction.* New York: Holt, Rinehart and Winston. An elegant treatment of the subject with excellent illustrations.

Cotton, F. A., and Wilkinson, G. 1966. *Advanced Inorganic Chemistry.* 2d ed. New York: Wiley-Interscience. A good discussion of the structure of matter and the nature of chemical bonding is given in part 1.

Evans, R. C. 1964. *An Introduction to Crystal Chemistry.* 3d ed. London: Cambridge University Press. An excellent nonmathematical review of crystal chemistry.

Fyfe, W. S. 1963. *Geochemistry of Solids.* New York: McGraw-Hill.

Mason, B., and Moore, C. B. 1982. *Principles of Geochemistry.* 4th ed. New York: Wiley.

Pauling, L. 1960. *The Nature of the Chemical Bond.* 3d ed. Ithaca, NY: Cornell University Press. The types of bonding in elements and compounds are discussed in detail. The covalent mechanism is emphasized.

Whittaker, E. J. W., and Muntus, R. 1970. Ionic radii for use in geochemistry. *Geochim. Cosmochim. Acta* 34:945–56. An extensive tabulation of ionic radii.

Mineralogic Variations

Determinative mineralogy is often unduly complicated by emphasizing minor differences among individual mineral species. A more fruitful approach is to look first at mineral groups with similar characteristics and then investigate why species within these groups differ. This chapter describes common mineralogical variations that allow us to group individual mineral species into families.

In general, mineralogical variations may be divided into chemical and physical categories. Chemical variations can occur without essential changes in the mineral structure; likewise, geometric variations in the structure of minerals can occur while chemical composition remains constant.

Chemical Variations

The growth units in a growing crystal have certain requirements for size, charge, and shielding efficiency that must be met before an ion can find a permanent place in its structure. The crystal may, however, not exercise much or any selectivity when confronted by two different ions with similar properties, and either may be accepted.

Solid Solution

Minerals always contain small amounts of foreign ions even when such ions are quite dissimilar to the formulary constituents. As they become more similar, the substitutional ability of the foreign ions increases accordingly. A significant amount of substitution can occur in ionic crystals when (1) the radius difference of the formulary and foreign ions is less than 20%, (2) their valences are within one unit of each other, and (3) their ionization potentials, I.P., do not differ by more than about 25% (about 40% for fluorides, 20% for silicates, and 10% for sulfides). Complete substitution can often occur when the radius difference is less than 15% of the smaller ion, the valences are equal, and the ionization potentials do not differ by more than 10%.

The word *small* in "small amount of foreign (nonformulary) ions" is a highly relative term. Crystal structure diagrams typically illustrate a unit cell, a *very* small portion of the mineral structure. A tetragonal unit cell with dimensions of $5 \times 5 \times 10$ Ångstroms (approximately those of chalcopyrite, $CuFeS_2$) has a volume of only 250 Å^3. A cubic millimeter of chalcopyrite thus contains $(10^7)^3$ or 10^{21} Å^3 or 4×10^{18} unit cells. If each cell contains four formula units, there are 1.6×10^{19} copper atoms present in the little cube. Should only one in every million of these atoms be replaced by a silver atom, the cube will contain 1.6×10^{13} atoms of silver—a number that makes the national debt look tiny!

Ionic substitution may be simple or coupled. **Simple ionic substitution** is the process by which a single ionic species substitutes for a formulary ion. For example, Ni^{2+}

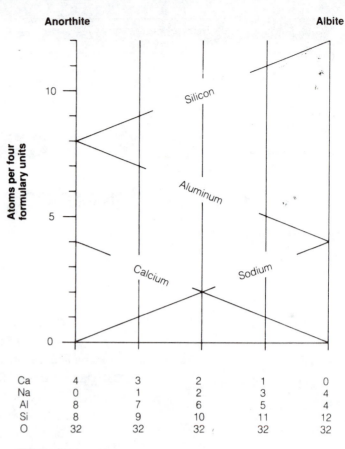

Figure 5.1
Plagioclase solid solution series.

(0.77Å, I.P. 7.63) may substitute for Mg^{2+} (0.80Å, I.P. 7.64). **Coupled ionic substitution** takes place when ions substitute by pairs whose total size, valence, and shielding are equivalent to those of the pair being replaced. Such paired substitution is exemplified by the series of plagioclase feldspars in which $(Ca + Al)$ substitutes for $(Na + Si)$. The formulas for the **end members,** or pure components, of the plagioclase series are $Na(AlSi_3O_8)$ and $Ca(Al_2Si_2O_8)$, and all intermediate ratios of $(Na + Si)$ to $(Ca + Al)$ can occur (see figure 5.1). For example, a compound midway between the two end members (see figure 5.1) would have the formula $Na_{0.5}Ca_{0.5}$ $(Al_{1.5}Si_{2.5}O_8)$ or $NaCa(Al_3Si_5O_{16})$. The sum data for valence, radius, ionization potential, and electronegativity for the substituting pairs in plagioclase are given in table 5.1. The close similarities indicate that the paired substitution is easy in this mineral series.

Because the size of the proxy ion must be nearly that of the formulary ion, the substitution of one atomic species for another should have little effect on the geometry of the structure and thus on the crystal form. Mineral series showing a continuous change in composition without significant change in form are said to be **solid solutions;** such minerals have the same crystal structure and similar but not identical formulas.

Table 5.1	Crystal Chemical Parameters for Ion Pairs in Plagioclase			
	Valence Sum	Radius Sum	Ionization Potential Sum	Electronegativity Sum
(Na + Si)	5+	1.58 Å	13.29eV	2.7
(Ca + Al)	5+	1.67 Å	12.09eV	2.5

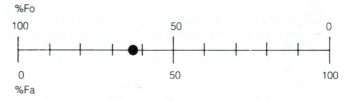

Figure 5.2
Two-component diagram for the olivine series.

*S*olution in its most often used sense describes the dispersal of a solid in a liquid, such as a solution of salt in water. More broadly, however, the term describes the dispersal of any solute in a solvent: oxygen in air, carbon dioxide in beer, and, in mineralogy, foreign elements in a mineral solid.

When extensive substitution is possible, the resulting continuous solid-solution series are often arbitrarily divided into several different species of minerals, a procedure that tends to obscure the close relationships within such a series and adds unnecessary names. A more exact way to designate the intermediate compounds in a series is to describe them in terms of the proportions of the end members present. For example, olivine, $(Mg,Fe)_2(SiO_4)$, is an intermediate member of the solid-solution series from forsterite, $Mg_2(SiO_4)$, to fayalite, $Fe_2(SiO_4)$, but this formula for olivine does not indicate the relative proportions of magnesium to iron. We only know that Mg, being written first, predominates over Fe. If the ratio of Mg to Fe atoms was known to be 63:37, however, a precise formula such as $(Mg_{0.63}Fe_{0.37})_2(SiO_4)$ could be written. This olivine could be considered either as a mixture of the two end members or as a compositional intermediate. In either case, the notation $Fo_{63}Fa_{37}$ (forsterite 63 mole percent, fayalite 37 mole percent) exactly describes the mineral composition. Diagrammatically, this olivine may be represented as a point on a line in a two-component diagram (figure 5.2).

The substitution for magnesium in olivine is not restricted to iron. Manganese sometimes competes successfully for this site, thus forming minerals with the general formula $(Mg, Fe, Mn)_2(SiO_4)$. Perhaps the best way to visualize the solid-solution relations in such a three-component system is to make a triangular diagram in which each corner represents 100% of a constituent. Figure 5.3 shows such a diagram representing various proportions of Mg, Fe, and Mn in the olivine group of minerals. Points 1, 2, and 3 represent forsterite, fayalite, and the olivine $(Fo_{63}Fa_{37})$ discussed earlier. Point 4 is a magnesian-manganoan fayalite; point 5 represents a manganoan fayalite; and point 6, the manganese end member, tephroite. All the individual minerals mentioned are related by solid solution, and intermediate mineral types must be expected to occur.

Where a continuous series of compositions can be found between end members, the solid solution is said to

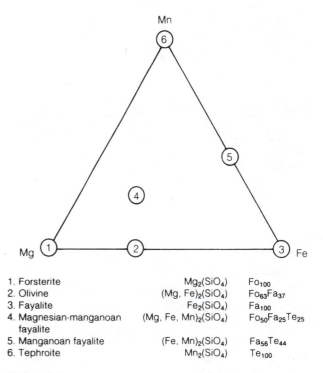

1. Forsterite		$Mg_2(SiO_4)$	Fo_{100}
2. Olivine		$(Mg, Fe)_2(SiO_4)$	$Fo_{63}Fa_{37}$
3. Fayalite		$Fe_2(SiO_4)$	Fa_{100}
4. Magnesian-manganoan fayalite		$(Mg, Fe, Mn)_2(SiO_4)$	$Fo_{50}Fa_{25}Te_{25}$
5. Manganoan fayalite		$(Fe, Mn)_2(SiO_4)$	$Fa_{56}Te_{44}$
6. Tephroite		$Mn_2(SiO_4)$	Te_{100}

Figure 5.3
Three-component diagram for the olivine series.

be **complete.** In many instances, solid solution, although present, is incomplete because of the inability of the structure to accept more than some fixed proportion of substituting ions without exceeding a limit imposed by energy considerations. The substitution of nonidentical ions always causes a local increase in strain, and accumulation of such local strain may eventually surpass the strength of the bonds. Obviously, crystals cannot form if the tendency for forming bonds is opposed by greater forces, and sharp limits to the amount of replacement may be imposed by dissimilarities of ions. An example of incomplete or **limited solid solution** is the clinopyroxene series shown in figure 5.6, in which no natural members having either Ca or Fe in excess of Mg exist. Complete solid solution at high temperatures is, however, found within the shaded portion of the diagram.

Solid solution is not restricted to the interchange of cations, although this is its most common expression. Anions also may substitute for one another, especially $(OH)^-$ for F^- and Se^{2-} for S^{2-}. Further, an unoccupied structural site (a vacancy is sometimes represented in

Box 5.1

Construction and Use of Triangular Composition Diagrams

It is often useful to express the chemistry of a mineral of variable composition in terms of the pure end members. For example, in figure 5.2 the composition of an olivine is illustrated in terms of two end members, forsterite and fayalite. In figure 5.3, olivines of various compositions are represented in terms of three end members, forsterite, fayalite, and tephroite. To develop such ternary diagrams, consider the general case given in figure 5.4 with the end members A, B, and C. The apices represent 100% of the components A, B, and C, and the sides opposite each of the apices represent 0%. To plot a composition on such a diagram, we need to know the concentrations of the end members normalized to 100%. For example, consider a mixture of components A, B, C, D, and E in the following proportions:

A	30%
B	20%
C	35%
D	10%
E	5%
	100%

To plot the composition in terms of A, B, and C, we must first extract those components and normalize them to 100%.

$A + B + C = 30 + 20 + 35 = 85$
$100/85 = 1.176$
$A = 1.176 \times 30 = 35.3$
$B = 1.176 \times 20 = 23.5$
$C = 1.176 \times 35 = \underline{41.2}$
100.0

The result is plotted as point X in figure 5.4. Note that only A and B are actually needed for plotting because the normalized concentration of C is necessarily fixed.

Feldspar can be described in terms of its end members orthoclase (Or), $K(AlSi_3O_8)$; albite (Ab), $Na(AlSi_3O_8)$; and anorthite (An), $Ca(Al_2Si_2O_8)$. Consider a feldspar with 68 mole percent Or, 32 mole percent Ab, and 2 mole percent An. The composition could be expressed as $Or_{68}Ab_{32}An_2$ and the result plotted as X in figure 5.5. For practice, replot this feldspar on a copy of figure 5.4 where the apices A, B, and C are Or, Ab, and An, respectively. Then plot the following feldspars:

	Or	Ab	An
Anorthoclase	20	60	20
Albite	1	95	4
Oligoclase	5	70	25
Andesine	3	60	37
Labradorite	2	33	65
Bytownite	3	15	82
Anorthite	0	3	97

Similar plots are also useful for representing compositions within ternary phase diagrams. Such diagrams are shown in Chapter 6.

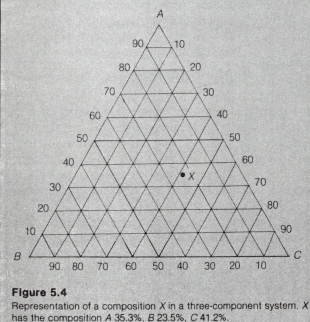

Figure 5.4
Representation of a composition X in a three-component system. X has the composition A 35.3%, B 23.5%, C 41.2%.

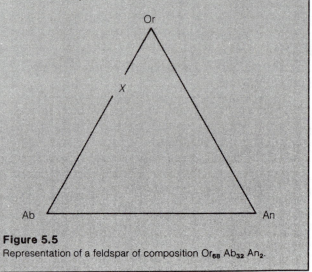

Figure 5.5
Representation of a feldspar of composition $Or_{68}\ Ab_{32}\ An_2$.

mineral formulas by an open square □) may occasionally interchange with an atom, as in pyrrhotite, (Fe, □)S, or, more usually, $Fe_{1-x}S$.

Our discussion of solid solution to this point has dealt only with the abundant, rock-forming elements that are the principal constituents of the earth's crust. Only eight elements compose well over 90% of the crust, leaving the other eighty-four naturally occurring elements of the periodic table in very low concentrations. These elements, often referred to as **minor** and **trace elements,** play an important role, however. Many of them are of considerable economic importance; others are diagnostic in the solution

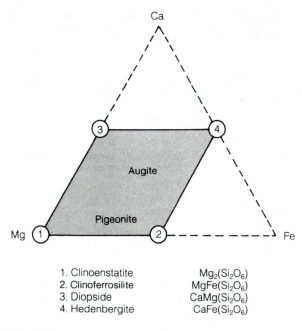

1. Clinoenstatite	$Mg_2(Si_2O_6)$
2. Clinoferrosilite	$MgFe(Si_2O_6)$
3. Diopside	$CaMg(Si_2O_6)$
4. Hedenbergite	$CaFe(Si_2O_6)$

Figure 5.6
Limited solid solution in the clinopyroxene series.

of geologic problems. In any case, they may be grouped into two principal categories. Those that do not easily substitute into the structures of the rock-forming minerals are the **incompatible elements,** and those that resemble a major element to the point that they may proxy for it in rock-forming minerals are the **compatible elements.**

The solution of a compatible element in a mineral can be described in geochemical terms as **capture** when a proxy element has similar size but greater charge, as **admission** when the proxy element has similar size but lower charge, and as **camouflage** when the proxy element has similar size and the same valence. Examples of capture include the common substitution of Ba^{2+} for K^+ in potassium feldspars or the substitution of Cr^{3+} for Mg^{2+} in ferromagnesian silicates. Admission is exemplified by the proxy of Li^+ for Mg^{2+} or by the very common substitution of Al^{3+} for Si^{4+} in silicates. The proxy of Ni^{2+} for Mg^{2+} and Ga^{3+} for Al^{3+} are good examples of camouflage. In some instances, the similarities between certain elements are so great that the proxy element is always found within the major element's minerals. These elements are referred to as **dispersed,** and good examples are Rb^+ for K^+, Cd^{2+} for Zn^{2+}, and Hf^{4+} for Zr^{4+}. In fact, these dispersed elements are always found in K, Zn, and Zr minerals, respectively, and only rarely form minerals of their own.

Generalizations concerning the ability of one element to substitute for another may be made by reference to the periodic table (see figure 4.4). Because valence is constant within a column and size increases as ionization potential decreases rather slowly down a column, replacing ions tend

to have adjacent positions in the table. Replacement pairs such as Sr-Ba, Nb-Ta, S-Se, and Ag-Au will thus arise.

Transition within the *A* subgroup of the periodic table limits the higher valences of these elements, and such geochemically cohesive groups as V-Cr-Mn-Fe-Co-Ni, Os-Ir-Pt, or the rare earths, are found. Occasionally, the position of an *A* subgroup element, being farther down the table, brings its size in line with that of a short-period element in a numerically earlier column. The pair Fe-Mg is an example.

Substitutions of elements between the *B* subgroup and either the early columns in the short periods or the *A* subgroup, although possible based on size and charge, generally do not occur. The barrier is the difference in bonding mechanisms as expressed by ionization potential or electronegativity. Cu^+ and Na^+ ions have the same charge and very similar radii, but their respective ionization potentials of 7.7 and 5.1 electron volts preclude significant substitution.

The foregoing discussion of solid solution is concerned only with minerals whose bonding is sufficiently ionic that the electrical charge of the reacting constituents is an important consideration. For covalently and metallically bonded compounds, the criterion for replacement is controlled not by charge and ionization potential but by size and the ability of the proxy element to form quantized bonds similar to those of the formulary element. Mineralogically, the compounds of sulfur (sulfides and sulfosalts) are the most important minerals in this category. Since these minerals are metal-sulfur compounds, solid solution will be found whenever two metals of similar covalent radius can satisfy the electronic requirements of the metal-sulfur bond. Sulfur is poorly shielded and lacks two electrons (in $3p$) for completion of an eight-electron outer shell. It will, therefore, bond covalently with any element that can supply this deficiency. Such elements are found in the central portion of the long periods in the periodic table—columns VIA to VB. As an example of the amount of solid solution that is possible, consider the fact that Cu, Fe, Mo, Sn, Ag, Hg, and Cd all replace Zn in sphalerite, ZnS.

The phenomenon of solid solution among the sulfides has considerable practical meaning to the mineral explorationist or metallurgist because proxy elements may represent considerable value. Cases in point are the bismuth content of galena, gold values in arsenopyrite, cadmium in sphalerite, rhenium in molybdenite, and nickel in pyrrhotite.

Solid-solution relations in metallocovalent structures are not limited to metal-for-metal replacements. Replacements of metals by vacancies are common, and replacements of sulfur by selenium or tellurium should be expected. Figure 5.7 suggests the wide range of substitutional styles that may be exhibited in minerals.

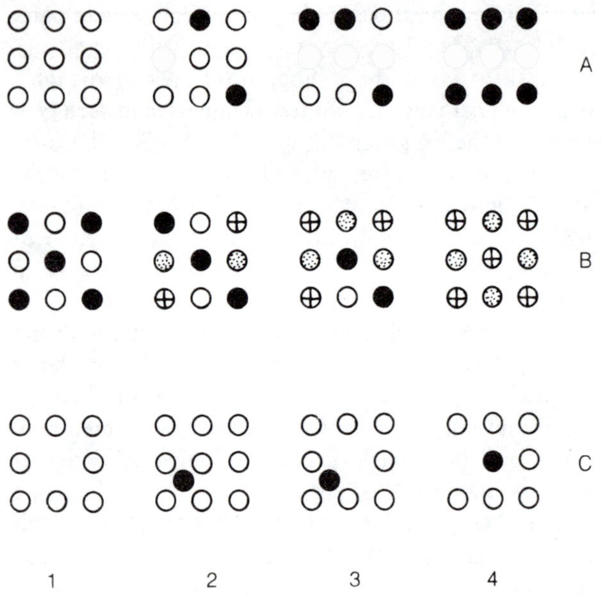

A. Simple substitution

B. Paired substitution

C. Miscellaneous

   1. Vacancy for atom

   2. Interstitial atom (stuffing)

   3. Paired interstitial atom and vacancy

   4. Paired atom and vacancy

Figure 5.7
Kinds of solid solution.

Calculations Based on Mineral Chemistry

In mineral formulas ions with similar structural sites are usually grouped together, and their relative proportions are shown by subscripts. Thus, in biotite, $K(Mg,Fe)_3$ $(AlSi_3O_{10})(OH)_2$, magnesium and iron occupy interchangeable positions in the octahedrally coordinated layer, as do aluminum and silicon in the tetrahedrally coordinated silica sheets. Potassium and hydroxyl have discrete and recognizable sites between and within the sheets. Chemical analyses, on the other hand, typically present the weight percent of the oxides present, as in MgO, FeO, Al_2O_3, SiO_2, K_2O, and H_2O. A structural formula gives proportions of the elements or atomic ratios, whereas chemical analyses give the elements in terms of atomic or molecular weights.

Conversion of chemical analyses to mineral formulas can be accomplished by any of several procedures that converts atomic or molecular weight to atomic or molecular proportion. The atomic weights of elements are given in appendix A, and molecular weights are simply the sum of the weights of the molecule's constituent atoms.

Table 5.2	Derivation of the Mineral Formula from the Mineral Analysis			
1 Element	**2** Atomic Weight	**3** Weight Percent Reported in the Analysis	**4** Relative Number of Atoms	**5** Atomic Ratios
Cu	63.55	63.2	0.933	$4.96 \approx 5$
Fe	55.85	11.2	0.200	$1.00 \approx 1$
S	32.07	25.6	0.798	$3.99 \approx 4$

Consider first an analysis that is given in weight percent of the elements (see table 5.2).

Column 4 gives the atomic proportions, or the weight percent (column 3) divided by the atomic weight (column 2).

Column 5 results from the division of each value in column 4 by the lowest value; for example, $0.993/0.200 = 4.96$.

The structural formula can thus be written as Cu_5FeS_4 and the mineral identified as bornite.

When the analysis is given in weight percent oxides, the weight percent of the elements can first be calculated, after which the procedure is like that given in table 5.2. These calculations for potassium-rich alkali feldspar are shown in table 5.3. The following comments apply:

Column 2: Atomic weight of cation divided by molecular weight of the reported compound; for example, $28.09/60.1$ for silicon in SiO_2.

Column 4: Column 2 times column 3.

Column 5: Column 4 divided by atomic weight of the element.

Column 6: Normalize the ionic proportions by dividing each by the sum, 4.749, obtained in column 5.

Column 7: Calculate an oxygen factor by dividing the number of oxygens in the formula by the normalized ionic proportion of oxygen; for example, $8/0.614 = 13.029$. Now multiply each of the normalized ionic proportions in column 6 by the oxygen factor. For example, for Si, $0.2276 \times 13.029 = 2.965$.

Column 8: Elements which are found in the same structural positions are grouped to give the formula.

The structural formula for this mineral, potassium-rich alkali feldspar, can thus be written (K,Na,Ca) $[(Al,Si)_4O_8]$.

Wide variations in feldspar chemistry are possible (see figure 21.6), but the ratio of atomic proportions of $[K + Na + Ca]:[Al + Si]$ is always 1:4. Thus we can express the actual composition of the feldspar as follows:

1. Sum the normalized ionic proportions (column 6) for the structural position containing K, Na, and Ca; that is, $0.0524 + 0.0232 + 0.0012 = 0.0768$.

Table 5.3

Table 5.3 Calculation of the Structural Formula from Weight Percent Oxides

1 Oxide	2 Fraction of Cation in Oxide	3 Weight Percent Oxide from the Analysis	4 Weight Percent Cation from the Analysis	5 Ionic Proportions	6 Normalized Ionic Proportions	7 Factored Ionic Proportions	8 Proportions in Formula
SiO_2	0.47	64.66	30.39	1.081	0.2276	2.965 ⎫	
						⎬ 4.026 ≈ 4	
Al_2O_3	0.53	19.72	10.45	0.387	0.0814	1.061 ⎭	
CaO	0.71	0.34	0.24	0.006	0.0012	0.016 ⎫	
Na_2O	0.74	3.42	2.53	0.110	0.0232	0.302 ⎬ 1.001 ≈ 1	
K_2O	0.83	11.72	9.73	0.249	0.0524	0.683 ⎭	
			53.34				
Weight percent oxygen by difference			46.66	2.916	0.6140		8 = 8
			100.00	4.749			

Table 5.4 Calculation of a Structural Formula Directly from the Oxides

1 Oxide	2 Molecular Weight	3 Weight Percent by Analysis	4 Relative Number of Molecules	5 Molecular Ratios
SiO_2	60.1	39.7	0.661	5.95 ≈ 6
Al_2O_3	102.0	33.7	0.331	2.98 ≈ 3
CaO	56.1	24.6	0.438	3.95 ≈ 4
H_2O	18.0	2.0	0.111	1

2. Divide each individual ionic proportion by the sum. For Ca, $0.0012/0.0768 = 0.02$; for Na, $0.0232/0.0768 = 0.30$; for K, $0.0524/0.0768 = 0.68$. The feldspar composition may be expressed as $(K_{0.68} Na_{0.30} Ca_{0.02})[(AlSi)_4 O_8]$.

It is also possible to calculate a structural formula directly from the oxides given by a chemical analysis (see table 5.4).

Column 4: Weight percent divided by molecular weight.

Column 5: Values in column 4 divided by 0.111, the smallest relative number of molecules.

The oxide proportions are $H_2O \cdot 4CaO \cdot 3Al_2O_3 \cdot 6SiO_2$, which may be rearranged by separating the cations and oxygen as $H_2Ca_4Al_6Si_6O_{26}$. Further simplification is accomplished by dividing the subscripts by a least common denominator and finally collecting the results into a probable structural form, $Ca_2Al_3(Si_3O_{12})(OH)$. Unfortunately, other arrangements are possible, and the correct structural formulation of this mineral, zoisite, is $Ca_2Al_3(SiO_4)(Si_2O_7)(O, OH)$.

Exsolution

Solid solution may occur at elevated temperatures when a mineral's tolerance for foreign ions is greatly increased by the additional energy from the thermal agitation of the atoms about their structural sites. The number of foreign ions accepted under such circumstances is often greater than can be tolerated in the structure at some lower temperature. In these circumstances the crystal structure must either be disrupted or must cast out the excess of foreign ions on cooling. The latter phenomenon is called **exsolution** or **unmixing,** the separation of a single homogeneous mineral into two or more distinct minerals *in the solid state.* This process is analogous to the formation of two immiscible liquids from some homogeneous liquid mixture as it cools, such as the separation of fat from soup. The process takes place without overall chemical change and without complete disappearance of the original phase. Separation into distinct solid phases without physical disruption is necessarily controlled by the structures of the resultant phases. These structures must have at least partial contact at their mutual boundary, and the energy localized in the boundary layer must be less than that required for rupture. The resulting mineral intergrowths are controlled in their orientation by the host structure and may appear as irregular masses (blebs), needles, plates, or equidimensional bodies, depending on whether the solute and host have zero, one, two, or three atomic planes in common. For example, a star sapphire results from the exsolution of rutile needles oriented parallel to the *a*-axis directions.

A very common mineral example of exsolution is **perthite,** an alkali feldspar of the composition $(K,Na)(AlSi_3O_8)$ that has exsolved $Na(AlSi_3O_8)$ on cooling as a separate albitic phase (see figure 5.8). This process is reversible, and heating of perthite to about 560° will regenerate the homogenous solid mixture.

Figure 5.8
Perthite with lighter exsolved lamellae of sodic feldspar in darker potassic feldspar. (Courtesy University of Windsor. Photograph by D. L. MacDonald.)

The physical chemistry of the exsolution process can be represented by a diagram as in figure 5.9. Complete solid solution can exist in crystals at high temperatures but not at lower temperatures (the shaded area). As temperature falls, a solid solution of some composition X arrives at the exsolution boundary (called the *solvus*) at 1. At that point the mineral separates into two minerals of composition 1 and 2. The process continues with falling temperature, for example, to points 3 and 4, and the result is a homogeneous mineral that has separated in the solid state into a host parent mineral and bands or blebs of a daughter mineral within it. The better the structural match of the two minerals, the more regular the geometry of the exsolved phase.

The general phenomenon of exsolution can be described graphically by use of a three-component diagram like the one used for solid solution. Assume that the region of complete solution of A, B, and C at some elevated temperature is represented by Field 1 in figure 5.10. Further, assume that at some lower temperature the mutual solution of these components is restricted to Field 2. A compound of 40% A, 30% B, and 30% C ($A_{40}B_{30}C_{30}$, point a) can exist as a single phase at the elevated temperature, but is unstable at the lower temperature and exsolution occurs. Assume the stable composition of the host is $A_{20}B_{60}C_{20}$ (point b). In order to reach point b from point a, the concentrations of both A and C must be decreased, and therefore A and C or $A + C$ will be the exsolved product(s). Should an AC compound be a stable product, its composition is readily determined by extrapolating the line ba to the edge AC (point c). The composition of $A + C$

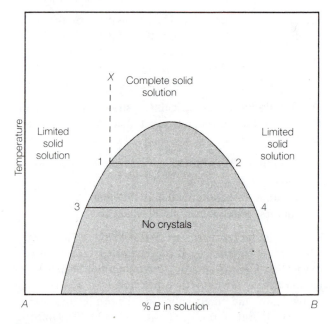

Figure 5.9
Schematic temperature-composition diagram for the system AB, which exhibits complete solid solution at high temperatures but exsolves to separate species at lower temperatures.

is thus determined to be $A_{60}C_{40}$. The entire exsolution reaction may now be written as

$$A_{40}B_{30}C_{30} \longrightarrow A_{20}B_{60}C_{20} + A_{60}C_{40}.$$

Exsolution is best developed in minerals with structures that have matching planes of atoms to serve as the

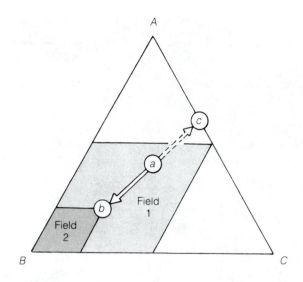

Figure 5.10
Exsolution in a three-component system.

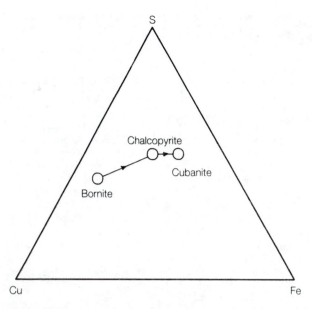

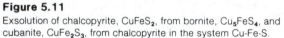

Figure 5.11
Exsolution of chalcopyrite, $CuFeS_2$, from bornite, Cu_5FeS_4, and cubanite, $CuFe_2S_3$, from chalcopyrite in the system Cu-Fe-S.

interface between the two exsolved phases. In this regard the process is very similar to the phenomenon of twinning, which is described later in this chapter.

Exsolution is particularly common in sulfide and sulfosalt minerals. It is best observed in specimens cut and polished to a mirror finish and examined with a reflected light microscope. For example, the amount of iron that can be held in the structure of bornite, Cu_5FeS_4, at high temperatures is in excess of that possible at low temperatures and chalcopyrite, $CuFeS_2$, is exsolved (see figure 5.11). Similarly, chalcopyrite rids itself of excess copper by exsolving cubanite, $CuFe_2S_3$. The exsolved mineral may be present as crystallographically oriented lamellae, as in figure 5.12, or as random blebs.

Isostructure and Isotype

Isostructural minerals are those that have an identical arrangement of atoms but totally dissimilar compositions, such as the cubic crystals galena, PbS, and halite, NaCl. The idea of isostructure is especially useful in describing a mineral by reference to a well-known structure. This practice reduces to a minimum the number of geometric configurations that must be learned.

Isotypic minerals are those with both isostructural characteristics and chemical similarity. No rigid requirements for crystallographic identity or compositions are implied by the term. An example of an isotypic pair is tridymite, SiO_2, and nepheline, $Na(AlSiO_4)$. These minerals are based on the same silicon-oxygen structural framework and thus have structural similarity. Aluminum occupies the same kind of structural site as the Si of both tridymite and nepheline. Therefore, Al = Si except for the charge discrepancy. The presence of Na in nepheline balances the charge, and Na occupies holes in the framework and does not disturb it. Another such pair is quartz, SiO_2, and eucryptite, $Li(AlSiO_4)$. Here, however,

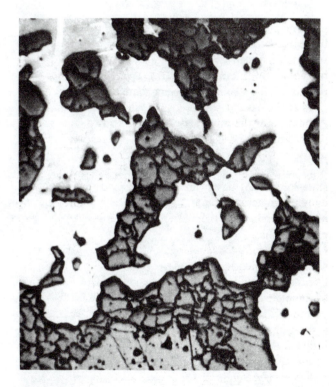

Figure 5.12
Exsolution lamellae of chalcopyrite (white) in bornite (light grey). Dark grey material with higher relief in quartz. Madeleine copper mine, Gaspé, Quebec. (Courtesy University of Windsor. Photograph by I. M. Samson.)

Box 5.2

The Effect of Crystal Fields

Although ionic size and charge are generally adequate to explain solid-solution phenomena in purely ionic substances, important exceptions may arise when the atoms have significant covalency in their bond character. Elements in the first transition series, scandium to zinc, in which $3d$ electrons participate in bond formation, are a case in point. The discussion of the spatial distribution of bonds in chapter 4 pointed out that electrons in different states occupy different probability volumes. More specifically, electrons occupy subshells or orbitals following Hund's rule by entering each of the available $3d$ orbitals separately before any pairing within an orbital occurs. This process results in parallel orientation of spins and minimizes interelectron repulsion in the shell. The particular configurations of $3d$ orbitals are illustrated in figure 5.13. The orbitals of one group, with lobes directed along three orthogonal axes, are termed the e_g orbitals and consist of d_{z^2} and $d_{x^2-y^2}$ orbitals. Orbitals in a second group, termed t_{2g} orbitals and consisting of d_{xy}, d_{xz}, and d_{yz} orbitals, have lobes projecting between the orthogonal axes.

Should an ion of a transition metal be placed at a particular site in a crystal, its electrons interact with those in its surroundings. At any instant, this surrounding **crystal field** can be regarded as a collection of point negative charges about the ion. The crystal field destroys the spherical symmetry of an isolated ion, and the magnitude of the change depends on the type, position, and symmetry of the coordinated anions or dipolar groups.

If a transition-metal ion is placed into an octahedrally disposed crystal field, two energetically different sets of orbitals arise. This **crystal field splitting** is due to the differential repulsion of electrons in the d orbitals by negatively charged anions, with those in the two e_g orbitals being more repulsed than those in the three t_{2g} orbitals. The energy separation between the e_g and t_{2g} orbitals is the actual octahedral crystal field splitting and may be designated as Δ_o. As shown in figure 5.14, the t_{2g} orbitals are energetically lowered by $2/5\Delta_o$, whereas the e_g orbitals are raised $3/5\Delta_o$ relative to the mean energy of the ion. Therefore, each electron in a t_{2g} orbital stabilizes a transition metal by $2/5\Delta_o$, and each electron in an e_g orbital diminishes the site stability by $3/5\Delta_o$. The enhancement or decrease in stability is measured in relation to a hypothetical nontransition element of the same size and charge, and the resultant net stabilization energy is called the **crystal field stabilization energy** (CFSE).

In an octahedral field, ions that possess one, two, or three d electrons, for example, Ti^{3+}, V^{3+}, or Cr^{3+}, can have only a single electronic configuration with each d electron in a different t_{2g} orbital, and their spins are parallel. However, ions with four to seven d electrons have a choice of electronic configurations, because if the crystal field is weak they

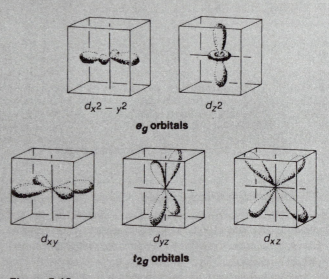

Figure 5.13
Spatial configuration of $3d$ orbitals.

may use both e_g and t_{2g} orbitals. However, in strong fields it is energetically more favorable for the t_{2g} orbitals to be filled first. Ions with eight, nine, or ten d electrons can have only one configuration, and the t_{2g} orbitals are completely filled.

In tetrahedral fields, the situation with respect to orbital-group stability is reversed in that the e_g subgroup is stabilized with respect to the t_{2g} subgroup, as shown in figure 5.15.

There are certain ions that are not stabilized by a regular octahedral field but, rather, are destabilized. One of these is Cu^{2+}, whose d electron configuration is such that two electrons are concentrated along the z-axis and one is in the xy plane. The electron in the xy plane thus shields the nucleus from negatively charged anions less effectively than those centered on the z-axis. Ions in the xy plane will then be attracted by a larger apparent nuclear charge than those along the z-axis. The system of ions may thus be expected to rearrange itself into an equilibrium array in which the interionic separation in the xy plane is lessened. This distortive effect is always present when an ion contains an odd number of electrons in the e_g orbitals. It may be predicted that Mn^{3+}, Cu^{2+}, Ni^{3+}, and Cr^{2+} will be distorted in an octahedral field, whereas Cr^{3+}, Mn^{3+}, Ni^{2+}, and Cu^{2+} will be distorted in a tetrahedral field.

Crystal field splitting may be measured in solids from infrared spectra of minerals and synthetic glasses. From these measurements, it is possible to calculate stabilization energies for octahedral (Δ_o) and tetrahedral (Δ_t) sites and the

---

because interstitial holes are smaller, the smaller ion Li^+ is included. Another isotypic pair is nitratine, $NaNO_3$, and calcite, $CaCO_3$. In both of these minerals triangular radicals alternate with cations to make up the structure. The members of a solid-solution series are isotypic but the reverse is not necessarily true.

Derivative Structures

A successive suppression of symmetry elements produces a related series of crystal structures. The structures of lower symmetry are thus derived from those of higher symmetry and are termed **derivative structures.** Diamond, C, to sphalerite, ZnS, to chalcopyrite, $CuFeS_2$, to stannite, Cu_2FeSnS_4, is such a sequence of derivatives. Zn and

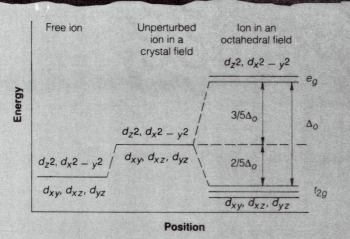

Figure 5.14
Schematic diagram of crystal field splitting in an octahedral field.

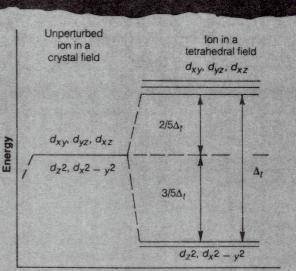

Figure 5.15
Crystal field splitting in a tetrahedral field.

Table 5.5	Crystal Field Stabilization and Preference Energies			
Number of d Electrons	Ion	Octahedral Stabilization (Kcal/mole)	Tetrahedral Stabilization (Kcal/mole)	Octahedral Preference (Kcal/mole)
0	Ti^{4+}	0.	0	0
1	Ti^{3+}	20.9	14.0	6.9
2	V^{3+}	38.3	25.5	12.8
3	Cr^{3+}	53.7	16.0	37.6
4	Mn^{3+}	32.4	9.6	22.8
5	Mn^{2+}	0	0	0
5	Fe^{3+}	0	0	0
6	Fe^{2+}	11.9	7.9	4.0
7	Co^{2+}	22.2	14.8	7.4
8	Ni^{2+}	29.2	8.6	20.6
9	Cu^{2+}	21.6	6.4	15.2
10	Zn^{2+}	0	0	0

Data from Burns, R. G. 1970. *Mineralogical Applications of Crystal Field Theory.* Cambridge: Cambridge University Press.

octahedral preference energies $(\Delta_o - \Delta_t)$ for various ions, as shown in table 5.5.

As an example of the use of these concepts, consider the spinel structure, in which cations occur in sites having both octahedral and tetrahedral coordination. Spinels may occur as normal spinels, with the divalent metal ions (M^{2+}) in the tetrahedral fields and the trivalent metal ions (M^{3+}) in the octahedral fields, or as inverse spinels, with the tetrahedral sites occupied by $\frac{1}{2}M^{3+}$ ions and the octahedral sites containing $\frac{1}{2}M^{3+} + M^{2+}$ ions. The substitution of a transition-metal ion into a spinel is thus related to the preference energy for octahedral sites for that element. For example,

Cr^{3+}, with a high octahedral preference energy, occurs only in normal spinels. On the other hand, Co^{2+} has a fairly low octahedral preference energy and occurs mainly in inverse spinels. Fe^{3+}, with no octahedral or tetrahedral preference energy, is found in both normal and inverse spinels, depending on the other cations in the crystal.

Crystal chemical parameters such as size, valence, ionization potential, and electronegativity predict that Cu^{2+} could substitute easily into mineral structures in much the same way that Ni^{2+} does. The crystal field destabilization of Cu^{2+} weighs against this, however, and copper tends not to proxy for other elements. Instead, copper is found in its own minerals.

S atoms in sphalerite occupy sites comparable to the C atoms in diamond; Cu and Fe atoms in chalcopyrite are found in sites equivalent to the Zn atoms in sphalerite; finally, the Sn atoms in stannite occupy sites equivalent to alternate Fe atoms in chalcopyrite. The suppression of symmetry in such a series can be seen in the increase in identity period, shown by the shaded portion of table 5.6.

Table 5.6	Suppression of Symmetry in Derivative Structures									
C	C	C	C	C	C	C	C	C	C	Diamond
Zn	S	Zn	S	Zn	S	Zn	S	Zn	S	Sphalerite
Cu	S	Fe	S	Cu	S	Fe	S	Cu	S	Chalcopyrite
Cu	S	Fe	S	Cu	S	Sn	S	Cu	S	Stannite

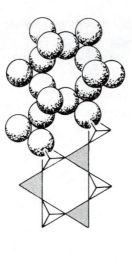

| Cristobalite | Tridymite | Quartz |

Figure 5.16
The principal polymorphic forms of SiO_2. The oxygen ions (spheres) are centered on the corners of shared tetrahedra arranged as shown in the lower portion. Shading indicates a tetrahedron pointing downward.

Geometrical Variations

Polymorphism

Polymorphs, as implied by the name (meaning "many forms"), are compounds that have the same chemical makeup but nevertheless are capable of crystallizing in more than one structural arrangement. In certain cases, these various forms are also capable of being readily transformed from one structural arrangement to another.

Fundamentally, polymorphism in ionic structures rests on (1) the possible existence of more than one distinct grouping of constituents with fixed radius ratios and (2) different groupings that may arise if the radius ratio varies. The possibility of combining a few constituents into several different structures may be likened to the possibilities that arise when a mason lays a brick wall. Certain fundamentals of good bricklaying must be observed, but within such limits a number of patterns may be developed.

Consider, for example, the polymorphs of SiO_2, in which the "bricks" are tetrahedral units composed of a silicon ion surrounded by four oxygen neighbors. The oxygen ions on the corners of each tetrahedral unit are in all cases shared between (common to) two adjacent tetrahedra, and all SiO_2 minerals are some arrangement of the resulting network of linked tetrahedra. Several of the various ways in which these linked tetrahedra can be arranged in space are shown in figure 5.16. The various arrangements are quite different geometrically, and, as we will see, one arrangement cannot be transformed into another without complete disruption of the structure. The situation is similar to that represented by the two stacks of lumber in figure 5.17. The transformation of one stack to the geometry of the other would require a complete rebuilding.

Because of the necessity of tearing down and rebuilding, a large energy barrier, represented by the energy required to break the chemical bonds, separates such polymorphic forms. The rearrangements are necessarily sluggish, commonly requiring the presence of catalysts. Transformations of this kind are called **reconstructive polymorphs.**

Another, somewhat more subtle, kind of reconstructive polymorphism occurs in mirror-image structures that are identical in all other respects except their mirror-image relationship. An analogy is the geometrical relationship between the left and right hands. Organic compounds related by this phenomenon have the prefixes *levo-* (left) and *dextro-* (right). A mineral example is quartz, in which rows of atoms in one plane may be succeeded in the next layer by rows displaced by 60° either clockwise or counterclockwise. The diagram of the quartz structure in figure 5.16 shows a clockwise or right-handed arrangement. Figure 5.18 shows the essentials of right- and left-handed polymorphs with the rows of atoms simplified to square rods.

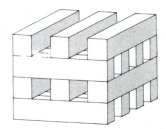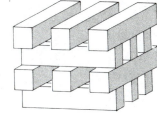

Figure 5.17
Polymorphic forms related by a reconstructive transformation.

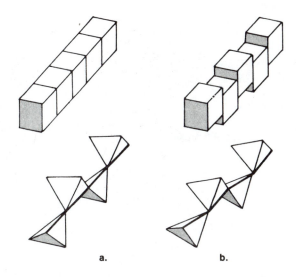

Figure 5.19
Displacive transformation. (a) Tetrahedra having a straight row of atoms (darker line) and represented by the row of blocks above may be displacively transformed to the situation represented in (b) without breaking of bonds.

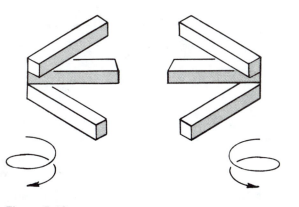

Figure 5.18
Right- and left-handed structures.

Each of the polymorphic forms of a particular compound may be further modified by slight displacements of the structural units in such a fashion as to change the symmetry but not disrupt any chemical bonds. This type of polymorphic transformation is illustrated in figure 5.19, which shows how square rods, here representing rows of tetrahedra, may themselves be modified. The energy difference between two such forms is very small since no bonds are broken, and transformation occurs very readily under an appropriate set of physical conditions. Such **displacive transformations** are characterized by rapid accomplishment and reversibility.

The various polymorphic forms connected by displacive transformation are called **high** and **low** (temperature) **forms.** These are often given Greek letter prefixes, as in α- or β-quartz, but there is no consistent usage as to which symbol represents high and which means low.

In the silica group of minerals that show reconstructive polymorphic transformations, displacive transformations of each of the mineral species also occur. These are characterized by different basic structures. Quartz has high and low forms with a transformation temperature at about 573° C; cristobalite has high and low forms that

transform between 198° C and 273° C; and tridymite has high, middle, and low forms transforming at 117° C and 163° C, respectively.

The interrelationships among the various forms of silica are illustrated in figure 5.20. Solid horizontal lines represent the stable forms. Horizontal dashed lines represent the metastable forms that are found because of the sluggish nature of the reconstructive transformations. Several bypass routes are shown as vertical dashed lines. These exist because of the sluggish transformations, such as the formation of vitreous silica rather than tridymite from high quartz at 1,410° C, or the crystallization of high tridymite from vitreous silica held at 1,670° C.

Combinations of ions whose radius ratio is close to the critical coordination values (see table 4.6) provide still another source of polymorphs. A very small change in the energy of the structure, either from thermal, mechanical, or chemical sources, such as contamination by substitution, can affect the average size or shape of the ions enough to dictate a new coordination and thus a new structural arrangement. The polymorphism of $Ca(CO_3)$ is an example. The radii of Ca^{2+} and O^{2-} are, respectively, 1.08Å and 1.30Å, and their radius ratio is thus 0.83. This is very close to the critical value separating structures in which the coordination polyhedron contains six or eight anions. $Ca(CO_3)$ can thus be expected to occur in two polymorphous forms. If calcium is in 6-fold coordination, the mineral is calcite; calcium in higher coordination is found in aragonite.

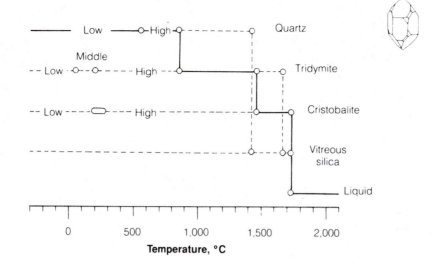

Figure 5.20
Principal polymorphs of silica and their transformations indicated by the heavy line. Metastable forms are shown as horizontal dashed lines and bypass transformations as vertical dashed lines. (After Sosman, 1965.)

Polytypism

A special kind of polymorphism in which successive layers in a sheet structure show regular variations in their orientation is called **polytypism.** This phenomenon is especially common in the micas and clay minerals whose monoclinic cells may be stacked so as to continue or alternate the orientation of an inclined edge. Figure 5.21 is a much simplified representation of polytypic stacking for square-based monoclinic cells which illustrates some of the complexities that can occur. In sheet structures with hexagonal symmetry of their basal plane, such as clays, micas, and sphalerite, many regular alternation sequences are common. Figure 5.22 summarizes the basic geometry of sheet structure polytypes and the resulting symmetry.

Order-Disorder

Masons who build a wall from bricks of two different colors have the choice of using bricks at random or laying them in a regular pattern. Growing crystals also are confronted with this kind of choice during growth.

Reacting ions tend to seek the structural sites in which their presence causes the greatest reduction in free energy. These sites are necessarily positions of regular pattern and spacing. On the other hand, both thermal agitation and rapid growth promote random occupation of all the possible sites, leading to irregular patterns. As an example of the interaction of these factors, assume a certain number of sites are available on a crystal surface and that some lesser number of ions will occupy those sites. When reaction is complete, the occupied and vacant sites will have

a regular pattern if the choice is entirely up to the individual ions that seek to minimize the free energy, or they will have some random arrangement if thermal agitation predominates or the rain of atoms on the surface has been rapid. In minerals the regular pattern is known as an **ordered arrangement** or **order.** It is the preferred arrangement since it provides the lowest free energy for the structure. The random or irregular arrangement is known as **disorder.** A completely random arrangement is one of complete disorder, and all intermediate degrees between complete order and complete disorder are possible. Disorder can occur between any of the structural components—ions and ions, ions and vacancies, or between radicals. Figures 5.7 and 5.23 show examples of ordered and disordered arrangements.

The possibilities for disordering do not cease with the formation of a crystal, since interchanges within the solid are possible. Crystal structures are typically composed of an anionic framework of relatively large anions within which one or more kinds of smaller cations are distributed. The channels connecting one interstitial site with another might seem too small to allow the passage of the cations, but a consideration of thermal motion suggests otherwise. The volume ascribed to a given ion is actually that swept out by the thermal vibration of the ion around its structural site and is consequently larger than the ion itself. Such volumes are usually spherical, but they can be ellipsoidal and will increase in size with a rise in temperature. Only at absolute zero is the rest volume and structural or packing volume of the ion the same. Figure 4.15 shows the volumes of thermal influence of ions in a structure. It can be readily seen that a somewhat oversized

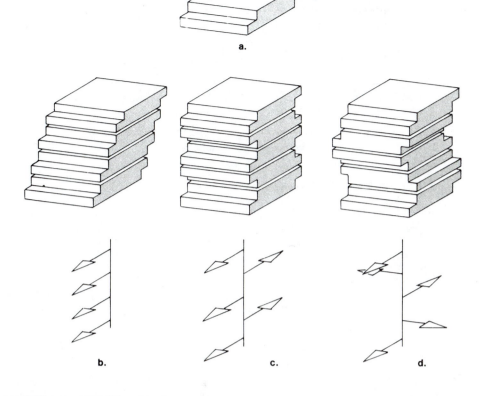

Figure 5.21
Polytypic stacking possibilities of a monoclinic cell having a square base. (a) Primitive cell (shown with step to facilitate visualization). (b) Stacking in common orientation 1-cell repeat, monoclinic. (c) Alternate cells rotated through 180°, 2-cell repeat, tetragonal. (d) Successive cells rotated 90°, 4-cell repeat, tetragonal.

cation can pass the anion barrier by moving at an appropriate time, much as a spy can pass a sentry line when the sentinels are at opposite ends of their beat. Because of thermal motion ions can migrate through a structure and will do so with increasing ease as the thermal energy to break bonds and enlarge packing volumes increases.

When an ion changes position, it may go to an unoccupied site or interchange with another ion, usually through a complex chain of moves. In either instance, the original orderly arrangement is gradually destroyed until site occupancy is purely a statistical matter (see figure 5.24). The change from complete order to complete disorder accelerates with increasing temperature to a critical temperature above which no regularity in position exists.

A characteristic of the disordering process is the reduction in identity period of the crystal. To illustrate, oxide ABO, containing two different cations, might have the following repetitions in the ordered state:

$$A \quad O \quad \underline{B} \quad O \quad A \quad O \quad \underline{B} \quad O \quad A \quad O$$

On heating, the A and B ions interchange rapidly and continuously in **dynamic disorder** so each A site is statistically occupied by both A and B ions, as is each B site. The repetition distance is thus halved:

$$
\begin{array}{cccc}
A & A & A & A \\
\; O & O & O & O \\
B & B & \underline{B} & \underline{B}
\end{array}
$$

As examples, consider the reduced-cell dimensions of disordered as compared with ordered minerals in table 5.7 and figure 5.23. Disorder may be retained in the structure as **static disorder** if the crystal is rapidly cooled or quenched. Slow cooling or annealing allows the recovery of order, although often by domains rather than throughout the crystal (see figure 5.25).

Order-disorder phenomena are widespread, and the concept may be used to develop models for many mineralogical relations, such as crystallization (ordering of a liquid), twinning, and polymorphism. Only in rather rare instances, however, is the effect sufficiently profound to warrant the designation of distinct mineral species. One case in which polymorphic forms do result in distinct species is in the alkali feldspar group. The potassic subgroup

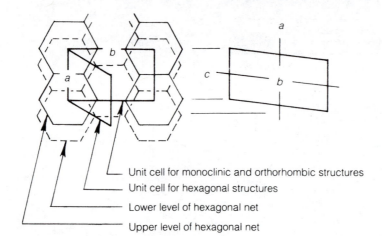

— Unit cell for monoclinic and orthorhombic structures
— Unit cell for hexagonal structures
— Lower level of hexagonal net
— Upper level of hexagonal net

Cell orientation	Layer shift	Layer repeat	Crystal class	Polytype
		1	Monoclinic	1*M**
		2	Orthorhombic	2*O*
		2	Monoclinic	2*M*₁
		2	Monoclinic	2*M*₂
	or	3	Hexagonal (trigonal)	3*T*
	or	6	Hexagonal	6*H*

* The notation $1M_d$ indicates disordered stacking.

Figure 5.22
Geometry and terminology of polytypes.

contains three minerals (sanidine, orthoclase, and microcline) of identical formula, $K(AlSi_3O_8)$, that differ fundamentally because of various order-disorder combinations between aluminum and silicon. Of the four possible sites needed to accommodate $AlSi_3$, occupancy is completely ordered in microcline and completely disordered in sanidine. In orthoclase, two sites are occupied by ordered Si ions and two by disordered Al-Si.

Twinning

A **twin** is a rational, symmetrical intergrowth of two or more individual crystals of the same species. *Rational* means the two portions of the twin have a point, line, or plane in common, and *symmetrical* means the twinned crystals are related by some crystallographic symmetry operation, such as reflection, translation, or rotation.

Table 5.7	Cell Dimensions of Ordered and Disordered Equivalents		
	Ordered (low form)	Transformation Temperature	Disordered (high form)
Chalcocite, Cu_2S	Monoclinic a 15.22Å b 27.323Å c 13.491Å β 116.35°	103° C	Hexagonal a 3.89Å c 6.68Å
Bornite, Cu_5FeS_4	Tetragonal a 3.89Å c 21.88Å	228° C	Isometric a 5.50Å

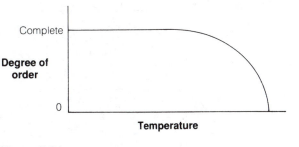

Figure 5.24
Relation of the degree of order to temperature.

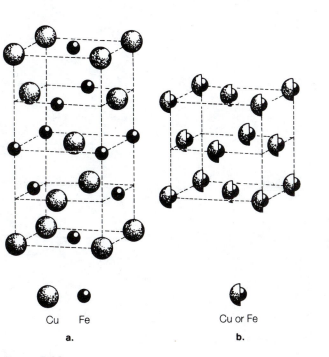

Figure 5.23
Site occupancy order-disorder. (a) Ordered arrangement of copper and iron ions in chalcopyrite, $CuFeS_2$. (b) Disordered arrangement; statistical occupancy of metal ion sites; c-axis length halved.

Mechanics of Twinning   Twins may arise in several different ways, and may be classified according to their origin as growth twins, transformation twins, or gliding twins.

Growth twins are always the result of an accident during the early stages of crystal growth. They are perpetuated in the growing crystal as distinct individuals with a mutual boundary or **composition plane.** A common accident that results in a twin configuration rather than a continuation of the normal structure can be visualized by considering the possibilities confronting the first ion to arrive in the situation, (see figure 5.26). This ion, say, the first to arrive to start a third layer, has a choice of two kinds of sites, A or B, over a first layer hole or a first layer ion, respectively. Site A is preferred since its occupation provides the larger decrease in free energy, and an ion attachment here represents a normal continuation of the

structure. The energy difference between sites A and B is, however, very slight, and it is possible that thermal agitation may be insufficient to move an ion initially arriving at B over the energy barrier to site A. In this case, the position of all future atoms in this layer is fixed, and the structure is continued as a twin of the earlier structure with a twin boundary between layers two and three. A similar accident could occur if the rain of reacting ions was so heavy that the first atom to arrive does not have time to adjust from the twin to the continuation position before it is hemmed in by new arrivals. On the other hand, if crystallization is sufficiently slow and there is competition for the lower energy state, growth twins will not occur.

Growth twins are usually restricted by energy considerations to a single twinning episode. This is because a very small crystal has a very high surface energy, and additions of more ions in either twin or continuation positions cause significant diminution of this energy. On the other hand, addition of ions in twin positions at a later stage of crystal growth increases the surface energy of the crystal and, thus, is not tolerated. This situation is illustrated in figure 5.27. The decrease in surface energy of a crystal nucleus (1) is rapid for either continuation (2) or for twin (3) increments. However, later twin configurations (4) are unlikely since their presence requires an energy increase.

Growth twins, because of their manner of formation, can usually be recognized in mineral specimens by the single, often sinuous, boundary separating twin components with slightly different orientations or by the interpenetration of the two individuals composing the twin. Nonparallel twin boundaries are also evidence for growth twinning.

The mechanism of displacive polymorphic transformation is also a mechanism in the formation of twins. In the high temperature form of a crystal, for example, the atoms may be vibrating rapidly between two possible sites that give the atoms an average position of high symmetry (figure 5.28a). Should thermal vibration cease, the atoms cannot occupy the average position but must assume one of the less symmetrical terminal arrangements (figure 5.28b).

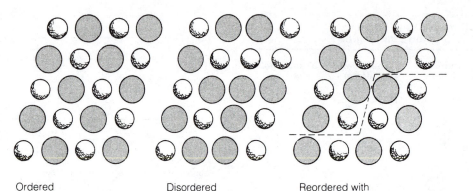

Ordered   Disordered   Reordered with
                                      domains in twin
                                      configuration

Figure 5.25
Ordering by domains.

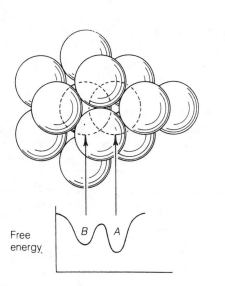

Free
energy

Figure 5.26
Choice of sites for a twin or continuation structure. *A* is the lower
energy site and results in structural continuity. An ion in site *B*
causes a twin to form.

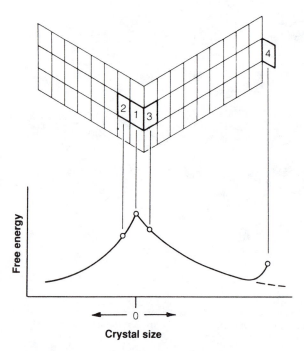

Crystal size

Figure 5.27
Restriction of growth twinning to a single episode. Reduction in free
energy by an ion attaching at either a continuation site (2) or a twin
site (3) is large at the nucleation stage. A second twinning episode
(4) is unlikely because of the energy increment needed.

When crystals go through a displacive transformation, all regions of the crystal may not arrive simultaneously at the transformation temperature because of
various crystal imperfections and consequent differences
in thermal conductivity. Two different areas may thus
arrive independently at the transformation temperature
and transform with different orientations. As transformation continues, the orientation of the original patch is
continued by adjacent areas. Eventually, the two transformed regions grow to a mutual boundary—a composition plane separating the two parts of the twin. This
sequence is illustrated in figure 5.29. Structural transformations with different orientations may thus spread out
from different centers, and when transformation is complete, adjacent areas of the mineral may be in twin configuration. In minerals, a large number of transformation
centers may develop, and the end result of transformation
twinning is often a mosaic of twins. Transformation twins

may also arise in a similar way as structures become ordered with a fall in temperature.

Crystals can sometimes be deformed by the application of mechanical stress, as in figure 5.30a. Assume that
the aggregate of atoms is stressed and yields by a slipping
along the plane separating rows 2 and 3. If the amount of
slippage is less than an identity period (as shown), row 2
will bear a twin relationship to row 3, as in figure 5.30b.
On the other hand, **slip** or **gliding** for any integral number
of identity periods deforms but does not twin the structure. The situation illustrated is only one of a number of
possible ways to produce twins or deform crystals by mechanically induced translational movements. Try drawing

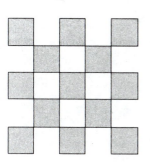

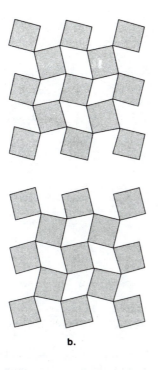

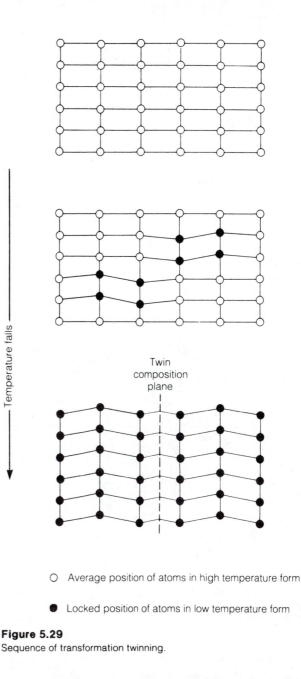

Figure 5.28
Possible mechanism for twin formation by displacive transformation. A higher symmetry due to thermal vibration (a) is lost on cooling. Two orientations of the lower temperature form can exist as shown in (b).

Temperature falls

Twin composition plane

○ Average position of atoms in high temperature form

● Locked position of atoms in low temperature form

Figure 5.29
Sequence of transformation twinning.

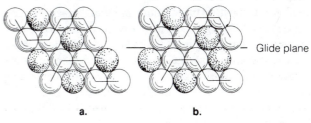

Glide plane

a.   b.

Figure 5.30
Twin formation by glide deformation. (a) Before deformation. (b) Twin formed by gliding.

a few layouts similar to figure 5.30 to see the results of gliding in monatomic aggregates, different geometric arrangements of binary compounds, and a ternary compound.

Glide twins form most easily if adjacent sites of the continuation and twin configuration are separated by a low energy barrier, as illustrated in figure 5.31. The energy barrier separating the sites for continuation and twin development in figure 5.31a is low and would readily allow the displacement of an ion from the normal to the twin location. The energy barrier in figure 5.31b, however, would prohibit displacement because the energy required for slip equals the energy for rupture.

The direction of gliding movement may be dictated by the presence of low-energy troughs leading away from the normal site. For example, in figure 5.32 an ion in position *A* can migrate readily to position *B* but may not continue to position *C*. The topography of the energy surface with its barrier hills and valley troughs is, of course, directly related to a mineral's structure.

Twin gliding in a uniformly stressed crystal begins by the localization of movement due to fortuitous circumstances such as a flaw, a foreign atom, or thermal motion. The twin lamellae grow laterally because of the high energy associated with their boundaries. Growth proceeds until blocked by an atomic arrangement that favors the start of a new twin lamella in preference to continuing the earlier twin. Figure 5.33 illustrates a series of episodes in the formation of glide twins. Crystals that have undergone twinning of this nature are characterized by the

presence of numerous parallel twin lamellae or by bands with linear boundaries, such as those characteristically present in plagioclase. Twin gliding is readily induced in calcite on $\{01\bar{1}2\}$ (see figure 5.34) and in many metals.

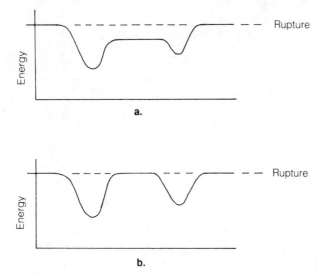

Energy

a.

Energy

b.

Figure 5.31
Energy profiles for gliding.

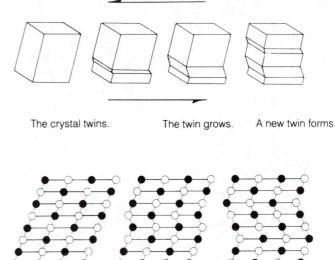

The crystal twins.　　The twin grows.　A new twin forms.

Figure 5.33
Sequence of events in glide twinning.

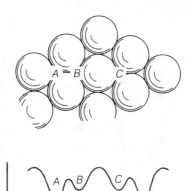

Free energy profile along line *ABC*

Figure 5.32
Restrictions to direction and amount of glide.

Symmetry of Twinning　Twins are rational, symmetrical intergrowths of two or more individual crystals of the same species. Intergrowths that do not conform to this definition can and do occur both between crystals of the same species and between crystals of different species. Such intergrowths, however, are not twins. The component parts of a twin crystal are symmetrically related to each other by a reflection plane, as in a **twin plane,** by a proper rotation axis as in a **twin axis,** or simultaneously by both if the crystal contains an inversion center. Twinning planes and axes have simple and rational relations to the crystallographic axes. Twin planes are parallel to possible crystal faces.

Certain planes and axes are forbidden for twinning simply because they are already elements of symmetry of the individual crystal, and no operation of these elements can produce a new geometrical arrangement of the parts. No plane of symmetry in the individual crystal can be a twin plane in its compound crystal, and no 2-, 4-, or 6-fold axis in the individual can become a twin axis in the compound crystal.

Combined twin planes and axes are present in all twin crystals in crystal classes having a center of symmetry ($m \cdot i \cdot A_{180} = 1$; see chapter 2). Either a twin plane or a twin axis may be present when the crystal class lacks a center of symmetry. The plane on which twinned crystals are united is the **composition plane.** This plane often coincides with the twin plane.

The operation of the elements of symmetry in the development of twinned crystals is illustrated in figures 5.35, 5.36 and 5.37. An octahedron (figure 5.35a) that contains a center of symmetry can be twinned by reflection across the plane outlined by dashed lines or by 180° rotation to produce a twinned octahedron (figure 5.35b). In this case the twin and composition planes are identical. Twinning of a tetrahedron as the result of the operation of a twin axis only is shown in figure 5.36. Right- and left-handed individuals possess neither a plane nor a center of symmetry. They can, therefore, twin only by reflection across a twin plane. Left- and right-handed individuals of tartaric acid are shown in figure 5.37a and could be combined into a single crystal with higher symmetry as shown in figure 5.37b.

The mechanisms of twin formation treated here do not always allow twins to be readily classified, since a given mechanism may give rise to different symmetrical relationships. It is often more useful to employ descriptive rather than genetic terminology in describing twins in mineral identification studies.

Figure 5.34
Two stages of glide twinning in calcite from mylonitized Grenville-age marbles. Kaladar-Actinolite area, Ontario. (Specimen courtesy of P. E. Holm. Photograph by D. L. MacDonald.)

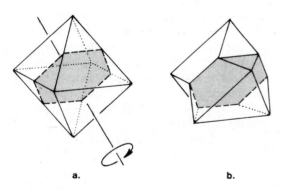

a.   b.

Figure 5.35
Contact twins (spinel law) related by rotation or reflection. Composition plane shaded. (After Ford, 1932.)

Figure 5.36
Penetration twins related by rotation. (After Ford, 1932.)

Twins may be descriptively classified as contact, penetration, or repeated. **Contact twins** are united to each other by the composition plane. **Penetration twins** are those in which two or more crystals seem to pass through each other. **Repeated twins** may have the individuals in parallel orientation, as in **lamellar** or **polysynthetic** twins, or in nonparallel orientation, as in **cyclic** twins. Cyclic twinning tends to produce circular forms. Examples of these different kinds of twins are shown in figure 5.38.

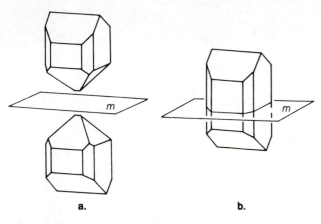

Figure 5.37
(a) Left (above) and right (below) crystals of tartaric acid showing their mirror symmetry. (b) The result of twinning.

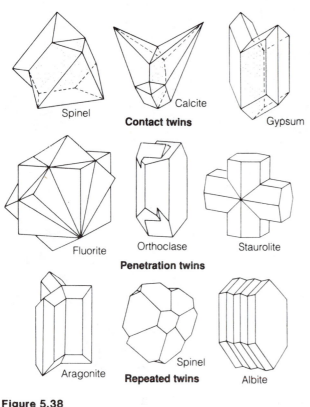

Figure 5.38
Types and examples of twins. (After Ford, 1932.)

Concluding Remarks

In the last two chapters we have drawn on concepts of physics, chemistry, and crystallography to investigate the nature of minerals at the atomic scale. Through the concepts of crystal chemistry, we saw how atomic structure directly controls the variation of properties and external forms of minerals. Mineral species show a wide range of variation in their chemical content and geometrical arrangements. These variations, however, are usually of a minor nature and are not sufficient to warrant the designation of a new species.

Chemical variations arise through solid solution as like components are substituted for formulary constituents and through exsolution phases which separates component phases. Geometrical variations arise from the ability of the same atoms to assume different arrangements, creating polymorphs, from the regularity of position of atoms as exemplified by their degree of order, and from intergrowth of individuals of the same species as twins.

We have now investigated the geometric and chemical controls on the forms and variability of minerals. In chapter 6 we will consider the energetic constraints on mineral stability and reaction.

Suggested Readings

Bloss, F. D. 1971. *Crystallography and Crystal Chemistry, an Introduction.* New York: Holt, Rinehart and Winston. A clear, understandable, and well-illustrated book written from a mineralogical point of view.

Bragg, L.; Claringbull, G. F.; and Taylor, W. H. 1965. *Crystal Structure of Minerals.* Vol. 4, *Crystal Structures of Minerals.* Ithaca: Cornell Univ. Press. Description and illustration of the structures of many minerals.

Buerger, M. J. 1945. The genesis of twinned crystals. *American Mineralogist* 30:469–82. Distinguishes and describes growth, transformation, and gliding twins.

Burns, R. G. 1970. *Mineralogical Applications of Crystal Field Theory.* London: Cambridge Univ. Press. An excellent review of crystal field theory with application to mineralogic problems.

Evans, R. C. 1966. *An Introduction to Crystal Chemistry.* 2d ed. London: Cambridge Univ. Press. A systematic and nonmathematical treatment of crystal chemistry.

Putnis, A., and McConnell, J. D. C. 1980. *Principles of Mineral Behaviour.* New York: Elsevier. Clear atomistic and microstructural descriptions of solid-state processes in minerals.

Wyckoff, R. W. G. 1968. *Crystal Structures.* Vol. 4, *Miscellaneous Inorganic Compounds, Silicates, and Basic Structural Information.* 2d ed. New York: Wiley-Interscience. Mineral structure descriptions and illustrations. Extensive bibliography.

Mineral Stability and Mineral Reactions

The essence of all natural processes is their progress toward physical and chemical equilibrium, that is, toward the state in which there is no tendency for further change. One often observes matter attaining equilibrium, usually by a change in position, under the influence of a gravitational, magnetic, or electric field. Chemical aggregates also adjust their atomic arrangement or composition in response to changes in the composition, temperature, and pressure of their environment. Physical and chemical equilibrium, then, in the most general sense, cannot be separated because the equilibrium state of matter in relation to its surroundings calls for both physical and chemical adjustments.

The formation and change of minerals and the rocks they compose is a complex and important aspect of geology. Understanding these changes requires an appreciation of the reactants, the processes, and the products. Minerals represent collections of atoms whose aggregation was dictated by the energy and composition of the environment in which they formed. In a general way, minerals (and rocks) represent a state of balance between the disruptive effects of thermal motion and directed pressure and the aggregative tendencies of uniform pressure and chemical-bond formation. Crystallization, solution, and similar physical and chemical changes do not always, however, occur instantaneously under the prescribed conditions. Crystallization may require supersaturation or seeding, and dissolution cannot take place until an excess of energy sufficient to break the chemical bonds is available. Because of the latter phenomenon, minerals commonly exist as unaltered species far outside the temperature conditions required for their formation. Such **metastable** minerals are, of course, susceptible to alteration if the energy necessary to disrupt their bonding becomes available, usually through chemical activity.

If only those minerals that are stable under conditions existing at the earth's surface were available for study, we would have a short list indeed!

Approaching geological processes through the equilibrium concept requires the use of some fundamental property that can be shown to differ in equilibrium and nonequilibrium situations. Energy is such a property, and in this chapter we will introduce energy or thermodynamic considerations into the study of mineralogy.

Physicochemical Equilibrium

Thermodynamics is a phenomenological discipline concerned with describing the macroscopic or gross properties of a system. The **system** is a selected part of the physical world, specifically, the portion of matter and its immediate surroundings chosen for study. An **open system**

can exchange mass, heat, and work with its surroundings. In a **closed system,** no exchange of matter, and therefore no exchange of mass, can take place between the system and surroundings. Transfers of heat and work are still possible, however. An **isolated system** has no interactions with its surroundings.

The thermodynamic properties of a system describe macroscopic observations over a period of time that is long compared to the periods of atomic motion. These properties are of two types: **intensive properties** which are independent of the mass of the system, such as temperature and pressure; and **extensive properties,** such as volume, which are dependent on the mass of the system; that is, their values are the sum of the values of the parts.

A **heterogeneous system** is one composed of two or more separable homogeneous regions called **phases.** As an example, consider a solid coexisting with a liquid of the same composition—an ice cube in water. Because there are obvious discontinuities in structural ordering of the matter across the boundary between solid and liquid in this case, we say that a solid phase coexists with a liquid phase, and the boundary between them is a phase boundary. An analogous situation arises if the system contains solid homogeneous regions of different chemical composition or of the same composition but of different structural arrangement. Thus, a rock is a heterogeneous system if it is composed of one or more homogeneous phases or minerals. A **homogeneous system,** on the other hand, would have only one phase. An example is pure water or a quartz grain.

A system may be further described in terms of its **components.** The number of independent components of a system is the minimum number of chemical substances from which the system can form. In mineralogy, chemical components are usually designated as the oxides of particular elemental species. The procedure is arbitrary, however, and purely a convenience, because oxygen is the most abundant element in the earth's crust and the bulk of minerals are oxygen bearing. Further, until recently, the chemical determination of oxygen in rocks and minerals was a difficult task, and it was convenient to assume a stoichiometric (electronic and elemental) balance for oxygen by taking the major chemical components as oxides.

Equilibrium in Mineralogic Systems

In general, the time required for a mineralogic system to attain chemical equilibrium depends on the nature of the reactants and on such externals as temperature and pressure. Conditions in mineralogic systems are often such that equilibrium has not yet been attained, even in geologic time, and the various reactants persist as a metastable assemblage. For systems that have attained physicochemical equilibrium, generally any change in the factors that describe the equilibrium will cause the system to change in such a way as to neutralize the effects of the change. This rule, which was summarized by the French chemist Henri Le Chatelier (1850–1936), can be applied either qualitatively or quantitatively to understand better the nature of chemical equilibrium. As an example of the qualitative use of the Le Chatelier principle, consider a reaction possibly involved in the deposition of cassiterite, SnO_2, from a gaseous hydrous phase containing tin as SnF_4:

$$SnF_4 + 2H_2O = 4HF + SnO_2$$
gas   gas   gas   solid

Certain types of tin deposits, such as those in Cornwall, England, or that at Mount Pleasant, New Brunswick, show geologic evidence of having been deposited from hot, watery (hydrothermal) solutions. These deposits typically contain fluorite, CaF_2, in association with the cassiterite. Thus, it is possible that the solutions carried tin in some sort of soluble fluorine compound.

Avogadro's law states that at the same temperature and pressure equal numbers of gas molecules of any kind occupy the same volume. Thus, if the tin-bearing system is initially in equilibrium, an increase in confining pressure would cause the reaction to go in the direction that counteracts the effective pressure change, that is, in the direction of reduced volume. The reaction goes to the left, then, as this will allow a reduction of four moles of gaseous HF to yield two moles of H_2O and one mole of SnF_4, a reduction of four to three moles and thus a reduction in volume. By this reasoning cassiterite would be deposited if pressure is released.

The rule of Le Chatelier is also readily applied to the effect of heat on a chemical equilibrium, because low temperatures promote heat-producing (exothermic) reactions whereas high temperatures favor heat-absorbing (endothermic) reactions. In both cases, the reaction attempts to maintain a constant quantity of heat in the overall system. A well-known exothermic reaction at standard conditions of 18° C and one bar of pressure is the slaking of quicklime:

$$CaO + H_2O$$
$$= Ca(OH)_2 + 236 \text{ kilocalories/gram}$$

The efficiency with which calcium hydroxide is generated decreases as the temperature of the system increases; reciprocally, sufficient heating of $Ca(OH)_2$ results in abundant CaO and water.

The crystallization of minerals from magmas to form igneous rocks takes place in response to falling temperatures. Thus, igneous crystallization is generally exothermic and can generate local heating of the magma that may affect or even reverse the crystallization trend. On the other hand, mineralogic changes during metamorphism are often due to rising temperatures, and these reactions are endothermic.

Mineralogic reactions can have significant effects on the environment. For example, the argillaceous carbonates (marls) of the Green River Formation of Colorado and Utah contain large amounts of hydrocarbon that can be recovered by retorting, or heating the marl to drive off petroleum. The retorting process is endothermic and results in a spent material rich in unstable, calcium-bearing compounds with high heat content. When these spent materials are placed back in the environment as fill, they combine with water and form new stable minerals by exothermic reactions. These reactions raise soil temperatures to levels that prohibit plant life and thus hinder mine-site restoration.

It is a common occurrence that when two or more substances combine they may react to form a new chemical compound. Further, this compound may be broken down by a reverse reaction into its component chemical parts:

$$A + B \rightarrow AB \text{ and } AB \rightarrow A + B$$
$$AB + C \rightarrow A + BC \text{ and } A + BC \rightarrow AB + C$$

etc.

These processes take place because reactants are unstable with respect to the products and transform spontaneously to the more stable product or assemblage. However, this statement has no meaning unless the factors that control the stability are understood. These factors relate to energy differences and will be discussed more fully later in the chapter. The basic reason for chemical reactions is that the volume and heat content of the compounds on each side of an equation are different.

Chemical reactions do not go to completion unless one of the products, such as a gas, is removed completely from contact with the other reactants. In the general case, forward and back reactions occur simultaneously. At a given temperature and pressure, the velocity of the forward reaction at first outstrips that of the reverse process; but eventually a state arises in which both forward and back reactions are taking place at equal rates, and the concentrations of the reacting substances become constant. The rate at which equilibrium is attained is related to the activities of the particular chemical species involved, and these activities, in turn, are proportional to their active masses represented by their vapor pressure. In effect, reactions cannot "go" unless atoms can exchange, and the number of free atoms available per unit of time controls the degree of reactivity.

The **law of mass action** states that the rate of a reaction is directly proportional to the concentration of each reacting substance. It provides a means of relating the velocities of a reaction and the activities of its reactants. A general chemical reaction

$$mA + nB = pC + qD$$

has A, B, C, and D as reactants and m, n, p. and q as small integer coefficients. The equal sign represents an equilibrium condition. From the law of mass action we know that the rate of the forward reaction (R_F) will be proportional to the activities of the reactants, for example,

$$R_F = k_F[A]^m[B]^n$$

where k_F is a constant of proportionality. Similarly, for the reverse reaction rate, R_R, will be

$$R_R = k_R[C]^p[D]^q$$

Note that the activities of the reactants are raised to the power of their coefficients. At equilibrium, there will be no tendency for change, and the rates of forward and reverse reactions must be equal. Therefore,

$$k_F[A]^m[B]^n = k_R[C]^p[D]^q$$

or

$$\frac{[C]^p[D]^q}{[A]^m[B]^n} = \frac{k_F}{k_R} = K$$

where K is the **equilibrium constant** for the reaction.

These principles can be applied to our speculated reaction for tin deposition as follows:

$$\frac{[HF]^4\,[SnO_2]}{[SnF_4]\,[H_2O]^2} = K$$

Because K is a constant for a particular temperature, any change in concentration of a chemical species involved in the equilibrium must be reflected by changes in the other species. Inspection of this relationship shows the HF concentration to have the greatest effect on the equilibrium, and should HF be lost at the reaction site by fissuring, the reaction goes rapidly to the right, depositing cassiterite. This reaction would be very rapid due largely to the presence of mobile and reactive gaseous species. The partial pressures of solids and liquids are much lower, and the effects of pressure changes on their equilibria will be correspondingly smaller.

Thermodynamic Considerations

Fundamental Concepts

The physicochemical processes of nature tend toward a uniform distribution of energy among the various parts of a system. Because energy cannot be destroyed, changes in the system represent redistribution of energy within the system plus those gains or losses due to interaction with the surroundings. Energy should be considered as a participating "substance" in any physical or chemical reaction. Like other substances, it may be transformed in the reaction, and like the chemical elements, it is not destroyed. This is the **First Law of Thermodynamics:** *The various forms of energy are equivalent and energy is indestructible.*

The total energy a system gains or loses during the course of any process or change is called the change in **internal energy,** ΔE or dE. A change in internal energy depends only on the initial and final states of the system, and therefore all systems in the same state have the same internal energy regardless of their paths to that state. The change in internal energy of a system (dE) during any closed process is defined as the gain in energy due to the heat absorbed by the system (dq) plus the work done on the system (dw):

$$dE = dq + dw$$

Furthermore, mechanical energy or work is the product of pressure times change in volume ($dw = -PdV$):

$$dE = dq - PdV \tag{6.1}$$

Internal energy, or E (U in many texts), is an abstract quantity that includes the total energy of the system—kinetic, potential, and chemical. The kinetic and potential forms of energy are manifested, respectively, in the intensive variables temperature and pressure. Obviously, thermodynamic systems are complex and involve a variety of energy forms. Analyzing such systems is easier if we use variables that restrict the analysis to one energetic form, such as kinetic energy.

Kinetic energy in thermodynamic systems is the result of atomic motion. Such motion disappears at **absolute zero** (0° Kelvin or −273.15° C). As heat is added to the system, atomic vibrations increase with an accompanying gain in kinetic energy. Thus, kinetic energy can be likened to thermal energy, or the heat content of the system as manifested in temperature. It is convenient to consider thermal energy as **enthalpy,** H, where

$$H = E + PV \tag{6.2}$$

On differentiation,

$$dH = dE + PdV + VdP$$

and at constant pressure, $dP = 0$, so that

$$dH = dE + PdV = dq$$

Now consider the reaction

$$A + B \longrightarrow C + D$$

$$\text{reactants (Re)} \qquad \text{products (Pr)}$$

Equation 6.2 can be written for each of the species involved as follows:

$$H_A = E_A + PV_A$$

$$H_B = E_B + PV_B$$

etc.

The enthalpies and internal energies of the products (Pr) and reactants (Re) may be added together so that if pressure is constant,

$$H_{Pr} = E_{Pr} + PV_{Pr}$$

and

$$H_{Re} = E_{Re} + PV_{Re}$$

Subtraction gives

$$H_{Pr} - H_{Re} = \Delta H = \Delta E + P\Delta V = \Delta q$$

where ΔE is the change in internal energy and ΔV is the volume change accompanying the reaction. Thus, for a reaction at constant pressure, ΔH is the heat of reaction and is often called the **heat content** of a system or the **latent heat** of reaction.

A process or reaction acting on a system is **reversible** when the system is so balanced that an infinitesimal change in conditions will cause the reaction to proceed in the opposite direction. This type of reaction takes place under the equilibrium conditions. On the other hand, a reaction that proceeds spontaneously and at a finite rate is **irreversible** and takes place under nonequilibrium conditions.

Consider now the concept of entropy, or S. Entropy is not easily defined, because the concepts of thermal energy, disorder, and probability call come into play. **Entropy** is best described as the possible ways to combine the properties of individual particles on the microscopic scale to produce the observable macroscopic properties of the system. By this description, large increases in entropy are involved in taking a substance from the solid through the liquid to the gaseous state. The solid state involves a rigid structure of atoms or molecules with little chance for changing configurations. At the other end of the scale, the particles of gases have extreme mobility, allowing many configurations without a change in the macroscopic appearance of the system.

Entropy, like internal energy, is a function of the system and is thus dependent only on the initial and final states of the system for any change. Further, entropy can be defined quantitatively for a reversible change as the heat absorbed by the system during a process at a particular temperature, or

$$\Delta S = dq/T \tag{6.3}$$

However, for an irreversible reaction

$$\Delta S > dq/T \tag{6.4}$$

holds true, because in an irreversible process the entropy change may be due to changes other than the absorption of heat by the system. In fact, the total entropy of all systems increases during an energy change in an irreversible or natural process, and equation 6.4 can be taken as a statement of the **Second Law of Thermodynamics**: *two bodies at different temperatures will exchange heat such that heat flows from the hotter to the colder body.* Rearranging and incorporating equation 6.1,

$$\Delta E = T\Delta S - P\Delta V$$

or

$$dE = TdS - PdV$$

for reactions involving only pressure-volume work.

Energy Distribution

To approach these energy relationships in a slightly different way, consider that the total energy in a system is the sum of all kinetic (motion) and potential (position) energies of all the atoms present; it therefore depends on temperature, pressure, and position in force fields (gravity, magnetic, electrical). This total energy may be considered as composed of (1) a certain quantity of bound energy that is present as thermal motion of the atoms and not available for mechanical work and (2) a quantity of free energy that is available for such work:

total energy = bound energy + free energy

Two further quantities, the *Helmholtz Free Energy* (A) and the *Gibbs Free Energy* (G), allow description of energy changes within a system:

$$A = E - TS \tag{6.5}$$

and

$$G = H - TS \tag{6.6}$$

On differentiation,

$$dA = dE - TdS - SdT$$

and

$$dG = dH - TdS - SdT$$

Many geologic reactions are considered to have taken place under conditions of constant temperature (**isothermal,** $dT = 0$) and pressure (**isobaric,** $dP = 0$). Thermal and mechanical energies are thus held constant, and the Gibbs Free Energy best describes the changes in chemical energy. From equation 6.6

$$G = H - TS$$

or from 6.2

$$G = E + PV - TS$$

For a reaction at constant T and P,

$$dG = dE + PdV - TdS$$

and if the reaction is reversible,

$$dE = TdS - PdV$$

and

$$dG = 0$$

If the reaction is spontaneous and irreversible, then

$$dE < TdS - PdV$$

and

$$dG < 0$$

Thus, there must be a decrease in the Gibbs Free Energy for any spontaneous reaction at constant temperature and pressure.

In a similar manner, it can be shown that spontaneous reactions must involve decreases in the variables E and A. Therefore, the thermodynamic potentials E, A, and G will allow prediction of whether or not a reaction should take place, and if it should, in which direction it will go. Because

$$G = E + PV - TS$$

differentiation gives

$$dG = dE + PdV + VdP - TdS - SdT$$

Also, because

$$dE = TdS - PdV$$

then

$$dG = VdP - SdT \qquad (6.7)$$

Equations analogous to 6.7 can be written for the free energy changes of both products and reactants:

$$dG_{Pr} = V_{Pr}dP - S_{Pr}dT$$

and

$$dG_{Re} = V_{Re}dP - S_{Re}dT$$

Subtraction gives

$$dG_{Pr} - dG_{Re} = d\Delta G = \Delta VdP - \Delta SdT \qquad (6.8)$$

If equilibrium is maintained as P and T are changed, then ΔG must equal zero at all times. Thus, $d\Delta G$ must also remain zero, and equation 6.8 can be written

$$\Delta VdP = \Delta SdT$$

or

$$\frac{dP}{dT} = \frac{\Delta S}{\Delta V} = \frac{\Delta H}{T\Delta V} \qquad (6.9)$$

This is the **Clausius-Clapeyron equation,** which holds for all transitions where ΔS and ΔV do not equal zero simultaneously.

Using the equations just derived it is possible to examine the stabilities of minerals in nature more rigorously. Given enough information about a particular system, one should be able to calculate the temperatures and pressures critical to the formation and destruction of minerals and mineral assemblages. Thermodynamic concepts also help us develop criteria for mineral stability and equilibration and ultimately understand the processes that shape the earth. As powerful as the thermodynamic approach is, however, little standard thermodynamic data is available for materials at the temperatures and pressures common to igneous and metamorphic processes. Further, the data that are available have sizable errors. In most cases, useful inferences can be drawn, but these should be treated with due caution.

Mineralogic Reactions

A few examples demonstrating the application of thermodynamic principles to mineralogic phenomena are given in the following paragraphs. In all examples, the systems are **isochemical,** or closed; the number and kind of atoms in the reactants and products are the same. Descriptions of systems that interact chemically with their surroundings, although certainly important, are beyond the scope of this book. Discussions of open systems can be found in the references at the end of this chapter.

Polymorphic Transformations As described in chapter 5, the same atoms may be arranged in more than one way, and the different minerals resulting are called polymorphs. The stable polymorphic form under any set of external conditions must have the lower free energy. Since volume changes will be very small, they can be disregarded and the Helmholtz Free Energy expression (equation 6.5) can be used to describe the thermodynamic relationship of a pair of polymorphic forms 1 and 2. Thus, for a polymorphic reaction

$$A = E - TS$$

and at equilibrium

$$E_1 - TS_1 = E_2 - TS_2$$

At any temperature above $0°$ Kelvin (where the entropy term vanishes), the random vibration of the atoms about their structural sites is a significant part of the energy of the system. The distribution of energy for a particular form at various temperatures might be that shown in figure 6.1a, whereas the distribution for another form might be that shown in figure 6.1b. When combined in the same diagram, figure 6.1c, it becomes apparent that the form having the lowest free energy is not the same at all temperatures. Transformation from form 1 to form 2 should occur where the free-energy curves intersect, and this point is the transformation or inversion temperature.

Also, the second form has a higher total energy (E_2) than the first form does at all temperatures. When the mineral transforms, this difference in total energy must be made up by the addition of heat to the mineral. The heat that enters (or leaves) a substance in order to balance the energy budget is the latent heat (L or ΔH). Changes in latent heat accompany polymorphic transformation, melting, dissociation, and crystallization.

The transformation temperature in the example could be determined experimentally by careful monitoring of the temperature of form 1 while it was being steadily heated. The temperature would rise smoothly until the transformation point was reached, at which time the curve would

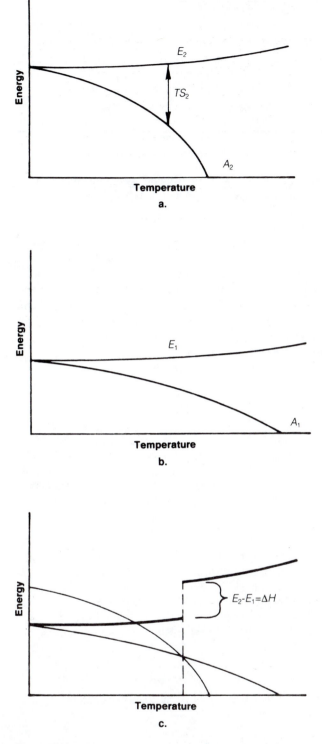

Figure 6.1
Energy relations of polymorphs as a function of temperature. *E* is the internal energy and *A* is the free energy. (After Buerger, 1961.)

flatten while the latent heat was supplied. When the necessary amount of latent heat had been added, the temperature would rise again. The opposite pattern would be generated on cooling, with the transition point being marked by a constant temperature as the latent heat was dissipated.

Because both the transformation temperature and the amount of heat generated or absorbed will differ with the particular reaction and the mineral species, differential thermal analysis (see chapter 11), can be used to identify many minerals.

Graphite-Diamond Consider the polymorphic transformation of graphite (Gr) to diamond (Di), ignoring the effect of pressure on the specific volumes. This assumption allows us to use the standard thermodynamic data for the minerals at 25° C (298° K), as given in table 6.1. The enthalpy, entropy, and volume changes for the reaction graphite → diamond are as follows, where the superscript ° refers to data at 25° C and 1 atmosphere:

$$\Delta H_T^\circ = \Delta H_{298}^\circ = H_{Di}^\circ - H_{Gr}^\circ = 453 \text{ cal/mole}$$

$$\Delta S_T^\circ = \Delta S_{298}^\circ = 0.568 - 1.372 = -0.804 \text{ cal/deg-mole}$$

$$\Delta V_T^\circ = \Delta V_{298}^\circ = 3.417 - 5.298 = -1.881 \text{ cm}^3/\text{mole}$$

The change in Gibbs Free Energy for the transition may be written

$$\Delta G_T^P = \Delta H_{298}^\circ - T\Delta S_{298}^\circ + P\Delta V_{298}^\circ$$

$$= 453 + 0.804T - \frac{1.881P}{41.8}$$

$$= 453 + 0.804T - 0.045P \qquad \textbf{(6.10)}$$

because 1 calorie = 41.8 bar cm³.

Now, at equilibrium $\Delta G_T^P = 0$, and the equilibrium pressure may be calculated at any temperature. For example, at 25° C

$$0 = 453 + (0.804)(298) - 0.045P$$

and

$$P = \frac{453 + 239.5}{0.045} \approx 15,400 \text{ bars}$$

Further, using the Clausius-Clapeyron equation (6.9),

$$\frac{dP}{dT} = \frac{\Delta S_T^\circ}{\Delta V_T^\circ}$$

$$= \frac{-0.804 \text{ cal/deg-mole}}{-1.881 \text{ cm}^3/\text{mole}} \times 41.8 \frac{\text{bar cm}^3}{\text{cal}}$$

$$= 17.9 \text{ bar/deg}$$

These calculations give one point and the slope of the simplified graphite-diamond equilibrium curve shown in figure 6.2 as line 3. When the temperature effect on ΔH and ΔS and the pressure effect on ΔV are taken into account, the equilibrium curve is as shown by line 2. The experimentally determined curve (line 1) is also shown along with the normal geothermal gradient. It is obvious from equation 6.10 and figure 6.2 that graphite remains stable at atmospheric pressure no matter how high the temperature—as long as oxygen is not present, of course!

Table 6.1	Standard Free Energies, Enthalpies, and Entropies for Selected Minerals			
	Enthalpy, $\Delta H°$ (Kcal/mole)	Entropy, $S°$ (cal/deg-mole)	Free Energy, $\Delta G°$ (Kcal/mole)	Molar Volume, $V°$ (cm³)
Graphite, C	0	1.372	0	5.2982
Diamond, C	0.453	0.568	0.693	3.4166
Calcite, Ca(CO₃)	−288.6	22.15	−269.908	36.934
Aragonite, Ca(CO₃)	−288.7	21.18	−269.678	34.15

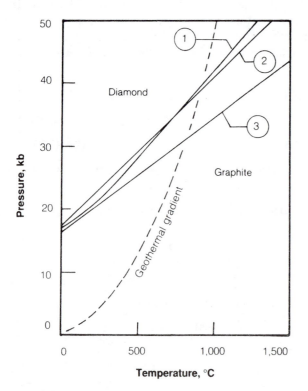

Figure 6.2
Stability relations for graphite-diamond. Curve 1, experimentally determined; curve 2, calculated considering temperature effects on $\triangle H$ and $\triangle S$ and the pressure effect on $\triangle V$; curve 3, calculated in text. The continental geothermal gradient is shown as a dashed line.

Examination of equation 6.10 also shows that an increase in pressure while temperature holds constant will reduce ΔG_T^P. At the pressure where $\Delta G_T^P = 0$, graphite and diamond can coexist. At higher pressures where $\Delta G_T^P < 0$, diamond will be the stable polymorph. Figure 6.2 shows that although the simplified calculations made here only approximate the experimental results, the inferences drawn from either case would be similar. The curve resulting from the more complex calculations (line 2), where the temperature dependence of ΔH and ΔS plus the temperature and pressure dependence of ΔV are taken into consideration, is an obvious improvement.

Although diamond is not stable at room temperature and atmospheric pressure, it obviously persists under these conditions. Diamond is said to exist *metastably* within the stability field of graphite, the stable polymorph. The reasons for the metastable persistence are within the realm of chemical kinetics and, although a rigorous treatment is beyond the scope of this book, more will be said on this in chapter 7.

Calcite-Aragonite Although aragonite is unknown in rocks older than Upper Carboniferous ($\approx 300 \times 10^6$ years), this mineral as well as the other Ca(CO₃) polymorph, calcite, may form under surficial temperature and pressure conditions. However, all aragonitic material will revert to calcite given enough time, and it is thus of interest to estimate the relative stabilities of the Ca(CO₃) polymorphs to understand further the inversion process. Using the data of table 6.1 to investigate the reaction

$$Ca(CO_3) \longrightarrow Ca(CO_3)$$
$$\text{aragonite} \qquad\qquad \text{calcite}$$

we see that

$$\Delta G°_{298} = G°_{Cal} - G°_{Ara}$$
$$= -269908 + 269678 = -230 \text{ cal/mole}$$

and also

$$\Delta V°_{298} = 36.93 - 34.15 = 2.78 \text{ cm}^3$$

What then is the pressure at which aragonite and calcite would be in equilibrium (P_{eq}) at 25° C, that is, the pressure at which $\Delta G_T^P = 0$?

We know that

$$\Delta G_T^P = \Delta G_T° + P\Delta V$$

or

$$\Delta G_{298}^P = \Delta G_{298}° + P\Delta V°_{298}$$

Because temperature is being held constant, at equilibrium

$$0 = -230 + P_{eq}(2.78)$$

or

$$P_{eq} = \frac{(230)(41.8)}{2.78} = 3,460 \text{ bars}$$

(Remember that 1 cal = 41.8 bar cm³.)

Thus, at 25° C aragonite and calcite are not in equilibrium until a pressure of 3,460 bars is attained. Further, using standard entropy data from table 6.1 and assuming

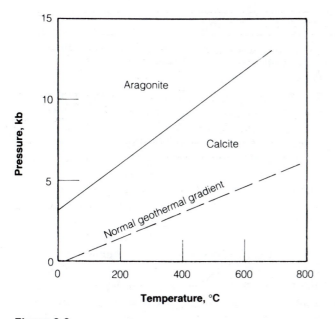

Figure 6.3
Calculated stability relations for calcite-aragonite.

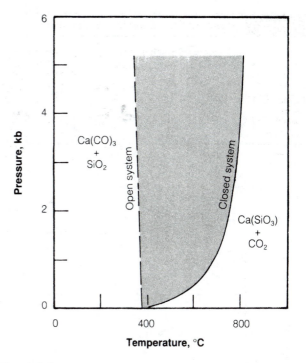

Figure 6.4
P-T relations for $Ca(CO_3) + SiO_2 = Ca(SiO_3) + CO_2$. (After Harker and Tuttle, 1956.)

that ΔS and ΔV are unaffected by temperature and pressure, then by the Clausius-Clapeyron equation

$$\frac{dP}{dT} = \frac{\Delta S^\circ_{298}}{\Delta V^\circ_{298}} = \frac{(0.97)(41.8)}{2.78} = 14.6 \text{ bar/deg}$$

The equilibrium curve for the transition aragonite → calcite can now be plotted as in figure 6.3. Also shown in this diagram is a dashed line representing the normal geothermal gradient. This figure and the previous calculations show that aragonite is never stable under atmospheric pressure no matter what temperature is attained, nor is it stable at any depth if the crustal geothermal gradient is normal. Thus, aragonite formed at the earth's surface is metastable and, in time, will revert to the calcite form. The reason for aragonite precipitation under surficial conditions is again a kinetic problem. Let it suffice to say that the rate of nucleation of aragonite relative to calcite is enhanced by the presence of other cations (e.g., magnesium) or by poorly understood organic reactions. For example, certain marine fauna and flora preferentially secrete aragonite test and shell materials.

When compared to the normal geothermal gradient, the equilibrium curve for the calcite-aragonite transition precludes the formation of aragonite under normal geologic conditions. Aragonite is found, however, in some metamorphosed carbonate-bearing rocks of orogenic belts. Thus, the aragonite-bearing assemblages of these rocks must be explained by a special type of high-pressure metamorphism characterized by a particularly low geothermal gradient.

The Effect of Gaseous Phases Reactions in which mechanical energy (PV) is important often involve a gaseous phase. The release of carbon dioxide from a carbonate in response to heating or chemical action, as in the following equations, are cases in point.

$$Ca(CO_3) + heat = CaO + CO_2$$
$$\text{solid} \qquad\qquad \text{solid} \quad \text{gas}$$

$$Ca(CO_3) + 2HCl = Ca^{2+} + 2Cl^- + H_2 + CO_2$$
$$\text{solid} \qquad \text{liquid} \qquad \text{liquid} \qquad\qquad \text{gas}$$

The reactions will proceed to the right and liberate gas so long as the carbon dioxide is free to escape (the Gibbs Free Energy of the right-hand side is less than that of the left). If, however, the system is closed, the internal pressure of CO_2 will build up until an equilibrium between the amount of gas being evolved and the amount of gas reacting with CaO or Ca^{2+} to form solid $Ca(CO_3)$ is reached.

The reaction of calcite and silica to form wollastonite is an important metamorphic reaction for which adequate thermodynamic data are available. The equilibrium pressure-temperature conditions for the reaction

$$Ca(CO_3) + SiO_2 = Ca(SiO_3) + CO_2$$
$$\text{calcite} \qquad \text{quartz} \quad \text{wollastonite} \quad \text{gas}$$

in a closed system from which CO_2 cannot escape are given by the solid line in figure 6.4. At pressures and temperatures to the right of the curve, for example, at 800° C and 1,000 bars, dG is less than zero, and the reaction will spontaneously proceed to the right. To the left of the curve, say, at 600° C and 4,000 bars, wollastonite and carbon dioxide react to form calcite and quartz. On the curve, $dG = 0$ and all phases are in equilibrium. Note that the

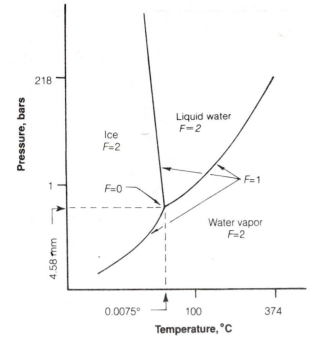

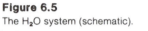

Figure 6.5
The H_2O system (schematic).

reaction may be made to go at constant pressure by a change in temperature or at constant temperature by a change in pressure.

If the system is open and allows the escape of CO_2, the equilibrium is shifted, as shown by the dashed line. Because it is geologically unlikely that a gas can be completely retained by rocks, but also that its escape will be slowed by the necessity of diffusion through minute openings, the probable boundary in natural systems lies within the shaded area. The loss of CO_2 largely removes the effect of pressure on the reaction; and the formation of wollastonite, $Ca(SiO_3)$, in a siliceous limestone, SiO_2 + $Ca(CO_3)$, implies the attainment of a temperature of about 500° C.

The Phase Rule

In an equilibrium situation there is a distinct relationship between the number of phases and the number of chemical and physical variables present in a system. This relationship is summarized by the **Phase Rule,** developed by the American chemist J. Willard Gibbs in the latter part of the nineteenth century. His original statement can be derived from first principles, but the derivation is lengthy and involved. Instead, we will attempt to induce it using examples from natural systems.

One- and Two-Component Systems

A One-Component System: H_2O   The phase relations of water at different temperatures and pressures are shown in figure 6.5. Three curves pass through a common point

called a *triple point*. The boundary curve between water and water vapor is the vapor pressure or boiling curve. This curve ends at 374° C and 218 bars, the **critical point** of water beyond which the physical distinction between liquid and vapor disappears. Boundary curves are also present between water and ice (the melting curve) and between water vapor and ice (the sublimation curve).

In this particular system, there is one chemical component and there are two physical variables, temperature and pressure. Consider a point within the area designated as liquid water. Obviously, temperature and pressure within the system can be changed independently so long as a boundary curve with another phase is not touched. Because two variables can change without changing the number of phases, we say the system has two **degrees of freedom.** Consider now a point on the boundary curve between water and vapor. Here two phases are in equilibrium and there is only one degree of freedom, because if the pressure on the system is fixed, the temperature is necessarily fixed. Further, if temperature changes, then the pressure must necessarily change to maintain the equilibrium. Finally, at the triple point, three phases occur together—water, ice, and vapor. Here both temperature and pressure are fixed and there are no degrees of freedom.

Figure 6.6 shows phase diagrams for two other one-component systems of mineralogic interest, SiO_2 and $Al_2(SiO_4)O$. These systems warrant a careful investigation. The figure shows that in a single-component system there are two degrees of freedom for one phase, one degree of freedom for two phases, and no degrees of freedom for three phases. In other words, the number of degrees of freedom or the **variance** (F) must equal the number of components (C) plus the physical variables, minus the number of phases (P), or

$$F = C + 2 - P \qquad (6.11)$$

This is the Phase Rule of Gibbs.

As we will see, the Phase Rule is applicable to multicomponent systems as well. In any case, the maximum number of phases (P) possible in a system of components (C) may be present only when the system is invariant, that is, when both temperature and pressure are fixed. In this case,

$$F = 0$$

and

$$P = C + 2$$

During the process of mineral formation, temperature and pressure are usually variable over large intervals that represent the stability conditions of the particular mineral. Thus, in many geologic situations there are two degrees of freedom ($F = 2$), and under these circumstances,

$$P \leq C$$

This is the **mineralogic phase rule** of V. M. Goldschmidt.

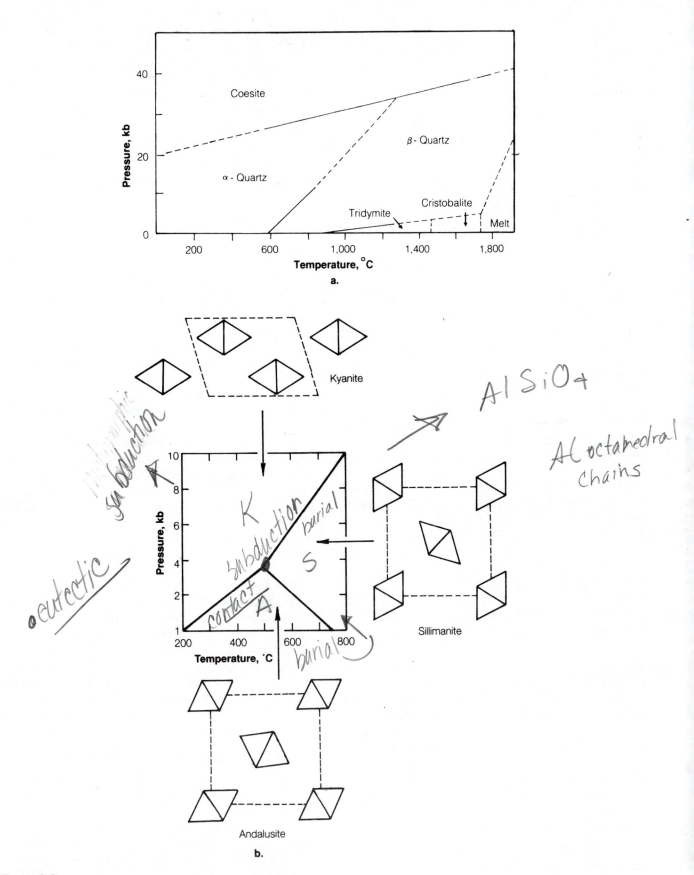

Figure 6.6
One-component systems of mineralogic interest. (a) SiO₂. (After
Boyd and England, 1960.) (b) Al(SiO₄)O, phase diagram. (After
Holdaway, 1971.) Structure diagrams showing the disposition of Al
octahedral chains. (After Putnis and McConnell, 1980.)

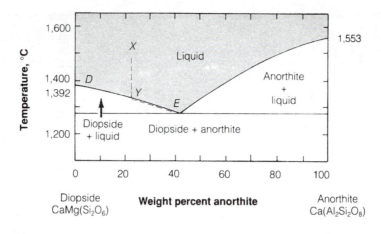

Figure 6.7
The two-component system diopside-anorthite. (After Osborne, 1942.)

A Two-Component System: Diopside-Anorthite The diopside-anorthite system is presented in figure 6.7 as a temperature-composition diagram. Obviously, to present components and both temperature and pressure, a three-dimensional diagram is necessary. Figure 6.7 is an isobaric section with composition (given in weight percent anorthite) as the abscissa and temperature as the ordinate.

Note the melting (or crystallization) temperatures of 1,392° C and 1,553° C for pure diopside and pure anorthite, respectively. With mixing of the components, the crystallization temperature is decreased in proportion to the concentration of the admixed component. This is analogous to the depression of the freezing point of water by addition of ethylene glycol (antifreeze). The two melting curves decrease toward the center and meet at a common point. This point (42% anorthite) is the composition of the lowest melting temperature (1,270° C) within the system and is called the **eutectic** point.

The diopside-anorthite system as shown in figure 6.7 is a simple eutectic system having complete miscibility in the liquid and complete immiscibility in the solid. In actuality there is a small amount of solid solution of anorthite in diopside, but this small amount can be ignored for the purpose of our discussion here.

Consider a melt of composition *X* in this system. Both temperature and composition can vary within bounds, and no alteration of the number of phases will occur. Thus, there must be two degrees of freedom within the liquid field since according to equation 6.11

$$F = C + 1 - P$$
$$= 2 + 1 - 1$$
$$= 2$$

Note that the physical degrees of freedom are reduced by one since pressure is held constant.

As the melt cools, no changes occur until the melting curve of diopside is encountered at *Y.* Here diopside will begin to crystallize in equilibrium with the liquid. There are now two phases (diopside and liquid) in equilibrium, and according to the Phase Rule there is only one degree of freedom. Thus, further cooling must be accompanied by the crystallization of diopside and changing composition of the liquid. The melting point for any further crystallization must be on the diopside melting curve *DE.* When the liquid composition reaches point *E,* the eutectic point, anorthite begins to crystallize. There are now three phases in equilibrium: liquid + diopside + anorthite. Thus, according to the Phase Rule, the system must be invariant, and temperature and composition must hold constant until crystallization is complete. An analogous path would have been followed by a melt rich in the anorthite component.

A Two-Component System: Leucite-Silica Keeping in mind the simple eutectic system just described, consider what happens when the two end members react to form a third compound. For example, in the system *X–Y,* if some compound XY_2 is formed, we could have two two-component systems, $X-XY_2$ and XY_2-Y, as shown in figure 6.8. In this case, the compound XY_2 is said to melt **congruently,** since the liquid formed also has the composition XY_2. The situation becomes more complex, however, if the intermediate compound undergoes decomposition below the melting point. Such is the case in the system leucite-silica (see figure 6.9). Here there is an intermediate compound, K-feldspar, that decomposes at 1,150° C to form leucite and melt. This type of melting is called **incongruent.**

The melt *X* in figure 6.9 has the composition of K-feldspar. On cooling, it arrives at the phase boundary between the fields of melt and melt plus leucite. Leucite will begin to crystallize, and with further cooling, the melt composition must change along the leucite-melt phase boundary since there is only one degree of freedom. At

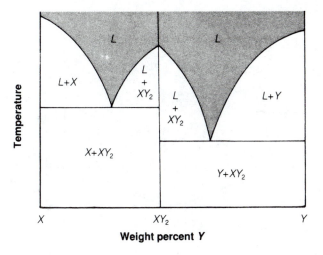

Figure 6.8
Two-component system with congruently melting compound XY_2.

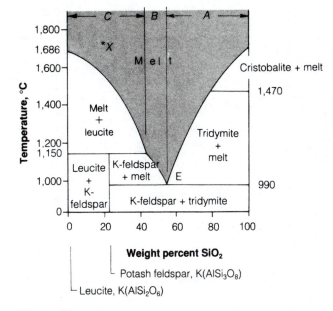

Figure 6.9
Leucite-silica system. (After Schairer and Bowen, 1947.)

1,150° C leucite begins to dissolve with simultaneous precipitation of K-feldspar. There are three phases in equilibrium, and the Phase Rule decrees that the system has no degrees of freedom. Since the original melt had a K-feldspar composition, the final product must also have this composition if the system remains closed. The temperature will hold constant at 1,150° C until all the leucite has been transformed to K-feldspar. Indeed, the last portion of melt will be used up in the transformation of the last grain of leucite to K-feldspar.

The type of invariant point just described is called a **peritectic.** It differs from the other invariant point, the eutectic, in that a solid phase is lost by reaction (reactive point). A eutectic point, on the other hand, involves no loss of a solid phase but only the precipitation of solid phases (subtractive point).

Melts that are more siliceous than K-feldspar but within the region *C* on figure 6.9 will continue to precipitate K-feldspar following the dissolution of all the leucite at the peritectic point. On the loss of leucite, one degree of freedom is gained, and the melt composition will proceed to change along the melt-feldspar phase boundary until the eutectic *E* is reached. Here K-feldspar and tridymite will crystallize together. On the other hand, melts less silica-rich than K-feldspar will use up all the melt at the peritectic before the leucite totally disappears. The final assemblage will be a mixture of K-feldspar and leucite. Melts within the region *B* will crystallize K-feldspar first and proceed to the eutectic, and those in region *A* will crystallize a silica mineral first on their path to the eutectic. Notice that another invariant point occurs within region *A* because of the polymorphic transformation of cristobalite to tridymite at 1,470° C.

A Two-Component System: Albite-Anorthite If the compounds of the system under study form a solid solution, a further complication arises. This is the case for the plagioclase feldspars represented by the albite-anorthite system shown in figure 6.10. Although recent work implies that the continuous chemical variation of the system as shown may not be quite complete (see figure 21.7), this discussion will assume it to be so. Several features of this type of diagram are notable. For example, the one-phase liquid field is bounded by a curve that represents the temperature of stability for various melt compositions. This is the **liquidus** curve above which crystals are not found. Further, the **solidus** curve serves as the upper boundary of the solid-phase field. A two-phase field, melt + feldspar, lies between these two curves. At this point it would be instructive for you to go back and identify the liquidus and solidus boundaries in figures 6.7–6.9.

In figure 6.10, consider a melt of composition *X* (60% anorthite) and examine its path of crystallization under conditions that allow equilibrium to be maintained between liquids and solids. This situation usually exists under conditions of slow crystallization and continuous contact between the reacting phases. As the cooling melt reaches the liquidus curve at point 1, crystals begin to form. The composition of the crystals formed at this temperature can be determined by drawing a line from point 1 parallel to the composition axis and intersecting the solidus at point 2. Such a line joining the compositions of phases in equilibrium, here an An_{60} liquid and an An_{86} solid is called a **tie line.**

Because we have a two-component system that is isobaric, and because there are two phases in equilibrium, there is only one degree of freedom. Thus, any further

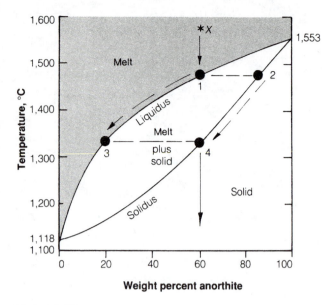

Figure 6.10

Temperature-composition relations of plagioclase. (After Bowen, 1913.)

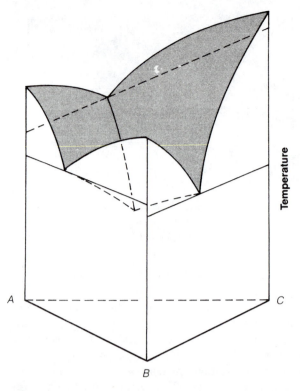

Figure 6.11

A three-component system *A-B-C* with a ternary eutectic. The binary systems *A-B*, *B-C*, and *A-C* all have simple binary eutectics. The liquidus surface is shaded.

cooling requires that the liquid composition change along the liquidus. As this happens, the crystals continually react with the liquid to become more albitic. The process ends when the last drop of melt produces a plagioclase feldspar of composition equal to the original melt (60% anorthite, point 4). Note that the last drop of liquid is quite sodic, with an anorthite content of only 19%.

The process involves equilibrium crystallization, where melt and crystals are in continuous contact and cooling is no faster than the equilibration reactions. A more common situation in natural systems happens when cooling and crystallization proceed more rapidly than the liquid-crystal equilibration reactions. This process, called **disequilibrium** or **fractional crystallization,** results in compositionally zoned plagioclase crystals with cores that are more anorthite-rich than the outer layers, which become increasingly albitic. Indeed, the outermost layer would have to be more albite-rich than the original melt (see figure 8.2).

Three-Component Systems

The use of one- or two-component systems can contribute significantly to our understanding of phase equilibria. These simple systems are handicapped, however, in that they inherently must portray extremes of both composition and conditions. To approximate more closely the mineralogic equilibria found in natural igneous and metamorphic rocks, it is necessary to increase the number of components. Although adding complexity, the expanded systems present more realistic mineral assemblages.

In the discussion of two-component systems, we saw that the equilibria could be represented in three-dimensional pressure-temperature-composition (*P-T-X*) space. Further, the systems were normally simplified to temperature-composition (*T-X*) planes by holding pressure constant. A similar procedure can be followed in the presentation of ternary systems. Figure 6.11 presents a three-component system in terms of temperature. The temperature-composition volume is bounded by binary eutectic systems, and the upper surface of the model is the ternary liquidus. Note that the liquidus surface descends toward the interior of the model and terminates at a ternary eutectic.

Three-dimensional models such as that given in figure 6.11 are easy to visualize but difficult to use for data plotting. Normally, three-component systems are viewed by looking directly down at the liquidus surface, compositions are given in triangular coordinates, pressure is held constant, and temperature is presented by contouring the liquidus surface. The topography of the liquidus surface is described graphically in a manner analogous to a topographic map, where location is given in latitude and longitude and elevations above sea level are contoured.

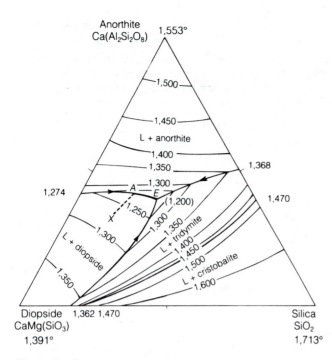

Figure 6.12
The contrived ternary system anorthite-diopside-SiO₂. The system is not actually known experimentally but is estimated from known binary joins. (Suggested by Osborne and Tait, 1952.)

A Three-Component System: Anorthite-Diopside-SiO₂
The edges of the ternary system prism in figure 6.11 are the binary systems diopside-SiO₂, SiO₂-anorthite, and diopside-anorthite described earlier in this chapter. Note that all three of these systems have simple binary eutectics, and combining them results in a ternary eutectic system. Such systems are normally presented as liquidus diagrams contoured in temperature, as shown in figure 6.12.

Consider a liquid of composition X in figure 6.12. There are three components and pressure is held constant. Thus, the Phase Rule (equation 6.11) is given by

$$F = 1 + C - P$$
$$= 1 + 3 - P$$
$$= 4 - P$$

If the initial temperature is above the liquidus surface, there is only one phase, liquid, and $F = 3$. If the temperature then falls to the liquidus, crystallization begins producing diopside. Two phases, liquid and diopside, now coexist, and the system variance must necessarily decrease to 2. Further, as diopside crystallizes, the composition of the liquid is depleted in diopside components and will move directly away from the diopside composition along line XA. At A, anorthite will begin to crystallize along with diopside, the system becomes univariant, and further cooling will be accompanied by melt compositional changes along the anorthite-diopside phase

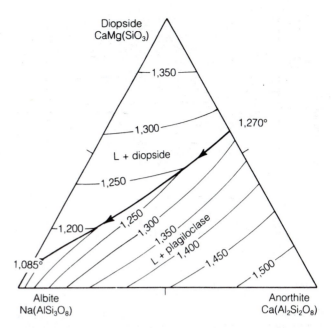
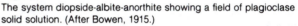

Figure 6.13
The system diopside-albite-anorthite showing a field of plagioclase solid solution. (After Bowen, 1915.)

boundary AE. At the ternary eutectic E, anorthite, diopside, and a silica mineral all crystallize together from the melt. Four phases now coexist and the system becomes invariant. Thus, the temperature must hold constant until the melt is exhausted.

Even in the very simple system portrayed in figure 6.12, a variety of igneous rocks can be generated. For example, if the diopside could be separated from the melt as it was formed, its accumulation would result in a special type of monomineralic rock called a pyroxenite. The assemblage anorthite-diopside approximates the plagioclase-pyroxene association typical of basalts, gabbros, and diorites. Further, the eutectic assemblage pyroxene-plagioclase-quartz could easily represent a granodiorite, the most abundant rock of orogenic batholiths. The implication here is that knowledge of the equilibrium mineral assemblage in the rock and of the available experimental data on analogous mineral systems can often lead to reasonable estimates of the physical conditions of rock genesis. Remember, however, that the experimental systems are chemically simple and that normal igneous magmas will contain other components, such as iron and sodium, in significant amounts. Extra components in the liquid will further depress liquidus temperatures, and thus the temperatures described in our simple system must represent maxima.

Ternary systems are extensions of the binary systems described in more detail earlier in this chapter. Thus, they can contain solid solution series (figure 6.13) or peritectics and eutectics (figure 6.14) to represent natural mineral equilibria. Intensive study of these or more complex systems is, however, left to courses in petrology and geochemistry.

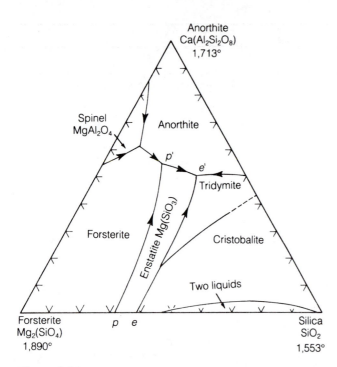

Figure 6.14
The system silica-forsterite-anorthite showing eutectics at e and e' and peritectics at p and p'. (After Anderson, 1915.)

Mineral-Aqueous Systems

The application of thermodynamic principles in mineralogy is not limited to the study of assemblages formed at elevated temperatures and pressures in water-free systems. Many reactions of mineralogic importance, and especially those taking place near the earth's surface in aqueous solution, may be usefully described in terms of the energetics of their exchange of electrons and the activity of free hydrogen ions. Diverse chemical species such as atoms, ions, or radicals in aqueous solution tend to exchange outer electrons, and the ease of electron loss or gain varies considerably from one species to another (see chapter 4). When an atom or ion gains an electron, its valence is decreased and the element is said to be reduced. In contrast, the term oxidation is used to describe the loss of electrons. Such reduction and oxidation reactions are often referred to as **redox reactions.**

The ease with which the loss or gain of an electron takes place is expressed as the standard potential, or the energy required to dislodge an electron from an outer shell. The standard reduction-oxidation potentials, $E°$, for a number of reactions at 25° C are given in table 6.2. The reaction $2H^+ + 2e^- = H_2$ is arbitrarily set at zero, and oxidizing ability increases downward; that is, elements higher on the list will lose electrons to those below them.

Values for the standard potential can be calculated using the equation $\Delta G° = 23.060 E° x$, where $\Delta G°$ is the standard free energy change in kcal/mole of a redox reaction and x is the number of electrons involved in the reaction.

Table 6.3 provides some selected data for $\Delta G°$. Other values of $\Delta G°$ can be found in the reference literature. For a given reaction, $\Delta G°_{SUM}$ is found by adding up the $\Delta G°$ for the formation of the products and subtracting this from the $\Delta G°$ values for the formation of the reactants. In the reaction

$$2FeFe_2O_4 + H_2O = 3Fe_2O_3 + 2H^+ + 2e^-$$
$$\text{magnetite} \qquad\qquad \text{hematite}$$

two atoms of ferrous iron are oxidized in the presence of water. For this reaction,

$$\Delta G°_{Pr} = 3(-177.1) + 2(0) = -531.3 \text{ kcal}$$

$$\Delta G°_{Re} = 2(-242.4) + (-56.69) = -541.49 \text{ kcal}$$

$$\Delta G°_{SUM} = \Delta G°_{Pr} - \Delta G°_{Re} = -531.3 - (-541.49)$$

$$= 10.19 \text{ kcal}$$

and

$$E° = \frac{10.19}{2(23.06)} = +0.221 V$$

In table 6.3 we can see that the presence of hematite, Fe_2O_3, is indicative of oxidizing conditions at the time of its formation when compared to magnetite, $FeFe_2O_4$. This is consistent with the occurrences of these minerals. Hematite is characteristically formed in surficial environments such as weathered zones in the presence of abundant electron-acceptor oxygen. Magnetite is formed under more reducing conditions typical of igneous and metamorphic environments.

The activity of a substance in dilute solution approximates its molar concentration. This approximation may, however, deviate significantly because of nonideal interactions, especially at higher concentrations.

Because redox reactions take place in aqueous solutions, we must take into account not only the redox reaction itself but also the effects of the dissociation of water:

$$H_2O = H^+ + OH^-$$

The dissociation product for this reaction at 25° C and 1 bar pressure is the concentration of H^+ times that of OH^- divided by the concentration of H_2O, or

$$\frac{[H^+][OH^-]}{[H_2O]} = 10^{-14}$$

The activity of a solution in terms of the negative logarithm of the hydrogen ion concentration is related to pH as follows:

$$pH = -\log_{10}[H^+]$$

Thus, if the concentration of H^+ is 10, $pH = -\log 10 = -1$, and if $[H^+]$ is 10^{-3}, $pH = 3$. When the concentration

Table 6.2 Redox Potentials for Some Reactions at 25° C

Potentials in Acidic Solutions

Reaction	Standard Potential, $E°$	Reaction	Standard Potential, $E°$, Volts
$Li^+ + e = Li$	−3.05	$Ni^{2+} + 2e = Ni$	−0.24
$K^+ + e = K$	−2.93	$Sn^{2+} + 2e = Sn$	−0.14
$Ba^{2+} + 2e = Ba$	−2.90	$Pb^{2+} + 2e = Pb$	−0.13
$Sr^{2+} + 2e = Sr$	−2.89	$2H^+ + 2e = H_2$	0.00
$Ca^{2+} + 2e = Ca$	−2.87	$Sn^{4+} + 2e = Sn^{2+}$	+0.15
$Na^+ + e = Na$	−2.71	$Cu^{2+} + e = Cu^+$	+0.16
$Mg^{2+} + 2e = Mg$	−2.37	$Cu^{2+} + 2e = Cu$	+0.34
$Th^{4+} + 4e = Th$	−1.90	$Cu^+ + e = Cu$	+0.52
$U^{3+} + 3e = U$	−1.80	$I_2 + 2e = 2I^-$	+0.54
$Al^{3+} + 3e = Al$	−1.66	$Ag^+ + e = Ag$	+0.80
$Zr^{4+} + 4e = Zr$	−1.53	$Hg^{2+} + 2e = Hg$	+0.85
$Mn^{2+} + 2e = Mn$	−1.18	$Pd^{2+} + 2e = Pd$	+0.92
$V^{2+} + 2e = V$	−1.18	$Br_2 + 2e = 2Br^-$	+1.08
$Zn^{2+} + 2e = Zn$	−0.76	$O_2 + 4H^+ + 4e = 2H_2O$	+1.23
$Cr^{3+} + 3e = Cr$	−0.74	$Cl_2 + 2e = 2Cl^-$	+1.36
$U^{4+} + e = U^{3+}$	−0.61	$Au^{3+} + 3e = Au$	+1.50
$Fe^{2+} + 2e = Fe$	−0.41	$Mn^{3+} + e = Mn^{2+}$	+1.51
$Co^{2+} + 2e = Co$	−0.28	$Au^+ + e = Au$	+1.68
$V^{3+} + e = V^{2+}$	−0.26	$Co^{3+} + e = Co^{2+}$	+1.82

Potentials in Basic Solutions

Reaction	Standard Potential, $E°$, Volts
$Mg(OH)_2 + 2e = Mg + 2(OH)^-$	−2.69
$UO_2 + 2H_2O + 4e = U + 4(OH)^-$	−2.39
$Al(OH)_4^- + 3e = Al + 4(OH)^-$	−2.32
$Mn(OH)_2 + 2e = Mn + 2(OH)^-$	−1.55
$Zn(OH)_2 + 2e = Zn + 2(OH)^-$	−1.25
$Sn(OH)_3 + 2e = Sn + 3(OH)^-$	−0.91
$Fe(OH)_2 + 2e = Fe + 2(OH)^-$	−0.89
$H_2O + 2e = H_2 + 2(OH)^-$	−0.83
$VO(OH)_2 + H_2O + e = V(OH)_3 + (OH)^-$	−0.64
$Fe(OH)_3 + e = Fe(OH)_2 + (OH)^-$	−0.55
$Pb(OH)_3 + 2e = Pb + 3(OH)^-$	−0.54
$S + 2e = S^{2-}$	−0.44
$Cu_2O + H_2O + 2e = 2Cu + 2(OH)^-$	−0.36
$Cu(OH)_2 + 2e = Cu_2O + H_2O + 2(OH)^-$	−0.08
$MnO_2 + 2H_2O + 2e = Mn(OH)_2 + 2(OH)^-$	−0.05
$HgO + H_2O + 2e = Hg + 2(OH)^-$	+0.10
$Co(OH)_3 + e = Co(OH)_2 + (OH)^-$	+0.17

of H^+ equals that of OH^-, the dissociation product is 10^{-7}, and the solution is neutral at pH = 7. The relations of H^+ activity and pH are shown in figure 6.15, which includes the extreme and common range of pH values observed in natural waters.

Redox potential, Eh, and pH are related by the equation

$$Eh = E° - 0.059pH$$

Eh-pH diagrams that show the stability fields of various minerals in aqueous environments are constructed in four steps:

1. List all of the oxide and hydroxide compounds of a given element.

Table 6.3 Selected Data for $\Delta G°$

Substance	$\Delta G°$ kcal/mole	Substance	$\Delta G°$ kcal/mole
H_2	0	Cu	0
H^+	0	Cu^+	12.0
O_2	0	Cu^{2+}	15.53
H_2O	−56.69	CuO	−30.4
C	0	Cu_2O	−34.98
Fe	0	CuS	−11.7
Fe^{2+}	−20.30	Cu_2S	−20.6
Fe^{3+}	−2.52	$Cu_2(CO_3)(OH)_2$	−216.44
Fe_2O_3	−177.1	Mn	0
$FeFe_2O_4$	−242.4	Mn^{2+}	−54.4
$Fe(OH)^{2+}$	−55.91	Mn^{3+}	−19.6
$Fe(OH)_2$	−115.57	MnO	−86.8
$Fe(OH)_3$	−166.0	MnO_2	−111.1
FeS_2	−36.0	$Mn(CO_3)$	−195.7
$Fe(CO_3)$	−161.06	MnS	−49.9

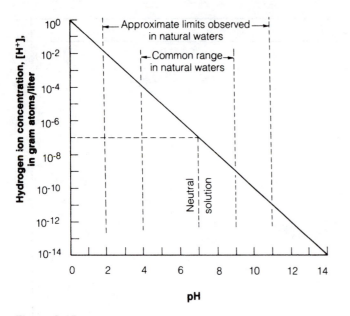

Figure 6.15
Relation of pH to hydrogen ion activity.

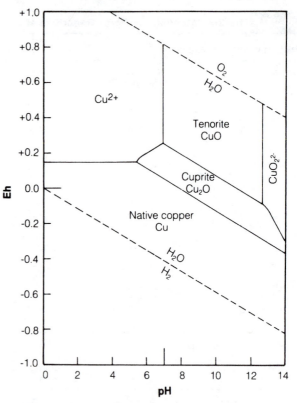

Figure 6.16
Stability relations among copper minerals in the system $Cu-H_2O-O_2$ at 25° C and 1 bar. (After Garrels and Christ, 1965.)

2. Write balanced reactions of the most reduced substance to each of the other compounds using only H_2O, H^+, OH^-, and electrons to balance the reaction, for example:

$$Fe° + 2H_2O = Fe(OH)_2 + 2H^+ + 2e^-$$

3. Calculate $E°$ for the reaction:

$$E° = -0.047V$$

4. Determine Eh:

$$Eh = -0.047V - 0.059pH$$

Thus, for any redox reaction in an aqueous environment, a series of Eh values can be found for corresponding values of pH.

Figure 6.16 provides an example of the Eh-pH stability fields for copper minerals in the system $Cu-H_2O-O_2$ and should be compared with figure 6.17 which shows the additional phases present as the result of adding geologically realistic amounts of carbon dioxide and sulfur to the system. Figure 6.18 is a similar diagram for manganese compounds. The diagonal lines at the top and bottom of each figure mark the oxidation of water to O_2 and its reduction to H_2. In nature, Eh values lie generally between -0.5 and $+0.6$ volts and pH between 4 and 9 (see figure 6.15). Together, these two parameters provide a particularly useful means of defining the stability fields of individual minerals and indicating those that can be expected to coexist. Figures 6.17 and 6.18 demonstrate the following:

1. The Phase Rule holds, and two phases may coexist in equilibrium only under the Eh-pH conditions at a phase boundary, for example, in native copper–cuprite or rhodochrosite-manganite pairs.

2. Because the diagrams represent phase equilibria possible for a specific temperature and pressure, the presence of incompatible phases in a mineral assemblage, as in malachite and chalcocite, indicates that the phases formed under different conditions at different times.

3. Compatible phases from different systems can be identified by the superposition of the diagrams. For example, alabandite can exist in equilibrium association with chalcocite but not with malachite. Further, mixing systems can add additional phases.

4. The course of alteration of a mineral under changing conditions of Eh or pH is indicated. Chalcocite oxidizes successively to native copper, cuprite, and malachite; malachite dissolves with lowered pH; and native copper alters to cuprite with increased alkalinity of the fluid.

Mineral Assemblages

The existence of a particular mineral species implies that its formation took place within a particular range of physicochemical conditions. When the mineral is accompanied by one or more other species that have formed with it, the conditions must necessarily embrace the entire suite represented, thereby restricting the size of their common field

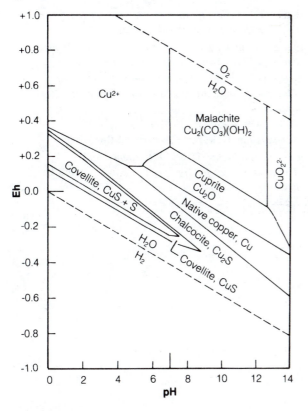

Figure 6.17
Stability relations among some copper minerals in the system
$Cu-H_2O-S-CO_2$ at 25° C and 1 bar. $P_{CO_2} = 10^{-3.5}$, total dissolved
sulfur species = 10^{-1}. (After Garrels and Christ, 1965.)

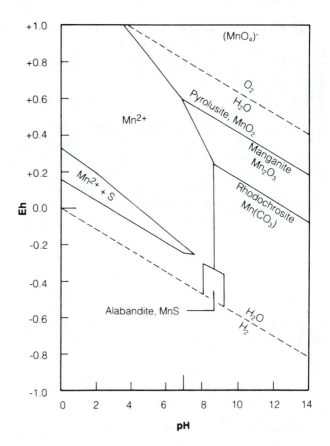

Figure 6.18
Stability relations of some manganese minerals in water at 25° C
and 1 bar. Total dissolved sulfur species = 10^{-1}, $P_{CO_2} = 10^{-4}$.
(After Garrels and Christ, 1965.)

of equilibrium. The conditions for mineral formation are incompletely known but can be generally assessed for the principal rock types and geologic processes. Table 2 in Appendix B lists some of the more common and representative minerals associated with a number of geologic processes and rock types.

Mineral assemblages as seen in the field or laboratory may be one of three kinds:

1. Minerals formed together at the same time and under the same physicochemical conditions
2. Minerals formed successively during a mineralizing event in which the chemical or physical parameters underwent continuous change (the order of crystallization usually being determined by observations of superposition, replacement, and cross-cutting relationships)
3. Minerals formed at separate times under different physicochemical conditions

Minerals that react to form intermediate compounds are, of course, incompatible and will not be found together unless they represent two periods of mineral formation separated in time. The common rule that forsteritic olivine and quartz do not occur together in primary association in igneous rocks is simply due to the fact that they react to form enstatite under igneous conditions.

The dominant mineral or minerals in an assemblage is often accompanied by small amounts of other compatible species that have been limited to the role of accessory minerals by the nonavailability of their components in the system. Such minerals, however, are often an important source of information as to the genesis of the assemblage or may be of considerable economic importance. For example, the presence of geikielite (magnesian ilmenite) in a mafic igneous rock signals its high pressure origin, and an ounce per ton (approximately 0.0035%) of gold in a quartz vein is rather rich ore.

Concluding Remarks

In this chapter we have examined the physicochemical basis for the stability of minerals. Our discussion provided only a first glance at this approach to mineralogic problems. Natural systems obviously are more complex than those presented here because they commonly must be described in terms of more than three components, some of which may be controlled by the chemistry of the surroundings. An understanding of the techniques presented, however, is basic to the solution of more involved problems of chemical mineralogy and to the understanding of natural mineral assemblages.

Suggested Readings

Edgar, A. D. 1973. *Experimental Petrology*. Oxford: Clarendon. A survey of the principles and techniques used in experimental mineralogic studies.

Ehlers, E. G. 1972. *The Interpretation of Geologic Phase Diagrams*. San Francisco: W. H. Freeman. Totally oriented toward systems of mineralogic interest, this book presents the geometric interpretation of phase diagrams with many examples.

Ernst, W. G. 1976. *Petrologic Phase Equilibria*. San Francisco: W. H. Freeman. Excellent descriptions of phase diagrams of mineralogic interest.

Faure, G. 1991. *Principles and Applications of Inorganic Geochemistry*. New York: Macmillan.

Garrels, R. M., and Christ, C. L. 1965. *Solutions, Minerals, and Equilibria*. San Francisco: Freeman.

Nordstom, D. K., and Muñoz, J. L. 1985. *Geochemical Thermodynamics*. Menlo Park, Cal.: Benjamin-Cummings.

Powell, R. 1978. *Equilibrium Thermodynamics: an Introduction*. New York: Harper and Row.

Wood, B. J., and Fraser, D. G. 1976. *Elementary Thermodynamics for Geologists*. Oxford: Oxford Univ. Press.

7

Crystallization, Growth, and Habit

Crystals are an orderly, solid arrangement of matter. Their constituents are atoms or ions present as discrete structural units or combined into radicals or molecules. These units of matter are accretion units that must initially be assembled into a crystal nucleus and then added in a regular manner to its surface.

How does an accreting crystal go about the process of selecting or rejecting ions from the nutrient reservoir? Of all the possible cation-anion combinations, what determines which are to be used in transforming some portions of the disordered matrix into an orderly array of some of its constituent ions? The critical parameters appear to be the availability of appropriate ions and the energetics of the system.

Energy Considerations

Common minerals are common because they are lowest-energy configurations of readily available elements. With an atypical set of natural abundances, a feldspar might well have the formula $Rb(GaGe_3O_8)$ because these cations are readily exchangeable with K, Al, and Si, respectively, in terms of the crystallochemical properties.

Generally, the particular proportions and arrangement of ions that nucleate and grow by accretion are those energetically most favored in the entire system. For crystallization from a cooling magma, for example, the loss of heat favors the energy change from the kinetic energy of free-moving ions to the potential energy of ions occupying structural sites. Because this is essentially a loss of heat from the system, the energy vector points toward crystallization as the system cools.

The ionic arrangements (phases) that form during crystallization are the ones that provide the greatest reduction of energy for a particular set of conditions. These arrangements, however, are not necessarily those that provide the greatest energy reduction under another set of conditions. Early-formed crystals can thus become unstable and dissolve as their environment changes, their components becoming available for incorporation in new, energy-favored structures.

A growing crystal represents an energy discontinuity in the physical and chemical system. Its surface is an interface between organized crystalline matter on the one hand and a source of accretion or growth units on the other. The growth units may be derived from a liquid, solid, or gaseous reservoir. A potential is obviously required to bring them to the point of nucleation or to the surface of a growing crystal, and some mechanism must be active at the growth surface to cause the units to be attached in a regular manner.

Crystals represent a state of matter for which the free energy of the particular solid is less than that of the system in which it occurs. At the incipient nuclear stages of the crystal's development, the crystal represents a marked discontinuity in the distribution of energy within the system. This discontinuity tends to be evened out either by dissolution of the nucleus or by transfer of atoms from the surroundings to the crystal. During this very early stage of development, random combinations of atoms form and disperse until eventually a stable configuration develops. Once formed, the surface of a nucleus is an energetically more favorable site for atoms than the surrounding medium, and atoms move to it down a chemical, thermal, or pressure gradient. Some nuclei are naturally favored over others, and gradients differ with both the kind of nucleus and its environment.

Crystal growth is generally not an accelerating process because of the negative feedback that occurs when an ion bonds to the crystal surface. For example, during exothermic crystallization (see chapter 6), latent heat is released and thus reduces the thermal gradient. Further, the concentration of growth units in the surrounding medium is reduced each time one attaches to the crystal surface. Therefore, for crystal growth to continue, either heat must be lost from the vicinity of the growing crystal, or new units must migrate into the local region, or both. Under these constraints, the growth of crystals tends to be rather slow and only local equilibration occurs. In rocks, a small thermodynamically equilibrated domain will surround a given mineral grain, but a grain of the same mineral a few centimeters away may have a slightly different growth environment and thus differ slightly in its chemical makeup.

The growth rates of crystals may be greatly accelerated under certain conditions but usually with a loss in perfection. Typical growth rates for synthetic quartz are around 1 mm per day, and since there are 10^7 Å per mm, 86,400 seconds per day, and the diameter of an oxygen ion is 2.6 Å:

$$\frac{(1 \text{ mm})}{(\text{day})} \times \frac{(10^7 \text{ Å})}{(\text{mm})} \times \frac{(1 \text{ day})}{(86,400 \text{ sec})}$$

$$\times \frac{(1 \text{ layer})}{(2.8 \text{ Å})} \approx 45 \frac{\text{layers}}{\text{sec}}$$

At such growth rates, it is not at all surprising that foreign atoms could be incorporated; that extraneous solids, liquids, or gases could be entrapped; or that perfect registry between adjacent parts of the crystal would not be attained.

Despite the long study of crystals and crystallization, little unanimity exists as to the exact nature of the process. Details of the growth mechanism have, for example, been variously ascribed to (1) functions of interface energy and face energy, (2) coordination of the accreting unit, (3) degree of satisfaction of surface bonds, and (4) interplanar spacing within the crystal. Our understanding of these mechanisms is not yet complete, but a basic explanation is in order because crystal growth is an important aspect of interpreting the end form (habit) of a mineral, one of the more provocative areas of mineralogic study.

Nucleation

Atoms dispersed in a fluid medium undergo random collisions that can result in temporary unions. When thermal motion is large, these groups disperse at the same rate as they form so that no net aggregation occurs. Such a steady state, however, exists only so long as the free energy of the system remains constant. Any changes favoring a transfer of energy from kinetic motion to interatomic bonds will promote a greater amount of aggregation among the atoms. Decrease in temperature or increase in pressure or concentration favors aggregation, and with changes of this kind clumps of atoms arise that constitute a nucleus. The nucleus, if stable, is then a site for further growth of a crystalline phase. Although most nucleation processes involve a change in the system leading to greater orderliness, as in the formation of a solid from a liquid, the reverse is also observed. For example, the formation of a bubble in a liquid is certainly a process of nucleation, yet the system loses order. It is preferable then to regard **nucleation** generally as the birth of a small volume of substance within a foreign matrix.

Nucleation does not occur exactly at the anticipated conditions. For example, we are told that under atmospheric pressure water will freeze at 0° C, but this is not quite true. In fact, water must be cooled below 0° C (supercooled) if it is to freeze. Reasonably clean water will probably have to be cooled to −5° C for nucleation to occur, whereas ultrapure water may have to go to −15° C. This being the case, there must be an energy barrier to nucleation. The presence of impurities plays a significant role in the nucleation process. Indeed, the prior presence of impurities, especially particulate matter, will greatly enhance nucleation. Two types of nucleation may thus occur: heterogeneous, in which discontinuities are present in the system, and homogeneous, in which nucleation is not dependent on the presence of foreign bodies.

The energy barriers to homogeneous nucleation are principally the thermal energy that keeps any two particles from coming together and the energy required to form the new surface on a nucleus. When the atoms or molecules do come together, they form chemical bonds that give back some of the energy to the system. As more particles are added to the cluster, the system's energetics change, and there is a critical point at which the process becomes spontaneous and irreversible, as shown in figure 7.1. Nuclei with a radius larger than critical (R_c) are stable and will persist. Those smaller are unstable and will disperse.

To be more specific, the energy of nucleation can be described as

$$\Delta G = -Ar^3 + Br^2 \qquad (7.1)$$

where r is the radius of the nucleus formed. The first term represents the change in cluster energy caused by the lower energy of interaction as each atom or atomic grouping attaches itself. A represents the forces favoring nucleation and is negative in this case because the process is

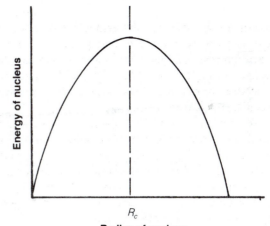

Figure 7.1
Critical radius, R_c, for a stable crystal nucleus.

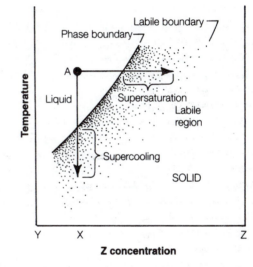

Figure 7.2
Crystallization in the labile region of a phase.

exothermic. The second term represents the energy increase due to the formation of the new surface. B is then the surface energy of the material being formed and acts against nucleation.

Using such energy considerations, the size of the critical nucleus has been calculated for many materials and is always close to a 10Å radius regardless of the material or type of phase change. This means that about 200 atoms, ions, or vacancies must be involved in most nuclei. It also means that the minimum dimensions of a crystal are about 10Å.

Nucleation does not occur exactly at the equilibrium phase boundaries described in chapter 6. To see where it does take place, it is instructive to examine a hypothetical phase diagram such as figure 7.2 in which there are two

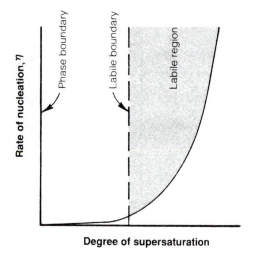

Figure 7.3
Nucleation rate as a function of supersaturation.

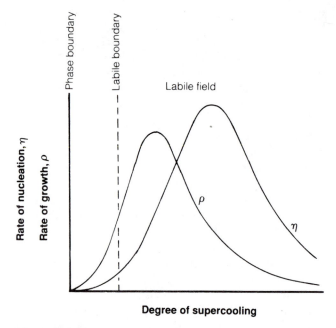

Figure 7.4
Rates of nucleation and growth as related to supercooling.

distinct ways to enter the phase field of the solid, one by supercooling and the other by supersaturation of one of the components. Generally, nucleation will take place only in that region described as **labile,** which is somewhat removed from the boundary.

Starting with a liquid of composition X at temperature T (point A), increasing the proportion of component Z causes the liquid to move isothermally and eventually to cross the phase boundary into the region of supersaturation. The rate of nucleation for various degrees of supersaturation is shown in figure 7.3, and it is obvious that significant nucleation does not take place between the phase boundary and the labile boundary. This phenomenon is easily demonstrated by starting with a brine solution saturated at room temperature and slowly increasing the salt concentration. Little salt crystallizes at low supersaturation, but large amounts are deposited when the labile region is entered.

A different pattern is observed on supercooling the same liquid A while holding the composition constant. This pattern is illustrated in figure 7.4, which shows that the rate of nucleation (η) increases rapidly on entering the labile field but drops off to zero at large degrees of supercooling. At small degrees of supercooling, although element mobilities are high, the critical nucleus dimensions are large and nucleation will not be favored (see equation 7.1 and figure 7.1). With large amounts of supercooling, the critical nucleus radii are much diminished, and a high rate of nucleation might be expected. Under these conditions, however, element mobilities are greatly lessened by increased fluid viscosities, and the atoms do not have time to rearrange themselves into crystalline order before complete solidification occurs, resulting in a glass.

At low degrees of supercooling, few nuclei are formed per unit time, and therefore further crystal growth must be concentrated on these limited centers. The rates of

crystal nucleation and crystal growth shown in figure 7.4 helps explain the grain size of igneous rock. In general, coarse-grained igneous rocks form at depths where the ambient temperature is high and the degree of supercooling is consequently low. In this case, there are lowered nucleation rates and augmented growth rates. On the other hand, volcanic rocks cooling on the earth's surface are severely supercooled with consequent high rates of nucleation and low crystal-growth rates, so the product is typically fine-grained. A natural glass such as obsidian exemplifies a case of extreme supercooling.

Nuclei, once formed, are minute volumes in which the atoms are arranged in an orderly way. The nuclei then serve as templates for further accretion, which may be either a relatively smooth plastering of the surface by successive atom layers or a point of departure for unidirectional spikes.

Crystal Growth

Secondary Nucleation

A useful rationale for many of the observed end results of crystal growth, such as planar faces, simple face indices, beveled edges, and lack of reentrant angles, can be developed from simple geometric considerations. If an accretion unit seeks an attachment site of maximum coordination on the surface of a nucleus, the unit provides the greatest reduction in the free energy of the system by its transfer to the crystalline phase.

Consider first the coordination differences for an atom on a flat surface and one against a step, as illustrated two-dimensionally in figure 7.5. Coordination in position 1 is

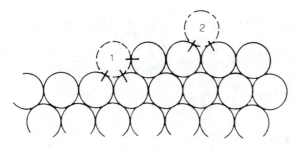

Figure 7.5
Coordination differences in accretion sites.

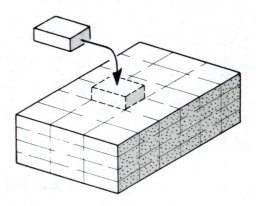

Figure 7.6
Crystal growth by secondary nucleation when no partially formed layers are present.

higher than in position 2, so the free energy drop will be greater there and accretion sites against a step are favored. In consequence, atoms joining a crystalline solid will be preferentially added in extension of developing layers, thus leading to the formation of planar faces with simple indices such as (100), (010), and (001). In three dimensions, the same arguments lead to the high probability of a linear advance of a growth step, since any reentrants in it provide increased coordination and would be preferentially filled. Because most minerals are not monatomic solids, it is unlikely that a growth step is as simple as has been illustrated. However, the mechanism for accretionary growth must, in general, be the same.

The initiation of a new layer may take place by **secondary nucleation** of new growth units upon a completed face. The greatest decreases in free energy, and hence most favored attachment sites, are away from an edge and on the face underlain by ions in closest spacing (see figure 7.6).

For discussion, assume ideal conditions exist with ions in a source uniformly distributed around a nucleus. Assume also that their rate of infall is sufficiently slow that each ion can find the attachment position providing the greatest free energy decrease for the system. The coordination at different kinds of possible attachment sites for ions on an existing crystal of a binary compound is shown in figure 7.7. In this case, the maximum coordination is four (interior ions), and accreting ions can attain coordinations of three in reentrant positions, two by attachment on faces, and only one on corners.

Because the formation of a chemical bond represents the transfer of free energy to bound energy in the system, the greatest reduction of free energy will occur when sites of higher coordination are occupied. The total energy associated with each attaching ion in figure 7.7a is represented by a large circle in figure 7.7b. The loss of free energy for different attachment sites is shown as the shaded portion. This illustration suggests that the preferred order of attachment for an ion as a function of free-energy drop is as follows:

1. In a reentrant with a step height greater than one translation

2. In a reentrant having a one-translation step height
3. On the face (and away from an edge) of the longer translation
4. On the face (and away from an edge) of the shorter translation
5. On a corner

This concept of preferential attachment as a function of the relative decrease in free energy can be extended to formulate the following rules which are consistent with the observed features of real crystals:

1. The preferential filling of reentrants leads to the development of planar crystal faces.
2. The growth rate of a crystal is reciprocal to the translation distances. Hence the slower-growing face becomes increasingly prominent, and the final shape or end form tends to be reciprocal to the shape of the structural cell.

Edges and corners have specific effects on the laying down of successive sheets of atoms. Consider the situations shown in figure 7.8, where growth units are simplified to cells and the influence of neighbors on a near-edge accretion unit is indicated by a circle. The shaded area of the circle represents the free-energy drop caused by the attachment of new units, shown as a dashed line. The situation shown in figure 7.8a is slightly favored over that in figure 7.8b. In this instance, a one-unit setback is energetically favored and succeeding layers will repeat the pattern, thus leading to a face or beveled edge of simple index (11). A similar preference pattern leads to the development of beveled corners.

The energy of a particular site as represented by the coordination that can be attained is not influenced solely by its nearest neighbors; those successively more distant contribute to some degree. The effects of second, third, and higher-order neighbors are much less important than those of first neighbors, but they cannot be disregarded.

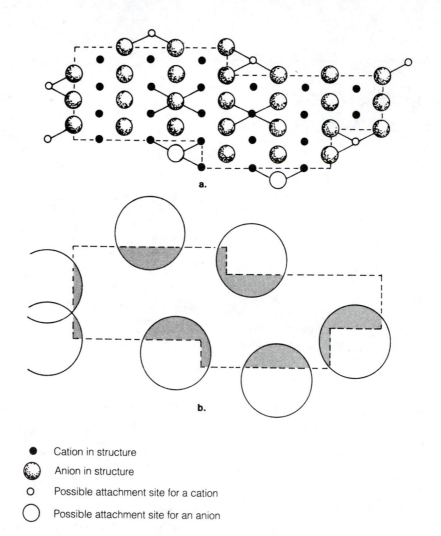

a.

b.

● Cation in structure

◍ Anion in structure

○ Possible attachment site for a cation

○ Possible attachment site for an anion

Figure 7.7
Coordination of accretion sites in a binary compound. The reduction in free energy for different sites is indicated by the shading in (b).

The amount of setback leading to edge and corner beveling depends on a variety of factors, such as field strengths, distribution of neighbors, and site geometry, which can be generalized as a volume of influence on atoms or growth units. As this volume increases, the amount of setback should increase from none to bevels of increasing, but always small, index. It is unlikely that bevel indices will be large simply because of the limited real dimensions of volumes of influence, but they should, and do, have simple small-number indices $(hk0)$, $(h0l)$, and $(0kl)$, where h, k, and l are usually 1, 2, or 3. Corner bevels also have simple (hkl) indices.

Steps against which growth can occur can also be provided by structural discontinuities known as **dislocations.** These are geometrically distinguished as **edge dislocations,** in which the discontinuity is across a plane, and **spiral dislocations,** in which the discontinuity has the form of a screw axis. Both are the result of misplacement of atoms in consequence of included impurities, mechanical stress, or growth discontinuities.

In edge dislocation, differential compression of a crystal may dislocate atoms across a glide plane, as shown

in figure 7.9a. The same number of atoms are present over the distances d and d', and some discontinuity of the structure must exist across glide plane g. Such an edge dislocation in a growing crystal would provide a preferred locus for growth. Dislocations resulting from the incorporation of an impurity atom in a structure are shown in figure 7.9b.

In spiral dislocation, the dislocation has a rotary component, as shown in figure 7.10. The dislocation step becomes the locus for accretion of growth units, and spiral sheets form as indicated by the arrow. Several differently oriented spiral dislocations may be simultaneously present in a structure, and each may provide a step against which growth units find maximum coordination. Figure 7.11 shows three mutually perpendicular spiral dislocations on a cube.

In polytypic minerals the regular repeat of complex layers many atoms thick is best explained by a spiral growth mechanism. Chance variations or imperfections on the face of the dislocation step are preserved as the ramp spirals outward from the nucleus and result in the regular repetition of the anomalous feature along the dislocation axis, as suggested by the sailboat motif in figure 7.12. This

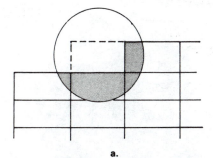

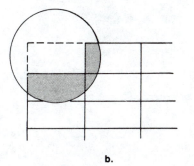

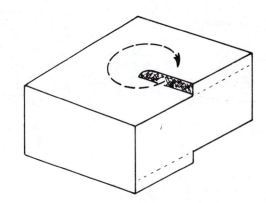

Figure 7.10
Spiral dislocation.

Figure 7.8
Edge effects on coordination. Newly attached units are shown dashed, and the free energy drop due to their attachment is indicated by shading.

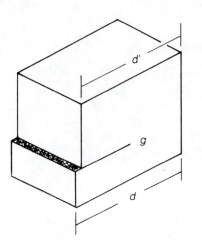

a.

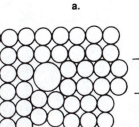

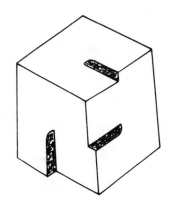

Figure 7.11
Spiral dislocations on a cube.

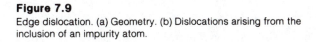

b.

Figure 7.9
Edge dislocation. (a) Geometry. (b) Dislocations arising from the inclusion of an impurity atom.

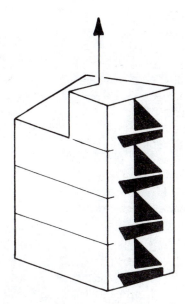

Figure 7.12
Repetition of a complex pattern (represented by a sailboat) along the axis of a spiral dislocation.

Figure 7.13
Dendritic growth.

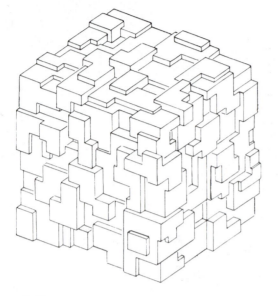

Figure 7.14
Lineage structure.

Figure 7.15
Galena cubes showing lineage structure. (Courtesy J. S. Hudnall. Mineralogy Museum, University of Kentucky. Photograph by T. A. Moore.)

mechanism provides a built-in memory that allows the crystal to exactly repeat complex atomic sequences along the growth axis. In some micas, for example, several dozen atomic layers may constitute a repeating motif along the *c*-axis. Such repetition is inexplicable by sequential single-layer additions to a (001) surface, but it is to be expected if the repeat is the same height as the dislocation step.

Crystal dislocations can sometimes be seen with the naked eye, but they generally require the use of a high-powered microscope. Etching of the crystal, which selectively dissolves the more soluble dislocation trace, or **decoration,** whereby other crystals are caused to nucleate and grow along the dislocation, enhances the visibility of the dislocation.

Dendritic Crystallization

The classic means of crystal growth is through the addition of successively younger layers of material, in which additions to the surface of a crystal are controlled by the steps formed in secondary nucleation or a dislocation. Another growth mechanism, analogous to framing a house, also occurs, however. This special mechanism for rapid crystal growth, which is probably more common than generally thought, is called **dendritic crystallization.** The term

comes from the rootlike habit of the crystals so formed, such as snowflakes or the dendrites of manganese hydrous oxides that are often mistaken for plant fossils. Dendritic forms themselves, shown diagrammatically in figure 7.13, are relatively rare, but because continuing growth fills in the interstices of the dendrite and removes the superficial evidence of its growth mechanism, this is not unexpected. The slight misalignments of the open dendritic lattice-work may be preserved, however, in the filled-in solid as dislocations and may give rise to a macroscopically recognizable **lineage structure,** as shown in figures 7.14 and 7.15.

The particularly favorable mechanism for dendritic growth, which suggests that this means of crystallization

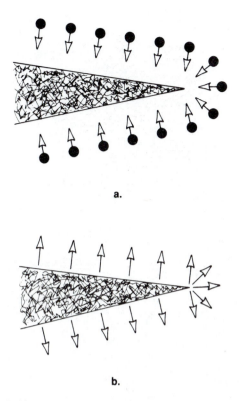

a.

b.

Figure 7.16
Growth conditions at a point. (a) Enhanced pointward accretion of a dendritic spike. (b) Enhanced heat loss from a point.

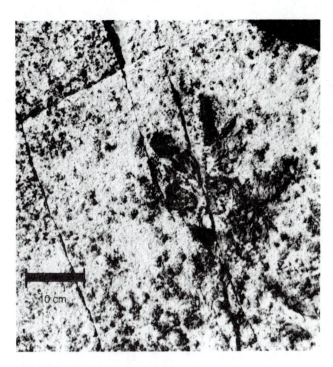

Figure 7.17
Large black dendritic augite crystals developed from a few nuclei in a gabbroic magma. The melt was thus locally enriched in the feldspar component and fine-grained plagioclase (white) crystallized from the supersaturated residuum. Nahant Gabbro, Nahant, Massachusetts. (Photograph by W. H. Dennen.)

will be exploited whenever possible, arises from the specialized chemical and thermal environment illustrated in figure 7.16a. Uniformly spaced atoms (of equal concentration) are located along a line of equal chemical potential with respect to the crystal surface and move at equal rates down the concentration gradient toward the crystal surface. Because a significantly greater number of atoms arrive at the point of the crystal per unit time, growth is dominantly pointward and at rates that may move the point continuously into regions not yet depleted in growth units. Similarly, the growth rate of a sharp point is enhanced by the more efficient removal of latent heat of crystallization from its immediate vicinity, as shown in figure 7.16b.

The history of crystal development through a dendritic stage would thus roughly follow this sequence: (1) nucleation and development of a favored accretion direction, (2) rapid growth of rodlike protrusions coupled with the development of crytallographically oriented side branches, and (3) backfilling of interstices. This last change leads to minor mismatching of the structure along the internal joins (dislocations) and the physical entrapment or inclusion of foreign matter.

Natural examples of dendrites include snowflakes, some gold and copper crystals, and deposits of hydrous manganese oxides found along rock fractures. Lesser known examples would include branching clusters of ferromagnesian silicates, such as olivine, amphibole, and pyroxene (see figure 7.17).

Crystal Faces

Certain observable features of crystals always arise regardless of the actual mechanism of crystal growth. So long as there is a contrast in composition across the interface and the face can develop without interference, the face will be planar. Rounded crystal faces develop only when minerals crystallize or recrystallize in environments very similar chemically to themselves. Some faces that seem to be rounded are actually made up of minute planar elements. The direction of crystal growth is normal to the accreting surface; that is, new faces parallel the old. This can sometimes be directly observed because the entrapment of impurities on early faces produces phantom crystals or zoning within a specimen. *Crystal shapes tend to be inverse to the shapes of their constituent unit cells and rapidly accreting faces tend to disappear.* Figure 7.18 shows that if the various faces grow at the rates indicated by the arrows the slowly accreting face *C* will soon predominate and faces *A* and *B* will become relatively less prominent. Appropriately oriented faces may thus be eliminated by their more rapid growth, whereas slow accretion may cause faces to appear. Figure 7.19 illustrates these phenomena. The successive faces are drawn for equal elapsed times. In figure 7.19a, the face 2–3 is growing much more rapidly than either 1–2 or 3–4, and eventually this face is lost at 0. In figure 7.19b, a possible face 2–3 begins to show at 0 when its growth rate falls below that of the adjacent faces. Eventually, this slow-growing face

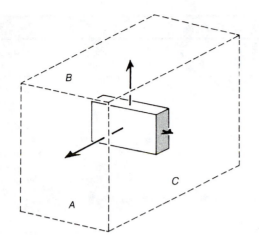

Figure 7.18
Increase in prominence of a slowly growing face. Rates of growth are proportional to the length of the arrows. Early stage solid, later stage dashed.

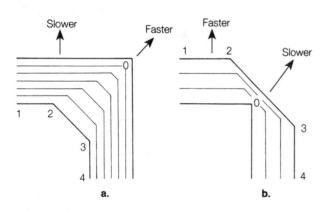

a.   **b.**

Figure 7.19
Effect of growth rates of adjacent faces on endform. Faces drawn for equal times.

will become very prominent on the crystal. The actual number of faces displayed by a crystal can thus be sharply limited by the interrelationship of growth rates of adjacent faces.

A simple analogy of the reverse situation is found in lawn mowing. Cuts of regular width on the sides of an irregular polygon (of more than three sides) eventually results in the elimination of the shortest side.

Not uncommonly an accreting crystal surface oscillates between its own orientation and that of an adjacent face. Such a face incorporates minute repetitions of the intersections of the adjacent faces into its own surface, and it appears to be ruled or scratched in grooves or **striations** that are always parallel to a crystal edge, as shown in figure 7.20. Some examples of striated crystals are sketched in figure 7.21. On the basis of their striated faces, the apparent 6-fold axis of quartz is demonstrated to be 3-fold, and the cube of pyrite is shown to lack a 4-fold

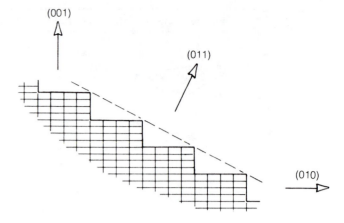

Figure 7.20
Striated face (011) as average of adjacent faces.

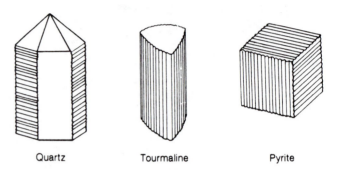

Quartz   Tourmaline   Pyrite

Figure 7.21
Examples of striated crystal faces.

axis. Remember, symmetrically equivalent faces must be identical in chemical and physical properties. It should also be noted that the striations described here are distinct in form from those that originate from parallel, multiple, or polysynthetic twins, which can be seen sometimes as shining lamellae on cleaved crystal surfaces.

Crystal Habit

Crystal **habit,** or its characteristic appearance, is determined by the dominant crystal forms developed and represents both the peculiarities of the mineral's structure and the influence of the environment during crystal growth. Habits can be broadly classified on the basis of the axial ratio using a_1/a_2 ($=$ unity) for the isometric system, c/a for hexagonal crystals, and $2c/(a + b)$ or $2c/(a_1 + a_2)$ for the other systems. Minerals with axial ratios near 1 are equant, stubby, or blocky; those with ratios <1 are prismatic or rodlike; and those with ratios >1 are tabular or platy. Some examples are listed in table 7.1.

In general, crystal habits are reciprocal to the shape of their unit cell. Often, the direct and essentially unmodified effect of a mineral's internal structure can be observed in its external form. A small sampling of mineral habits that persist in the face of wide variations in the environment of growth is given in table 7.2. However, two

Table 7.1   Relationship between Axial Ratio and Habit for a Variety of Minerals

Mineral	Crystal System	a	b (or a_2)	c	Axial Ratio	Habit
Magnetite	Isometric	8.39	—	—	1.0	Equant
Graphite	Hexagonal	2.46	—	6.71	2.7	Platy
Schorl	Hexagonal	16.03	—	7.15	0.4	Prismatic
Rutile	Tetragonal	4.59	4.59	2.96	0.6	Prismatic
Leucite	Tetragonal	13.07	13.07	13.74	1.1	Equant
Stibnite	Orthorhombic	11.23	11.31	3.84	0.3	Prismatic
Bournonite	Orthorhombic	8.16	8.75	7.81	0.9	Blocky
Biotite	Monoclinic	5.31	9.23	20.36	1.7	Platy
Orthoclase	Monoclinic	2.56	13.00	7.19	0.9	Blocky
Kyanite	Triclinic	7.12	11.82	5.56	0.7	Bladed
Wollastonite	Triclinic	7.94	7.32	7.07	0.9	Blocky

Table 7.2   Relation of Structure to Habit

Structural Type	Habit	Example
Sheets of linked Si-O tetrahedra	Platy	Micas
Mo-S sheets	Platy	Molybdenite
Sb-S chains	Bladed	Stibnite
Si-O tetrahedral chains	Rodlike	Tremolite
Si-O tetrahedral network	Blocky	Orthoclase
Si-O isolated tetrahedral units	Equidimensional	Garnet

isostructural minerals can also show different habits if their cell dimensions or radius ratios differ. Figure 7.22 shows the variations for some cubic crystals.

Inherent structural differences affect not only crystal habit but also the environment of growth. In general, the mineral preserves only those elements of symmetry common to both the crystal and the growth medium, and selective adsorption or activity differences between different faces, moving nutrients, attachment, or mutual interference may be reflected in the resultant habit. Figures 7.23 and 7.24 illustrate some examples of these situations. Obviously, each environment impresses its own characteristics on a growing crystal. Reciprocally, each mineral specimen, when fully interpreted, provides a key for unlocking the history of the mineral. The difficulty of such interpretations lies in the incompleteness of our present knowledge of geologic environments and growth mechanisms.

A large number of adjectives are commonly used to describe the appearance of an individual mineral or a crystalline aggregate. A habitually assumed form or habit is, of course, not a constant mineral attribute, but it is so often associated with a particular mineral species that it is one of the more reliable means of mineral identification. Most practicing mineralogists tend to make tentative identifications based on habit and color—two inconstant mineral properties!

A glossary of the terms used in describing habit together with some sketches of these habits are shown in table 7.3.

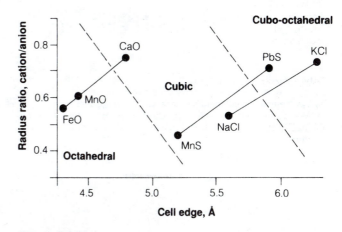

Figure 7.22
Variation in crystal habit with cell size and radius ratio. (Redrawn from Kostov, 1968.)

Crystal Imperfection

A perfect organization of atoms or ions into a crystal structure is an exception rather than a rule of crystal growth. Minor deviations from the ideal case can be seen in the most cursory examination of a crystal. On further investigation, more and more evidence of imperfection is usually found. Some examples of deviations from the ideal are (1) slight nonparallelism of paired faces, (2) the presence of inclusions of various kinds, (3) zoning, (4) lineage structure, (5) internal flaws, (6) ionic conduction, and (7) discrepancies between measured and computed strength. Such deviations from perfection should not cause dismay, however; rather, one should be surprised that crystal structures are so homogeneous. After all, a 1×4 mm crystal face when magnified until the constituent atoms have diameters of golf balls would be the size of Tennessee!

Departures from ideal crystallinity are found at different scales. At the atomic level, one kind of atom may substitute for another, an atom may be omitted, or nonformulary atoms may occupy interstitial positions. Such atomic phenomena necessarily alter the properties of

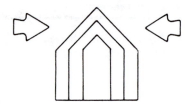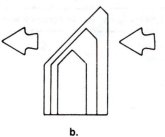

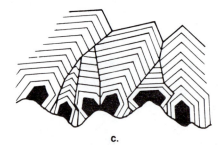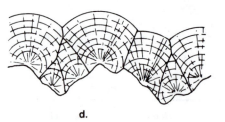

Figure 7.23

Effects of environment and interference on crystal habit.
(a) Symmetrical environment, nutrients uniformly supplied.
(b) Asymmetrical environment, nutrient solution moving from right to left. (c) Effects of crystal orientation and interference. (d) Mutual interference during spherulitic growth.

Figure 7.24

Misshapen quartz crystals deposited from a solution flowing from right to left. Maximum growth on upstream side. (Photograph by Melissa Steven.)

| Table 7.3 | Glossary of Terms Used to Describe Habit |

Terms Describing Single Crystals

Description	Appearance
Capillary, filiform acicular—hairlike, threadlike, or needlelike crystals.	
Stout or stubby—usually applied to pyramidally terminated crystals whose *c*-axis is short compared with its other axes.	
Bladed—crystals in elongate, flattened blades.	
Blocky—brick-shaped.	
Tabular, lamellar—booklike in shape.	
Columnar—columnlike crystals.	
Foliated, micaceous—easily separated into sheets or leaves, micalike.	
Plumose—featherlike arrangement of fine scales.	
Geometrical terms—various geometrical terms are used as applicable (e.g., cubic, tetrahedral, octahedral, prismatic, dodecahedral, scalenohedral).	

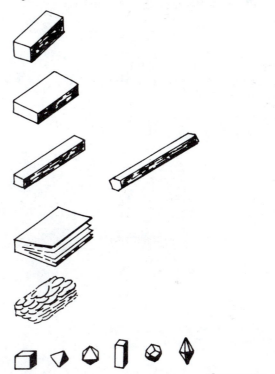

(Continued on next page)

a crystal and are especially important in modifying bulk density and those properties that are dependent on electrons.

Any variation from ideal crystallinity that involves atomic or ionic interchange alters the configuration of electrons at the interchange site. Properties such as electrical conductivity, which depends on electron flow, or color, which results from the interaction of electromagnetic waves and electrons, can be markedly changed. Semiconductor materials owe their remarkable properties of controllable electron flow and energy conversion to this kind of crystal imperfection. An extensive technology has developed to provide controlled additions of proxy atoms (dopants) to various semiconducting crystals.

A further kind of conduction involving movement of the ions themselves also occurs. When structures containing defects are under the influence of an external electric field, migration of interstitial ions and vacant sites (the latter equivalent to ionic movement in the opposite direction) occur, accounting for the conductivity often observed in some ionic crystals.

Larger-scale crystal imperfections can be seen in most crystals with the naked eye, a hand lens, or a microscope. Many crystals contain inclusions, either trapped during growth or exsolved later. Original inclusions may have random shapes and distributions or their orientations may be crystallographically controlled. Exsolved material

Table 7.3   *Continued*

Terms Describing Crystal Groups and Mineral Aggregates

Description	*Appearance*
Columnar—an aggregate of columnlike individuals.	
Divergent, radiated, stellated—individuals arranged in fan-shaped groups or rosettes.	
Bladed—an aggregate of bladed individuals.	
Colloform (botryoidal, reniform, mammillary, globular)—radiating individuals forming spherical or hemispherical groups. The various terms have been used to designate the extent and radius of the hemispherical surfaces developed. Colloform includes all other terms.	
Fibrous—an aggregate of capillary or filiform individuals.	
Reticulated—slender crystals arranged in a latticelike array.	
Dendritic—treelike or mosslike form.	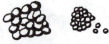
Pisolitic, oölitic—composed of rounded masses respectively the size of peas or BB shot.	
Granular—an aggregate of mineral grains.	
Banded—bands or layers of different color and/or texture.	
Massive—a compact aggregate without distinctive form.	
Concentric—onionlike banding.	

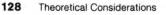

commonly reflects the symmetry of the host and is arranged on or along some crystallographic line or plane of simple index.

On the crystal surface, nonideality is evident in the fact that many crystal faces are not perfect geometrical planes. One type of plane takes the form of a mosaic of slightly disoriented blocks and probably reflects filled-in dendritic growth. Atoms in perfect registry can be traced from the center of the crystal along any branch to the surface, but adjacent branches are not in registry. Cubic crystals of galena often display this phenomenon. Other irregularities are seen in the presence of hopper-shaped crystals resulting from accelerated growth on edges, low mounds marking the surface position of a spiral dislocation, or pits, usually triangular, caused by selective solution.

Another external deviation from ideality is the presence of faces having large indices, which are nearly but not quite in their proper crystallographic location. Such **vicinal** faces may be due to dendritic growth, dislocations, or possibly other causes.

Intergrowths

In the cramped quarters in which most minerals grow, two or more individuals often manage to share the same limited volume. For example, often two crystals nucleate independently very close to one another. Further, one crystal may use another as a substrate or seed and grow upon it. With continued growth, the two crystals may develop a mutual boundary, or intergrow, or one may completely enclose the other.

If the two crystals are of the same species, the most stable configuration will be that of a twin (see chapter 5), but mineral grains of the same species can also intergrow or enclose one another without having a twinned relationship. Two individuals of different species cannot twin (by definition). The structural and geometrical relationships of inclusions of one mineral in another, however, are often more than accidental. This is probably because the anion-dominated structures of many minerals are grossly similar in packing, and pseudocontinuity from one mineral to another is often possible.

Foreign matter may be incorporated into a crystal during its growth. Such inclusions may be uniformly distributed through the host, but they are often oriented with respect to some favored crystallographic direction or restricted to some former growth surface. The latter situation suggests variation in the composition of the nutrients over time. Common results of such included matter are the phantom crystals of quartz with internal planes marked by fluid inclusions, the black cross found in andalusite (variety chiastolite) resulting from the pushing aside of carbon particles during growth, and the coloration of nominally light-colored or clear minerals.

Intergrowths, usually well oriented, also commonly develop because of the separation of solid phases by exsolution (see chapter 5). Common results of this process are the intergrowth of potassic and sodic alkali feldspar called perthite; the presence of cubanite, $CuFe_2S_3$, in chalcopyrite, $CuFeS_2$; the subopalescence or **schiller** exhibited by some pyroxene crystals containing oriented inclusions of rutile, TiO_2; and the segregation of pyrrhotite, $Fe_{1-x}S$ into domains richer and poorer in iron.

Apart from their complex geometric relationships, intergrowths and inclusions provide useful clues to the environment in which the particular assemblage formed. Since they and their host have a mutual history, the growth conditions must have been in the region where stabilities of both phases overlap.

Concluding Remarks

Crystallization is the ordering of such units of matter as atoms, ions, radicals, or molecules into a regular structural array. It begins when random motions and interatomic forces bring a sufficient number of growth units into a configuration that is energetically favored over the dissociative tendencies of the system. Crystals then grow by accretion of growth units on a nucleus, which serves as a template for their arrangement.

Growth units may be added to the surface either as successive coatings (secondary nucleation) or as spikelike extensions (dendritic growth) whose interstices are later filled.

The interaction of crystal-forming forces and the environment in which the crystal develops leads to characteristic end forms or habits. Their interpretation is one of the more provocative aspects of mineral study.

Variables such as different rates of growth, accidents during the crystallization process, mutual interference, and asymmetry of the environment lead to imperfections on both the atomic and macroscopic scales. These are represented by such phenomena as asymmetric crystals, solid solution, vicinal faces, exsolved phases, dislocations, color centers, and intergrowths.

Suggested Readings

Bollman, W. A. 1970. *Crystal Defects and Crystalline Interfaces.* Berlin: Springer-Verlag. Excellent, often mathematical, description of crystal imperfections. Well illustrated and includes Moire models.

Buckley, H. E. 1951. *Crystal Growth.* New York: John Wiley & Sons. Compilation of various theories and observations regarding the growth and habit of crystals. Old but valid.

Bunn, C. 1964. *Crystals, Their Role in Nature and Science.* New York: Academic. A clear, interesting, and reliable book for science students and interested laypersons.

Burke, J. 1965. *The Kinetics of Phase Transformations in Metals*. Oxford: Pergamon. An excellent presentation of the theory and mechanisms of nucleation and growth of crystalline materials.

Dekeyser, W., and Amelinckx, S. 1955. *Les Dislocations et la Croissance des Cristaux*. Paris: Masson et Cie. The various mechanisms of crystal growth are reviewed.

Holden, A., and Singer, P. 1960. *Crystals and Crystal Growing*. New York: Anchor Books. An elementary book about the nature of crystals and the means for growing them.

Putnis, A., and McConnell, J. D. C. 1980. *Principles of Mineral Behaviour*. New York: Elsevier. Atomistic and microstructural descriptions of mineralogic processes and reactions.

Geological Framework for Mineral Study

Minerals are normally found as members of associations or assemblages of several different species. For example, coarse-grained, light-colored igneous rocks typically contain quartz, alkali feldspar and plagioclase, plus possibly micas or an amphibole such as hornblende. Dark, fine-grained, igneous rocks, on the other hand, often have a mineral assemblage dominated by plagioclase and pyroxene, with the possible addition of olivine and magnetite. Mineral assemblages such as these that are repeated in time and space argue for the establishment of equilibria within groups of minerals. These equilibria in turn represent common origins for the minerals in the assemblage. Thus, it is often possible to characterize a geologic process from its resultant mineral suite or to anticipate the mineral suite from the geological environment. A thorough coverage of geological processes is beyond the scope of this book, but certain terms have genetic significance and are used in descriptions of individual minerals (see part 3). Therefore, in this chapter we will briefly review some typical materials, processes, and environments.

In general, no rock or mineral type has been excessively concentrated in the earth's crust over the whole of geologic time. It follows that the processes of geology are essentially cyclic. Such a cycle is illustrated in figure 8.1, in which the principal rock types are boxed and geological processes are shown on the arrows. This **rock cycle** provides a geological framework that controls the form and stabilities of minerals. Because of the bypasses and subcycles in the rock cycle, which can be seen even in this simplified representation, it may be useful to visualize the system as a network in which the incoming and outgoing matter and energy at any point are in balance. Broadly speaking, the rate of generation of the various products is equal to the rate at which they are removed or changed. Local situations in time or space, however, may dictate a net accumulation, and some of the "solid wastes" so formed, such as limestone or coal, represent important natural resources.

Igneous Rocks

Magma and Its Products

Portions of the lower crust and upper mantle of the earth can be partially melted if sufficient heat is accumulated. The resultant masses of liquid material, or **magma,** can never be directly observed, although a low pressure equivalent—**lava**—is often seen. Magma is a complex melt solution, and its nature can be inferred by studying the igneous rocks that form when it cools and by laboratory experiment.

Silicate magmas are composed principally of silicon and oxygen with varying amounts of the other elements in approximately the proportions of their crustal abundances (see table 1.1). The actual composition of the magma depends on the source material, the physical conditions of melting, and the stage of magmatic evolution. Magmas also contain volatile species. These are dominated by water but also include CO, CO_2, SO_2, H_2S, and the halogens. The water content, although small, has a very important impact on the cooling history of the magma. Water dissolved in a magma depresses the freezing or crystallization points of the various mineral products (see chapter 6). It also reduces magma viscosity and acts as a carrier of certain elements that do not easily enter the structures of the principal mineral products. The effect of the volatile constituents can be observed in (1) the alteration of previously formed minerals both within the cooled mass and at its borders; (2) the presence of gas bubbles, or **vesicles,** in volcanic rocks; (3) the development of localized accumulations of very coarse-grained rocks; or (4) the presence of late-stage veins. The volatile components can be directly observed as the gaseous emanations and sublimates from volcanoes and as the almost ubiquitous fluid inclusions in igneous minerals.

Magma is the starting material for all of the solid materials of the earth's crust. The processes whereby it separates into a relatively few compounds are of special interest to the geologist. In order to understand the differentiation of a magma, it is necessary to know something of the mineral compounds that may form and the way in which their stability is related to temperature and pressure. The potential intercombinations of the principal components of a magma are obviously very large, and no complete quantitative description of these interactions has been developed. Nevertheless, thoughtful observation of the cooled products, extrapolation from experiments on more limited systems, and application of the principles of physical chemistry have provided geologists with a semiquantitative understanding.

Oxygen and silicon dominate the magma composition and readily combine into chemically stable arrangements based on the silica tetrahedron. The strong chemical affinity of silicon and oxygen atoms allows them to form Si–O linkages, or polymers, in a magma in spite of high temperatures and consequent thermal motion. The polymerization of the SiO_4 tetrahedra into chains and sheets results in structures not far removed from those of solid silicates and thus increases viscosity. The increased viscosity reduces both the magma's ability to flow and the mobility of its constituent chemical species. Water and other volatiles can increase the fluidity somewhat by their tendency to break the polymer linkage. Even so, the very high viscosity of silicate magma is apparent in the rather small size of the individual crystals that form, implying that diffusion of only a few centimeters is possible in the perhaps million years needed to cool the mass.

Igneous crystallization takes place in response to loss of heat and consequent decrease in thermal motion, which allows bonds to form at a greater rate than that at which they are broken. The individual crystals that form do not

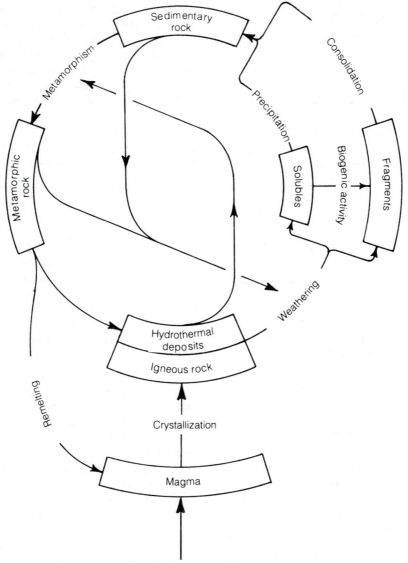

Figure 8.1

The rock cycle. The principal products are boxed and the processes are on the arrows.

usually attain a regular geometric external crystal form but are blunted and indented by interference with one another's growth. Earlier-formed minerals should show better developed crystal outlines than later-formed minerals because of the lower number of solids per unit volume. Resorption, however, often destroys this evidence of relative age. Equidimensional, columnar, tabular, and platy habits are common and express the tendency of minerals to grow preferentially in certain crystallographic directions.

The size of all crystals of the same species should be approximately the same under equilibrium conditions. Variations in size are thus clues to variations in cooling history. Generally speaking, the size of each crystal is dictated by the ability of atoms to move to it and become incorporated in its structure. This ability, in turn, will depend on the time available, the abundance of nutrient elements, the temperature, and the melt viscosity.

The discussions of nucleation and growth of crystals in chapter 7 showed that rapid cooling tends to increase the numbers of crystals while lowering growth rates. Slow cooling, on the other hand, promotes crystal growth at the expense of crystal abundance. Thus, igneous rocks that crystallized at depth typically will be coarser-grained than those that crystallized near or at the earth's surface. The presence of two distinct grain sizes in an igneous rock argues for two stages of crystallization.

Any fluid motions within the cooling magma tend to orient mineral grains that are not equidimensional. A record of these movements may be preserved as an alignment of elongate grains or as a common orientation of platy grains. The degree of orientation can, of course, range from

Table 8.1 Terminology for Describing Igneous Textures

Size Terms

Glassy	No grains
Fine-grained	Mean grain size < 1 mm
Medium-grained	Mean grain size 1–5 mm
Coarse-grained	Mean grain size > 1 cm
Porphyritic	Some crystals are distinctly larger than the average

Shape Terms

Euhedral	Crystals show all characteristic faces
Subhedral	Crystals show some characteristic faces
Anhedral	Characteristic crystal faces missing
Equant	Grains are equidimensional
Needles, rods	Grains have one dominant dimension
Plates	Grains have two dominant dimensions
Blades	Grains have three distinct but unequal dimensions

Arrangement Terms

Homogeneous	Random disposition of grains
Foliate	Parallel arrangement of tabular or platy grains
Lineate	Parallel arrangement of elongate grains

the barest suggestion to a highly ordered arrangement. Not uncommonly, minerals are both oriented and clustered, and if these groupings are drawn out by magma flow, the final rock may have a streaky appearance.

The size, shape, orientation, and arrangement of grains in a rock make up its **texture.** Size is generally described by a mean grain diameter, shape by the ratio of two (or more) dimensions, and arrangement by the degree of ordering in the position of the grains. Igneous rock textures are generally described using the terms in table 8.1.

Crystallization of Igneous Rocks

Each mineral species has a characteristic melting temperature determined by its constituent elements and their bonding. Thus, crystallization in igneous systems tends to take place in an orderly fashion with the mineral species appearing in inverse order to their melting temperatures. Consider a magma having an initial temperature of 1,200° C made up of the principal elements O, Si, Al, Ca, Mg, Fe, Na, and K, in that order of abundance. A few percent of H_2O and the rest of the elements in the periodic table may also be present in minute amounts. The greater proportion of the silicon and aluminum will probably be combined in simple $(SiO_4)^{4-}$ and $(AlO_4)^{5-}$ radicals. With falling temperature, the first minerals to crystallize are a magnesian olivine, $Mg_2(SiO_4)$, and a calcic plagioclase, $Ca(Al_2Si_2O_8)$. These early-formed crystals become unstable as temperature continues to fall and react with the magma to form new species stable under the new conditions. At lower temperature, plagioclase is able to accommodate increased amounts of sodium and silicon at the expense of calcium and aluminum (see figure 6.10). Thus,

later-formed and lower-temperature plagioclase are more albitic. The early-formed Mg-rich (forsteritic) olivine acts in a fashion similar to plagioclase and becomes more Fe-rich (fayalitic) at lower temperatures if allowed to react with the magma. Such **continuous** reactions are accomplished without significant structural changes in the minerals and are common to any of the solid-solution mineral series of igneous systems.

An analogy for the continuous reaction is a barber shop in which the fixed chairs represent the structural positions in the mineral and the transient clientele the compositional changes.

If the continuous reactions take place slowly while the crystals remain in constant contact with the melt, the final crystals will be compositionally homogeneous. In contrast, if the crystals grow rapidly enough or are displaced from that portion of the liquid in which they began to crystallize, they may be compositionally zoned. Plagioclase crystals are commonly zoned with anorthitic cores and more albitic rims (figure 8.2).

Crystallization, of course, removes the components of the solids from the magma and thus increases the concentration of residual elements in the melt. The chemical evolution of the liquid coupled with falling temperatures forces some minerals to the limit of their stabilities. Thus, even the more iron-rich olivines, formed by continuous reactions with the melt, may become unstable at lower temperatures. They then react with the more siliceous magma according to a reaction of the type

$$\underset{\text{solid}}{(Mg,Fe)_2(SiO_4)} \quad + \quad \underset{\substack{\text{from the} \\ \text{melt}}}{SiO_2} \quad = \quad \underset{\text{solid}}{2(Mg,Fe)(SiO_3)}$$

to form a new mineral, a pyroxene, that is stable. Such a reaction will take place at a peritectic (see figures 6.9 and 6.14) and is described as **discontinuous.** In igneous rocks, evidence for such a reaction commonly appears as rims of pyroxene around olivine grains.

As the magma continues to cool and evolve, the chemical constituents of the higher-temperature minerals are distributed to the solid phases. Components such as SiO_2, Na_2O, K_2O, and H_2O are, meanwhile, concentrated in the residual liquids where they become available for late-forming, lower-temperature minerals. At lower temperatures, the pyroxene may become unstable and transform by a similar discontinuous reaction into an amphibole, typically hornblende. Further reactions with the magma can generate biotite. The final consolidation of the major constituents may allow for the formation of muscovite, alkali feldspar, and quartz from the liquid residuum.

The minerals generated by the continuous and discontinuous reactions described above are the principal products from the cooling of a magma. They are the essential constituents of igneous rocks. Their equilibrium

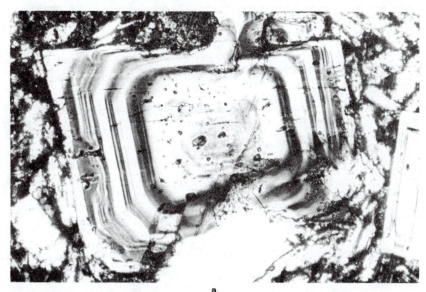

a.

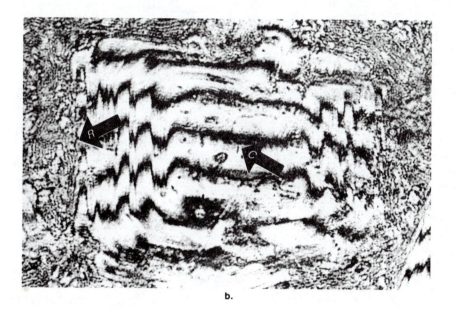

b.

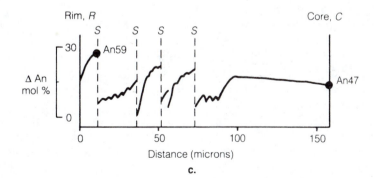

c.

Figure 8.2

Zoned plagioclase crystal. (a) Crossed-nicols transmitted light photomicrograph of a normal oscillatory-zoned plagioclase phenocryst from Volcan Popocatepetl, Mexico. (b) Narrow-fringe laser interferogram of the same crystal in (a); *R* is the rim and *C* is the core. Each fringe can be considered as a graph of mineral composition, providing an accurate visual representation of zoning profiles. One full shift of a fringe (equal to the distance between fringes) is equivalent to a change in composition of 30% An. Conventionally, interferograms are viewed with An% increasing upwards. (c) Interpreted zoning profile from the core *C* to the rim *R* of the interferogram in (b). This profile shows that each major zone begins with a sharp increase in An% at a resorption surface (*S*) followed by a steep decrease with minor oscillations. Baseline compositions can be determined optically or by microprobe analysis. (Sample supplied by A. Kolisnik. Laser Interference Microscope patented by T. H. Pearce and Queen's University, Laser Research Laboratory, Queen's University, Kingston, Ontario, Canada.)

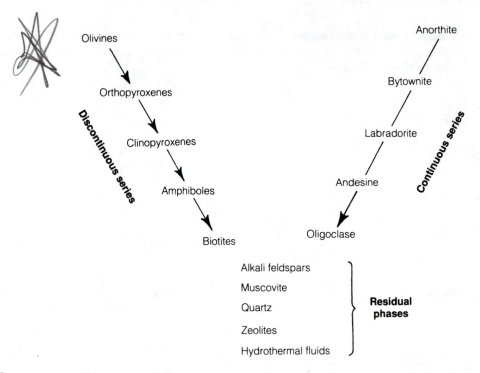

Olivines

Orthopyroxenes

Clinopyroxenes

Amphiboles

Biotites

Anorthite

Bytownite

Labradorite

Andesine

Oligoclase

Discontinuous series

Continuous series

Alkali feldspars

Muscovite

Quartz

Zeolites

Hydrothermal fluids

} **Residual phases**

Figure 8.3
Bowen's reaction series. Note that each mineral in the discontinuous series is a solid solution and thus may partake in a series of continuous reactions while it coexists with the magma.

physical-chemical relationships can be epitomized by **Bowen's reaction series** (figure 8.3), named for the Canadian petrologist N. L. Bowen, who in 1922 first described igneous fractionation and evolution in terms of mineralogic reactions. The various combinations of the minerals in the series provide a principal parameter for the classification of igneous rocks.

The cooling of a magma is not necessarily a quiet equilibrium process, and many deviations from the equilibrium course of cooling may be inferred from a careful study of the final products. Upset of the reaction sequences can occur fairly readily. For example, solids do not have the same density as liquids of similar composition, so crystals tend to float or sink in the magma. This is especially true in the early and less viscous stages. As a result, olivine is often enriched downward by crystal settling. Occasionally, this process forms concentrated crystal layers, but more often it simply depletes the upper and enriches the lower portion of a magma in iron and magnesium. The two portions continue crystallization but to different end products, for example, forming quartz-bearing upper and quartz-free lower portions.

Changes in containing pressure, liquid movement, or cooling rate also impress interpretable features upon the crystalline product. Changes in pressure affect the stabilities of minerals. In fact, they may change the order of igneous crystallization as predicted by the Bowen reaction series. For example, both the amphiboles and the micas contain water in the form of hydroxyl radicals (OH) incorporated in their structure. In a general way, the for-

mation of these hydrous minerals is promoted by high pressures on the magma coupled with decreased temperature, as shown below.

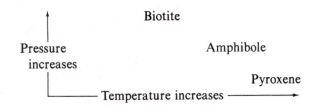

Pressure increases

Biotite

Amphibole

Pyroxene

Temperature increases ⟶

Rapid changes in pressure may cause either dissolution or quenching. Rapid movement of magma to the surface may result in fine-grained or even glassy products because of the rapid drops in both temperature and pressure. If partial crystallization has taken place, the early-formed crystals may be incorporated into the chilled phase as isolated, larger grains called **phenocrysts.** Rapid drops in pressure may allow gases held in solution to separate (exsolve) and froth, and thereby to vesiculate the rock. In contact with oxygen at the surface, these gases may undergo heat-producing (exothermic) reactions sufficient to temporarily raise the temperature of the lava, thus locally reversing the cooling effect of movement to the surface.

Many other means of magmatic fractionation that are effective in upsetting the course of equilibrium crystallization are known or suspected. As described earlier, the interaction of a solid phase with the magma may proceed in such a way that the reacting grain is overmantled or armored with the newly formed mineral, effectively separating the unreacted core from contact with the magma. The chemical interaction of magma with its containing rock walls (wall or country rock) may selectively remove some substances from them while adding others. This process can often be observed in the field in the transitional nature of igneous rock-country rock contacts. Another mechanism is the squeezing out of liquids from partially crystallized magma by movements of the enclosing rocks. Early-formed ferromagnesian minerals are thus retained, and a liquid enriched in silicon, aluminum, and alkalis is injected into the country rock, where it then crystallizes.

The formation of silicates is favored in a silicate magma. Oxygen, however, can also form compounds directly with some of the cations present, and small amounts of oxide minerals—notably magnetite, rutile, and ilmenite—are commonly found in igneous rocks. Other accessory compounds, which incorporate either negative ions other than oxygen or nonsilicate radicals, are also usually present in small amounts. Minerals such as apatite and monazite accommodate phosphorus. Sulfur is usually found in a compound with iron, such as pyrite or chalcopyrite.

Classification of Igneous Rocks

Igneous rocks can be classified according to their mode of emplacement. If the magma crystallizes at depth, the resultant mass is called a **pluton** and the rock, **plutonic.** Such rocks normally cut or intrude others. Thus, plutonic rocks are also intrusive. **Volcanic** igneous rocks can be called extrusive because they have crystallized from a magma (lava) that has been extruded onto the earth's surface. Some shallow intrusive rocks that have direct relationships to surface extrusions are also included in the volcanic designation.

Many classifications of igneous rocks are based on rock chemistry. One of the more common classifications employs the concentration of silica, SiO_2, by weight percent as follows:

Ultramafic	Mafic	Intermediate	Felsic
	45	52	66

Weight percent SiO_2

In addition, those rocks with total alkalis, $Na_2O + K_2O$, that are high relative to silica are designated **alkaline.** The term **mafic,** a contraction of *magnesium* and *ferrum,* includes minerals rich in iron and magnesium. Mafic minerals are usually dark-colored. Light-colored minerals including feldspar, feldspathoid, and quartz (silica) are referred to as **felsic.** The **Color Index** is the volumetric sum of the mafic minerals (see Box 8.1).

Chemical classifications are important because they may be the only means by which fine-grained or glassy rocks can be classified. The necessary chemical data are

not, however, generally available to most geologists. The most useful classification schemes are those based on the proportions of diagnostic, rock-forming minerals present in coarse-grained members of each rock-chemical family. One such classification is given as figure 8.5. To use such a scheme, the volume percents of each of the major mineral types are estimated and plotted on the diagram. The appropriate name is then assigned according to the average grain size. In figure 8.5 the heavy line separating the ferromagnesian minerals from the felsic minerals represents the Color Index.

Many semiquantitative field classifications of igneous rocks, such as that presented in figure 8.5, are available. The placement of boundaries is approximate in each case, and thus considerable latitude results in the naming of a rock. The classification of igneous rocks, however, is now standardized by the International Union of Geological Sciences (IUGS) Committee on Petrology.

Plutonic rocks are classified on the basis of their felsic mineral content. For rocks with a Color Index of less than 90, meaning that at least 10% felsic minerals are present, the classification is made according to the relative proportions of the mineralogic factors Q, A, P, and F, where

Q = quartz
A = alkali feldspar (sanidine, orthoclase, microcline, perthite, anorthoclase, albite)
P = plagioclase (oligoclase-anorthite), scapolite
F = feldspathoids

After estimating the volume percents of the individual felsic minerals, the proportions of Q, A, P, and F are normalized to 100%. The result is then plotted on the QAPF double triangle (see figure 8.6). Remember that quartz and feldspathoid are not stable together. Thus, Q and F will not appear simultaneously in the calculations.

A rock with a Color Index greater than 90 is referred to as **ultramafic** and can be classified according to the relative proportions of olivine, orthopyroxene, and clinopyroxene (see figure 8.7).

The fractional crystallization of the dominant components of the magma leading to the formation of igneous rocks is the principal part but not all of the complex history of this primary substance. Magma contains many constituents not previously emphasized that play a part in its cooling history. Some elements are very close in atomic parameters to the major elements in the magma and substitute readily for them in minerals. These elements are thus dispersed throughout the rock. Examples of such compatible substitutions are rubidium for potassium, gallium for aluminum, and manganese for iron.

Many elements, however, are sufficiently different from the major constituents that they are, in fact, rejected from the common mineral structures and are continuously concentrated in the residual fluids. Water, although it enters some mineral structures, is far from completely removed, and so the residual liquid becomes more and more

Box 8.1

Estimating Volume Percent of Minerals by Point Counting

If one makes the reasonable assumption that the areas of randomly oriented and dispersed minerals (or any other objects) exposed on a flat surface are directly proportional to the volumes of the minerals in the sample, the volume percent of a given mineral can be found by a procedure called **point counting.** This is done by measuring the length of a mineral along regularly spaced or randomly oriented lines as a fraction of the total length of the line. As usually performed, the sample is moved by regular increments, or points, on a microscope stage by a mechanical device. Point counting for coarse-grained rocks, however, can be done using a plastic scale and a hand lens.

The grid in figure 8.4 represents a slab of rock containing two minerals, one of which is shaded. The shaded mineral is distributed randomly over 20% of the slab area (20 of 100 squares). Numbers on the abscissa and ordinate give the percent of shading along regularly spaced east-west and north-south traverse lines, respectively; their average in both cases is 20%.

Alternatively, the number of points falling on randomly oriented lines of standard length may be used. Eight such lines are shown in figure 8.4 whose lengths each represent 100 points; the circled number on a line is the number of points falling on shaded squares. Taking the average of the results again gives a value of approximately 20%.

When applied to rocks, the procedure described here is called a **modal analysis** with the proportions of minerals given in volume percent. In contrast, a chemical analysis gives the chemical components in terms of weight.

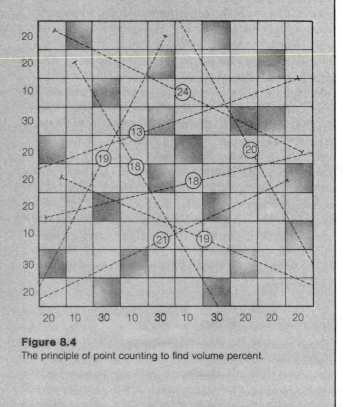

Figure 8.4
The principle of point counting to find volume percent.

hydrous and consequently less and less viscous. Eventually, this liquid will (1) react with early-formed minerals with which it may now be out of equilibrium (deuteric alteration), (2) react with the country rock (wall-rock alteration), or (3) be forcibly injected into the now crystallized igneous rock or adjacent walls. This hydrous residuum is rich in exotic elements including volatiles, such as fluorine, chlorine, sulfur, and boron; metals, such as copper, lead, uranium, beryllium, lithium, and tin; and the unused constituents of the late magmatic stage, such as silicon, aluminum, sodium, and potassium. Crystallization of this material leads to the formation of coarse-grained **pegmatites.** Any remainder may be emplaced as **hydrothermal veins.**

Weathering and Its Products

Igneous rocks represent materials that are stable in a physical and chemical environment that is distinctly different from surface conditions. When these rocks are brought to the surface by outpourings of lava, by deep erosion, or by mechanical movements, a whole family of physical and chemical reactions grouped under the general heading of **weathering** come into play.

Lowered pressure allows minute elastic expansions of the rock, and the tiny fractures that form weaken it mechanically. The fractures also provide a point of attack for the wedging actions of roots and the freeze-thaw cycle of water. Chemical reactions involving changes in volume help in the mechanical disintegration of the rock, and abrasion plays a role in reducing the size of particles.

The reduction in the physical size of surface rocks produces increased amounts of fresh surface upon which chemical reactions may take place. Chemical weathering, although generally less apparent than mechanical breakdown, is quantitatively dominant in rock decay. The earth's surface abounds in active reagents, such as water, carbon dioxide, free oxygen, and organic acids, that alone or in combination can attack and transform all but the most refractory of minerals.

The relative resistance of a mineral to chemical weathering is inversely proportional to its solubility in the aqueous environment. Quartz, for example, has a low solubility and tends to be concentrated in weathered products. In contrast, minerals with high solubilities, such as halite, are readily removed during weathering. The solution of minerals is congruent if the mineral dissolves as an entity; examples are halite, calcite, and quartz. Most

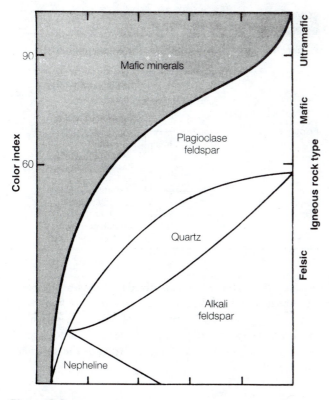

Figure 8.5
A general classification of igneous rocks using volume percentages of the rock-forming minerals. The heavy boundary is the color index (volume percent of dark-colored minerals).

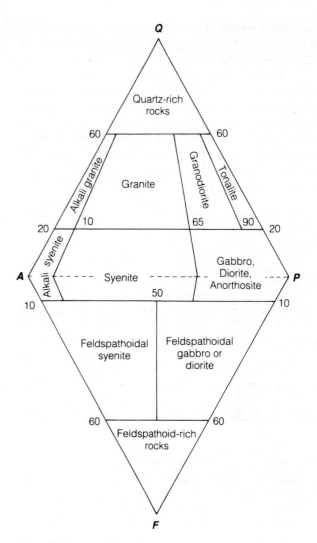

Figure 8.6
The *QAPF* diagram for the classification of medium- to coarse-grained igneous rocks. (Redrawn from Streckeisen, 1974.)

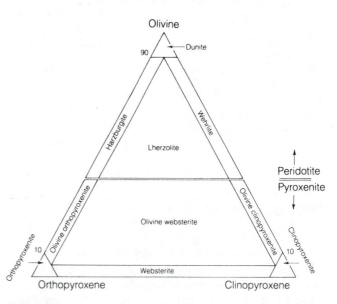

Figure 8.7
Classification of igneous rocks with Color Index greater than 90. (Redrawn from Streckeisen, 1974.)

minerals, however, dissolve incongruently and the alteration process produces a new solid phase. Consider the weathering of a potassium feldspar as an example. In the presence of water, feldspar readily alters to muscovite (most often a fine-grained, sericitic variety) according to the relationship

$$3KAlSi_3O_8 + 14H_2O = KAl_3Si_3O_{10}(OH)_2$$
$$\text{K-feldspar} \qquad\qquad \text{muscovite}$$
$$+ 6Si(OH)_4 + 2K^+ + 2(OH)^-$$
$$\text{solution}$$

The muscovite itself may also dissolve in the weathering environment to produce kaolinite and gibbsite:

$$2KAl_3Si_3O_{10}(OH)_2 + 10H_2O = Al_4Si_4O_{10}(OH)_8 + 2Al(OH)_3$$
$$\text{muscovite} \qquad\qquad \text{kaolinite} \qquad \text{gibbsite}$$
$$+ 2Si(OH)_4 + 2K^+ + 2(OH)^-$$
$$\text{solution}$$

and under extreme conditions the kaolinite can break down to form gibbsite:

$$Al_4Si_4O_{10}(OH)_8 + 10H_2O = 4Al(OH)_3 + 4Si(OH)_4$$
$$\text{kaolinite} \qquad\qquad\qquad \text{gibbsite} \qquad \text{solution}$$

Ferromagnesian minerals, such as olivine, augite, hornblende, and biotite, yield numerous products, among which are talc, chlorite, serpentine, and epidote. Calcite, quartz, and iron oxides and hydroxides are also common products.

The solid products of mineral alteration are, of course, either minerals in their own right or complex mixtures of minerals such as bauxite.

In general, resistance to alteration varies directly with the degree of structural complexity of the silicates and inversely with their temperature of formation. Thus, simple, high-temperature silicates such as olivine tend to break down faster under surficial conditions than do chain silicates, which in turn are less stable than the sheet or framework silicates. The weathering products of simple silicates tend to be sheet or framework silicates.

Incipient alteration often provides a useful clue to the identity of an unknown mineral because many alteration products have a characteristic color. For example, oxidized and hydrated iron compounds are red-brown to yellow-brown; the alteration products of copper-bearing minerals are usually blue or green; nickel minerals alter to green products; cobalt to pink; and uranium to bright green, yellow, or orange products.

The final products of the chemical decay of the common minerals are resistant primary solids such as quartz; secondary solids such as clays, aluminum hydrates, and iron oxides and hydroxides; and the dissolved ions Na^+, K^+, Ca^{2+}, and Mg^{2+}. A much simplified scheme for chemical decay is shown in table 8.2. The resultant solids are thoroughly mixed with organic debris and colloids at the earth's surface, resulting in a distinctive new product, the soil.

The earliest stages of chemical weathering are dependent on the local chemical environment and so may not be the same even for different parts of the same outcrop. The first effects in many instances are the oxidation of ferrous iron to the ferric state and the removal in solution of calcium and, to a lesser extent, magnesium. On the other hand, the long-continued activity of chemical weathering tends to be controlled by climate, with the result that the products converge to a relatively few kinds of widespread materials.

Both disintegration and decomposition yield solid particles and dissolved substances that can be more or less readily moved. Particles roll or creep down a hillside under the pull of gravity and are carried along by moving fluids—wind, water, or glacier ice. These geologic agents of transportation, especially wind and running water, are highly selective in the materials they will carry and are capable of sensitive discrimination on the basis of size, density, and shape of transported material. In consequence, like materials tend to travel together and to be deposited in the same area.

Dissolved substances are carried by surface and subsurface waters, but because conditions that cause the solution of a substance at one point may not hold at another, the dissolved load continuously varies. In essence, a balance exists between the amount and kind of material in solution and the rocks with which the water is in contact.

Table 8.2	Mineral Weathering	
Primary Mineral	**Principal Cations**	**Weathered Product**
Quartz	Si ⟶	**Primary solid** (resistate) Quartz, SiO_2
Feldspar	Si, Al ⟶ K, Na, Ca ⟶	**Secondary solids** Quartz, SiO_2 Muscovite, $KAl_3Si_3O_{10}(OH)_2$ Kaolinite, $Al_4Si_4O_{10}(OH)_8$
Olivine Augite Hornblende	Si, Al ⟶ Fe ⟶ Ca, Na, Mg ⟶	Hematite, Fe_2O_3 Goethite, $Fe(OH)_3$ **Dissolved ions** K^+, Na^+, Ca^{2+}, Mg^{2+}

Because the water with its dissolved load moves, this equilibrium is maintained only by continuous modification of the dissolved load. Precipitation of dissolved materials on the surfaces of sediment grains leads to a coalescing of the films and consequent cementation of the sediment into a rock.

Sedimentary Rocks

The products of weathering may be broadly grouped according to the processes that formed them in the surficial environment. **Resistate** materials, such as quartz, basically maintain their mineralogic integrity. **Hydrolysate** elements, such as aluminum, iron, and manganese, may be taken into solution but subsequently form insoluble oxides and hydroxides. **Soluble cations,** including the alkali and alkali earth elements, go into solution and may remain there until saturation limits for their mineralogic species are attained.

Sedimentary rocks are formed by the deposition or precipitation of materials derived from the surficial environment. The weathering products allow us to classify sediments broadly, as described in table 8.3. Sediments consisting of accumulations of minerals and rock fragments produced by the disintegration of older rocks are referred to as **clastic.** They are principally **terrigenous,** or derived from the land by subaerial weathering and erosion. If the minerals of the sedimentary rock originate by precipitation within a depositional basin, they are referred to as **precipitates.** A third large class of sediments, the **volcaniclastics,** includes fragmented material of volcanic origin, often deposited directly by explosive volcanism.

Clastic Sediments

Clastic sediments are the most abundant sediments in the earth's crust. They include conglomerates, breccias, sandstones, and mudstones. The classification of clastic sediments (see table 8.4) is made on the basis of both grain size and fragment composition as independent variables.

Table 8.3 | A Classification of Sedimentary Rocks

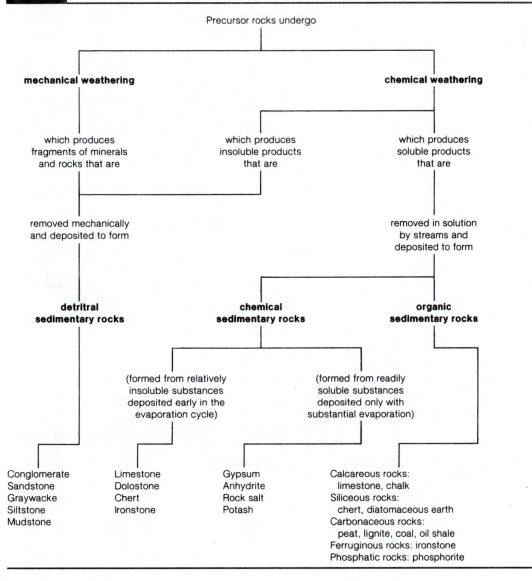

The mineralogy of clastic sediments can be very complex because the material represents the detritus of weathered rocks. However, the nature of weathering at the source and the conditions and longevity of transport greatly affect the final sediment mineralogy. Quartz is the most resistant of the common minerals, and because it is also abundant in terrigenous rocks, it is generally the principal constituent of clastic sediments. Detrital feldspar is also known but is restricted to sediments deposited close to their source. This is especially true for plagioclase, which is less resistant to weathering than alkali feldspar. Common also in terrigenous clastics, but normally minor in concentration, is a group of resistate minerals including zircon, topaz, rutile, and monazite. In exceptional cases, these may

Table 8.4 | A Classification of Clastic Sedimentary Rocks

Sediment	Grain Size	Rock Type
Gravel		Conglomerate
	2 mm	
Sand		Sandstone
	0.0625 mm (62.5 μm)	
Silt		Siltstone ⎫
	0.004 mm (4μm)	⎬ Mudstone
Clay		Shale ⎭

As examples of terminology, sandstones are further designated as follows:
Quartzose sandstone—dominantly quartz
Arkose sandstone—dominantly quartz and feldspar
Graywacke sandstone—mixture of rock and mineral fragments

be concentrated enough to form **placers,** deposits that can be economically extracted. Examples are gold, cassiterite, and rutile.

Clastic sediments generally undergo lithification by compaction, dehydration, or cementation. The cements are most commonly quartz and calcite, but hematite and zeolite cements are also known.

Mudstones are the most abundant clastic rocks in the crust. Their very fine-grained nature (less than 0.062 mm diameter) precludes easy visual identification, and x-ray diffraction techniques are most often employed. Shales can be distinguished from the other clastic types by their common fissility, or ability to split into thin beds.

The principal mineralogical components of shales are the clay minerals, formed by the breakdown of feldspars and ferromagnesian silicates during weathering, and very finely divided quartz. The common clay minerals kaolonite, muscovite, smectite, and chlorite may occur alone, but often there is a mixture or assemblage of these. A finely divided muscovite (illite) is the most common of the clay minerals in marine sediments and usually contains some iron and magnesium in its structure. In the sedimentary environment, the muscovite structure is highly disordered and contains excess water, but during diagenesis (lithification) and very low-grade metamorphism it becomes ordered. The ordering of muscovite, often referred to as illite crystallinity, can be correlated with coal rank and temperature and is an important geothermometer for the diagenetic environment.

Smectite-group minerals (montmorillonite is an important member) are the most reactive of the clays and readily invert to muscovite and chlorite. Similarly, kaolinite that forms in acidic environments holding bases such as K^+, Na^+, Ca^{2+}, Mg^{2+}, Fe^{2+}, and Fe^{3+} in solution normally reacts during diagenesis to form muscovite and chlorite. The overall tendency during burial and lithification, then, is the depletion of kaolinite and smectite group minerals to produce muscovite and chlorite.

Shales may also reflect their environment of deposition by including unoxidized organic material. The environment represented by these rocks is likely to be reducing because of the depletion of oxygen by anaerobic bacteria and decaying fauna. Such conditions promote the production of H_2S, which by reaction with iron can lead to the deposition of pyrite or, in more acidic conditions, marcasite. Similar processes in swampy, terrestrial environments, but with decaying plant material, may give rise to lignite or brown coal. Pyrite, marcasite, and the iron carbonate siderite are common accessories in coals and associated rocks.

Precipitates

Sedimentary rocks classified under the general term precipitates are widely variable in their source material, depositional environments, and mineralogy. A characteristic common to all, however, is that they consist of minerals precipitated from solution within the depositional area. Included are biochemical and biogenic deposits such as limestones and dolostones, chemical precipitates such as ironstones and evaporites, and organic deposits such as coal and oil shale.

The most common precipitate sediments are limestones deposited by either the action of carbonate-secreting organisms or by direct, inorganic precipitation from warm, shallow, marine waters. Calcite is the principal carbonate mineral formed if the crystallization takes place slowly from slightly supersaturated solutions. The calcite so formed is stable and will remain so indefinitely, implying that it is in equilibrium under surficial conditions. Aragonite, however, is often deposited first, especially if the carbonate deposition has been through the agency of organisms. Aragonite is unstable at the earth's surface and with time will transform spontaneously into calcite. Most modern carbonate rocks are mixtures of calcite and aragonite, but the latter is unknown in rocks older than Pennsylvanian since enough time has elapsed for its inversion. (Aragonite stability was discussed in chapter 6.)

Dolomite may replace the calcium carbonates in limestones by a reaction such as

$$2Ca(CO_3) + Mg^{2+} = CaMg(CO_3)_2 + Ca^{2+}$$

whereby some of the calcium in the carbonates is replaced by magnesium. Dolomite precipitation can take place directly from highly saline waters, but most is thought to be a replacement of existing $Ca(CO_3)$. The replacement can take place in near-surface, unconsolidated sediment by reaction with Mg-rich saline pore water or by the circulation of Mg-bearing waters in consolidated limestone sequences.

Most of the world's oceans are near saturation with respect to $Ca(CO_3)$. Colder waters may be somewhat undersaturated, but warmer waters tend to be supersaturated and precipitate calcite or aragonite. Higher evaporation rates can lead to the direct precipitation of dolomite or the deposition of the sulfate and chloride minerals normally associated with **evaporites.** In general, evaporite minerals occur in mixtures with sand, silt, clays, and carbonates. Some very pure and extensive deposits are known, however. These require rather special depositional environments with high evaporation rates and continued replenishment of the waters from the open ocean.

In normal seawater at 30° C, the deposition of gypsum, $Ca(SO_4) \cdot 2H_2O$, will begin when evaporation has raised the salinity to 3.4 times its normal value. Continued evaporation supersaturates the remaining brine with respect to $Ca(SO_4)$ and results in the precipitation of anhydrite. At even higher degrees of salinity, perhaps 10 times normal, halite will precipitate. Continued evaporation and increase in salinity may cause the precipitation of magnesium and potassium salts such as polyhalite, $K_2Ca_2Mg(SO_4)_4 \cdot 2H_2O$, or sylvite, KCl. The general rarity of deposits of these last minerals, however, reflects the extreme conditions necessary for their formation.

Minor constituents in evaporites often include magnesite, hematite, pyrite, barite, celestine, and carbonaceous matter.

Other precipitate rocks include phosphate deposits in which calcium phosphate has been deposited from seawater. Collophane, a cryptocrystalline variety of apatite, is a common but minor constituent of essentially all sedimentary rocks. It is, however, concentrated in rocks known as **phosphorites.** The origin of collophane is principally those biogenic reactions that form bones, teeth, and marine shells. Seawater, however, is close to saturation with respect to calcium phosphate, and much collophane may be of inorganic origin. It is, for example, a common cement in marine clastics.

Ironstones may be deposited in shallow marine conditions by precipitation of the ferrous iron minerals siderite and chamosite, a member of the chlorite group. If conditions are more oxidizing, hematite may be deposited. Siderite and chamosite may occur separately or together in limestones or mudstones. Chamosite often occurs as ooids, sometimes partially replaced by siderite. Both minerals are easily oxidized and may appear to be limonitic. Hematite and chamosite are commonly interlayed in oolites, and goethite is a common associate in these rocks.

An interesting and important group of residual deposits is formed under conditions of tropical weathering when mean annual temperatures are about 25° C, rainfall is abundant and seasonal, and rainwater infiltrates and percolates freely underground. Under these conditions, many elements, including silicon, go into solution and are removed, leaving resistant compounds behind. The resultant soil is called **laterite** (from Latin *later,* brick) and locally may be so enriched in a particular compound as to form exploitable ore. Laterites formed in the present or past in the tropics account for nearly all of the world's production of aluminum and iron ores and for a large proportion of the manganese and nickel produced.

Metamorphism and Metamorphic Rocks

Metamorphism takes place when rocks are displaced from the conditions of their formation. Such displacements can lead to weathering at the earth's surface or melting in the upper mantle or lower crust. However, metamorphism is limited to changes that take place between sedimentary diagenesis and melting, thus forming the last link in the rock cycle. Changes in the physical and chemical form and arrangements of rock materials that take place in response to changes in environmental conditions have previously been described for magmatic crystallization and for weathering. The conditions under which magma crystallizes to form intrusive igneous rocks are those of high but falling temperature and fairly high pressure. In these circumstances, the mineral phases that form are in equilibrium with the magma and themselves if cooling is sufficiently slow that the necessary atomic interchanges can

take place. However, minor disequilibrium is rather common, especially when the temperature falls to lower values. Weathering and diagenetic processes are carried on at relatively low temperature and pressure. The transformation of geological materials from one form and arrangement to another in such a circumstance tends to be rather sluggish and often does not reach or maintain equilibrium. The products of weathering, therefore, are often disequilibrium mixtures.

In contrast, mineralogical changes induced by rising temperature or pressure are relatively rapid and are usually able to remain in balance with the changes in their environment. Because of this, the study of such systems is especially rewarding, as it leads rather directly to an understanding of the geological conditions themselves. The minerals and their arrangement in each assemblage often represent a high water mark of change and can be interpreted in terms of the available material and energy.

The process by which rock constituents are rearranged without significant gain or loss (in closed systems) is called metamorphism (from Greek *meta,* change, and *morphos,* form). Processes that entail chemical loss or gain (in open systems) are called metasomatic (Greek *soma,* body). The rearrangements are usually recognized by the development of characteristic mineral species and distinctive rock fabrics. Bear in mind that the changes in metamorphism and metasomatism take place only in the solid. Gases and liquids, if present, may promote the reaction rates but are not a volumetrically significant part of the system.

The intensity (or grade) of metamorphism can be interpreted from the mineral assemblage produced. Metamorphism may also generate new arrangements of minerals or rock textures. **Prograde** metamorphism takes place with increasing temperature and pressure; **retrograde** metamorphism takes place during the relaxation of metamorphic conditions.

Types of Metamorphism

The three principal types of metamorphism are contact, dynamic, and regional. **Contact metamorphism** takes place in response to the distinct thermal disequilibrium generated by the intrusion of igneous rocks. The degree of metamorphism is controlled by the relative temperatures of the intrusion and the country rock and by the length of time the thermal differential is maintained. Typically, rocks that have undergone contact metamorphism are harder than their unmetamorphosed equivalents. They are also brittle and may retain many of their original structures and fabrics. Such rocks are normally referred to as **hornfels** (fine-grained) or **granofels** (medium- to coarse-grained).

Rocks produced by **dynamic metamorphism** are altered texturally by mechanical forces, but mineralogic changes are minimal. Dynamically metamorphosed rocks are limited in extent, being restricted to zones of intense deformation, such as faults, or intense shock, as occurs in meteorite impact craters.

In **regional metamorphism** rocks are produced by high pressures and temperatures during deep burial and are exposed in uplifted and eroded orogenic belts and in the Precambrian shields of the continental crust. Regionally metamorphosed rocks are developed in large volumes in the earth's crust. They often show the effects of directed stress by developing characteristic directional metamorphic fabrics. Such fabrics include **lineation,** an alignment of elongate minerals, and **foliation,** a parallel arrangement of the constituent minerals in planes.

The regional metamorphism of a shale illustrates the range of rock types that can develop. At very low grades, shale undergoes recrystallization to form the phyllosilicates muscovite and chlorite, which are arranged so that the crystal plates are subparallel with their broad surfaces perpendicular to the active stress. This imparts a foliation to the rock to produce the characteristic rock cleavage of **slate,** although there may be few other visual indications of metamorphism. At somewhat higher grades, continued recrystallization produces a larger grain size that results in the silky sheen typical of **phyllites.** Increased metamorphic intensity is accompanied by continued mineral growth resulting in strongly foliated **schists,** the cleavage planes of which are more irregular due to the growth of nonplaty minerals. Finally, at the highest grades, the minerals separate or differentiate into bands to produce a **gneiss.**

Metamorphic reactions may be of two principal types, solid-solid or solid-(solid + vapor). The actual reactions that take place may be quite complex and, at least over small volumes, they generally involve all the minerals present. Solid-solid reactions do not directly involve a vapor phase as a product or reactant. They include polymorphic transformations, such as kyanite-sillimanite or microcline-orthoclase inversions. Examples of vapor-producing reactions include the production of wollastonite:

$$Ca(CO_3) + SiO_2 = Ca(SiO_3) + CO_2$$

or the breakdown of muscovite to K-feldspar and sillimanite:

$$KAl_3Si_3O_{10}(OH)_2 + SiO_2$$
$$= KAlSi_3O_8 + Al_2SiO_5 + H_2O$$

Devaporization reactions occur when hydrated assemblages such as those typically found in sediments are converted to increasingly dehydrated assemblages by prograde metamorphism. Indeed, rocks representing the highest grades of metamorphism are typically anhydrous mineral assemblages similar to many igneous rocks.

Classification of Metamorphic Rocks

Chemical exchanges on a large scale, known as **allochemical** metamorphism or metasomatism, are sometimes observed in contact metamorphism, especially when the intrusion and country rock are chemically disparate. Much

metamorphism, however, seems to be isochemical, except for the loss or gain of volatiles such as water or carbon dioxide. Thus, it is often possible to recognize the original rock, or **protolith,** from the chemistry of the metamorphic rock regardless of its grade. The mineralogy of the metamorphic rock represents an equilibrium assemblage for the particular conditions of metamorphism and for a particular bulk chemistry. Thus, the identification of the assemblage and the estimation of the relative proportions of the minerals within it tell much about the nature of the original rock. For example, a *quartzite,* a metamorphic rock that is almost totally quartz, can only be the result of metamorphism of a quartzose sandstone or a quartz vein. Further discrimination can usually be accomplished using field observations. *Marbles,* or metamorphosed carbonate rocks, are almost certainly former calcareous sediments. Claystones, because of their highly aluminous nature, likely produce metamorphic assemblages rich in phyllosilicates, such as muscovite and chlorite. As a last example, consider an *amphibolite,* a schistose rock with roughly equal proportions of hornblende and plagioclase. Using the compositions of these phases along with their specific gravities, an estimate of the rock chemistry can be made (see Box 8.2). The calculations show that such a rock has a composition very similar to a basalt. Indeed, most amphibolites are metabasalts, although a very few may be generated by rare combinations of argillaceous and carbonate sediments.

On the basis of their mineral compositions, metamorphic rocks can be categorized into broad chemical groups which are described below.

Pelitic rocks have the general composition of shales or clay-rich mudstones. They are most often **peraluminous;** that is, they have an excess of Al_2O_3 over that needed to form the feldspars. Rocks of this type generally contain aluminous phyllosilicates and aluminum-rich minerals, such as chlorite, biotite, muscovite, andalusite, kyanite, sillimanite, garnet, staurolite, and corundum. Progressive regional metamorphism of pelitic rocks generates a typical suite from slate to phyllite to schist to gneiss.

Psammitic rocks are chemically analogous to the arenaceous sedimentary rocks, including quartzose sandstone through arkose and the wackes. Felsic igneous rocks, from diorite to granite, and their volcanic equivalents can also be included in this class because their chemistries and starting minerals may not be very much different from those of psammitic sediments. Indeed, it is often difficult to distinguish the metaigneous and metasedimentary rocks of this class on the basis of mineralogic and textural criteria. Chemical tests are sometimes successful discriminators. Psammitic rocks commonly contain quartz, feldspars, phyllosilicates such as muscovite and biotite, and sometimes Al-rich minerals like garnet or the aluminosilicates.

Table 8.5 A Classification of Metamorphic Rocks

Grain Orientation	Grain Size or Composition	Metamorphic Rock
Preferred orientation of grains apparent	Banded or eyed structure of blocky grains	Gneiss
	Laminated structure of coarse grains	Schist
	Laminated structure of very fine flakes; silky appearance	Phyllite
	Slaty cleavage; no grains visible	Slate
Nonoriented fabric	Fine-grained	Hornfels
	Fine-grained	Argillite
	Coarse-grained	Granofels
Coarse, sutured fabric	Predominantly quartz	Quartzite
	Pebbly, any composition	Conglomerite, metaconglomerate
	Calcite or dolomite	Marble
	Calcium silicates	Skarn, calc-schist

Mafic rocks include silicate-based rocks that are higher in iron and magnesium than normal sediments and felsic igneous types. The class includes basalts and mafic andesites and their plutonic equivalents, plus some magnesium-rich sediments. Volumetrically, basalts are the most important rocks of the crust. Thus, this chemical class is widely represented in metamorphic sequences and includes the greenstones and most amphibolites. Typical metamorphic minerals found in metabasalts include chlorite, actinolite, cummingtonite, hornblende, albite, epidote, hematite, and magnetite. Pyroxene may be present in the higher grades.

Calcareous rocks may occur either as relatively pure limestones or dolostones or as **marls** that have a silicate component, generally in the form of clays and quartz. Calcite and dolomite occurring alone or in combination are very stable throughout the regime of metamorphism. The metamorphism of these pure carbonate rocks produces marbles. The addition of silicate minerals greatly reduces the stability fields of the carbonates, producing a wide range of calc-silicate minerals. Minerals found in these **calc-silicate** rocks include calcite, dolomite, lawsonite, tremolite, grossular, quartz, chlorite, wollastonite, olivine, diopside, scapolite, brucite, and serpentine.

Ultramafic rocks include dunites, peridotite, and pyroxenites often tectonically emplaced in orogenic zones. They commonly have undergone regional metamorphism that first hydrated them at low grades and subsequently dehydrated them with increasing metamorphic intensity. At lower grades, these rocks generally include talc, chlorite, serpentine, and brucite. Olivine, anthophyllite, cummingtonite, enstatite, and spinel may appear at higher grades.

A simple classification for metamorphic rocks based on rock fabric and general composition is given in table 8.5.

Controls of Metamorphism

In response to the new conditions imposed by deep burial or the nearby intrusion of an igneous heat source, the mineral assemblages of the protolith change to form new assemblages representing equilibria at the new conditions. The assemblages change by reaction of existing minerals leading to the nucleation and growth of new minerals. (The processes and controls of nucleation and growth of minerals were reviewed in chapter 7.) These processes are not simple, and the formation of metamorphic minerals is further hindered by the nature of the solid-state medium. Nucleation and growth are to a large extent controlled by the diffusion of materials to and from the reaction site. Metamorphic reaction rates then are extremely slow even within a geologic time frame. They have, however, obviously taken place, and the reaction rates must have been enhanced by catalysts.

Diffusivities and, therefore, reaction rates can be increased by physical catalysts, such as temperature and shear stress, or by chemical catalysts, of which water may be the most important. Diffusion rates are largely controlled by thermal motion at the atomic scale, and temperature is a measure of the thermal motion or heat content. Thus, reaction rates should increase with rising temperature. Shearing stress enhances diffusion by leading to strain by gliding within minerals. Water acts as a chemical catalyst through its mobility and solvent capabilities. At the same time, however, confining pressure may in fact reduce reaction rates by limiting thermal motion and reducing porosity and permeability in the reaction medium. Thus, grade of metamorphism is often correlated with temperature, although this is not strictly so.

Box 8.2

A Method of Estimating Rock Chemistry

The determination of the bulk chemistry of a rock from its mineralogical composition might seem an insurmountable task considering the variability in mineral compositions and the complexity of their identification. The calculation is quite simple, however, and can provide a good estimate of bulk-chemical composition. It is only necessary to make reasonably good estimates of the minerals present and their proportions.

As an example, consider an amphibolite containing the following proportions of minerals:

Plagioclase	26.5
Hornblende	71.5
Quartz	2.0

The proportions are given in volume percent. Such an analysis (Box 8.1) is usually done by optical techniques on thin sections but can also be performed in the field using a squared grid on the specimen or the outcrop. In simple rocks, such as the one used here, fairly good estimates can often be made on hand specimens using only a hand lens. One cannot overemphasize the importance of visual recognition of common minerals, a skill that is acquired by practice.

Chemical analyses are most useful if given in terms of weight percent of the constituent species. Thus, our basic problem is (1) to convert the mineralogic volume data to a weight basis and (2) to determine the chemical constituents contributed by each mineral to the whole rock. The data in table 8.6 and the corresponding numbered notes for each step demonstrate the procedure.

1. Column A is an average hornblende calculated from end-member compositions. Column D is calculated for a plagioclase of composition $Ab_{50}An_{50}$ from pure end members, columns B and C.
2. Representative average specific gravities for the observed phases.

3. Volume percents (V_i) of the phases as determined by optical thin-section examination.
4. The weight percents of the constituent minerals are determined by first calculating the specific gravity of the rock (S_{rock}) according to

$$k_i = S_i V_i$$

and

$$S_{rock} = \frac{\sum_{i=1}^{n} k_i}{100}$$

where S_i and V_i are the specific gravity and volume percent, respectively, of each phase i. Then, the weight percent of each mineral (W_i) is determined by

$$W_i = \frac{k_i}{S_{rock}} \times 100$$

5. The chemical contributions made by each mineral to the whole rock are calculated by multiplying each constituent chemical component by its weight percent (expressed as a fraction). For example, the SiO_2 contributed by the hornblende is $46.27 \times 0.748 = 34.6$.
6. The rock composition is determined by adding the rows of the table generated in step 5. For example, SiO_2 in the rock is equal to 34.6 (from hornblende) + 13.2 (from plagioclase) + 1.7 (from quartz).
7. This column gives the actual rock composition as determined by chemical analysis. A comparison of the calculated and actual analyses shows reasonably good agreement. Discrepancies are probably due to the choice of mineral compositions. For example, the selection of a more aluminous but less iron-rich amphibole would afford a much more accurate result.

Many of the prograde reactions are dehydrating, and water is continually being driven from the reacting system. Normally, a sequence of metamorphic reactions proceeds until an equilibrium is established that represents the peak of metamorphic intensity for that particular rock. As the physical conditions ease (possibly because of uplift or cooling of the heat source), the rock is subjected to temperatures and pressures lower than those of the established mineral assemblage. This is the period of possible retrogression, when minerals may react to form new, lower-temperature assemblages. Retrograde products, although common, are not ubiquitous, and rarely do the reactions appear to have been completed. This is because of two important factors: (1) lower temperatures and the resultant loss of thermal motion and (2) loss of water from the rock

during the prograde period. Thus, to form the more hydrous, lower-grade minerals, water has to be reintroduced to the rock system.

Prograde metamorphism produces a series of minerals and mineral assemblages that represent equilibria for a particular bulk composition under particular temperature and pressure conditions. These mineral assemblages represent grades of metamorphism that can be evaluated in terms of temperature and pressure. The actual conditions at which a reaction might occur or under which an assemblage is stable can be estimated from the results of laboratory experiments. These results can be summarized in a P-T diagram, such as figure 8.8, in which fields are established by boundaries representing the average stability fields of diagnostic assemblages. An elaborate

The accuracy of the results is determined by (1) the accuracy of mineral identification, (2) the estimate of mineral proportions, and (3) the choice of mineral compositions used in the calculation. In the analysis shown here, mineral compositions were selected to be reasonable for the rock type under consideration.

However, good estimates of rock composition can sometimes be made using simple reference formulas such as those given in the mineral descriptions in part 3. The calculation of mineral compositions from their chemical formulas is described in chapter 5.

Table 8.6 **Calculation of Rock Chemistry from Mineral Data**

Weight Percent of Oxides

1		A Hornblende	B Albite	C Anorthite	D $Ab_{50}An_{50}$	E Quartz
	SiO_2	46.27	68.75	43.20	55.98	100.0
	TiO_2	1.21	0	0	0	0
	Al_2O_3	8.56	19.44	36.65	28.04	0
	Fe_2O_3	3.95	0	0	0	0
	FeO	12.98	0	0	0	0
	MgO	12.09	0	0	0	0
	CaO	11.71	0	20.15	10.08	0
	Na_2O	1.10	11.81	0	5.91	0
	K_2O	0.65	0	0	0	0
	H_2O	1.68	0	0	0	0
2	Specific Gravity	3.15	2.62	2.76	2.69	2.65
3	Volume % of Mineral	71.50	—	—	26.50	2.0
4	Weight % of Mineral	74.80	—	—	23.60	1.7

5		Hornblende × 74.8%	$Ab_{50}An_{50}$ × 23.6%	Quartz × 1.7%	6	7
	SiO_2	34.6	13.2	1.7	49.5	49.7
	TiO_2	0.9	0	0	0.9	0.6
	Al_2O_3	6.4	6.6	0	13.0	16.1
	Fe_2O_3	2.9	0	0	2.9	2.4
	FeO	9.6	0	0	9.6	8.0
	MgO	9.0	0	0	9.0	7.8
	CaO	8.8	2.4	0	11.2	10.2
	Na_2O	0.8	1.4	0	2.2	3.0
	K_2O	0.5	0	0	0.5	0.6
	H_2O	1.3	0	0	1.3	1.0

Note: Data in step 1, column A, from Deer, Howie, and Zussman, 1963. Data in step 3 and step 7 from Barth, Correns, and Eskola, 1939.

nomenclature has been established for describing metamorphic grade, but a rather simple system is used here: very low, low, medium, and high.

Three possible geothermal gradients, identified by the circled numbers, are shown in figure 8.8. Gradient 1 represents metamorphism via deep burial in low heat-flow regimes. Progressive metamorphism leads to assemblages in field *A* where glaucophane, jadeite, and aragonite are stable. The field labeled *A* represents a high-pressure–low-temperature metamorphism normally associated with the trench side of orogenic belts. Gradient 2 describes most regional metamorphic systems, and gradient 3 represents metamorphism in a region of high heat flow where contact metamorphism may be prevalent.

The curve *XY* in figure 8.8 is the minimum melting curve for granite, or the granite eutectic, under water-saturated conditions where $P_{H_2O} = P_{Total}$. In most metamorphic systems, water pressures are generally much less than the total pressure, and the granite melting curve will lie to the right of the one plotted. In any case, it is instructive to consider what will happen if metamorphism proceeds along gradient 2 in rocks of high H_2O content. Although at lower grades metamorphism takes place in the solid state, at higher grades partial melting generates granitic magma which will segregate and form layers within its source rock, creating a **migmatite.** The highest grades of metamorphism will not be attained in these rocks.

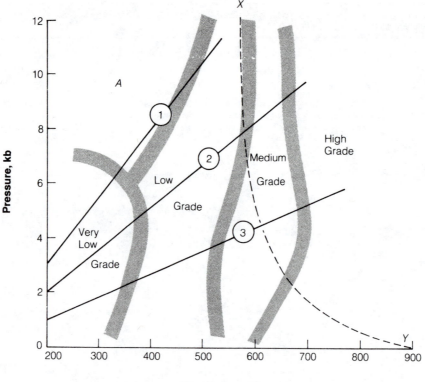

Figure 8.8
The approximate pressure-temperature boundaries for metamorphic grades (shaded bands). Curves 1, 2, and 3 are respectively low, average, and high geothermal gradients. Curve XY is the minimum melting curve for granite under water-saturated conditions.

Metamorphism can thus be described in terms of rock fabrics and mineral assemblages, the latter representing particular grades or intensities. The mineral assemblages produced are determined both by the physical conditions of metamorphism and by the bulk-rock chemistry. Table 8.7 shows mineral stabilities for the rock-chemical classes. Keep in mind that the presence or absence of a single mineral may not present much grade information. The presence of mineral assemblages can, however, be quite diagnostic.

Economic Mineralogy

Certainly the earliest interest in minerals was purely utilitarian. The raw materials of the earth have provided tools and decoration since the earliest times of human existence, and the development of civilization has been greatly influenced by the discovery, distribution, and technology of minerals.

Mineral resources are often thought of as including all natural substances that may be recovered profitably from the earth. However, minerals as we have defined them in this text are naturally occurring, crystalline, inorganic substances, and thus exclude organic fuels such as coal and petroleum. Mineral resources can be divided into metalliferous and nonmetalliferous types. **Metalliferous** deposits are exploited for the metals they contain. They include native metals such as gold and metal-bearing compounds like copper or nickel sulfides. **Nonmetalliferous** deposits are exploited for the minerals themselves because of some property that makes them useful to society. For example, kaolin, although high in aluminum, is extracted for its ability to be molded and fired as a ceramic. Diamond and graphite are not recovered for their carbon content but because of their distinct and useful characteristics.

The chemical elements that make up the earth's crust are widely variable in their concentration (see table 1.1) and widely dispersed as well. The overall crustal concentration of an element, however, does not tell us much about its economic usefulness. For example, one might compare the crustal concentration of an element, such as gold, to the volume of the crust to calculate a total crustal tonnage. The resultant value is of academic interest, but it does not imply that the gold is recoverable. For an element to be recoverable and therefore useful to us, it must be concentrated by some geologic process or a combination of processes. If the mineral is concentrated to a level at which it can be extracted profitably, the resource is called an **ore.** Table 8.8 presents data on the enrichments necessary to make ore for several elements. The enrichment values presented are necessarily approximate since they depend on both economic and geologic factors.

Ore deposits are generally classified according to their origin, as shown in table 8.9. Detailed reviews of each class

Table 8.7 **Stability of Minerals in Various Rock-chemical Classes (The elements are listed for each class in approximate order of abundance.)**

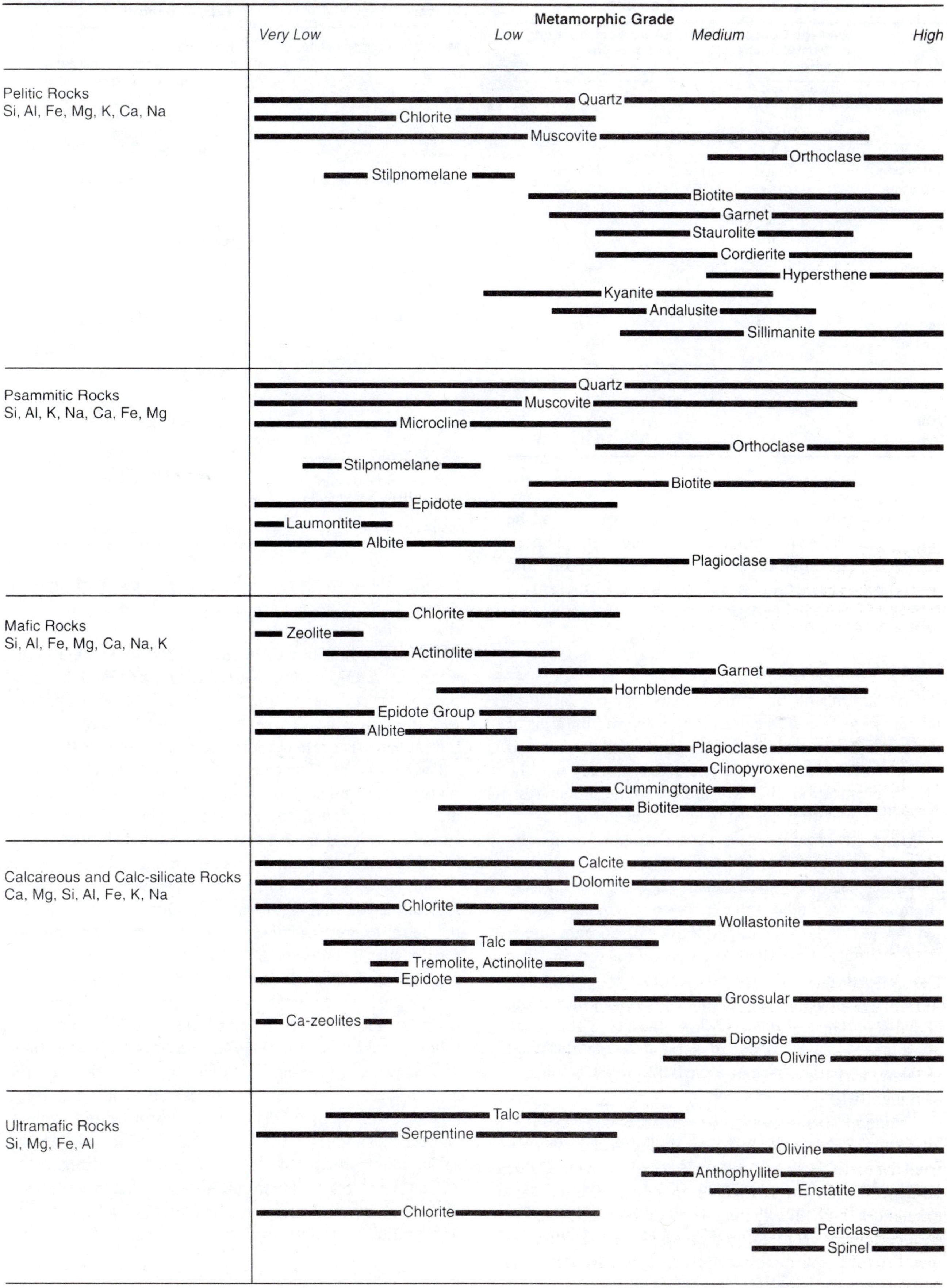

Metamorphic Grade: Very Low — Low — Medium — High

Pelitic Rocks
Si, Al, Fe, Mg, K, Ca, Na

- Quartz
- Chlorite
- Muscovite
- Orthoclase
- Stilpnomelane
- Biotite
- Garnet
- Staurolite
- Cordierite
- Hypersthene
- Kyanite
- Andalusite
- Sillimanite

Psammitic Rocks
Si, Al, K, Na, Ca, Fe, Mg

- Quartz
- Muscovite
- Microcline
- Orthoclase
- Stilpnomelane
- Biotite
- Epidote
- Laumontite
- Albite
- Plagioclase

Mafic Rocks
Si, Al, Fe, Mg, Ca, Na, K

- Chlorite
- Zeolite
- Actinolite
- Garnet
- Hornblende
- Epidote Group
- Albite
- Plagioclase
- Clinopyroxene
- Cummingtonite
- Biotite

Calcareous and Calc-silicate Rocks
Ca, Mg, Si, Al, Fe, K, Na

- Calcite
- Dolomite
- Chlorite
- Wollastonite
- Talc
- Tremolite, Actinolite
- Epidote
- Grossular
- Ca-zeolites
- Diopside
- Olivine

Ultramafic Rocks
Si, Mg, Fe, Al

- Talc
- Serpentine
- Olivine
- Anthophyllite
- Enstatite
- Chlorite
- Periclase
- Spinel

Table 8.8 | Average Crustal Concentrations and Enrichments Necessary to Form Ore for Some Elements

Metal	Average Concentration in Crustal Rocks (grams/tonne)	Enrichment Needed to Form Ore
Aluminum	81,300	4
Iron	50,000	5
Titanium	4,400	7
Manganese	950	380
Vanadium	135	160
Chromium	100	3,000
Nickel	75	175
Zinc	70	350
Copper	55	140
Cobalt	25	2,000
Lead	13	2,000
Tin	2	1,000
Uranium	1.8	500
Molybdenum	1.5	1,700
Tungsten	1.5	6,500
Mercury	0.08	26,000
Silver	0.07	1,500
Gold	0.004	2,000

Table 8.9 | A Classification of Ore Deposits

Process	Type of Deposit
Magmatic concentration	Early magmatic Disseminated crystallization Crystal segregation Late magmatic Residual liquid segregation Immiscible liquid segregation
Contact metasomatism	Contact metasomatic iron, copper, or gold
Hydrothermal processes	Cavity filling Replacement
Submarine volcanic exhalative	Submarine volcanic
Sedimentation	Sedimentary iron, manganese, or phosphate Evaporites Residual concentration by weathering Mechanical concentration by transport (e.g., placers)
Bacteriogenic	Bacterial products or reduction
Surficial oxidation and supergene enrichment	Oxidized, supergene sulfides
Metamorphism	Metamorphic deposits

of deposit are beyond the scope of this book but can be found in the pertinent references at the end of the chapter. Here we will examine the principal types of deposits and the minerals that occur in each. Table 8.10 presents a summary of exploitable deposits for some elements.

Magmatic Ore Deposits

The mechanisms of differentiation and crystallization of magmas were discussed earlier in this chapter in terms of their principal mineralogical components. Other chemical species are present, however, and although they may be in low concentration, they are subject to the same physicochemical laws that govern magmatic evolution.

Magmatic ore deposits are concentrations formed by the igneous processes of crystallization and segregation. In some cases, the desired mineral species are sparsely disseminated throughout the intrusion. Diamond distributed in a particular type of peridotite called kimberlite is a good example of a disseminated magmatic ore. In other cases, the ore minerals are concentrated by crystal settling. This is a particularly effective process in ultramafic and mafic magmas of relatively low viscosity. The layered chromite and platinum deposits found in the Bushveld Complex of South Africa and the Stillwater Complex of Montana are excellent examples.

Primary crystallization in mafic magmas can produce residual liquids rich in iron and titanium. The liquids, which may crystallize in the pore spaces among early-formed silicates, can squeeze out to form massive segregated deposits within the magma chamber or be injected into the wall rocks. Deposits of magnetite and ilmenite,

such as those in the Adirondacks of New York and at Allard Lake, Quebec, are thought to have originated in this fashion.

The residual liquids of more felsic magmas are enriched in SiO_2, Al_2O_3, Na_2O, K_2O, and H_2O by progressive crystallization and fractionation of primary rock-forming minerals. These highly volatile, low-viscosity liquids can segregate to form coarse-grained silicate rocks called pegmatites. Typically, these are rich in quartz, alkali feldspar, and micas, but often they also contain enrichments of the incompatible elements. These elements, because of their distinctive crystallochemical character, are fractionated away from the rock-forming silicates into the residual liquids. Thus, minerals such as beryl, spodumene, lepidolite, tourmaline, topaz, tantalite, columbite, and pollucite, as well as those of uranium, thorium and the rare earths, are commonly concentrated in pegmatites. Feldspar and mica are also exploited from pegmatites.

Because magmas are at least in part crystallographically structured, elemental species within them must follow the same rules of bonding and association that they do in crystals. In the case of liquids, we know that certain liquids, because of their structures and bonding characteristics, will not mix. Oil and vinegar provide an excellent domestic example of the **immiscibility** of liquids. High-sulfur mafic silicate magmas can exsolve an immiscible sulfide liquid phase. Upon solidification this liquid forms disseminated or massive Cu-Ni ores with pyrrhotite, pentlandite, and chalcopyrite as the principal minerals.

Table 8.10

Occurrence of Some Elements in an Exploitable Form

Minerals	Formula	Occurrence
Abundant Metals		
Iron		
Magnetite	Fe_3O_4	Igneous rocks
Hematite	Fe_2O_3	Sedimentary or laterite deposits
Goethite	$FeO(OH)$	Sedimentary or laterite deposits
Siderite	$Fe(CO_3)$	Sedimentary rocks
Iron silicates	Variable	Banded iron formations
Aluminum		
Gibbsite	$Al(OH)_3$	Bauxite
Boehmite	$AlO(OH)$	Bauxite
Diaspore	$AlO(OH)$	Bauxite
Nepheline	$Na(AlSiO_4)$	Igneous rocks
Kaolinite	$Al_2(Si_2O_5)(OH)_4$	Sedimentary rocks
Magnesium		
Magnesium	Mg^{2+}	From seawater
Magnesite	$Mg(CO_3)$	Metamorphic rocks; sedimentary rocks
Dolomite	$CaMg(CO_3)_2$	Sedimentary rocks
Titanium		
Ilmenite	$FeTiO_3$	Igneous rocks or placers
Rutile	TiO_2	Igneous or metamorphic rocks; placers
Scarce Metals		
Berylium		
Beryl	$Be_3Al_2(Si_6O_{18})$	Pegmatites
Chromium		
Chromite	$FeCr_2O_4$	Ultramafic igneous rocks
Copper		
Chalcopyrite	$CuFeS_2$	Porphyry copper deposits; sedimentary deposits
Chalcocite	Cu_2S	
Enargite	Cu_3AsS_4	
Bornite	Cu_5FeS_4	
Gold		
Native; tellurides	Au, AuTe	Volcanics; porphyry copper; placers
Lead		
Galena	PbS	Hydrothermal deposits; porphyry deposit by-product
Manganese		
Pyrolusite	MnO_2	Sedimentary or laterite deposits
Manganese nodules	MnO_2	Seafloor
Mercury		
Cinnabar	HgS	Epithermal deposits
Molybdenum		
Molybdenite	MoS_2	Porphyry deposits
Nickel		
Pentlandite	$(Fe,Ni,Co)_9S_8$	Mafic magmatic deposits
Garnierite	Ni(Co)-silicate	Laterites
Niobium		
Columbite	$(Fe,Mn)Nb_2O_6$	Pegmatites; syenites; carbonatites
Platinum		
Native	Pt	Ultramafic rocks; placers (continued on next page)

Table 8.10 Continued

Minerals	Formula	Occurrence
Scarce Metals		
Silver		
Acanthite	Ag_2S	Hydrothermal deposits; porphyry deposit by-product
Native	Ag	
Tantalum		
Tantalite	$(Fe,Mn)Ta_2O_6$	Pegmatites; syenites; carbonatites
Tin		
Cassiterite	SnO_2	Granitic rocks; placers
Vanadium		
V-oxides, silicates	Complex	V,U deposits in sandstones; carbonatites
Tungsten		
Scheelite	$CaWO_4$	Pegmatites
Wolframite	$FeWO_4$	Pegmatites
Zinc		
Sphalerite	ZnS	Hydrothermal deposits; copper porphyry by-product
Nonmetals		
Barium		
Barite	$Ba(SO_4)$	Sedimentary deposits
Boron		
Borax	$Na_2(B_4O_7) \cdot 10H_2O$	Nonmarine evaporites
Bromine		
Bromine	Br^-	Seawater evaporation
Chlorine		
Halite	$NaCl$	Evaporites, seawater evaporation
Chlorine	Cl^-	
Fluorine		
Fluorite	CaF_2	Hydrothermal deposits
Phosphorus		
Apatite	$Ca_5(PO_4)_3(F,OH)$	Marine sediments
Potassium		
Sylvite	KCl	Evaporites
Langbeinite	$K_2Mg_2(SO_4)_3$	
Carnallite	$KMgCl_3 \cdot 6H_2O$	
Kainite	$KCl \cdot MgSO_4 \cdot 3H_2O$	
Polyhalite	$K_2Ca_2Mg(SO_4)_4 \cdot 2H_2O$	
Sodium		
Halite	$NaCl$	Evaporites
Sulfur		
Native	S	Salt domes
Pyrite	FeS_2	Sulfide deposits
Pyrrhotite	$Fe_{1-x}S$	

Platinum, gold, silver, tellurium, and selenium minerals are often present as well. Deposits of this type include those at Insizwa, South Africa, and Sudbury, Ontario. Also of interest here are the immiscible, carbonate-rich magmas that form **carbonatites,** which are sources of many minor and rare incompatible elements.

Hydrothermal Ore Deposits

Underground water plays a crucial role in geologic processes and mineral formation. Water is a ubiquitous substance in the upper portions of the earth, filling the interstices between mineral grains and the cracks in rocks, dissolved in magma, and incorporated into hydrous minerals.

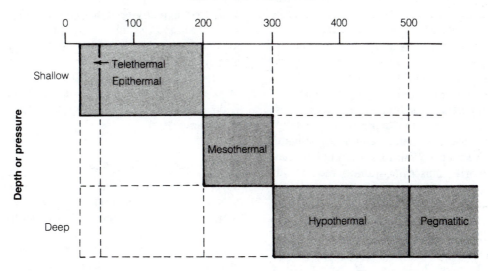

Figure 8.9
Categorization of hydrothermal solutions.

Near the surface, groundwater is active in the process of solution, transportation, and deposition of mineral matter. Solubility is enhanced as temperature goes up, and many elements are soluble as ions or complexes in hot water. This mobile fluid moves down a pressure gradient as a **hydrothermal** (hot-water) solution until conditions call for the deposition of the dissolved substances.

Hydrothermal solutions can be generated in magmas by the fractionation of water into residual liquids during crystallization of anhydrous minerals or by separation of a separate hydrous phase as pressure is released during the magma's rise toward the surface. Hydrothermal solutions can also be created by dehydration of hydrous minerals during metamorphism or by the activation of connate (trapped) water through sediment loading or tectonic movement. In any case, hydrothermal solutions are important carriers of many elements and the deposition of minerals from them will take place under appropriate physicochemical conditions. Deposition is normally in permeable host rocks, producing complex systems of veins and impregnations. Hydrothermal deposits can be described in terms of the depth and temperature of their formation (see figure 8.9).

The ore minerals commonly found in **hypothermal** deposits are gold, wolframite, scheelite, pyrrhotite, pentlandite, pyrite, arsenopyrite, löllingite, chalcopyrite, sphalerite, galena, cassiterite, bismuthinite, uraninite, and the arsenides of cobalt and nickel. Because these ores are deposited at considerable depths and are exposed at the surface only through deep erosion or orogenic processes, they are more abundant in older, often metamorphosed terrains. The Broken Hill District of Australia is an example of a major hypothermal ore deposit of lead, zinc, and silver. Minor amounts of cadmium, gold, antimony, cobalt, and copper are also produced.

Mesothermal deposits take many forms, including disseminated ores, veins, pipes, and irregular replacement masses. The most abundant ores found are of copper, lead, zinc, silver, and gold, with gangues of quartz, pyrite, and carbonate minerals. The ore-bearing solutions that form these deposits are typically out of chemical and thermal equilibrium with the country rocks. Thus, extensive alteration of the wall rocks is common. Muscovite, quartz, calcite, dolomite, pyrite, microcline, chlorite, and clay minerals are the most common alteration products.

The extensive variety of mesothermal deposits makes it difficult to characterize them with simple examples. However, the great disseminated copper deposits of orogenic zones, the copper porphyries, are economically significant. Typically these deposits formed above subduction zones in porphyritic granodioritic intrusions derived by partial melting of oceanic basalts and emplaced at depths of less than 2 km. As the magmas rose, the thermal regime changed drastically, internal fluids reacted with earlier-formed crystals to cause extensive alteration, and deep groundwater circulation was greatly modified. Alternating periods of silicification and fracturing provided space for ore deposition primarily in the form of chalcopyrite and bornite with minor sphalerite and molybdenite. Much of the primary ore of these deposits has been enriched by the oxidation and supergene enrichment processes that will be described later in this chapter. Excellent examples of these deposits are found at Chuquicamata, Chile; Bingham Canyon, Utah; and Cerro Colorado, Panama.

Molybdenum, from molybdenite, is also produced from porphyry deposits, either as a by-product of copper mining or in Cu-free molybdenum porphyries. The Climax and Henderson deposits of Colorado are good examples.

Epithermal deposits typically form at depths of less than one kilometer, and thus are characteristic of regions of recent igneous activity where erosion has not removed them. The deposits are generally in the form of veins, irregular fissures, or breccia pipes, and not uncommonly the fissure system has a direct connection to the surface. Some hot springs and steam vents, including submarine vents, are the surface expression of subsurface epithermal systems. Ore minerals of epithermal deposits include sulfosalts of silver, tellurides of silver and gold, stibnite, cinnabar, and native copper. Wall-rock alteration is often extensive, with chlorite, muscovite, alunite, zeolites, adularia, silica, and pyrite as characteristic products. An excellent example of an epigenetic deposit is the Almaden Mine of Ciudad Real Province, Spain, which has produced at last 240,000 tonnes of mercury. The ore is cinnabar with small amounts of native mercury in a gangue of pyrite, carbonate minerals, barite, and zeolites. Other epithermal deposits include the famous bonanza gold deposits at Goldfield, Nevada, and Cripple Creek, Colorado; deposits of "invisible" gold at Carlin, Nevada; the silver districts of Pachuca-Real del Monte and Guanajuato, Mexico; and the antimony-mercury-arsenic deposits in China.

Deposits formed by low-temperature solutions that have migrated far from an igneous source or that have no igneous source are called **telethermal**. In these deposits, mineralization seems to take place for the most part in open space from fluids with temperatures close to those of the host rocks. Examples of these deposits include the Mississippi Valley-type deposits named for the extensive lead-zinc mineralization of Missouri, Oklahoma, Kansas, and Wisconsin. Mineralization of this type is also observed in the south-central Appalachian district, as exemplified by deposits in eastern Tennessee and at Austinville, Virginia. The principal ore minerals in these deposits are sphalerite and galena, with minor wurtzite, marcasite, pyrite, and chalcopyrite. Mineralization is commonly at the borders of major sedimentary basins in brecciated, near-surface carbonate rocks.

The origin of these deposits remains controversial, but the source solutions may be mixtures of connate and deeply circulating meteoric waters. These waters leach metals from the sediments deep in the basin and carry them as metal chloride complexes along permeable horizons. The sulfate in the brines may be reduced by bacteria to form H_2S, which reacts with the metals under appropriate conditions, that is, where the brines mix with near-surface meteoric groundwater, to form sulfides. A basinal origin for these deposits is supported by their geologic placement and by the nature of their gangue mineralogy.

Many massive sulfide deposits are also caused by hydrothermal mineralization. Typically, these are large, roughly lenticular bodies of sulfides interlayered with marine volcanic rocks. Their origin seems to be in the accumulation of metal-rich muds at or near submarine hot springs where juvenile or circulating meteoric waters were exhaled between periods of volcanic activity. Sulfide minerals have actually been observed forming on the Pacific seafloor around volcanic vents. Mineralogical associations include pyrite-sphalerite-chalcopyrite in mafic to felsic volcanic rocks, pyrite-galena-sphalerite-chalcopyrite in felsic volcanics, and pyrite-chalcopyrite in mafic volcanics of ocean-floor affinity. Deposits at Noranda, Quebec; Kidd Creek, Ontario; and Kuroko, Japan are excellent examples.

Sedimentary Ore Deposits

Several examples of mineral concentration by sedimentary processes were mentioned earlier in the discussion of sedimentary rocks. These included phosphate deposits, sedimentary iron formations from the deposition of hematite and chamosite, and the evaporite deposits, which provide many minerals used for industrial or agricultural purposes. Also mentioned was the concentration of resistate minerals such as gold and diamond into placer deposits. We should note that a principal source of titanium is from rutile placers in beach sands.

There is also a group of copper-sulfide deposits that seem to have a sedimentary origin. These are thought to have been formed by the deposition of organic-rich sediments that presumably were anomalous in their copper content. The source of the copper is unclear, but it may have been derived by the weathering and erosion of nearby copper-rich rocks. Further, organic-rich sediments commonly contain abundant pyrite formed by the interaction of iron and H_2S. Thus, copper-bearing fluids generated by hot springs could have caused a secondary replacement of the pyrite by copper sulfides such as chalcocite. Examples of these deposits include the Precambrian White Pine District of Michigan, the widespread Permian Kupferschiefer (copper shale) of Europe, and the gigantic Precambrian Copper Belt deposits of Zambia and Zaire.

Weathering and Ore Deposition

Weathering, as noted earlier, may produce laterites by intense alteration of rock materials. Ores of several important metals can be formed in this manner. Iron ores of this kind are found at Cerro Bolivar, Venezuela; Carajas, Brazil; and in extensive deposits in West Africa and Western Australia. Aluminum ores (bauxite) are represented by deposits in Jamaica, in Guinea, at Weipa in northern Australia, and at Trombetas, Brazil. Nickel laterites are exploited in New Caledonia, in the southwest Pacific, and form significant reserves in Cuba and the Philippines. Manganese deposits are worked at Amapa, Brazil, and at Nsuta, Ghana.

The formation of bauxite is a particularly good illustration of laterite generation. The source rocks may be any aluminous rock low in iron and nickel. If these metals are present, iron or nickel laterites form. Source rocks with a high Al/Si ratio are also preferred. Thus, syenite would be more amenable than granite, and argillaceous limestone would be particularly well suited to the concentration of aluminum since the bulk of the carbonate-forming components are readily soluble. The extensive Jamaican deposits, for example, are formed on argillaceous limestones.

Under conditions of a tropical climate, the rocks undergo chemical weathering during which Na^+, K^+, Ca^{2+}, Mg^{2+}, and some Si^{4+} are taken into solution, leaving a cap of kaolinite. Important to the process is abundant rainfall, rocks and topography that promote infiltration and internal drainage, and acidic waters, which effect solution of silica. In tropical regions, the acidity of the waters is assured by organic acids derived from the decay of plant life. If the waters are very acidic, the kaolinite dissolves totally. On the other hand, if the waters are only slightly acidic, the kaolinite breaks down, the silica dissolves, and an assemblage of hydrated aluminum oxides, gibbsite, böhmite, and diaspore are left behind as bauxite.

Chemical weathering is also known to be responsible for the alteration and enrichment of some metallic ore bodies at or near the earth's surface. The most active region of chemical weathering is the **vadose** zone between the surface and the water table. Water moves downward through this undersaturated zone by infiltration into the water-saturated zone beneath the water table. The water table is not fixed with respect to the surface; it can move up with increased infiltration following atmospheric precipitation or down during times of drought and evaporation. The vadose zone is typically oxidizing, whereas the saturated zone below is normally reducing. Thus, it is of interest to examine the behavior of sulfides, minerals that are particularly sensitive to changes in a redox environment.

Many metallic sulfides oxidize readily in an environment containing oxygen and water, and the sulfide reforms as a sulfate or oxide. These oxides and sulfates may have very different solubilities. Insoluble compounds may remain where they formed, whereas soluble matter migrates downward with the groundwater.

For example, consider the series of reactions for chalcopyrite in the zone of weathering.

1. Chalcopyrite, $CuFeS_2$, is readily attacked by ferric sulfate and oxygen in the presence of water to yield iron oxide plus copper and sulfate ions in solution. The iron oxide remains in place giving a characteristic red color to the leached zone, or **gossan,** while copper and sulfate ions migrate downward.

2. In some types of rocks, new minerals may precipitate after the solutions have moved downward only a short distance, forming a zone of oxidized ore. For example, in carbonate rocks copper will often form malachite, azurite, or cuprite. If present, zinc and lead will precipitate as smithsonite or cerussite, respectively.

3. At greater depths and below the water table, the infiltrating waters encounter a reducing environment and the dissolved ions recombine to form new solids, such as chalcocite, which is free of iron and contains proportionately more copper than the original chalcopyrite:

$$2Cu^{2+} + SO_4^{2-} \rightarrow Cu^{2+}S + 2O_2$$

The process above, referred to as **supergene enrichment,** can lead to a substantial increase in the concentration of metals. Although many sulfide deposits exhibit this phenomenon, the typically low-grade porphyry-copper deposits have particularly benefited from supergene concentration.

Concluding Remarks

The subject matter of mineralogy covers or impinges on a broad spectrum of sciences and thus cannot be restricted simply to describing the naturally occurring substances defined as minerals. Aspects of chemistry, physics, and geology are important to understanding what minerals are, why they form where they do, and how they are used in our technological world. Thus, in this chapter we have examined the geological framework in which minerals are studied.

The controls on the number and kind of mineral species are of three types: geometric, chemical, and energetic. The geometric constraints on atomic arrangements place limits on the variability of chemical compounds. The very restrictive chemistry of the earth's crust further limits the number of mineral species that occur in substantial amounts. Energetic constraints in mineral formation not only determine the stability ranges of individual minerals but also sharply limit the stability of mineral associations.

Mineral associations allow classification of geologic materials and contribute data for interpreting geologic processes and the conditions of rock formation. Thus, the principles of mineralogy constitute a discipline that is basic to the earth sciences.

Suggested Readings

Barker, D. S. 1983. *Igneous Rocks*. Englewood Cliffs, N.J.: Prentice-Hall.

Best, M. G. 1982. *Igneous and Metamorphic Petrology*. San Francisco: W. H. Freeman.

Blatt, H. 1982. *Sedimentary Petrology*. San Francisco: W. H. Freeman.

Dennen, W. H. 1989. *Mineral Resources: Geology, Exploration, and Development*. New York: Taylor and Francis.

Fry, N. 1984. *The Field Description of Metamorphic Rocks*. New York: Halsted Press/Wiley.

Garrels, R. M., and Mackenzie, F. T. 1971. *Evolution of Sedimentary Rocks*. New York: W. W. Norton.

Guilbert, J. M., and Park, C. F. 1986. *The Geology of Ore Deposits*. New York: W. H. Freeman.

Hall, A. 1987. *Igneous Petrology*. New York: Wiley.

Hyndman, D. W. 1985. *Petrology of Igneous and Metamorphic Rocks*. 2d ed. New York: McGraw-Hill.

Jensen, M. L., and Bateman, A. M. 1981. *Economic Mineral Deposits*. 3d ed. New York: Wiley.

Mason, B., and Moore, C. B. 1980. *Principles of Geochemistry*. 4th ed. New York: Wiley.

Thorpe, R., and Brown, G. 1985. *The Field Description of Igneous Rocks*. New York: Halsted Press/Wiley.

Tucker, M. E. 1982. *The Field Description of the Sedimentary Rocks*. New York: Halsted Press/Wiley.

Tucker, M. E. 1991. *Sedimentary Petrology*. 2nd ed. Oxford: Blackwell.

Turner, F. J. 1980. *Metamorphic Petrology*. 2d ed. New York: McGraw-Hill.

Yardley, B. W. D. 1989. *An Introduction to Metamorphic Petrology*. New York: Wiley.

Part Two

Practical Mineralogy

The Physical Properties of Minerals

The beginning student of mineralogy may become bewildered by the mass of descriptive information covering all phases of the subject. Understanding the basic philosophy and practice that underlie the identification of minerals can make this material more comprehensible.

Identification of Minerals

Given a single property, it is possible to divide all minerals into at least two large groups; with two properties, each of these groups can be further subdivided; and with a half-dozen or so properties, minerals can be assigned to very small groups or individual species. It is, therefore, important to know thoroughly the more distinctive properties and imperative to use these properties correctly when identifying minerals. With experience, however, the mineralogist develops the ability to shortcut a laborious tabulation of properties and need not sift through all the possibilities to find a match. Snap identification of specimens is possible in time. The learning process for identifying minerals is a bit like getting to know people; early confusion is gradually replaced by familiarity as new acquaintances become old friends.

Many kinds of determinative tables, which classify minerals into groups according to various sets of properties, have been designed, and such tables have considerable value in developing a facility for mineral identification. Tables using the key properties of luster, cleavage, color, and hardness are given in appendix B. In addition to using these tables, students may also want to try designing their own identification tables.

The tentative identification of a mineral should always be followed by a confirming test. Such a test might be found by reading over the description of the mineral and checking the physical or chemical properties, especially those listed as diagnostic, that were not used in the original identification.

For many mineralogists, mineral identification becomes a matter of inspiration or intuition. While this might seem unscientific, especially to one trained in a scheme of analysis where the investigation always follows the same plan of attack, the proof of this method lies in the speed with which an accomplished mineralogist can correctly name specimens.

The technique of mineral identification by inspiration could be likened to an experienced physician's preliminary diagnosis of a patient.

A number of physical properties are applicable to the hand-specimen identification of a wide variety of minerals, and these properties are related to the internal structure of the mineral. Bear in mind that the properties of minerals discussed in this chapter are not only useful in identification but also have scientific and economic significance.

Manner of Breaking

Two characteristics of primary importance in mineral identification are the kind of surface produced when minerals rupture and the orientation of these surfaces with respect to crystallographic directions. The direction and ease with which the mineral can be broken are two of the best clues to its internal structure.

In a substance that is equally strong in all directions, the density and strength of the interatomic or interionic bonds is similar in all directions. Rupture of such material may occur in any crystallographic direction, the break being localized by some flaw in the crystal. The irregular surfaces produced are called **fracture surfaces,** and their qualities are often further described as follows:

1. **Even.** Nearly flat fracture surfaces, as in lithographic limestone.
2. **Uneven** or **irregular.** Rough and irregular fracture surfaces, as in rhodonite.
3. **Hackly.** Ragged fracture surface with sharp edges and points, as in copper.
4. **Splintery.** Fibers and splinters produced by fracture, as in pectolite.
5. **Conchoidal** (shell-like). Smoothly curving, ribbed fracture surface, as in glass or quartz (see figure 9.1).

Many minerals have planes across which there are fewer or weaker bonds than in other parts of the crystal. In figure 9.2a, for example, the number of bonds broken by the two fractures is different. The direction crossed by the fewest bonds will be the surface on which the mineral breaks. In figure 9.2b, the spacing of atomic planes parallel to a and b is different, and fractures of equal length must break four bonds across (01) for each six across (10). Obviously, the preferred break is along the (01) plane.

Many mineral structures have planes across which bond densities are markedly lower. This condition might arise, for example, if the mineral were composed of sheets of atoms held together by strong bonds within the sheets and weaker bonds between the sheets, or by a framework made up of an alternation of strongly bound radicals held together by weaker cation bonds. Two mineral examples are shown in figure 9.3. In general, mechanically weak planes arise either as a natural consequence of the atomic packing or because of strongly developed polarization. Whatever the cause, planes of low bond density occur periodically at small separations within the crystal structure and provide a natural locus for rupture. When a mineral breaks along this locus, the observable result is a planar surface, and the mineralogic property of breaking with

Figure 9.1
Quartz showing conchoidal fracture. (Courtesy Royal Ontario Museum, Toronto.)

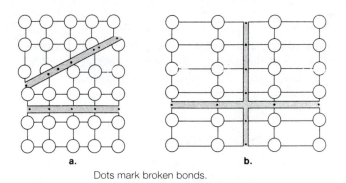

Dots mark broken bonds.

Figure 9.2
Preferred cleavage directions. (a) Bonds of equal length. The number of bonds per unit length of the inclined rupture that must be broken is greater than for the horizontal rupture; cleavage will occur preferentially on the horizontal (01) plane. (b) Bonds of unequal length. The number of bonds per unit length of a (10) rupture is greater than for (01). Breakage along (01) is expected.

Table 9.1	Shapes of Typical Cleavage Fragments	
Number of Cleavage Directions	Characteristic Fragment	Example
0		Quartz
1		Muscovite
2		Augite
		Orthoclase
		Hornblende
3		Halite
		Anhydrite
		Calcite
4		Fluorite
6		Sphalerite

such a surface is called **cleavage.** The cleavage planes are not necessarily true planes on an atomic scale but are macroscopically so. Mineral cleavage seldom yields a single perfect plane. More often, the surface is composed of many parallel planes connected stepwise by other cleavages or by fractures. Four factors should be noted when the property of cleavage is being considered—ease of breaking, perfection of surface, number of cleavage directions, and crystallographic directions (indices of the planes).

The number of possible directions of cleavage through a given mineral is restricted to planes parallel to actual or potential crystal faces and is thus always consistent with symmetry. The number of potential cleavage planes in a given direction is theoretically limited only by atomic spacing and hence is essentially infinite. The intersections of cleavage planes produce characteristic cleavage fragments. Some examples are shown in figure 9.4 and table 9.1. A list of minerals with one, two, and three or more good cleavages is given in table 9.2.

Under certain conditions, a mineral may break with a cleavage-like rupture that, on close examination, is found to be related to weak planes having a macroscopic spacing. **Parting** is the name given to this phenomenon. Parting and cleavage are superficially similar. The distinction in hand

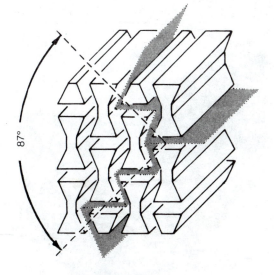

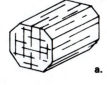

a.

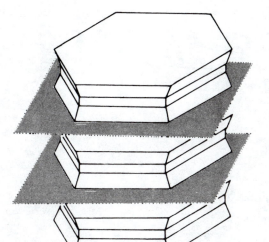

b.

Figure 9.3
Generation of cleavage along planes of weaker cation bonds.
(a) Pyroxene. (b) Mica.

specimens is usually based on the fact that the solid bounded by two adjacent planes of parting cannot be broken by further parting.

Parting, like cleavage, is related to weak bond planes within the mineral structure, but unlike cleavage, it is not

a necessary result of a given packing. Parting is usually related to weak planes arising from twinning or crystal deformation and is thus not necessarily present in all specimens of the same mineral. Planes of parting may occur in several different directions and may circumscribe polygonal solids as do planes of cleavage.

The beginner may experience some difficulty in distinguishing between plane surfaces arising from cleavage or parting and plane faces resulting from crystal growth. Careful examination of such surfaces, however, will usually show the delicate markings that distinguish one from the other. Crystal faces commonly show the results of accidents in crystal growth in such observable features as lineage structure and striation or pitting, whereas cleavage tends to produce geometrically plane surfaces with high reflectivity.

Hardness and Tenacity

The **hardness** of solids is measured by their resistance to deformation by an indenting tool or to abrasion or scratching. With minerals, the technique of hardness testing that uses the simplest equipment and is the most satisfactory for general mineralogical work is relative resistance to scratching.

The accepted scale of relative mineral hardness (scratchability) is given in table 9.3 along with some readily available tools and materials for testing. This mineral scale is known as Mohs scale of hardness after F. Mohs, the Austrian mineralogist who developed it in 1825.

a.

b.

c.

d.

Figure 9.4
Cleavage in some common minerals. (a) One perfect {001} cleavage in muscovite. (b) Two {110} prismatic cleavages at 56° and 124° in hornblende (left), and 87° and 93° in augite (right). (c) Three perfect cleavages in calcite form a {10$\bar{1}$1} rhombohedral fragment. (d) Four perfect {111} cleavages in fluorite form an octahedron. (Samples courtesy University of Windsor. Photographs by D. L. MacDonald.)

Minerals are described using the scale in terms of hardness numbers. For example, the hardness of chromite is 5.5 (H = 5.5). Chromite, therefore, will scratch apatite and be scratched by orthoclase. Minerals of the same hardness may scratch each other or may not affect each other.

The absolute spacing between the relative hardness values on Mohs' scale is not equal, and absolute mineral hardnesses cannot generally be determined with accuracy. There appears to be, however, an approximate exponential progression of absolute hardness for the minerals of Mohs' scale as shown by figure 9.5.

All the intrinsic properties of atoms as well as the variations possible in crystal structures must be called upon to obtain a full appreciation of mineral hardness. Some measure of bond strength is necessary, and this requires a consideration of charge, radius, and shielding efficiency. Bond density, which is a function of charge, coordination, and structure, must also be taken into account, and the distribution of bond density with respect to crystallographic directions is important. In addition, the presence of weak planes (e.g., cleavage) that intersect the test surface may markedly affect hardness determinations.

Table 9.3	Mohs Scale of Hardness	
Mineral	**Scale Value**	**Comparative Hardness**
		Wax
Talc	1	
Gypsum	2	
		Fingernail *2½*
		Aluminum
Calcite	3	
		Penny *3*
		Brass
Fluorite	4	
		Iron
Apatite	5	
		Knife blade
		Window
		glass *5*
Orthoclase	6	
		Steel file *5½*
		Streak plate
Quartz	7	
Topaz	8	
Corundum	9	
Diamond	10	

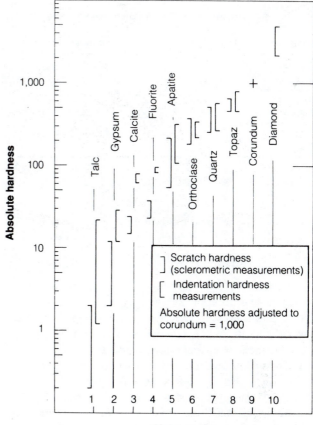

Figure 9.5
Relative versus absolute hardness of minerals.

Mineralogists have had some success in predicting the relative hardness of simple, ionically bonded binary compounds of the same structure type on the assumption that hardness varies as an inverse square with interionic distance and varies directly with valence. Thus, MgO should be, and is, harder than NaCl. Likewise, the hardness of the isostructural oxides BaO, SrO, CaO, and MgO increases from BaO to MgO as the cation radius decreases. The extension of these simple rules to more complex substances, however, is not valid. For example, the radii of cations and the hardness of the rhombohedral carbonates, as given in table 9.4, show no such progression. Other studies relate hardness to changes in interionic distance and "lattice energy" as measured by changes in volume.

Note the universally accepted misuse of the term lattice. An imaginary array of points cannot contain energy!

A somewhat more general approach to the relationship of hardness to crystal structure in ionic minerals has been made by correlating hardness with the number of bonds per unit volume in the structure. The bonding index so obtained takes account of the variable interionic distances and valences when more than two cations are

Table 9.4	Cation Size-Hardness Data for the Rhombohedral Carbonates		
Mineral	**Formula**	**Cation Size, Å**	**Hardness**
Calcite	$Ca(CO_3)$	0.99	2.5–3
Rhodochrosite	$Mn(CO_3)$	0.80	3.5–4.5
Siderite	$Fe(CO_3)$	0.74	3.5–4
Smithsonite	$Zn(CO_3)$	0.74	5.5
Sphaerocobaltite	$Co(CO_3)$	0.72	3–4
Magnesite	$Mg(CO_3)$	0.66	3.5–4.5

present. It does not, however, account for the uneven distribution of bonds, which affects hardness in certain crystals. In many minerals, uneven bond distribution is not the principal factor that controls hardness, and thus the bonding index correlates with mean hardness.

The unevenness of bond distribution is occasionally of sufficient magnitude to be detected even by such a relatively insensitive standard as Mohs' scale. For example, calcite has a hardness of 3 on its $\{10\overline{1}1\}$ cleavage face and 2.5 on its $\{0001\}$ base, and kyanite has a hardness of 4.5 parallel to the *c*-axis and 7 perpendicular to the *c*-axis. These differences are readily explained by the crystal

Table 9.5	Hardness of Some Polymorphic Pairs	
Mineral	**Formula**	**Hardness**
Calcite	$Ca(CO_3)$	2.5–3
Aragonite	$Ca(CO_3)$	3.5–4
Arsenolite	As_2O_3	1.5
Claudetite	As_2O_3	2.5
Graphite	C	1–2
Diamond	C	10

structure. The soft and hard directions in kyanite are, respectively, parallel and perpendicular to chains of Al–O octahedra, whereas in calcite the least hardness is on the plane of the CO_3 groups.

Hardness is very structure-sensitive, and the same atoms in two different arrays may have quite different hardnesses. This feature is readily seen in table 9.5 which gives the hardness of a few polymorphic pairs.

Hardness determinations should always be made on a fresh surface because of the possibility of soft-surface alterations. Also, make sure that an apparently positive test is not the result of powdering one specimen against the other or of the disruption of granules or cleavage fragments from the surface.

Tenacity is the manner in which a mineral ruptures or deforms under stress and is described by various terms:

1. **Brittle.** The mineral breaks or powders (e.g., quartz).
2. **Sectile.** The mineral deforms plastically under the stress of cutting, drawing, and hammering, respectively (e.g., copper and gypsum). Also called *ductile* or *malleable*.
3. **Flexible.** The mineral may be bent but does not return to its original form (e.g., talc and chlorite).
4. **Elastic.** The mineral, after bending, assumes its original form (e.g., muscovite).

The degree of brittleness or sectility can usually be observed when a scratch hardness test is performed. Brittle minerals, when scratched with a sharp point, have small particles cast out beside the scratch. The scratching of sectile materials, on the other hand, causes material to flow from the deformed area but not to be detached.

Specific Gravity

Density is the weight per unit volume of a substance in stated units. **Specific gravity** is the dimensionless ratio of the mass of a substance to an equal volume of water at 4° C.

Specific gravity is usually determined by weighing the substance first in air and then immersed in water (or an-

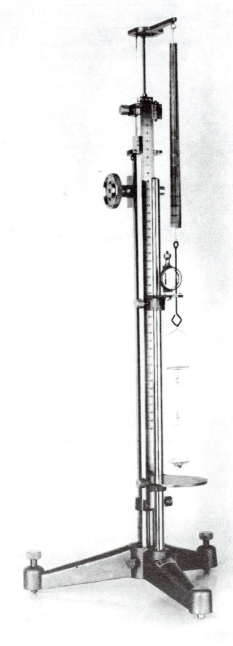

Figure 9.6
The Kraus-Jolly balance for determining specific gravity. (Courtesy of Eberbach Co., Ann Arbor, MI.)

other liquid of known density). Immersion displaces a volume of water equal to the volume of the test material. Thus,

$$\text{specific gravity} = \frac{\text{weight in air}}{\text{weight in air} - \text{weight in water}}$$

Specific gravity is commonly measured using an analytic balance with a support for a beaker of water or specially designed Kraus-Jolly or Berman balances (see figures 9.6 and 9.7). An alternative method, particularly useful for small amounts of granular material, employs a pycnometer (figure 9.8). This is a small bottle fitted with a ground

Figure 9.7
The Berman torsion balance for determining specific gravity. (Courtesy Bitronics, Inc., Bethlehem, PA.)

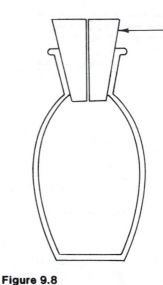
Ground glass stopper with capillary hole

Figure 9.8
Pycnometer used for measuring the specific gravity of small or pulverized specimens.

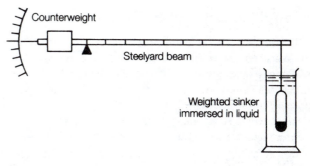
Counterweight
Steelyard beam
Weighted sinker immersed in liquid

Figure 9.9
Using a Westphal balance to measure the specific gravity of a liquid.

glass stopper having a capillary hole. The bottle is filled with water to the top of the capillary and weighed (B). A weighed amount of the specimen (S) is introduced, the capillary aperture is wiped off, and the filled bottle plus specimen is then reweighed (W). With this method,

$$\text{specific gravity} = \frac{S}{B + S - W}$$

A third way of finding the specific gravity of a mineral is to employ heavy liquids of known density, such as bromoform (2.89), methylene iodide (3.33), or Clerici solution (4.2). A mineral fragment is floated on the heavy liquid and a dilutent, usually acetone with a density of 0.79, is added drop by drop and stirred in. When the fragment neither sinks nor rises when it is displaced, its specific gravity is equal to that of the liquid, which may be measured with a pycnometer or by the sinker method using a Westphal balance (see figure 9.9). In this method a glass sinker of known weight and volume is immersed first in water and then in the heavy liquid. The formula is

$$\text{specific gravity} = \frac{W_w - W_l}{V} - 1$$

where W_w and W_l are the weights of the sinker in water and in the heavy liquid, respectively, and V is the volume of the sinker.

The techniques described here can be used to determine specific gravity with considerable precision. The accuracy of the measurement, however, is largely determined by the purity of the specimen. For best results, the fragment used in a test should be free of mineralogic inclusions and fracture planes along which air can be trapped.

The specific gravity of a mineral is the measurable consequence of the kinds of atoms present, which differ in atomic weight and the closeness of their packing. The latter is expressed as a **packing index:**

$$\text{packing index} = \frac{\text{volume of ions}}{\text{volume of unit cell}} \times 10$$

The effect of a change in constituent atoms on specific gravity can be seen in table 9.6, which lists the specific gravities and cation weights of the isotypic aragonite group minerals. The effect of packing on the specific gravity is shown in table 9.7, which lists the specific gravities of some polymorphic pairs.

Table 9.6	Relationship of Cation Weight to Specific Gravity		
Mineral	**Formula**	**Atomic Weight of Cation**	**Specific Gravity**
Aragonite	$Ca(CO_3)$	40.1	2.9–3.0
Strontianite	$Sr(CO_3)$	87.6	3.6–3.8
Witherite	$Ba(CO_3)$	137.3	4.2–4.35
Cerussite	$Pb(CO_3)$	207.2	6.4–6.6

Table 9.7	Relationship of Specific Gravity to Packing in Some Polymorphous Pairs

Mineral	Specific Gravity
Diamond	3.5
Graphite	2.1–2.2
Calcite	2.7
Aragonite	2.9–3.0
Quartz	2.65
Tridymite	2.3
Pyrite	4.9–5.1
Marcasite	4.8–4.9

Figure 9.10 presents an overview of the distribution of specific gravity within and among mineral groups. Because no regularity is seen as a function of the anionic components, the importance of cation weight and structural density is emphasized.

Interrelationships of Physical Properties

When matter is arranged in space in a regular way, there are consistent interrelationships among the various physical and geometrical properties of the aggregate. The arrangement and bonding of ions in minerals control such properties as cleavage, habit, hardness, and certain optical phenomena. Similarly, the nature, size, and number of atoms per unit volume correlates with cell dimensions and content, density, and the refractive index.

The bulk density of a crystal and the density of one of the cells that compose it are the same. Thus, D = cell mass/cell volume. D can be determined experimentally with considerable accuracy, so either cell mass or cell volume can be found if the other is known. The cell mass is the mass of the atoms in a formulary unit times the number of such units per cell. The cell volume can be found if its geometric parameters are known. Thus,

$$D = C/V = zM/V$$

where

C = cell mass

z = formula units per cell (cell content)

M = molecular weight

V = cell volume

D is usually expressed in grams per cubic centimeter, and the other variables should also be given in SI units. Thus, cell dimensions, normally in Ångstrom units, must be expressed in centimeters: $1Å = 10^{-8}$ cm, so $V = Å^3 \times 10^{-24} cm^3$. Avogadro's Number, 6.0225×10^{23}, gives the number of atoms per gram; that is, an atom of atomic weight 1.0000 weighs $1/(6.0225 \times 10^{23}) = 1.6602 \times 10^{-24}$ g. Thus,

$$D = \frac{z \times M \times 1.6602 \times 10^{-24}}{V \times 10^{-24}} = \frac{1.6602zM}{V}$$

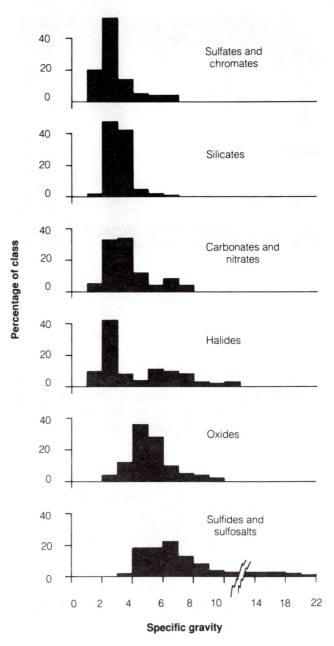

Figure 9.10
Distribution of specific gravity within and between mineral groups.

The volumes of cells can be computed from cell geometric parameters using the following formulas.

Isometric:	$V = a^3$
Tetragonal:	$V = a^2c$
Hexagonal:	$V = a^2c \sin 60°$
Orthorhombic:	$V = abc$
Monoclinic:	$V = abc \sin \beta$
Triclinic:	

$$V = abc\sqrt{1 - \cos^2\alpha - \cos^2\beta - \cos^2\gamma + 2\cos\alpha\cos\beta\cos\gamma}$$

Some sample calculations based on these relations follow.

1. Zircon, $ZrSiO_4$, is tetragonal, and ($a = 6.60$ and $c = 5.98$). It has a density of 4.65. From this information, we can find the number of formulary units per cell (z). The formula weight, FW, is Zr $(91.22) + Si(28.09) + 4\ O(4 \times 16.00) = 183.31$, and the cell volume is $a^2c = 260.49$ Å^3. Thus,

$$z = \frac{DV}{1.66 \times FW} = \frac{(4.65)(260.49)}{(1.66)(183.31)} = 3.98$$

The theoretical value of z can be taken as 4, and the small calculated discrepancy can be ascribed to small differences in D or V due to measurement errors or solid solution.

2. Magnetite, $FeFe_2O_4$, is isometric with a formula weight of 231.55. It has eight formula units per cell and a density of 5.20. Using this information, we can determine the cell edge dimension as follows:

$$V = \frac{1.66 \times z \times FW}{D}$$
$$= \frac{(1.66)(8)(231.55)}{5.20} = 591.34$$
$$a = \sqrt[3]{V} = 8.39 \text{ Å}$$

Properties Dependent on Light

Luster is the general appearance of a macroscopic fresh surface in reflected light. It depends quantitatively on the mineral's composition and qualitatively on its bond type, smoothness of surface, and size of the reflecting grains. For minerals with ionic and covalent bonding that are transparent or light-colored, luster is essentially dependent on index of refraction (n) in accordance with the relation

$$R = \frac{(n-1)^2}{(n+1)^2}$$

where R is the index of reflectivity with values from zero to one.

The index of refraction (n) is the ratio of the velocity of light in air to its velocity within a substance. Refractive index is further discussed in chapter 10.

The reflectivities of opaque minerals, usually those with metallic or metalloidal bonding, depend also on the degree of light absorption, and their reflectivities are given by the expanded relation

$$R = \frac{(n-1)^2 + n^2k^2}{(n+1)^2 + n^2k^2}$$

where k is an absorption coefficient.

Figure 9.11 shows the relations between R and the refractive index for metallic and nonmetallic minerals along with the terms used to describe mineral luster. The luster of minerals is also described by using terms descriptive of commonly known substances,—earthy, greasy, glassy, dull, waxy, silky, pearly, resinous, or pitchy.

Mineral **color** is the combined effect of a mineral's formulary constituents and their arrangement, chemical impurities, and crystal imperfections. As a rule, minerals whose colors are normally dark will not suffer color changes because of the presence of impurities or imperfections. On the other hand, impurities or imperfections in a mineral may interact strongly with light and markedly modify the colors of normally pale minerals. The amount of an impurity need not be large to produce a strongly colored substance; indeed, the amount of coloring agent is often measured in small fractions of a percent. Color should always be determined by examining a fresh surface because surface alterations and tarnishes can obscure the normal color.

Different wavelengths of light are perceived as different colors ranging through the visible spectrum from short-wavelength (λ) violet light ($\lambda \approx 3,900$ Å) to long-wavelength red light ($\lambda \approx 7,700$ Å). When all wavelengths are present the light appears white, but removal of some wavelengths by interference, selective scattering, or absorption gives the substance the color of the remaining wavelengths. Thus, red-colored minerals result from the selective absorption of the shorter blue wavelengths, and blue minerals are those that absorb red light. Should all the light be absorbed, a mineral will be opaque; and if none is absorbed in the visible region, the mineral is transparent.

Light rays passing through a mineral grain are refracted, reflected, and scattered by changes in internal physical conditions, such as fractures or inclusions. Simple scattering of all wavelengths causes translucency, and occasionally scattering from crystallographically oriented inclusions leads to asterism, as in the star of sapphire. Also, scattering from very tiny inclusions may be wavelength selective and the mineral assumes a color due to the scattering loss of part of the spectrum. Refraction and reflection get the wave trains out of phase and may be the cause of brilliant interference colors, as seen in some opal and the surface tarnishes of a number of minerals.

Solid solution involving the substitution of one ion for another in a crystal structure always causes some distortion of the local electronic configuration. Not uncommonly, later exposure of such substituted structures to ionizing radiation provides the energy for an electron to transfer from one ion to its neighbor, thus changing the effective valence of both. Localized electron transfer of this kind forms a **color center** where light is selectively absorbed. Because the mineral grain may contain many scattered absorbing points, the mineral assumes a color characteristic of the particular solid solution. The transition elements (Ti, V, Cr, Mn, Fe, Co, Ni, Cu), with partially filled d orbitals, and the lanthanide and actinide series elements, with partially filled f orbitals, are particularly active in color development because of the ease of

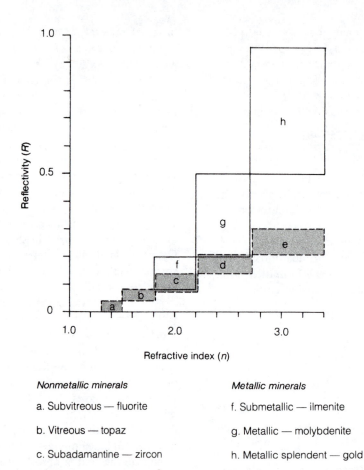

Nonmetallic minerals

a. Subvitreous — fluorite

b. Vitreous — topaz

c. Subadamantine — zircon

d. Adamantine — diamond

e. Adamantine splendent — cinnabar

Metallic minerals

f. Submetallic — ilmenite

g. Metallic — molybdenite

h. Metallic splendent — gold

Figure 9.11
Luster of minerals as a function of reflectivity, *R,* and refractive index, *n.*

electron jumps that absorb energy in the visible region. Predicting color is difficult, however, because the same element in different structural environments may have different transitional energy and thus cause different coloration.

The colors of the mineral quartz provide an excellent example of the varied mechanisms that cause mineral coloration. Normally, quartz is clear and colorless, but internal fractures or inclusions such as tiny bubbles can cause it to assume a milky appearance. Exsolution of submicroscopic needles, usually rutile, causes scattering of blue light and results in a rosy color, whereas coarser needles scatter red light and blue quartz results. Substitution of Al^{3+} for Si^{4+} followed by irradiation yields a color center that absorbs over the entire visible spectrum, creating a dark-colored or smoky quartz. Smoky quartz is bleached colorless if heated. Substitution of Fe^{3+} results in a selective color center that causes the purple color of amethyst. If amethyst is heated, the center is modified through reduction of Fe^{3+} to Fe^{2+}, and yellow citrine or, very rarely, green quartz results.

Streak is the color of the powdered mineral and has a much more constant color than does the bulk mineral. For example, a series of colored glasses, when powdered, all show a very similar white streak. Streak tests are usually made by rubbing the mineral across a piece of unglazed porcelain (streak plate) and examining the resultant powder smear against the white background of the plate. If no streak plate is available, the color of the mineral powder can be observed on white paper.

Many mineral properties that depend on light are sometimes observed, but because of their limited applicability to mineral identification we will describe them only briefly.

1. **Diaphaneity.** The property of transmitting light, described as transparent, translucent, or opaque.
2. **Play of colors.** Different radiant colors observed as the mineral is turned (e.g., precious opal, labradorite).
3. **Iridescence.** A play of colors caused by a thin coating on the mineral surface or partially opened cleavages (e.g., goethite).

4. **Opalescence.** Milky or pearly reflections from the interior of the specimen (e.g., moonstone).

5. **Chatoyancy.** Silky sheen resulting from fibrous structure (e.g., satinspar variety of gypsum, tiger eye).

6. **Tarnish.** Dull, but occasionally brilliant, thin surface coatings resulting from chemical reactions between the mineral and the air (e.g., copper, bornite).

7. **Aventurine.** Spangled surface with minute scales of mica or hematite (e.g., orthoclase, quartz).

8. **Asterism.** A six-pointed star of light seen in reflected or transmitted light (e.g., star sapphire, phlogopite).

9. **Pleochroism.** The appearance of various colors when the specimen is viewed in different directions as a result of selective absorption of light along different crystallographic directions (e.g., tourmaline, cordierite).

10. **Luminescence.** Emission of light from a nonincandescent material.

11. **Triboluminescence.** Luminescence of the mineral when it is scratched, rubbed, or crushed (e.g., fluorite, sphalerite).

12. **Thermoluminescence, pyroluminescence.** Luminescence is produced on heating (e.g., tourmaline).

13. **Fluorescence.** Mineral becomes luminous during exposure to x rays or ultraviolet light (e.g., scheelite).

14. **Phosphorescence.** Luminescence that continues after exposure to x rays or ultraviolet light (e.g., pectolite).

Light from Excited Atoms

The preceding discussion, as well as those in chapters 4 and 10, tacitly assumes that atoms are built up of electrons disposed in a configuration that places the electrons in the lowest possible energy state. Such an electronic configuration constitutes the **ground state** of an atom. However, atoms whose electrons are in states of higher energy can and do exist. Additions of energy to an atom from an external source may cause its electrons to go to states of higher energy, much as warming an inflated balloon increases the energy of the enclosed air. Atoms in which electrons have been raised to higher levels of energy are in an **excited state.** Such electron states are not stable, and an excited electron will return to its ground state within a very short but finite time after excitation ($\approx 10^{-8}$ seconds). At this time it will emit the same amount of energy as was previously absorbed. The effect would be as though the balloon had a delayed-action safety valve.

When an atom is subject to an external electromagnetic field, such as light, the various electron transitions that are allowed act as a series of oscillators with different

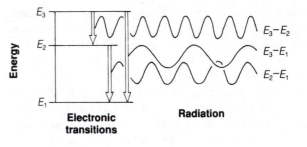

Figure 9.12
Electron transition and radiation in an excited atom.

resonant frequencies. Free atoms have only a discrete set of excited energy levels and can absorb only a discrete set of frequencies. Atoms bonded into various crystalline aggregates have a greater number of resonating frequencies, and in metals a continuum of excited levels exists.

The energy differences between any two states of an atom are controlled by quantized energy differences according to the relation

$$E_2 - E_1 = h\nu$$

where E_2 and E_1 are the energies of the higher and lower states, h is Planck's constant, and ν is the frequency. When the energy flow associated with an electromagnetic field provides an energy increment to an atom such that the energy difference between a lower and higher state is exactly matched, this energy is absorbed by the atom, held for a short time, and given back as radiation.

In general, the radiated light has the same frequencies as the absorbed light, but in certain situations the light returned from an atom does not have the same frequencies it had when it was absorbed. Consider the situation shown in figure 9.12, in which the energy of the ground state of an electron and two possible excited states are represented by horizontal lines. Absorption of energy sufficient to raise the energy level of the electron to level E_2 can be given back only as a single frequency, $E_2 - E_1$. Absorption of energy sufficient to reach level E_3, however, can be reemitted as a single frequency, $E_3 - E_1$, or as two frequencies, $E_3 - E_2$ and $E_2 - E_1$, by a stepwise return of the electron to its ground state. If the frequencies at which radiation occurs are all in the visible region, the quality of the reemitted light is not affected. If, however, one or more of the radiated frequencies are outside the limits of visible light, then some portion of the visible light is removed and colors arise by subtraction.

The change of electromagnetic frequencies through the mechanism of electron transitions is not required for an understanding of the general phenomena associated with the interaction of light and matter, but it is very useful in understanding certain associated phenomena. Many transitions between electron energy states, especially for outer electrons, cause energy to be radiated as light. Light itself may be the source of energy that starts the process.

Certain substances absorb higher frequencies of radiant energy (shorter wavelengths) than visible light and give back some portion of that energy in the visible region. The fluorescence of substances irradiated with x rays or ultraviolet light illustrates this phenomenon. Electrons will radiate energy whenever they go through a transition by which they lose energy. The energy that raised the electron to an excited step initially need not be electromagnetic in nature, however. Thermal luminescence and flame coloration occur through the input of thermal energy, and chemiluminescence occurs through the input of chemical energy, as does phosphorescence. Input of mechanical energy may give rise to triboluminescence.

Magnetic, Electrical, and Thermal Properties

The differing electrical and magnetic properties of minerals have limited use in hand-specimen identification of species because these properties are generally difficult to measure and the values overlap from species to species. Electrical and magnetic properties, however, are widely exploited in laboratory determinations and chemical analysis (see chapter 11), in the search for ore deposits and in large-scale separation of valuable minerals from waste (gangue) material.

Magnetic Properties

The magnetism of a material is an atomic property that depends on the summation of the magnetic moments of constituent electrons. Electrons possess magnetic moments because they are spinning and orbiting in an electric field. When the spins of two electrons are opposed, the net magnetic moment is zero, but parallel spins are additive (see chapter 4). In most ionic and covalent compounds where electrons are paired, the total magnetic moments are small, and the compounds are repelled by a strong magnetic field. Such substances are called **diamagnetic.** Examples are gypsum, halite, and quartz. Structures containing parallel electron spins are attracted to a magnet. Such **paramagnetic** substances include garnet, biotite, and tourmaline. If interatomic forces that maintain the magnetic moments of many atoms parallel to each other are present (a high-spin state), the magnetic moments of this domain are correspondingly large and the substance exhibits **ferromagnetism.** Examples are magnetite, pyrrhotite, and native platinum. As might be predicted from the rules governing electronic configurations, iron is responsible for most of the observed magnetic properties of minerals, as shown in figure 9.13, in which the magnetic susceptibility of a number of paramagnetic minerals is plotted against their iron content. The magnetic properties of minerals can be employed for physical separation of species from pulverized mixtures using an apparatus like that shown in figure 9.14.

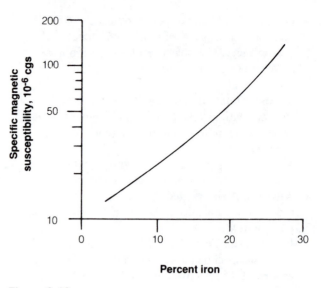

Figure 9.13
Effect of iron content on magnetic susceptibility of paramagnetic minerals.

Electrical Properties

The electrical properties of a mineral can be separated into conductive, static charge, and thermoelectric phenomena. The ability of a substance to conduct electricity depends on the presence within the material of charge carriers that are free to migrate. In perfect crystals, such carriers can only be electrons which in turn are free to migrate only in structures with metallic or semimetallic bonds, such as native metals and the sulfides. Electronic conduction through such structures takes place readily whenever a potential gradient is established. Relatively small amounts of charge may be transferred by the migration of ions or vacancies through an imperfect or defective structure.

Poor conductors may be electrically charged by any of several means and will retain this charge for a period of time. If the mineral does not have some symmetry element relating faces on opposite sides of the crystal (for example, if it lacks a center of symmetry but has a 2- or 6-fold rotoreflection axis), unlike charges appear on the opposed faces. The presence of these charges can be proved by dusting the electrified specimen with a mixture of powdered red lead and sulfur, which are attracted to negatively and positively charged surfaces respectively. The crystal may be electrified by friction, by heating or cooling (pyroelectricity), or by pressure (piezoelectricity). The relationship of the latter two properties in the crystal classes is given in table 9.8.

Thermal Properties

The thermal properties of minerals are not readily used in identification but are very important in the theoretical aspects of mineralogic reactions (see chapter 6). Thermal properties fall naturally into two groups, those related to

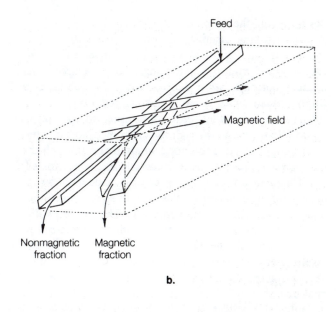

Figure 9.14

(a) The Franz isodynamic magnetic separator, which uses magnetic susceptibility for mineral separation. Semiquantitative determinations of mineral magnetic susceptibilities can also be made. (Courtesy S. G. Franz Co., Trenton, NJ.) (b) Operating schematic. Roll and pitch of the split channel and magnetic field intensity can be varied.

the transfer of heat and those related to changes in heat content occurring with mineralogic reactions.

Heat passes through matter both by conduction due to thermal vibration of the constituent atoms and by transmission through interatomic space. These mechanisms are called **thermal conduction** and **thermal transmissibility,** respectively. Thermal conduction through matter differs with both the material and the crystallographic direction. The different conductivity in different directions can be observed by coating a crystal or cleavage face with paraffin and observing the shape of the melted area when the end of a hot wire is placed against the surface. The wax will melt in a circle on faces with 3-, 4-, and 6-fold symmetry and melt in an ellipse on those that are 1- or 2-fold.

Thermal transmissibility is analogous to optical diaphaneity. Materials that are transparent to heat are termed **diathermanous,** whereas those that are opaque to heat are **athermanous.** This property, like thermal conduction, is controlled by symmetry.

The increased thermal agitation of atoms that occurs when minerals are heated causes each atom to effectively occupy a larger volume of space, and **thermal expansion** of the whole crystal follows. This expansion is not uniform in all directions for most structures but is controlled by the symmetry of the mineral and can be divided into isotropic and anisotropic classes. Thermal expansion is usually expressed as the **coefficient of thermal expansion,** which is the change in volume per unit volume for a temperature change of 1° C.

Temperature, as distinguished from heat, provides one very useful marker in mineral identification. This is the temperature at which the crystallinity of the structure is

Table 9.8	Piezoelectricity and Pyroelectricity in the Crystal Classes	
Class	**Piezoelectric**	**Pyroelectric**
1	+	+
$\bar{1}$	−	−
2	+	+
m	+	+
2/m	−	−
222	+	−
2mm	+	+
2/m 2/m 2/m	−	−
3	+	+
$\bar{3}$	−	−
3/m	+	−
32	+	−
3m	+	+
$\bar{3}$ 2/m	−	−
432	−	−
$\bar{4}$3m	+	−
4/m $\bar{3}$ 2/m	−	−
4	+	+
$\bar{4}$	+	−
4/m	−	−
422	+	−
4mm	+	+
$\bar{4}$2m	+	−
4/m 2/m 2/m	−	−
6	+	+
6/m	−	−
622	+	−
6mm	+	+
$\bar{6}$m2	+	−
6/m 2/m 2/m	−	−

destroyed and the mineral decomposes, vaporizes, or fuses. A useful scale of fusibility is given as table 9.9.

The mineralogical blowpipe is a simple tool for the production of intense local heating. The fuel source is any hydrocarbon-rich flame, such as that of a candle or a Bunsen burner with the air intake closed. Oxygen from the lungs or some other source of compressed air is introduced into the flame through the blowpipe, and a narrow, hot jet is produced. Too strong an airstream through the blowpipe will cause the jet to flutter and should be avoided.

The intense heat at the tip of a blowpipe jet or a propane torch is sufficient to cause the fusion, transformation, or sublimation of many minerals. The absolute value of this temperature varies with the individual and equipment but is usually between 1,200 and 1,500° C. For best results the blowpipe should be calibrated for an individual user using the data in table 9.9. This can be done by melting a sharp edge or point of a mineral fragment by a few minutes' application of the blowpipe jet. A hand lens is useful for examining the fragment before and after heating. Melting temperature is a very useful property and can be obtained by this method with very little practice.

The manner in which a mineral fuses may also be distinctive and should be noted whenever minerals are fused. Four types of fusion are characteristic:

1. **Simple melting.** The test fragment changes passively from a solid to a liquid (e.g., stibnite).
2. **Intumescence.** The test fragment swells and bubbles because gases (such as H_2O and CO_2) are being liberated from the molten assay (e.g., lepidolite).
3. **Exfoliation.** The test fragment protrudes small branches (e.g., stilbite).
4. **Decrepitation.** The test fragment snaps or explodes because of the rapid evolution of gases such as steam, CO_2, or SO_2.

The presence of hydroxyl ions or water in a specimen can be readily ascertained by heating the powdered substance in a closed tube over a Bunsen burner. The closed tube is simply a test tube or a 10–15 cm piece of 5–7 mm soft glass tubing sealed at one end. Water, if present, is driven out of the powdered mineral and condenses as droplets on the colder upper portion of the tube.

An input of thermal energy to atoms or ions can raise their electrons to excited levels. For the atoms of many elements, this amount of thermal energy is available in an ordinary Bunsen burner flame, and the wavelength of the emitted radiation is in the visible portion of the spectrum. As a consequence, introducing such elements into a flame gives the flame a color characteristic of the element.

Flame coloration can be observed if a platinum wire, cleaned by heating, is dipped into a solution of the specimen and then held in the flame. Specimens to be tested in this manner should be powdered, placed on a watch glass, and moistened with concentrated HCl. The solution can then be used to moisten the platinum wire.

Another technique, which does not require dissolving the specimen, is to introduce the powdered mineral into

Table 9.9	Melting Points and Fusibility Numbers		
Fusibility Number	Mineral	Temperature (° C)	Metal
	Diamond	>3,550	
		3,410	Tungsten
	Corundum	2.980	
	Periclase	2,800	
	Zircon	2,550	
	Forsterite	1,910	
	Rutile	1,840	
		1,772	Platinum
7	Cristobalite	1,713	
	Leucite	1,686	
	Hematite	1,565	
	Anorthite	1,551	
		1,535	Iron
	Fayalite	1,503	
White Heat			
6	Beryl	1,410	
	Diopside	1,391	
	Fluorite	1,360	
5	Rhodonite	1,323	
Incipient White Heat			
4	Cuprite	1,235	
	Actinolite	1,200	
	Molybdenite	1,185	
	Pyrite	1,171	
	Cassiterite	1,127	
	Galena	1,114	
	Albite	1,100	
Yellowish-Red Heat			
		1,083	Copper
	Wulfenite	1,065	
		1,064	Gold
3	Almandine	1,050	
	Nickeline	968	
		961	Silver
Bright Red Heat			
	Halite	801	
2	Chalcopyrite	800	
Dark Red Heat			
		660	Aluminum
		631	Antimony
1	Stibnite	550	
Dull Red Heat			
	Chlorargyrite	455	
		420	Zinc
		327	Lead
	Orpiment	300	
		271	Bismuth
	Carnallite	265	
		232	Tin
	Sulfur	113	
	Mirabilite	32	
	Ice	0	
		−39	Mercury

Table 9.10　Flame Coloration

Element	Flame Color	Remarks
Barium, Ba	Yellow-green	Readily obtained
Bismuth, Bi	Pale greenish white	—
Boron, B	Yellow-green	—
Calcium, Ca	Orange-red	Readily obtained
Copper, Cu	Emerald green	If the mineral is moistened with HCl, the flame will be azure blue; readily obtained
Lead, Pb	Pale skim-milk blue	Usually observed on a block of charcoal
Lithium, Li	Crimson	—
Molybdenum, Mo	Faint yellow-green	—
Phosphorus, P	Pale bluish green	Better results with H_2SO_4 than with HCl
Potassium, K	Pale violet	Purplish-red when observed through cobalt glass; may be necessary to decompose mineral with a flux of gypsum to obtain a flame color
Selenium, Se	Indigo blue	Horseradish odor
Sodium, Na	Intense yellow	Readily obtained; often obscures less sensitive elements
Strontium, Sr	Crimson	Readily obtained; residue is alkaline, which differentiates it from lithium
Tellurium, Te	Grass green	—
Zinc, Zn	Bluish green	Usually appears as bright streaks in the flame

the air hole at the base of a Bunsen burner. This procedure can be used whenever specimens are being ground in a mortar. A few quick plunges of the pestle into a mortar containing finely powdered material yields a fine, airborne dust that is easily entrained into the burner intake. The flame colors that are characteristic of certain elements are given in table 9.10.

Concluding Remarks

The physical properties of minerals, like the physical characteristics of people are readily recognized once they become familiar. The physical properties of minerals are based on the kinds of atoms, their arrangement, and the nature of their chemical bonding in different species. Aside from their use in mineral identification, physical properties are important in all aspects of mineral exploration, exploitation, and use.

The most useful properties for identifying minerals are manner of breaking, scratch hardness, specific gravity, and interaction with light, particularly in the way of luster and color. The determinative tables in appendix B are arranged using these parameters. Many minerals break on planar cleavage surfaces, resulting in fragments of characteristic shapes. Mineral hardness has a wide range, and dividing minerals into those harder or softer than test items such as quartz, a knife blade, or a copper penny provides useful groupings. Determining specific gravity requires

testing apparatus but is a particularly useful means for discriminating between species. Mineral luster and color arises from interaction of the mineral substance with light. Light properties that are particularly useful include the distinction between metallic and nonmetallic lusters and the characteristic colors of many mineral species.

Suggested Readings

Correns, C. W. 1967. *Introduction to Mineralogy.* 2d ed. New York: Springer-Verlag. Minerals and mineralogy in the larger context of petrology. Determinative tables.

Frye, K. 1981. *The Encyclopedia of Mineralogy.* New York: Van Nostrand Reinhold.

Kostov, I. 1968. *Mineralogy.* English ed. Translated by P. G. Embry and J. Phemister. Edinburgh: Oliver and Boyd. Mineral descriptions include crystallographic, crystal chemical, morphologic, and physical properties. Profusely illustrated. Extensive bibliography.

Nassau, K. 1980. The causes of color. *Scientific American* 217: 222–38.

Read, H. H. 1970. *Rutley's Elements of Mineralogy.* 26th ed. London: Thomas Murby. General mineralogy text including chapters on the chemical, optical, and physical properties of minerals and on crystallography. Contains mineral descriptions arranged by principal cations.

Sinkankas, J. 1972. *Gemstone and Mineral Data Book.* New York: Winchester. Extensive tabulation of mineralogic data.

10

Interaction of Light and Matter

The refractory nature of most geologic materials has often caused geologists to turn to physical methods of analysis that do not require dissolving the sample. Such methods typically employ some portion of the electromagnetic spectrum because when electromagnetic radiation of any sort passes through or is reflected by matter, the two must interact. The result is detectable by comparing the qualities of the incident and exit beams.

The Nature of Light

Light is electromagnetic radiation with wavelengths between about 3,900 and 7,700 Ångstroms, or 390–770 nanometers (see figure 11.9) whose wave motion is transverse to the direction of its propagation. Because of its wave nature, the passage of light causes an electrical disturbance that varies sinusoidally with time. Thus, any point through which a light ray passes is subjected to a fluctuating electrical field in which a charged particle, such as an electron, is set into forced vibration.

Because atoms (and ions) consist of an electrically positive nucleus in a cloud of negatively charged electrons, these differently charged parts tend to become separated as a light wave passes. Protons and neutrons are about 1,800 times heavier than electrons. Thus, the nucleus has most of the mass and hence the inertia of an atom. Because of this, the oscillating electrical field of passing light has little effect on the atom's nucleus. The electrons, however, are strongly affected. Under ordinary circumstances the electrical potential of a light wave is insufficient to remove an electron from the intense electrostatic field of the nucleus, but the electrons do slosh back and forth as the field oscillates from positive to negative and back again. The result is a rapidly oscillating atomic dipole as shown in figure 10.1.

The periodic asymmetry of an atom's nucleus and electron cloud sets up a small, fluctuating electrical field of its own that opposes the field of the light waves. This field puts a load on the light wave and slows it down. An oscillating atomic dipole is only one of the many combinations of electrified units that may be set into forced vibration by the passage of light. Each positive and negative pair resonates with a characteristic frequency, and in general the load on a light wave is the sum of all loads from the different oscillators.

A mechanical analogy for the slowing of a light wave by its interaction with other fields is a weighted spring hanging vertically. This spring will execute simple harmonic motion if it is displaced and released. The frequency of the motion can be decreased by adding more weight to the end of the spring, increasing the inertia that must be overcome.

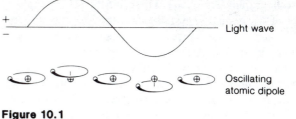

Figure 10.1
Oscillating atomic dipole.

Table 10.1	Refractive Index and Luster of Some Isotropic Minerals	
Mineral	**Luster**	**Refractive Index**
Fluorite	Subvitreous	1.43
Analcime	Vitreous	1.49
Halite	Vitreous	1.54
Spinel	Vitreous	1.75
Andradite	Subadamantine	1.87
Sphalerite	Adamantine	2.37
Diamond	Adamantine	2.42
Cuprite	Adamantine splendent	2.85

Refractive Index

The loading of a light wave by electrical oscillators decreases the velocity of light in a material medium below that of its velocity in free space. The constant of proportionality between these two velocities is known as the **index of refraction** (n), which is the ratio

$$\frac{\text{velocity in free space}}{\text{velocity in the material medium}}$$

and is hence always greater than unity. The refractive index of air is only 1.00029 and is usually used as a reference.

In general, the greater the contrast between light velocities within and outside a substance (the larger is n), the more flash and brilliance the substance shows. The variations in n are often grouped under the term *luster* (see chapter 9 for further discussion of luster). For examples of variation in luster with refractive index, compare the appearance of the minerals in table 10.1.

In transparent gases, liquids, amorphous solids, and isometric crystals, the kind and arrangement of atoms is such that every direction in the crystal presents the same aspect to passing light. Such substances are **isotropic.** In certain other substances, the arrangement of atoms is such that the electrical load on a ray of light is different in different directions. Light traveling through such materials must then move with different velocities along different crystallographic directions. These substances are **aniso-tropic.** Only minerals crystallizing in the isometric system and amorphous substances are isotropic; minerals crystallizing in all other systems are anisotropic.

The refractive index of a substance depends on many things, one of the more important being the nature of the constituent atoms or ions and their arrangement. As a rule, the heavier the atoms and the more tightly they are packed, the more they will slow passing light and the higher will be the index of refraction. The simple relationship is

$$K = \frac{n - 1}{\rho}$$

where K is the specific refractive energy of the substance, n its average refractive index, and ρ its density.

For complex substances, K is the weighted sum of the contributions of the various components

$$K = kx_1 + kx_2 + kx_3 + \ldots$$

where k represents the specific refractive energy and x the weighted molecular proportion of each component. Specific refractive energies for a number of common elements and oxides together with their respective atomic or molecular weights are given in table 10.2.

As an example of the use of these relations, consider the calculations for orthoclase, which is anisotropic with refractive indices of 1.518, 1.526, and 1.524 and has a density of 2.56. Thus,

$$K = \frac{n - 1}{\rho} = \frac{\sqrt[3]{(1.518)(1.526)(1.524)} - 1}{2.56}$$

$$= 0.20$$

Further, the proportions of the components of orthoclase, (see table 10.3), $K(AlSi_3O_8)$ or $K_2O \cdot Al_2O_3 \cdot 6SiO_2$, are equal to the molecular weights of the constituent oxides divided by the molecular weight of the compound. Taking values of k from table 10.2,

$$K = (kx)_K + (kx)_{Al} + (kx)_{6Si}$$

$$= (0.19)(0.17) + (0.19)(0.18) + (0.21)(0.65)$$

$$= 0.20$$

The index of refraction of tiny crystalline fragments can be determined accurately with a microscope by the **oil immersion method,** in which the grains are successively immersed in oils of known refractive index. In practice, the grains are placed on a microscope slide and covered with a cover slip. Then an oil drop placed at one edge is drawn between the plates by capillarity. Since the fragments must be thinner on their edges than in their middles, they act as tiny lenses that are concave or convex depending on the relative indices of oil and crystal. In a liquid of lower index, the lens action concentrates a beam of parallel light as does a convex or positive lens. Thus, an upward focus of the microscope tube causes a bright, fringing line of light, called the **Becke line,** to move inward.

Table 10.2	Specific Refractive Energies of Some Mineral Components	
Species	**Approximate** k	**Species Weight**
C	0.15	12.01
Cl	0.40	35.45
F	0.04	18.99
O	0.20	15.99
S	1.00	32.06
Ag_2O	0.15	231.74
Al_2O_3	0.19	101.96
B_2O_3	0.22	69.62
BaO	0.13	153.34
BeO	0.24	25.01
CO_2	0.22	44.01
CaO	0.23	56.08
CoO	0.18	74.93
Cr_2O_3	0.27	151.99
CuO	0.19	79.55
FeO	0.19	71.85
Fe_2O_3	0.31	159.70
H_2O	0.33	18.02
K_2O	0.19	94.20
Li_2O	0.31	29.88
MgO	0.20	40.31
MnO	0.19	70.94
MoO_3	0.24	143.94
Na_2O	0.18	61.98
Nb_2O_5	0.30	265.82
NiO	0.18	74.71
P_2O_5	0.19	141.94
PbO	0.14	223.19
SiO_2	0.21	60.09
SnO_2	0.15	150.69
SrO	0.14	103.62
Ta_2O_5	0.13	441.89
TiO_2	0.40	79.90
V_2O_5	0.43	181.88
WO_3	0.13	231.85
ZnO	0.15	81.37
ZrO_2	0.20	123.22

Table 10.3	Proportions of the Components of Orthoclase	
Oxide	**Molecular Weight**	**Proportion**
K_2O	94.20	0.17
Al_2O_3	101.96	0.18
$6 SiO_2$	6(60.09)	0.65
	556.70	

The opposite is observed if the crystal index is less than that of the oil. Because index oils can be accurately calibrated, indices of refraction can be measured to ± 0.0005 under ideal conditions, although in practice a precision of ± 0.001 is more likely.

A related method for measuring refractive index is to shadow part of the microscope field by inserting a card below the stage and noting which side of the fragment is shaded. If the particle is shaded on the side facing the darkened field, its index of refraction is lower than the

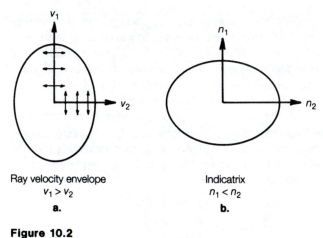

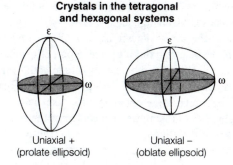

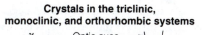

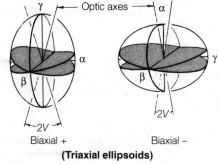

Figure 10.2
Reciprocal relationship of the ray velocity envelope and optical indicatrix.

immersion medium; and vice versa. Colored shadings occur when the grain and medium are close in index; blue appears away from the field shading and red-brown facing it.

The Optical Indicatrix

The velocity with which light travels along any path through a solid is controlled by the drag it experiences as its transverse wave motion is damped by atoms lying in the ray path. Both the packing density of constituent ions and their structural arrangement will affect the passage of light. Imagine the wave front of a light flash emitted from a point source that is embedded in a crystal. If the density of atoms is the same in all directions, the loading on the light will be radially symmetrical. An instant after the light source is emitted and while the wave front is still within the crystal, the wave front is in the form of a sphere. If, however, the density of atoms varies for different directions, as it would in nonisometric substances, the light will travel different distances along different crystallographic directions and the wave front will be ellipsoidal (see figure 10.2a). This is a useful concept but not a practical one because the velocity of light cannot be measured under these circumstances.

The ratio of light velocities, however, is the reciprocal of the ratio of the indices of refraction:

Index of Refraction, n	Velocity of Light, km/sec
1.00 (free space)	299,792.5
1.50	199,861.7
2.00	149,896.3
2.50	119,917.0
3.00	99,930.8

Light in a substance of $n = 1.50$ has its velocity reduced by one-third, at $n = 2.00$ the speed is halved, and at $n = 3.00$ the speed is only one-third of that in free space.

Figure 10.3
The optical indicatrix. Circular sections are shaded.

Because of this relationship, the velocity ellipsoid can be transformed into a refractive index ellipsoid (see figure 10.2b). Relative and absolute values of the refractive index can be readily determined for different vibration directions within a transparent substance. To represent the differential aspects of light speed in different crystallographic directions, it is conventional to use a three-dimensional figure that shows the refractive index for light waves in their direction of vibration (not propagation). This representation (figure 10.3), known as the **indicatrix,** is a spherical or ellipsoidal surface having the different indices as axes.

For isometric crystals, crystallographic axes are all the same length, and the atomic aspect is the same in three mutually perpendicular directions. Consequently, the refractive indices in these directions are the same and the indicatrix is a sphere.

Crystals in the tetragonal and hexagonal systems have a unique crystallographic axis normal to the plane of two or three equal axes. Indices for light vibrating in this plane (traveling parallel to c) are the same and larger or smaller than the index for vibration in the c-direction. The indicatrix is thus a uniaxial ellipsoid having a circular section in the ab plane. Minerals crystallizing in the tetragonal or hexagonal systems are thus referred to optically as **uniaxial.** An index in the circular section of the indicatrix is designated as ω and the index along the plane normal, the **optic axis,** is designated as ϵ. The ellipsoid is prolate or positive $(+)$ if $\epsilon > \omega$, and oblate or negative $(-)$ if $\epsilon < \omega$ (see figure 10.3b). The difference in index between ϵ and ω is called **birefringence.**

For all other crystal systems, $a \neq b \neq c$ and the indicatrix must be a triaxial ellipsoid. Such ellipsoids have two circular sections in which indices are equal and two optic axes normal to them. This type of indicatrix is thus **biaxial.** The acute angle between optic axes is called the **optic angle,** $2V$. The lengths of the axes of the biaxial indicatrix are proportional to the three refractive indices, with α being the smallest, β the intermediate, and γ the greatest. In orthorhombic crystals, the axes of the indicatrix coincide with the crystallographic axes $a, b,$ and c. In monoclinic substances, one indicatrix axis coincides with b; the other two are in the ac plane but independent of a and b directions. In triclinic crystals, none of the axes of the indicatrix coincide with crystallographic axes. Birefringence in biaxial crystals is the difference between the greatest and least indices, or $\gamma - \alpha$.

Crystals with a layered structure typically have an oblate indicatrix; that is, they are optically negative. This is because of greater damping (higher index) in the direction of the densely packed layer planes. Exceptions are found, especially in hydroxides in which the oxygens in $(OH)^-$ groups are polarized by H^+ perpendicular to the layer. Chain structures, on the other hand, are generally optically positive with the highest index in the chain direction.

Dispersion and Absorption

The charged components of matter that oscillate under the influence of a passing light wave have mass and consequently inertia. For this reason, they do not react instantaneously to the electromagnetic field of light but lag to some extent behind it. The amount of lag depends on the frequency of the light; and as the frequency increases, the amount of lag increases. As a consequence, there is a variation or **dispersion** of the index of refraction with frequency, as shown in figure 10.4a. This curve is only a portion of a more extended curve shown in figure 10.4b. The sudden drop in the curve represents a resonant frequency at which light is strongly absorbed by the material because of energy transfer from the light to some resonant oscillator.

The frequency of light is an intangible property, and it is customary to use the readily measured parameter of wavelength when dealing with ordinary light. Light velocity equals frequency times wavelength, and hence frequency is 1/wavelength.

The absorption of radiant energy from electromagnetic radiation is not restricted to wavelengths in the visible region. Most materials have several regions of anomalous dispersion in which absorption occurs. For example, quartz is opaque to certain infrared wavelengths to which stibnite is perfectly transparent.

White light is an incoherent mixture of all visible wavelengths which, if separated into its components, would be perceived as a sequence of colors constituting the visible spectrum. These colors and their wavelengths in Ångstroms are as follows:

Violet	3,900–4,300
Indigo	4,300–4,600
Blue	4,600–5,000
Green	5,000–5,700
Yellow	5,700–5,900
Orange	5,900–6,500
Red	6,500–7,700

White light can be separated into its component colors by several means. Colors can be removed by the selective absorption of filters (including minerals). All colored substances owe their color to this or a related action; their color is white minus the removed wavelength. For example, the removal of wavelengths above 5,000Å by absorption results in a bluish color.

White light can also be resolved into its component colors by passage through a prism. Because the different wavelengths have different refractive indices, each is bent through a different angle on entering and leaving the prism. Figure 10.5 shows a beam of white light decomposed through a prism into its component colors.

The angle through which a light ray is bent on passing from one medium to another, the **angle of refraction,** is dependent on the contrast in refractive index of the two materials. For a particular wavelength the relations are given by **Snell's Law,** (formulated in 1621 by the Dutch mathematician Willebrod Snell):

$$n = \sin i / \sin r$$

where i is the angle of incidence of light arriving at the interface in medium 1 and r is the angle of refraction of light in medium 2. The refractive indices of the two media have the relation

$$n_1/n_2 = \sin i / \sin r$$

where $n_1 < n_2$ (see figure 10.6a). Holding $\sin i$ constant and knowing either index, we can find the other index by measuring the angle of refraction. Accurate measurements of the angle of refraction of liquids can be made using a **refractometer** in which the index of the unknown liquid is compared to that of air.

As the angle of incidence and $\sin i$ increase, there must come a point at which the angle of refraction is $90°$, where $\sin r = 1$. Under this condition the refracted ray lies along the interface (see figure 10.6b). This angle of incidence is

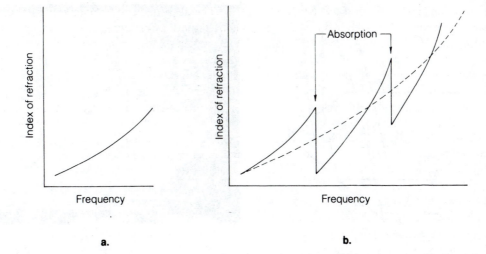

a.

b.

Figure 10.4
Dispersion (a) and absorption (b) of light versus the index of refraction.

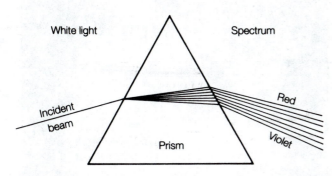

Figure 10.5
Dispersion of white light into a color spectrum by refraction through a prism.

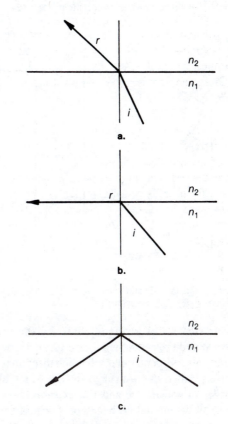

Figure 10.6
Refraction and total internal reflection where $n_1 > n_2$.

known as the **critical angle.** Beyond the critical angle, the ray is totally internally reflected and no light escapes at the point of incidence (see figure 10.6c). For fluorite ($n = 1.43$) the critical angle is about 44°, whereas for diamond ($n = 2.42$) it is only 24.4°.

The flash and brilliance of a well-cut diamond is due to its high index of refraction, the internal reflection of light entering the stone through facets and the table on top, and strong dispersion effectively separating the spectral colors, as shown in figure 10.7.

Polarization of Light

An ordinary light ray is comprised of a heterogeneous group of light waves having different wavelengths and vibrating in different planes perpendicular to the direction in which the ray is traveling. If all the wavelengths are the same as they are for monochromatic or laser light, the ray is **coherent,** and if the vibrations are constrained to a single direction by reflection, refraction, or differential absorption, the light is **polarized.**

When incident light strikes a crystal, part is reflected and part refracted, and both portions of the beam are polarized. Within anisotropic crystals, the refracted light is further divided into two rays, polarized at right angles to each other, that travel with different velocities along different paths. One ray vibrates in a plane defined by the direction of ray propagation and a normal to the c-axis,

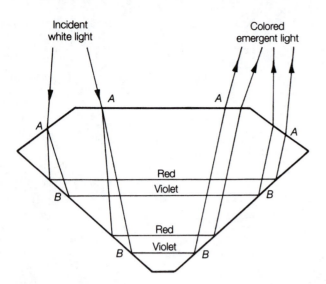

Figure 10.7
How a cut diamond works. Refraction takes place at table and side facets (A); total internal reflection occurs at bottom facets (B).

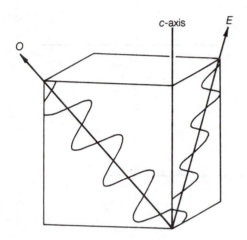

Figure 10.8
Vibration planes of ordinary (O) and extraordinary (E) rays.

and the other vibrates in a plane defined by the propagation direction and the c-axis (see figure 10.8). These two rays have mutually perpendicular vibration directions and are respectively called the **ordinary ray** (O), which obeys the usual rules of refraction, and the **extraordinary ray** (E), which exhibits anomalous behavior. Because the ordinary and extraordinary rays proceed through the crystal with different velocities, they have different refractive indices.

The separation of refracted light into two rays is called **double refraction.** The amount of double refraction is usually not sufficient to be detected without a properly equipped microscope, but in the case of calcite (see figure 10.9) a doubling of images can easily be seen when viewed through a cleavage rhomb.

When a thin slice or small fragment of a crystal is observed on the stage of a polarizing microscope, the transmitted light is constrained to vibrate parallel to a cross

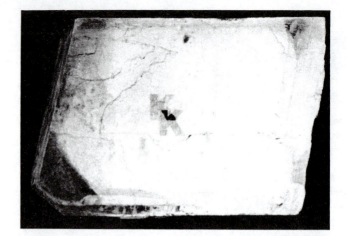

a.

b.

Figure 10.9
Double refraction in calcite. The rhomb in (b) has been rotated 90° clockwise from its position in (a). (Photographs by T. A. Moore.)

hair after passing through the substage polarizer. The crystal then resolves this ray into components of the E and O rays which vibrate in the single planes allowed by the substage polarizer P and the superstage analyzer A, as shown in figure 10.10. Only those components in the direction of the analyzer vibration, E' and O', are transmitted to the eye. In such a position, the crystal appears bright against a dark field.

If the stage is rotated, the E and O vibration directions can be brought into coincidence with the vibration directions of P and A. In such positions, no light is transmitted; the crystal is dark and is said to be **extinguished.** Four positions of extinction will be present per stage revolution (see figure 10.11).

No extraordinary ray is formed by the passage of light through isometric crystals because of their uniform symmetry in three dimensions; they are thus extinguished in

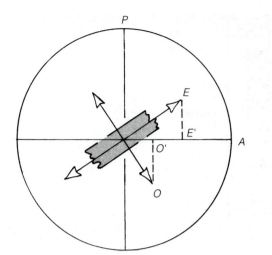

Figure 10.10
Resolution of *E* and *O* rays. A crystal in general orientation viewed down the axis of a petrographic microscope has its *E* and *O* vibration directions resolved to *E'* and *O'*. *P* is the vibration direction of the polarizer; *A*, the vibration direction of the analyzer.

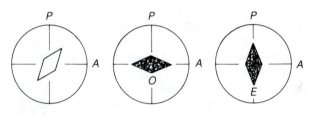

Figure 10.11
Extinction of light when the *E* and *O* vibration directions coincide with the vibration directions of the polarizer, *P*, and analyzer, *A*.

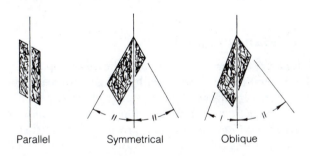

Parallel   Symmetrical   Oblique

Figure 10.12
Geometries of extinction.

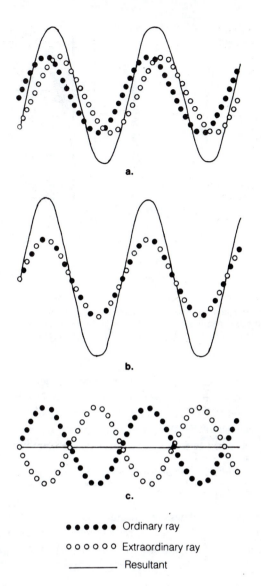

• • • • • Ordinary ray
○ ○ ○ ○ ○ ○ Extraordinary ray
————— Resultant

Figure 10.13
Interference of waves. In the general case (a), two out-of-phase waves of the same wavelength (color) interfere constructively with a resultant of the same wavelength and increased amplitude. For waves exactly in phase (b), the resultant amplitude (intensity) is at a maximum. When exactly out of phase (c), the waves cancel and no light is seen.

all positions. The symmetrical aspect of crystals in the tetragonal and hexagonal systems viewed along their *c*-axes is comparable to that of isometric crystals, and they are extinguished in all positions of rotation when their *c*-axes are perpendicular to the stage.

Crystals with an edge or face parallel to the *c*-axis are extinguished when such a face or edge parallels a cross hair. Rhombohedra, pyramids, and domes are extinguished when the bisector of a silhouette angle parallels a cross hair. Other forms extinguish at some angle to the cross hairs, as shown in figure 10.12. Extinction angles

between the traces of cleavage, faces, or similar crystallographically controlled elements are important physical parameters of minerals and can be easily measured using the scale on the microscope stage.

Retardation and Color

The two rays emerging from an anisotropic crystal are constrained by the analyzer to vibrate in the same plane. However, any initial phase separation of the rays is preserved and the rays travel at different speeds through different path lengths in the crystal. In consequence, their vibrations are generally out of phase and the rays mutually interfere, as shown in figure 10.13. This leads to the

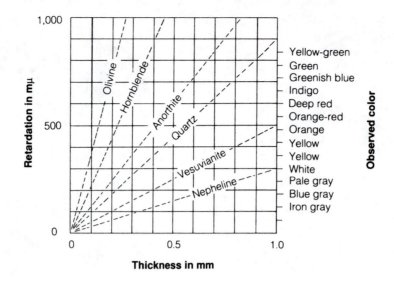

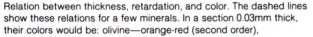

Figure 10.14

Relation between thickness, retardation, and color. The dashed lines show these relations for a few minerals. In a section 0.03mm thick, their colors would be: olivine—orange-red (second order), hornblende—indigo, anorthite—yellow, quartz—white, vesuvianite—blue gray, nepheline—iron gray.

subtraction of certain wavelengths from the incident white light, and the mineral appears to be colored. The color depends on the amount of retardation of one set of waves with respect to the other and is a function of the crystal thickness and the refractive indices of the O and E rays.

Retardation Δ is given by the expression

$$\Delta = t(n_2 - n_1)$$

where t is the thickness in millimicrons (1 mμ = 10^{-6} mm = 10Å), and n_2 and n_1 are, respectively, the greater and lesser indices of refraction in a particular orientation. The phase difference (ϕ) of the emerging rays is

$$\phi = \frac{\Delta}{\lambda} = \frac{t(n_2 - n_1)}{\lambda}$$

where λ is the wavelength.

Interference colors vary with the kind of mineral, its orientation, thickness, and the light used. The relation between retardation and thickness is shown in figure 10.14. Since the maximum difference in n_2 and n_1 (double refraction) is approximately constant for a given mineral, a characteristic interference color appears for a particular thickness of the specimen section. The O and E refractive indices for quartz are, respectively, 1.5442 and 1.5533. Thus, $n_E - n_O = 0.009$, and at the standard section thickness of 0.03 mm or 30,000 mμ, the retardation is 30,000 × 0.009 or 270. In this situation, a quartz grain appears white between crossed polars when not at extinction. Since quartz is easily recognized, this color is often used to determine section thickness.

The direction of vibration of the fast and slow rays in a mineral can be determined because the two rays have different refractive indices with the one of greater index being the slow ray. At extinction the rays are vibrating

parallel to the cross hairs in the ocular. When the stage is turned by 45° they are brought to their position of maximum interference. If now an oriented accessory plate of mica or gypsum or a quartz wedge is inserted into the optical path with its slow direction parallel to a vibration direction in the mineral, the retardation is changed. The mineral color will change to a higher color (see figure 10.14) if the slow ray directions of the mineral and the plate coincide, and to a lower color if the fast ray of the mineral and the slow ray of the plate are parallel.

Interference Figures

The nature and orientation of the indicatrix or optic sign of a mineral can be ascertained by using **interference figures.** These are formed by the passage of convergent polarized light through a mineral which causes a range of retardations. Convergent light is provided by an auxiliary condensing lens below the microscope stage. A Bertrand lens is inserted in the optical path above the specimen to bring the interference figure into focus at the ocular. Good figures, although of small size, can sometimes be seen without a Bertrand lens by removing the ocular.

In convergent light, all rays of equal inclination to the microscope axis have the same retardation. Thus, the interference colors appear as rings, and because light is not resolved in the vibration directions of the polarizer and analyzer, these directions appear dark. For uniaxial crystals (tetragonal and hexagonal systems) with *c*-axes parallel to the microscope axis, the result is a black cross or **isogyre** superimposed on colored rings called **isochromes** (see figure 10.15). If the crystal axis lies somewhat off the microscope axis, the figure will appear off center and will move with stage rotation, as illustrated in figure 10.16.

Figure 10.15
Uniaxial interference figure. The closely spaced isochrome rings arise because the crystal is highly double refractive. (From E. Weinschenk and R. W. Clark, *Petrographic Methods.* © 1912 McGraw-Hill Book Company.)

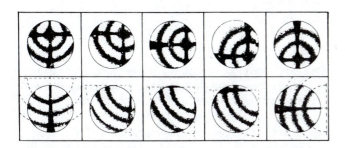

Figure 10.16
Uncentered uniaxial interference figures. (From E. Weinschenk and R. W. Clark, *Petrographic Methods.* © 1912 McGraw-Hill Book Company.)

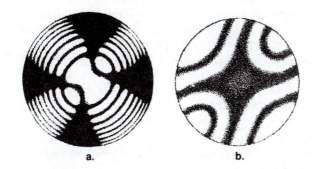

Figure 10.17
Biaxial interference figures. The crystal in (a) has a small optic angle and large double refraction; the opposite is true for (b). (From E. Weinschenk and R. W. Clark, *Petrographic Methods.* © 1912 McGraw-Hill Book Company.)

In biaxial crystals (orthorhombic, monoclinic, and triclinic systems), the two optic axes complicate the interference figure. If the optic angle $2V$ is small, the interference figure approximates the uniaxial case (see figure 10.17). With increasing $2V$, the isogyres separate to form arcuate bands, as in figure 10.17.

The optic sign of a mineral, that is, the shape of its indicatrix, can be determined from an interference figure by inserting an accessory plate or wedge at 45° to the direction of the polarizers. The resulting color changes due to retardation follow those previously described, and some cases are illustrated in figure 10.18.

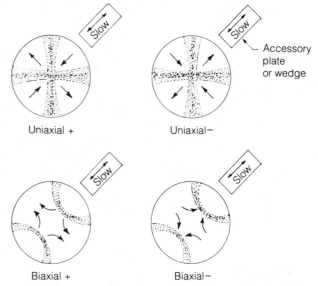

Figure 10.18
Determination of optic sign using an interference figure. Colors rise as indicated by the arrows with the insertion of an accessory plate or quartz wedge whose slow direction is as indicated.

The Petrographic Microscope

Much of the foregoing material would be of little but academic interest to the ordinary mineralogist except for the fact that many optical features of minerals are easily measured using a **petrographic microscope** (see figure 10.19). This specially equipped microscope is widely used in mineralogic and petrographic studies. The materials studied are usually either finely crushed mineral powders or very thin rock slices (0.02 to 0.03 mm) called **thin sections.** Some examples are shown in figure 10.20.

The microscope is probably the most widely used tool of the geologist and deservedly so, since it combines great versatility with relatively low cost and ease of manipulation. Aside from certain obvious advantages in magnification, the various auxiliary test systems allow measurement of such optical properties as

Refractive index

Birefringence (difference between indices)

Pleochroism (color effects arising from differential absorption)

Optic sign (character of the indicatrix)

$2V$ (angle between optic axes in a biaxial indicatrix)

Extinction angles

Relief (contrast of mineral's refractive index with that of the mounting medium)

Color

Obviously, mineral characterization using this microscope is highly reliable because many different properties can be measured with considerable precision. Indeed, it is

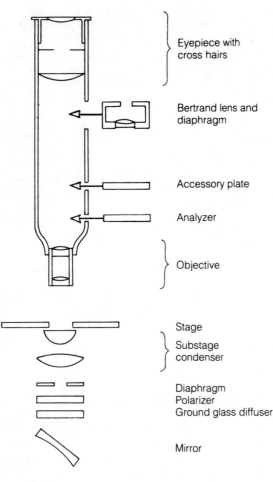

Eyepiece with cross hairs

Bertrand lens and diaphragm

Accessory plate

Analyzer

Objective

Stage

Substage condenser

Diaphragm
Polarizer
Ground glass diffuser

Mirror

Figure 10.19
Principal components of a petrographic microscope.

often possible to characterize the chemical composition of an intermediate member of a solid-solution series. In addition to aiding mineral identification, the microscope allows us to obtain much useful information concerning the shape, size, and interrelationships of grains, either isolated or within an aggregate. Petrographic microscopy is a complex and important field, and numerous texts exist on the subject. The interested student should investigate these for further information.

Concluding Remarks

Crystal structure and chemical composition are closely associated with the physical properties of a mineral. In this chapter we have examined optical properties such as refraction, reflection, luster, color, and luminescence. These properties are determined by the crystallographic and chemical features of minerals and are an important means of mineral identification.

Much optical examination of minerals is done using a petrographic microscope, a basic tool of the practicing mineralogist or petrologist. The microscope is widely available, relatively inexpensive, and, with continued use, a ready source of quality data. Students are encouraged to become proficient in using the petrographic microscope. They will find it a very powerful tool for the identification of minerals and the interpretation of mineral associations.

Suggested Readings

Bloss, F. D. 1961. *An Introduction to the Methods of Optical Crystallography.* New York: Holt, Rinehart and Winston. An excellent treatment of the concepts of optical crystallography with many illustrations.

Ehlers, E. G. 1987. *Optical Mineralogy.* Vol. 1, *Theory and Techniques.* Vol. 2, *Mineral Descriptions.* Palo Alto: Blackwell.

Hutchison, C. S. 1974. *Laboratory Handbook of Petrographic Techniques.* New York: Wiley. An excellent and wide-ranging compilation of techniques including detailed outlines of optical procedures and measurements.

Kerr, P. F. 1977. *Optical Mineralogy.* 4th ed. New York: McGraw-Hill. Part 1 covers the theory and practice of optical mineralogy; part 2 provides mineral descriptions with emphasis on optical properties.

Larsen, E. S., and Berman, H. 1934. *The Microscopic Determination of the Nonopaque Minerals.* 2d ed. U.S. Geological Survey Bulletin 848. Washington, DC: U.S. Government Printing Office.

Nesse, W. D. 1986. *Introduction to Optical Mineralogy.* New York: Oxford Univ. Press.

Shelley, D. 1985. *Optical Mineralogy.* 2d ed. New York: Elsevier.

Stoiber, R. E., and Morse, S. A. 1972. *Microscopic Identification of Crystals.* New York: Ronald. A straightforward, practical text on optical mineralogy using the petrographic microscope.

Wahlstrom, E. E. 1969. *Optical Crystallography.* 4th ed. New York: Wiley.

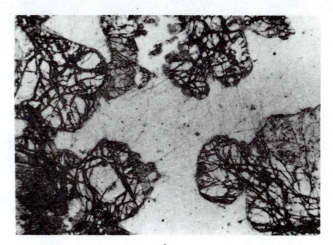

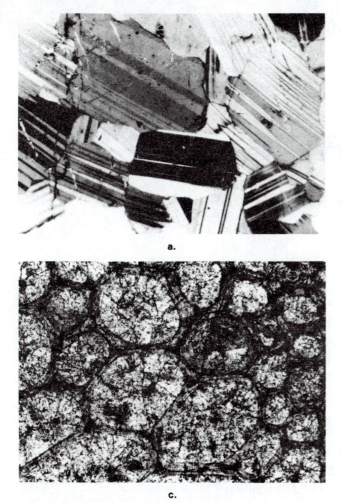
a.

b.

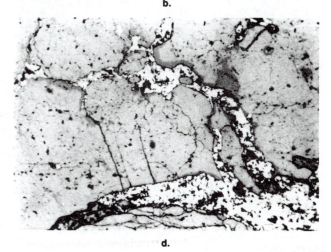
c.

d.

Figure 10.20

Examination of rocks with the petrographic microscope.
(a) Anorthosite showing tabular crystals of plagioclase with polysynthetic twinning. Crosséd polarizers. Field of view is 3 × 2mm. (From Choudhry, 1981.) (b) Gabbro with crystals of olivine (fractured) enclosed in plagioclase (light gray). Thin rims of orthopyroxene (dark gray) separate the olivine from the plagioclase. Plane polarized light. Field of view is 3 × 2mm. (From Choudhry, 1981.) (c) Cell structure in a mollusc shell. The darker gray rim around each cell is calcite representing the original cell material. Coarser (lighter gray) calcite fills the interior of each cell as a secondary cement. Larger cells are 0.5mm in diameter. (Photograph by I. S. Al-Aasm.) (d) Polished section in reflected light showing chalcopyrite (white) filling fractures in quartz (gray). Magnification 50×. (Courtesy A. E. Williams-Jones. Photograph by I. M. Samson.)

Mineral Analysis

The goal of mineral analysis may be either to identify a particular mineral compound or to determine some or all of its chemical constituents. In **qualitative analysis** the aim is to determine only the presence or absence of a particular mineral species or chemical component. In **quantitative analysis** the concentration of the component is determined.

The most common instrumental method of mineral identification is by the qualitative observation of optical properties using a petrographic microscope (see chapter 10). Quantitative proportions of minerals in a rock can be determined optically by a modal analysis as described in chapter 8. The qualitative identification of a mineral and analysis of its structure can both be accomplished using x-ray diffraction analysis. This important technique is widely used in mineralogy and is the subject of chapter 12.

The kind and amount of chemical constituents in a mineral, as distinct from species identification, can be assessed by a variety of chemical and electromagnetic means. Chemical analysis based on reaction-specific reagents is useful when the mineral is readily soluble in water or acids, but many minerals are not easily dissolved. Also, because the reactions are difficult to quantify, physical analysis is more often used.

The number of analytic instruments that are used for the identification and chemical characterization of minerals is large and growing. Table 11.1 lists some of the more common systems and devices, which are discussed in this text.

The suffixes *-scopy*, *-graphy*, and *-metry* refer to the means of observation of recording that are used with analytic instrumentation. Direct observation by the eye is implied by *-scopy*, as in microscopy or spectroscopy. The suffix *-graphy* is used when the instrumental output is recorded on a photographic plate or a paper chart recorder, as in polarography or spectrography. The suffix *-metry* simply means measurement. Spectrometry involves measurement by the means of various electronic devices, as in gamma-ray spectrometry or infrared spectrometry.

Differential Thermal Analysis

Differential thermal analysis (DTA) of minerals is usually conducted in a differential system that allows the specimen to be steadily heated. Only those increases or decreases in temperature that correspond to some physical change in the specimen, such as a polymorphic transition, loss of a volatile constituent, alteration, or dissociation, are recorded.

The apparatus (see figure 11.1) consists basically of a pair of thermocouples, one embedded in the powdered specimen and one in thermally inert alumina (α-Al_2O_3)

powder placed in the proximity of the specimen. Thermocouples are joined wires of different composition that generate an electrical current when their junction is heated. The thermocouples are commonly made of chromel-alumel or platinum-platinum with 13% rhodium. The specimen and alumina powders are placed in a furnace designed to heat them uniformly, and the flow of electrical current on heating is nullified by the use of a common thermocouple leg between them so that no signal is received at the recorder.

When a temperature is reached at which some heat-absorbing (endothermic) or heat-evolving (exothermic) reaction takes place in the specimen, the specimen thermocouple generates more or less current than does the alumina thermocouple. Current then flows through the circuit and is recorded. The actual temperature of the change can be found by adding a thermocouple to monitor the furnace temperature, shown as a dashed line in figure 11.1.

Conventionally exothermic reactions are recorded as positive deflections (above the baseline) and endothermic reactions as negative. The area under the curve is approximately proportional to the heat of transition.

The interpretation of thermograms is difficult because so many phenomena cause thermal changes in minerals, and usually supplemental information is required. One related means of obtaining further data is through **thermogravimetric analysis** (TGA), in which the weight of the specimen is continuously monitored as it is heated.

A sensitive weighing system with a specimen and thermally inert reference material initially in balance will show displacement if, on heating, a thermal change in the specimen is related to such phenomena as oxidation or loss of volatiles (but not to a polymorphic transformation). Differential thermographic curves can thus be compared with DTA thermograms to discover changes not accompanied by weight changes.

Infrared Spectrometry

Solids absorb visible and infrared wavelengths of light by transitions involving quantized energy changes in the interlocked assemblage. The nature of particular atoms is of no greater importance in this instance than are the structural linkages that hold them together.

The energy changes involved in the bending and stretching of chemical bonds lie within the frequency (energy) spectrum of infrared light, about 10^{-6} to $10^{-3.5}$ m or 1 to 150 μm (figure 11.9). Consequently, solids illuminated by infrared radiation (IR) show strong resonant absorptions at wavelengths characteristic of particular bonds.

The nature and degree of resonance is principally related to the kind of structural bonding present in the mineral. Isodesmic structures resonate as a whole and show

Table 11.1 **Analytic Systems and Devices**

Instrument or Analytic System	Use and Advantages	Limitations	Remarks
Mineral Identification			
Petrographic microscope	Major tool for identification of minerals as grains or in thin section	Cannot identify clay-size material	A principal tool; objects magnified and optical parameters measured by appropriate attachments
Differential thermal analysis (DTA)	Identifies temperature of mineral transformations	Not characteristic for most minerals; used principally for clay minerals	Measures thermal changes arising from mineralogic changes
Thermogravimetric analysis (TGA)	Quantitative mineral identification in some instances; used in conjunction with DTA	Not diagnostic for most minerals	Measures weight change as a function of heating
X-ray diffraction (XRD)	Major tool to identify minerals and determine their structure	Mineral mixtures sometimes present problems	Diffraction pattern of impinging x radiation is characteristic of crystal structure
Infrared spectrometry (IR)	Mineral identification and bonding studies	Grain size must be uniform; complex spectra of aggregates may have many interferences	Specimen irradiation by infrared light causes characteristic resonances
Chemical Analysis			
Wet chemistry	Determination of presence and amount of any element, usually above 0.1%	Requires specimen dissolution, procedures vary with element to be determined	Used in all older work; now usually employed with nonsilicates and in spot tests
Blowpipe analysis	Simple equipment, spot testing for many elements	Qualitative only	A traditional mineralogic technique now little used
Neutron activation analysis (NAA)	Excellent sensitivity (ppb-100%) and precision for a wide range of elements	Specimen must be activated by thermal neutrons, usually in a nuclear reactor	Radioisotopes formed by neutron bombardment decay with characteristic half-lives, measured by γ-ray spectrometry
γ-ray spectrometry	Best technique for measuring natural or induced radioactivity; used to determine U, Th, and K	Limited to active γ-ray emitters	γ rays from decaying radioisotopes cause scintillation in special crystals; flashes counted by photomultiplier tube
Polarography	Quantitative measure of elements in solution	Specimen must be in solution; element must be reducible	Electrical potential needed to remove electrons from ions in solution characteristic of the ionic species

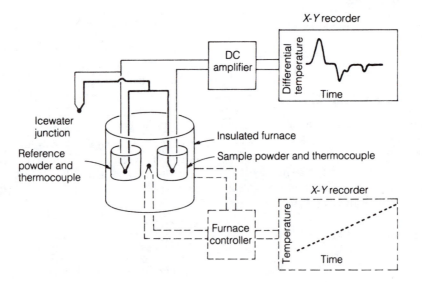

Figure 11.1
Differential thermal analysis (DTA) apparatus. The icewater thermocouple serves as a temperature datum.

Instrument or Analytic System	Use and Advantages	Limitations	Remarks
Chemical Analysis—Continued			
Optical emission spectrography (OES)	Excellent sensitivity (ppm-100%) for determination of cations; rapid multielement measurement	Lower precision than XRF, NAA or AAS	Valence electrons, excited by a flame arc, or spark will emit light characteristic of the atomic species.
Atomic absorption spectrometry (AAS)	High sensitivity (ppb-100%) and precision	Normally specimen is in solution; single element measurement	Light of characteristic wavelength emitted by an element-specific cold cathode tube is absorbed by a flame containing the same element
X-ray fluorescence spectrometry (XRF)	Good sensitivity and precision for quantitative determination of a wide range of elements	Best sensitivity for elements above atomic number 12	Very widely used for major and trace element determinations in rocks and minerals
Electron microprobe	Study of chemical composition of very small areas ($\approx 1\mu m^2$)	High initial and maintenance costs	Electron-beam-excited secondary x rays have energies characteristic of the atomic species
Spark source mass spectrometry	Chemical determination of low-level constituents	Equivalent results obtainable from less complex instruments	Ions accelerated by an electrostatic field pass through a magnetic field to separate them by mass and charge
Physical Constants			
Mass spectrometry	Determination of relative atomic mass; employed for isotope analysis	Lengthy sample preparation	Accelerated ions deflected by a magnetic field as a function of their mass
Nuclear magnetic resonance spectrometry(NMR)	Molecular structure studies, mainly of organic compounds	Nuclear properties limit which molecules can be examined	Resonant absorption of radio-frequency energy by atoms in a strong magnetic field
Mössbauer spectrometry	Examination of bond symmetry; identification of valence state of iron	Limited to about thirty nuclei	Reestablishes resonance of γ radiation being emitted and absorbed by relative motion of source and specimen

broad and weak absorption bands at low frequencies in highly ionic structures (e.g., halite, NaCl). Minerals with a degree of covalent bonding (e.g., quartz, SiO_2) show complex spectra. Anisodesmic structures involving radicals have strong absorption bands at high frequencies accompanied by low-frequency body resonances. There are no fundamental vibrational-absorption bands in metallically bonded minerals.

The kinds of bonds, their length and angular relations, along with the radii, coordination, and relative masses of the constituent atoms, give rise to characteristic frequencies, intensities, and shapes of the absorption bands. Because no two minerals yield exactly the same spectrum, these absorption bands can be used for mineral identification. Figure 11.2 shows some typical IR spectra.

Infrared spectrometry has long been used in the structural analysis of organic materials, but the fundamental vibrations associated with mineral bonding in the 2.5 to 50 μm range have been less used. Aircrafts and satellite-based photographic systems using the infrared wavelength range, however, are now routinely used to obtain information about minerals, soils, and rocks.

An IR spectrometer employs a source of infrared radiation, such as a hot wire or Globar (water-jacketed incandescent lamp with a carbon-rod or platinum filament), an absorption section where the specimen is placed, a spectrographic system to disperse the radiation, and a means for measurement of the signal. Because the characteristics of absorption over a spectral region are usually sought, the spectrograph employs a scanning device, such as a rotating prism, that causes the spectrum to move past a fixed sensing point or vice versa. Sensing is accomplished with a sensitive thermometer (bolometer), thermocouple, radiometer, or photoconductive cell.

Remote-sensing applications take advantage of the different sensitivities of photographic films in cameras or television scanners. Appropriate filters allow observation of narrow spectral regions.

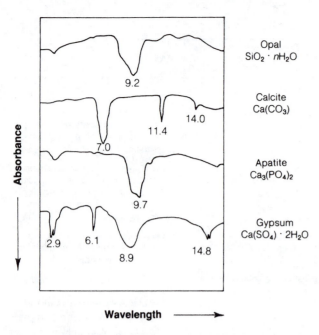

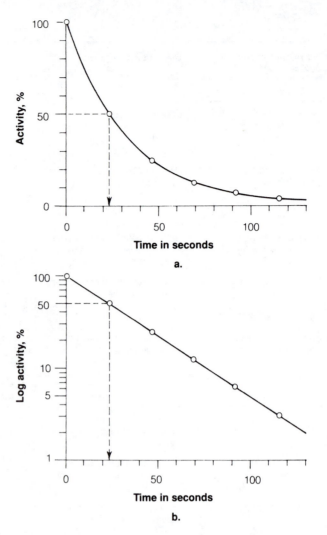

Figure 11.2

Examples of IR spectra. The wavelengths in micrometers (μm) of some of the strong absorption peaks are given beneath the spectra.

Neutron Activation Analysis

Radioactivity can be induced in stable atomic nuclides if they are subjected to a flux of particles, such as neutrons, protons, or deuterons, or to high-energy x or γ radiation. The radiodecay products are characteristic in kind and energy of the decaying nucleus, and their measurement serves as an analytic tool. For chemical analysis, the most used activator is the thermal neutron flux produced in quantity in nuclear reactors (10^{12} to 10^{15} neutrons/cm^2/s).

Thermal neutrons are those that have been slowed down to thermal equilibrium with their surroundings by interaction with such moderators as graphite or heavy water. At 20° C, their energy is a fraction of an electron volt and their velocity is on the order of 2,200 m/s. Because neutrons are not charged, they can interact with nuclei even at very low energies.

Most elements have at least one stable isotope that can be transformed by the absorption of a thermal neutron into an isotope whose mass is one greater. In many cases, the new isotope is radioactive. For example, in the transition $^{75}As + \text{neutron} \rightarrow {}^{76}As + \gamma$, the product ^{76}As is radioactive and decays to selenium by emission of β^- (beta particles = electrons) and gamma radiation with a half-life of 26.4 hours: $^{76}As \rightarrow {}^{76}Se + \beta^-, \gamma$.

Neutron activation analysis (NAA) involves measuring the production of radioactive nuclides from stable nuclides subjected to a neutron flux as well as the γ-ray activity of the radioactive species and the determination of their half-lives. Gamma rays are used in preference to other decay products to identify the decaying nuclide because they exhibit a definite energy spectrum. Further,

Figure 11.3

Decay of radioisotope. Both plots are based on a half-life of 23 seconds. The linear plot (a) best illustrates the exponential nature of the decay; the semilog plot (b) makes for simpler measurement.

they can be measured more easily than either alpha or beta particles. Both qualitative and quantitative analysis can be performed by this means, often with sensitivity in the range of 10^{-7} to 10^{-11} grams. Other than particular precautions to guard against contamination, sample preparation consists of putting weighed specimens (and standards) into airtight aluminum, silica, or polyethylene containers for placement in a reactor.

Radionuclides are activated exponentially and also decay exponentially. The half-life for a decaying species can be obtained either by calculation or graphically. The exponential decay of a typical element is shown in figure 11.3 as linear and semilog plots. The half-life, by definition, is the time required for the activity to decrease to one-half of its initial value. If, for example, a radionuclide has a half-life of 23s and an initial measurement at time t_o gives an activity N_o of 100%, the activity will drop to 50% of the original at $t = 23$s. Similarly a measurement made at $t = 40$s is exactly twice that at $t = 63$s, one half-life later.

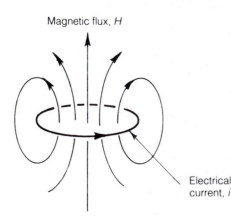

Figure 11.4
Fundamental relations of magnetic flux and electrical current.

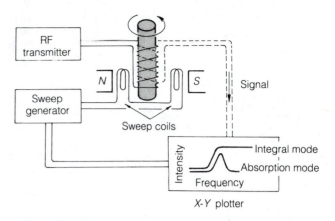

Figure 11.5
Nuclear magnetic resonance (NMR) apparatus. The sample is spun inside magnetic coils (*N-S*), sweep coils, coils to apply RF energy, and signal-sensing coils.

Mathematically, these relations can be expressed as

$$N_t = N_o e^{-kt}$$

or

$$\log(N_t/N_o) = -0.4343\lambda t$$

where N is the total number of atoms of a radionuclide, N_t is the number of undecayed atoms after a time interval t, N_o is the original number of atoms, and λ is the decay constant. After one half-life, $N_t:N_o = 1:2$ and the half-life is $0.693/\lambda$.

Semiconductor detectors are usually used for detection and measurement of γ-ray spectra. Such detectors are often fabricated from a cylinder of germanium metal into which lithium ions have been drifted electrostatically. The lithium ions are introduced to compensate for impurities in the germanium. Ge(Li) detectors are capable of high resolution and allow the discrimination of photopeaks of nearly the same energy.

Scintillation detectors are used in some specialized applications where high resolution is not necessary but where efficiency for the detection of γ radiation is required. This type of detector consists of a crystal of NaI activated (drifted) with thallium, which shows a flash of light, or scintillates, when it absorbs a γ ray. The flashes are monitored by a sensitive photomultiplier. Scintillation detectors are particularly useful for measuring naturally radioactive isotopes such as ^{235}U, ^{238}U, ^{232}Th, and ^{40}K.

Photoenergies deposited by γ radiation in the detectors are passed to a linear amplifier that transforms them to square wave pulses. These pulses are fed to a multichannel analyzer that converts the analog information from the amplifier to a series of digital signals. The digital signals are linearly related to pulse height and thus to the energies deposited in the detector. The digital signals are accumulated in the memory channels of the analyzer from which they can be recovered for data reduction and calibration.

The advantage of neutron-activation analytic techniques in mineralogic and petrologic problems is sensitivity. Thirty to forty elements can be determined with detection limits in the part per million or part per billion range. Thus, the technique is particularly applicable to the determination of elements at very low concentrations and to analyzing small amounts of material. Examples of the application of NAA include the direct determination of oxygen in rocks, minerals, and meteorites, the determination of rare earth elements, and the determination of trace elements that may serve as guides for ore exploration.

Nuclear Magnetic Resonance

Atomic nuclei have electrical charge, and like electrons they may be thought of as spinning on an axis. This spinning requires them to have an associated magnetic field, because a spinning charge is equivalent to an electrical current flowing in a conductor loop (see figure 11.4). The magnetic dipole that arises can be interpreted in terms of the nuclear spin quantum number m_s. When $m_s = 0$, there is no magnetic moment; when $m_s = \frac{1}{2}$, the spinning charges are spherical; and when $m_s > \frac{1}{2}$, the spinning charges are nonspherical.

Nuclear magnetic resonance (NMR) studies investigate the quantum energy gaps between nuclear states of different energy. The NMR method generally involves the absorption of applied radio frequency (RF) energy by nuclei exposed to a strong magnetic field (see figure 11.5). Most earlier studies dealt with the spherical, spinning hydrogen atom, and the most important applications were in elucidating molecular structure, especially in organic compounds. Recently, however, NMR studies concentrating on the nature of the silicon-29 and aluminum-27 atoms have proven very useful in examining the subtleties of silicate mineral structures. Investigations in feldspars, spinels, and layered silicates have provided new information on the nature and extent of Al–Si order-disorder, the number of crystallographically distinct Si sites, the estimation of Si-O-(Si,Al) bond lengths and angles, and the distribution of Al in 4-fold and 6-fold coordination sites.

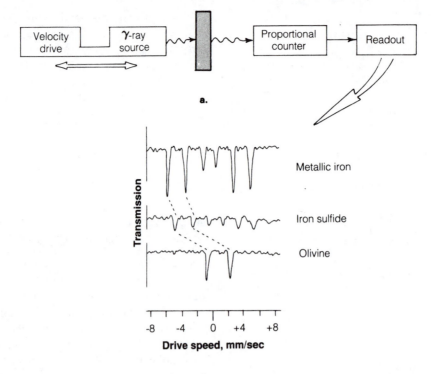

Figure 11.6
Mössbauer spectrometry. The basic components of the apparatus are shown in (a). Mössbauer absorption spectra of some iron-bearing substances are shown in (b).

Mössbauer Spectrometry

When γ radiation is absorbed or emitted by an atomic nucleus, the momentum of the system must be preserved. The recoil of a nucleus emitting γ radiation thus changes the energy (frequency) of the emitted radiation, and a Doppler broadening of the emitted line occurs. Under special circumstances, however, as developed by Mössbauer in 1958, recoil-free absorption and emission takes place with sharp resonance of the absorbed and emitted energy. This can occur for certain nuclei if both the source and absorbing nuclei are fixed in a crystalline structure. In this case, a large fraction of the recoil is distributed over the entire crystal.

If the absorbing nucleus is in a structural environment different from that of the emitting nucleus, the slight difference in their nuclear energy levels is sufficient to destroy their resonance. Resonance can be restored, however, if movement of the emitting nucleus with respect to the absorber provides the needed small energy changes. With the appropriate apparatus, diagramed in figure 11.6, a resonance spectrum can be obtained relating the γ-ray flux passing through the absorbing specimen to the relative velocity of the source and specimen. The shift of the resonant absorption bands can then be interpreted in terms of the structural environment of the absorbing nucleus.

When a Mössbauer atom is in tetrahedral or octahedral coordination, its nuclear levels are degenerate and only a single resonance is observed. For other coordinations, the nuclear energy levels are separated into $m_s +$ 1/2 levels for half-integral spins (1/2, 3/2, 5/2, . . .) or into $2m_s$ levels for integral spins (1, 2, 3, . . .). Thus, the spectrum provides an immediate measure of the symmetry of the bond distribution at a particular crystalline site.

Most Mössbauer studies have examined phenomena related to ^{57}Fe, whose spin states are 1/2 and 3/2. As a result of these studies, we now have the ability to distinguish between Fe^{3+} and Fe^{2+} on the basis of the relative velocity of source and specimen. Resonance for Fe^{3+} occurs at a relative velocity significantly different from that for Fe^{2+}.

Mass Spectrometry

Different ions and isotopes have characteristic charges and masses and can be separated on this basis by properly designed instruments. The usual method is to accelerate the charged particles as an ion beam that passes through a magnetic field in which the particles are differentially deflected.

The kinetic energy given to an ion by an accelerating potential is

$$\tfrac{1}{2}mv^2 = eV$$

where m is the ion mass, v is the ion velocity, e is the charge

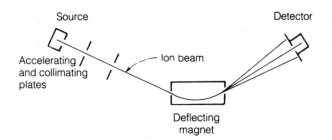

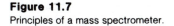

Figure 11.7
Principles of a mass spectrometer.

on the ion, and V is the accelerating potential. If the accelerated ion now enters a magnetic field oriented perpendicular to its direction of travel, it is subjected to a deflecting force:

$$f = Hev \qquad (11.1)$$

where H is the magnetic field strength. This force continues to act as the ion is deflected along a circular path and the centrifugal force is

$$f = mv^2/r \qquad (11.2)$$

where r is the radius of path curvature. Combining equations 11.1 and 11.2,

$$Hev = mv^2/r$$

or

$$mv = rHe$$

Combining this relation with that for kinetic energy, the mass/charge ratio for an ion is

$$m/e = r^2H^2/2v \qquad (11.3)$$

Mass spectrometers used for isotopic studies in geology are designed to shift from one mass to another by rapidly changing either the accelerating potential or the magnetic field strength. This allows for great sensitivity and precision in measuring the relative abundances of ions of different mass in the ion beam.

The design of a mass spectrometer is shown in figure 11.7. The sample may be admitted as a gas that is ionized and accelerated through the instrument. Alternatively, solid sources can be made by drying a sample solution on a filament. In the instrument, the filament is heated, and the sample is driven off as ions. Ions produced in the source are accelerated and collimated by a series of charged plates, then pass through a magnetic field where they are separated according to their mass/charge ratio into different beams. At an appropriate combination of magnetic field strength and accelerating potential, a selected beam falls upon a detector and is recorded. An analogy to the action of a mass spectrometer is a stream from a garden hose composed of droplets of different size that are deflected by the field of gravity. Droplets of a particular size will then fall into an appropriately placed bucket.

The principal applications of mass spectrometry in geology are in determining isotope ratios for studies of the age, formational temperature, or history of a sample. The system, however, has also been developed for chemical analysis as a **spark source mass spectrometer.** In this instrument, powdered specimens are vaporized by a spark, and the resultant ions are accelerated sequentially through an electrostatic field and a magnetic field to sort them into particle beams whose strength may then be measured electronically.

Polarography

Many kinds of ions in solution can be reduced, or gain electrons, by the application of a modest potential gradient (see chapter 4). In a specimen solution containing reducible ions, increasing the negative potential causes a stepwise rather than linear change in current as successive ionization levels are reached. A polarograph is an apparatus used to obtain information as to the kind and amount of the particular reducible-ionic species that are present in a test solution. The device graphs a current-voltage curve for interpretation.

The fundamental part of the apparatus is a glass capillary tube attached to a mercury reservoir. The capillary allows tiny drops of mercury to form and fall through the test solution every three or four seconds. The falling drops become the cathode of an electrolytic cell, and the mercury pool that forms on the bottom of the test solution container serves as the anode. The reason for the dropping mercury electrode is to continuously provide a clean surface to which cations in the solution migrate and become attached. A schematic of the apparatus is given in figure 11.8a.

The applied voltage is gradually increased, and a small increase in current (residual current) occurs because of an increased migration of ions. When the potential needed to remove electrons from a particular ionic species is reached, however, ions diffuse rapidly to the cathode to replace those that have been reduced, and a sharp increase in current occurs. The current continues to increase until all ions arriving at the cathodic drops are reduced (see figure 11.8b). The half-wave potential is characteristic of the reducible species, and the distance between the residual and limiting currents is a measure of their concentration.

Analysis Using Electromagnetic Radiation

The concepts of the interaction of light and matter as discussed in chapter 10 can be extended to electromagnetic radiation of both shorter and longer wavelengths, that is, into the ultraviolet (x-ray) and γ-ray regions at shorter wavelengths and the infrared region at longer wavelengths (see figure 11.9). The interaction of electromagnetic radiation with atoms or ions excites the inner

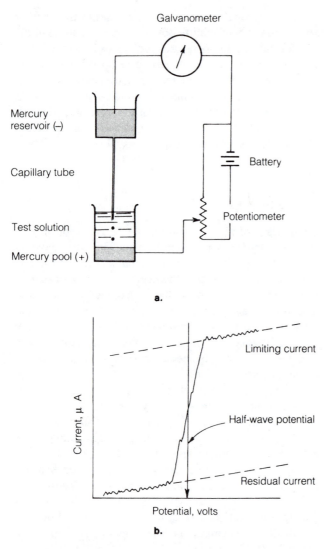

Galvanometer

Mercury reservoir (−)

Capillary tube

Battery

Potentiometer

Test solution

Mercury pool (+)

a.

Current, μ A

Limiting current

Half-wave potential

Residual current

Potential, volts

b.

Figure 11.8
Polarograph. A schematic of the apparatus is shown in (a), and the change in current flow when an ionic species is reduced as voltage increases is illustrated in (b). The half-wave potential is characteristic of the ionic species.

electrons in x radiation and γ radiation and valence electrons in ultraviolet, visible, and near-infrared radiation. Interaction with long wavelengths of far-infrared radiation bends and stretches the bonds but does not cause electronic transitions.

Because of the electronic structure of atoms (see chapter 4) the energies and hence the frequency or wavelength associated with particular electronic transitions differs from element to element. The basic components of all electromagnetic analytic devices are thus (1) a means of specimen excitation, (2) a means of sorting out or resolving the radiant output into a characteristic energy or wavelength spectrum, and (3) a means of recording the spectral data.

X-ray Fluorescence Spectrometry

Appropriate excitation of a specimen can produce electromagnetic radiation in the x-ray region (see figure 11.9). The x-ray spectrum produced can be used to identify an element by measuring its characteristic emission wavelength. Further, the quantity of the element can be determined by measuring the emission-line intensity and relating this to concentration.

If an atom is bombarded with electrons or primary x-ray photons, an inner orbital electron may be removed, leaving the atom in an excited state. The vacancy due to the emission of the photoelectron may be filled by an outer orbital electron of initial potential energy E_i. When the outer electron falls to the inner orbital, an x-ray photon is emitted, the energy of which is directly related to the initial and final states of the transferred electron. In each atom, every electron has a characteristic binding energy. Thus, although a given atom may emit many x-ray photons, the energies of these are always characteristic of the parent atom.

X rays can be described in terms of both energy and wavelength:

$$E_i - E_f = h\nu = hc/\lambda = 12.4/\lambda$$

where E_i and E_f are the energies of the initial and final states of the electron; ν is the frequency; λ is the wavelength in Ångstrom units; h is Planck's constant (6.625×10^{-34} joule · sec); and c is the velocity of light.

As the English physicist H. G. J. Mosely showed in 1913, the wavelengths of characteristic lines are related to atomic number (Z) according to the equation

$$1/\lambda = K(Z - \sigma)^2$$

where k and σ are constants for a particular line series.

It follows that if the characteristic spectral lines for an element can be resolved the element can be identified. Further, measurement of photopeak intensity reveals the concentration of the particular element.

A typical x-ray fluorescence (XRF) spectrometer system is shown in figure 11.10 and diagramed in figure 11.11. Bombardment of the specimen by x rays produced in a sealed x-ray tube causes emission of secondary x rays (fluorescence). A portion of the x-ray photons are allowed to pass through a collimator and fall on a crystal that is cut and oriented to present to the x rays an interatomic spacing (d). The spectrum can then be resolved according to the Bragg equation (see also chapter 12)

$$n\lambda = 2d \sin \theta$$

In addition, a detector can be rotated into position to measure a particular photopeak. Note that the x rays are diffracted by the crystal through an angle 2θ by the diffracting planes of atoms in the crystal, which are spaced at a distance d. A second collimator is placed at an angle 2θ to the primary collimator, and the x rays pass through

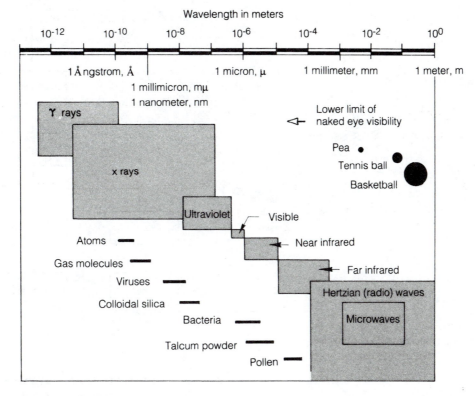

Figure 11.9
The electromagnetic spectrum.

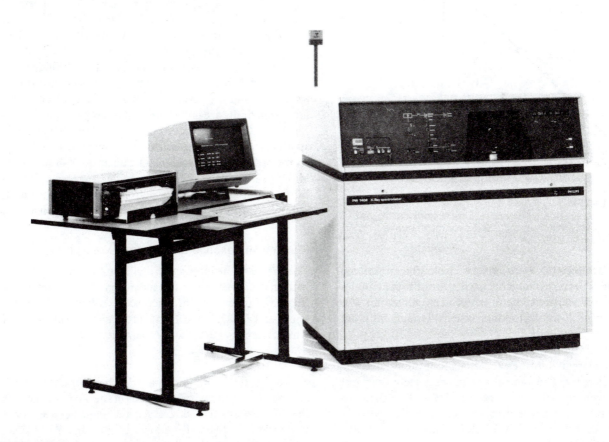

Figure 11.10
A typical x-ray fluorescence spectrometry system. (Courtesy Philips
Electronic Instruments, Mahwah, NJ.)

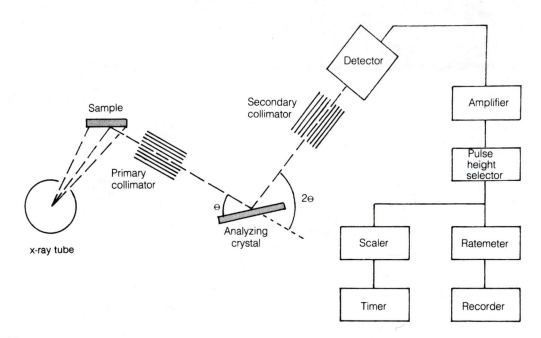

Figure 11.11
Components of an x-ray fluorescence (XRF) spectrometer.

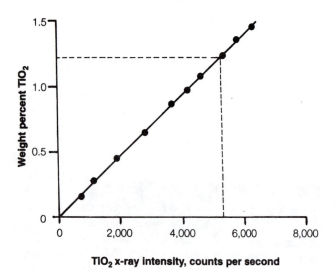

TiO₂ x-ray intensity, counts per second

Figure 11.12
Calibration curve for x-ray fluorescence spectrometry of TiO₂. An unknown giving an intensity of 5,300 cps corresponds to a concentration of 1.23%.

this to the detector. A goniometer enables the crystal and detector to be moved together so they are always at angles of θ and 2θ, respectively, to the primary collimator. The geometry of x-ray diffraction systems is more fully described in chapter 12.

The diffracted x-ray beam presents to the detector photons generated by the excitation of the atoms of a particular element. The number of photons detected per unit time is proportional to the concentration of the element in the sample. Thus, if a series of standards with known concentrations is available, the XRF system can be calibrated. Such a calibration curve is shown in figure 11.12 where concentrations of standard with known amounts of

TiO_2 are plotted against their x-ray intensities. Once the calibration curve is established, an unknown sample can be measured using the same instrumental conditions. As shown in figure 11.12, the unknown gives an x-ray intensity for TiO_2 of 4,600 counts per second. The concentration of TiO_2 in the sample, 1.23%, can then be read off the calibration curve. Several interatomic effects can affect the measured x-ray intensity, but these are beyond the scope of this brief description of the XRF technique.

A modern x-ray spectrometer is capable of measuring all elements in the periodic table above oxygen ($Z = 8$). Detection limits are on the order of a few parts per million for elements above argon in the periodic table but rise substantially for lighter elements. Instrumental precision is on the order of 0.1% and analytic accuracies in the range of 0.2 to 1% relative. The electron microprobe, discussed below, is basically an x-ray spectrometer as described here, but it uses an electron beam to excite the specimen atoms.

Electron Microscope and Electron Microprobe

The nature of the interaction of electromagnetic radiation with matter is strongly dependent on the wavelength of the radiation employed. Microscopists have long known (since Abbe, 1873) that the theoretical resolving power of a microscope is governed by the relation

$$R = 0.61\psi/n \sin\alpha$$

where R is the minimum separation between distinguishable points, n is the refractive index of the medium in which the objective is situated, and ψ is the angle between a ray along the objective axis and one to the edge of the field. Since $n_{air} = 1.00029$ and the maximum value of α is 90°, the maximum numerical aperture of the system, $n\sin\alpha$, is

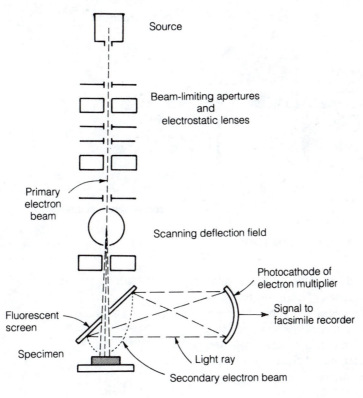

Source

Beam-limiting apertures
and
electrostatic lenses

Primary
electron
beam

Scanning deflection field

Photocathode of
electron multiplier

Signal to
facsimile recorder

Fluorescent
screen

Specimen

Light ray

Secondary electron beam

Figure 11.13
Schematic of a scanning electron microscope (SEM).

near unity. Thus, no microscopic system can resolve the details of objects separated by less than approximately one-half of the wavelength of the illuminating radiation. This is about 3,000Å for an ordinary light microscope, 2,000Å if oil immersion is used, or 1,000Å if ultraviolet light is used. Obviously, a very high-resolution system requires the use of radiation of significantly shorter wavelength than visible light.

In 1924, Louis de Broglie demonstrated that particles, particularly electrons, have associated wavelengths that vary with their velocity. For example, an electron accelerated by a potential of 60,000 volts has a wavelength of about 0.05Å. This discovery led in the 1930s to the development of the first electron microscopes, in which an electron beam was collimated and focused by magnetic coils in a manner analogous to the lens system in a light microscope. The image is formed by the impingement of electrons on a fluorescent screen similar to that of a television monitor. The resolving power of modern instruments is of the order of 10Å, or about five atomic diameters. Several kinds of electron microscopes have been developed of which only the transmission electron microscope (TEM) and the scanning electron microscope (SEM) are widely employed.

The transmission electron microscope is used to examine thin specimens in a manner analogous to thin-section study with an ordinary light microscope. It is an exact electronic counterpart of the light microscope whose essential components are (1) the illuminating system in which the electron beam is produced and focused on the specimen by magnetic or electrostatic lenses, (2) the imaging system that focuses the electrons transmitted by the specimen into a highly magnified image, and (3) the image-transforming system that converts the electron image into a form that can be perceived by the human eye, either on a fluorescent screen or in a photograph.

Electron microscopes, like ordinary ones, can function in either a transmission or reflection mode. The popular scanning electron microscope (see figure 11.13) is a reflection instrument that allows observation of an image produced by secondary electrons released by electron bombardment of the specimen. The 30–60 kV electron beam causing the excitation is about 100Å in diameter. Activated by a deflecting magnetic field, it rapidly scans the specimen in the manner of the picture tube of a television set. An image is formed as secondary electrons are accelerated back into the instrument and focused on a fluorescent screen. This image can then be picked up by a photoelectric tube, amplified in a conventional manner, and sent to a facsimile recorder.

The de Broglie relation is $\lambda = h/mv$ where λ is wavelength, h is Planck's constant, m is mass, and v is velocity.

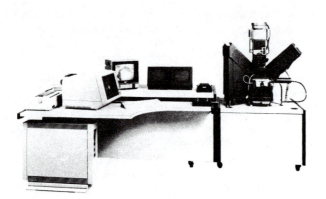

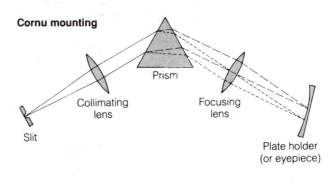

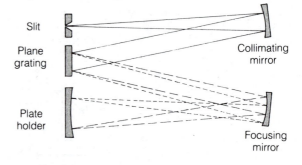

Figure 11.14

An electron microprobe. (Courtesy Cameca Instruments, Inc., Stamford, CT.)

If the electrons employed are sufficiently energetic, they will excite the inner electrons of atoms in the specimen to produce a characteristic x radiation. This secondary radiation can then be resolved and measured using an analyzing crystal and a counter in a manner analogous to an x-ray fluorescence spectrometer. Such a procedure is employed in the electron microprobe, an example of which is shown in figure 11.14.

Because the electron beam can be focused on a very small spot, typically two micrometers in diameter, the elemental content of very small portions of the specimen can be determined. Typical applications include studies of the fine detail of crystal intergrowths, mineral inclusions, and the nature of compositional zoning in minerals. Indeed, the electron microprobe is the most important analytic innovation in mineralogy since the introduction of x-ray diffraction spectrography.

Optical Emission Spectrography

Optical emission spectrography uses a flame, arc, spark, discharge tube, or inductively coupled plasma as a means of sample excitation. Such sources excite only valence electrons whose transitions produce radiation in or near the visible region. Thus, recording is done visually, photographically, or with a photoelectric system.

The typical spectrograph consists of a source operated by an appropriate power supply, an optical bench carrying devices to manipulate the incident light, and a spectrograph or camera. This last is furnished with a narrow entrance slit, an optical train consisting of focusing and collimating lenses or mirrors as needed, the dispersing element (prism or grating), and a means of observing or recording the images of the slit as spectral lines formed by the internal optical system. The optical dispersal systems for two common spectrographs are illustrated in figure 11.15.

Qualitative emission spectrography is based on the fact that valence electrons of different elements have energetically different excited states. The return of p electrons to the ground state in silicon, for example, involves

Figure 11.15

Some optical arrangements of emission spectrographs.

a different energy change than for p electrons of aluminum. A different change in energy must lead to differences in the radiated frequency, since $\Delta E = hv$.

The source provides a mixture of frequencies (or wavelengths, $v = 1/\lambda$) resulting from the very large number of different electron transitions. These are resolved by the spectrograph as follows. Divergent light from the illuminated slit is brought parallel by a collimating lens or mirror and dispersed by a prism or grating. In a prism, different wavelengths of light are refracted through characteristic angles and those of identical wavelength exit as parallel rays (see figure 11.15). The exiting rays are then focused to provide images of the slit at different positions on a photographic film or plate or on a battery of photoelectric tubes. The position of a spectral line is thus characteristic of the particular transition and hence of the element excited in the source.

Quantitative emission spectrography is based on the principle that the number of photons of characteristic energy emitted from a source is directly related to the number of particular electronic transitions and thus to the number of atoms of a particular kind present in the specimen. The number of photons received at a particular line position is represented by the amount of blackening of a photographic emulsion or the integrated signal from a photocell. When such variables as amount of specimen, dominant matrix, excitation conditions, and exposure time are held constant, the complex interrelationships can be reduced to $I = kC^n$ where I is light intensity at a particular wavelength, C is the concentration, and k and n are

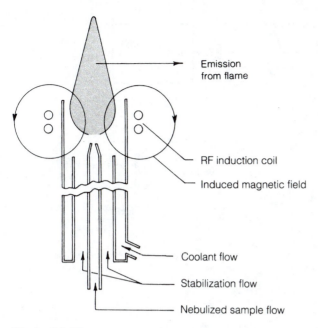

Emission
from flame

RF induction coil

Induced magnetic field

Coolant flow

Stabilization flow

Nebulized sample flow

Figure 11.16
Schematic of an inductively coupled plasma (ICP) torch.

position and slope constants. Specimens can thus be compared with themselves or with standards by determining relative intensity values.

A recent advance in emission spectrography has been the development of an inductively coupled plasma (ICP) torch for excitation. This source has advantages over arcs, sparks, and flames in the higher temperatures attained, because of its innate stability, lowered spectral interferences, and improved detection limits for many elements.

The ICP discharge is initiated by a pilot spark in flowing argon gas which then becomes electrically conductive. The discharge is maintained by heating the gas by a torus-shaped magnetic field generated by a radiofrequency coil (see figure 11.16).

Samples are introduced as powders or liquids carried by argon gas through the center of a set of coaxial quartz tubes. Annular flow in the surrounding tubes provides for flow stabilization and cooling.

Atomic Absorption Spectrometry

X-ray fluorescence and emission spectrometric methods are based on the direct excitation of atoms or ions and are restricted in sensitivity by the limited efficiency of this process. For most elements, this efficiency is only a few percent with ordinary sources, and sensitivity would be improved markedly if the larger number of neutral atoms could be measured. Absorption spectrographic methods are designed to exploit this situation by measuring the reduction of beam intensity in passing through a region of neutral atoms. Specifically, neutral atoms in the sample are excited by incident radiation of appropriate wavelength, but they do not necessarily reradiate in the direc-

tion of the spectrographic slit. Incident radiation is thus absorbed in accordance with the Beer-Lambert Law:

$$\log_{10}(I_0/I) = KC$$

where K is a constant for a given wavelength and absorbing path length, C is concentration, and I_0 and I are the incident and transmitted light intensities. Determination of I as a function of concentration by the use of standards thus provides a working analytic system.

To determine the concentration of elements by absorption spectrometry, the elements must be isolated from the energetic interference they experience in liquid or solid substances. A gaseous absorption region can quite easily be produced by a flame aspirating the dissolved specimen, and this procedure is typically employed.

An absorption spectrometric system for measuring atomic concentrations, or **atomic absorption,** thus fundamentally consists of a source emitting a line spectrum of the element(s) of interest, a flame that serves as a gaseous absorption region, and a spectrometer to disperse the entering light and provide a means of recording light intensity. Such a system is shown schematically in figure 11.17. The energy source is a sealed tube filled with an inert gas at low pressure and having a cathode containing the element to be determined. An applied potential causes atoms to sputter from the cathode. Subsequent collisions with the inert gas atoms cause emission of the characteristic line spectrum of the cathode element.

To produce the sample atomic vapor, a liquid sample is sprayed by aspiration into a flame of temperature sufficiently high to decompose the molecular species and produce ground-state atoms of the element under study. In some systems, solid rather than liquid specimens can be used, and these can be atomized by cathodic sputtering or by vaporization by electric arc in a graphite crucible.

The light emitted from the hollow cathode source is focused at the sample atomization region. It is then refocused on the entrance slit to a spectrometer where the wavelength for a particular resonance line is selected. Calibration of the system for the measurement of unknowns is done in a manner analogous to that described for x-ray fluorescence analysis. Here, however, concentration is calibrated to light absorption or, inversely, its transmissivity.

Basic Statistical Concepts

The measurement of a mineral feature, whether chemical or physical, is inherently subject to a certain amount of variance. Repeated measurements clearly show the inability of the analyst and apparatus to exactly reproduce a particular value. This variability in repeated measurements is referred to as **precision** and can be assessed if a sufficient number of readings are plotted against the number of tries. The histogram in figure 11.18a represents such a series of measurements and is readily transformed into the bell-shaped curve in figure 11.18b. The

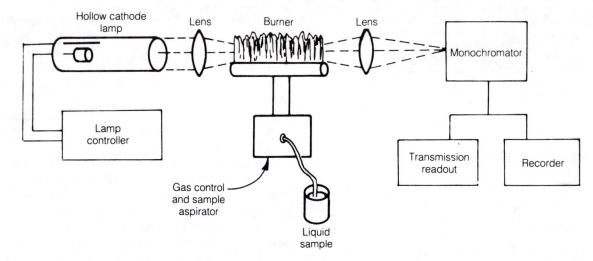

Figure 11.17
Atomic absorption spectrometry (AAS) apparatus.

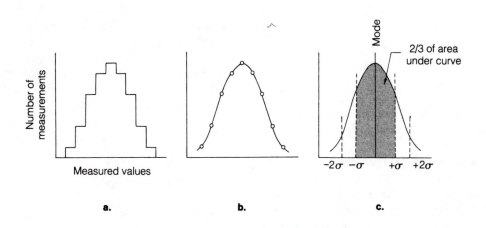

a.   b.   c.

Figure 11.18
The bell curve. The histogram (a) can be converted to a bell curve (b) by connecting the centers of the tops of the histogram bars.

Two-thirds of the area under this curve lies between the limits of $\pm \sigma$ and 95% of its area lies between $\pm 2\sigma$ (c).

mathematical properties of this curve were investigated by Carl Friedrich Gauss (1777–1855), and the curve is often referred to as a Gaussian curve. The curve is symmetrical about some central value termed the mode, and two-thirds of the area under the curve lie between the limits shown in figure 11.18c. As the curve also shows, 95% of the area lies at twice these distances from the mode. Thus, in a series of measurements the probability is that 66.7% of the values will lie between the inner limits and 95% between the outer limits.

Different degrees of variability are represented by different curves (see figure 11.19) showing the precision of measurement as high or low. This precision can be evaluated by assigning a numerical value to the limits defining the 66.7% probability zone. This value is expressed as sigma (σ) and is related to the number of individual measurements (n) and their individual differences (d) from the average of the set by the equation

$$\sigma = \pm \frac{\Sigma\, d^2}{\sqrt{n-1}}$$

For example, the series of readings given in column A in table 11.2 has an average ($\bar{x}$) of 19. The differences of the

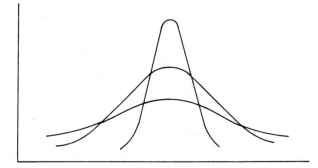

Figure 11.19
Bell curves representing different statistical variances.

individual readings (d) from this value are given in column B and d^2 values in column C. For this set of ten readings,

$$\sigma = \pm \frac{20}{\sqrt{10-1}} = 1.49$$

and 66.7% of the measured values lie between $\bar{x} \pm \sigma$ or 19 ± 1.5. In addition, 95% of the values lie between $\bar{x} \pm 2\sigma$ or 19 ± 3.0.

Table 11.2	Example of a Variance Calculation	
A Readings	**B** Differences, d	**C** d^2
17	2	4
21	2	4
19	0	0
18	1	1
20	1	1
21	2	4
18	1	1
19	0	0
17	2	4
20	1	1
$n = 10$ $\overline{x} = 19$		$\Sigma\, d^2 = 20$

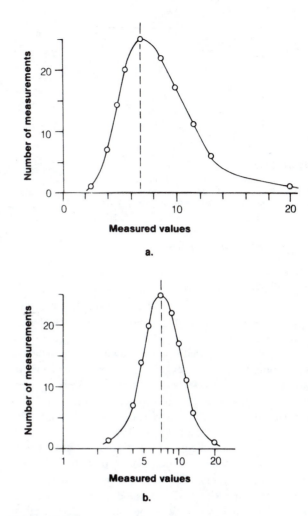

Figure 11.20
Lognormally distributed data. The linear curve (a) and the semilogarithmic curve (b) are plotted for the same data.

This analysis assumes the variance to be linearly distributed, which is often true. For many phenomena, however, and especially for the chemical distribution of the elements in nature, a different distribution is present. Curves reflecting this distribution using linear and logarithmic abscissas are shown in figure 11.20. Obviously the logarithmic (lognormal) distribution better represents the true nature of the variance:

$$\sigma = \pm\, \log \frac{\Sigma\, d^2}{\sqrt{n-1}}$$

It is often useful in assessing precision to extend the values obtained by repeated runs on a particular specimen to the probable variance for a set of related samples analyzed by the same instrument and analyst. The **coefficient of variation**, (C), serves this purpose by transforming absolute sigma values to a percentage of the sample mean:

$$C\,(\%) = \frac{\sigma}{\overline{x}} \times 100$$

The nonrepeatability of measured values by the same operator-instrument combination is usually greater when data sets obtained by different operators and instruments are compared. Each operator-instrument combination, or mode, generates a characteristic distribution of repeated data in the way of $\overline{x}$ and σ. Differences in the data sets are called **bias.**

In the discussion to this point we have not concerned ourselves with the true value of the parameter being measured. Although a correct determination can sometimes be established, very often this value is only approximate. This is especially true in chemical analysis. **Accuracy** is the statistical term for the correctness of a determined value.

The relations among accuracy, precision, and bias are illustrated simply in figure 11.21, which shows bullet holes in a target. The bull's eye represents the correct or accurate value and the scatter of holes represents the precision. The separation of average hole location from the bull's eye represents the bias.

To understand better the nature of precision and bias, try this exercise. Determine the percentage of some mineral on a polished slab of moderately coarse-grained rock such as granite. This can be accomplished using a plastic scale with 1 mm divisions and the procedure for point counting described in chapter 8. First, count the number of marks of the scale lying on the mineral chosen along a randomly oriented 5 cm line. Figure the mineral percent:

$$\text{Mineral percent} = \frac{\text{number of intersections}}{50} \times 100$$

Do this for two other randomly chosen lines and determine the mineral percentage for each. Then, find the precision for this type of measurement for which $n = 3$.

To continue, compare your grand average for the mineral percent for the 150 counts with that of a colleague to find the analytic bias.

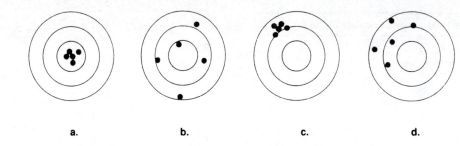

a.   b.   c.   d.

Figure 11.21
Relations between accuracy, precision, and bias shown by the grouping of hits on a target. (a) Accurate and precise. (b) Accurate but imprecise. (c) Precise but inaccurate. (d) Imprecise and inaccurate. Bias is shown by the separation of the center of one group from another. (After Dennen and Moore, 1986.)

Concluding Remarks

We have presented in this chapter brief reviews of some of the instrumental methods used by mineralogists today. The history of mineralogy and our knowledge of mineralogic structure and chemistry reflect the history of technology. With each advance in instrumentation, data that were hitherto unobtainable become available to support or refute older ideas, refine existing knowledge, and generate new hypotheses. One instrumental method that has been in use since the early part of this century and is still very important for mineralogic studies is x-ray diffraction. This technique is described in chapter 12.

The results of chemical analyses on rocks and minerals are often used for mineralogic, petrologic, or geochemical calculations. The results of any calculation can be no better than the quality of the data used. The adage of our computer-based society—garbage in, garbage out—is apt. Basic statistical analysis allows us to estimate data quality. The student should bear in mind that any measurement will have an associated error, and an analytic or measurement method should be selected so that error is minimized.

Suggested Readings

Fairbridge, R. W., ed. 1971. *The Encyclopedia of Geochemistry and Environmental Sciences.* New York: Van Nostrand Reinhold.

Jeffrey, P. G., and Hutchinson, D. 1981. *Chemical Methods of Rock Analysis.* 3d ed. Oxford: Pergamon.

Johnson, W. M., and Maxwell, J. A. 1981. *Rock and Mineral Analysis.* 2d ed. New York: Wiley.

May, L. 1971. *Introduction to Mössbauer Spectroscopy.* New York: Plenum.

Potts, P. J. 1987. *A Handbook of Silicate Rock Analysis.* Glasgow, Scotland: Blackie.

Reeves, R. D., and Brooks, R. R. 1978. *Trace Element Analysis of Geological Materials.* New York: Wiley.

Wolff, E. A., and Mercanti, E. P. 1974. *Geoscience Instrumentation.* New York: Wiley.

Zussman, J., ed. 1967. *Physical Methods in Determinative Mineralogy.* London: Academic.

X rays in Mineralogy

The relatively long wavelengths of visible light (3,900–7,700Å) interact with the macroscopic features of mineral structures and provide an essentially statistical view of the mineral. X rays and γ rays, on the other hand, have wavelengths of the same order of magnitude as the size of the individual atoms, a few Ångstroms, and are a particularly delicate means of probing the details of an atomic array.

Physics of X rays

X rays and γ rays have the shortest wavelengths in the electromagnetic continuum (see figure 11.9). Those in the longer wavelength portion of the range are sometimes referred to as soft x rays and those of the shorter wavelength as hard γ rays, especially if derived from nuclear reactions. Hard or soft in this context refers to more or less energetic radiation as represented by differing wavelength or frequency.

The energy of the radiation is directly proportional to frequency and inversely proportional to wavelength. Wavelength and frequency are related by the formula

$$c = \lambda \nu$$

where λ is wavelength, ν is frequency, and c is the speed of light, approximately 3.00×10^{10} cm/sec. Further,

$$\nu = \frac{\Delta E}{h}$$

where ΔE is the energy of the x-ray quanta and h is Planck's constant. Thus,

$$\Delta E = \frac{ch}{\lambda} = h\nu$$

X rays are diminished by absorption by all matter through which they pass. The absorption is of two types. Electrons lying in the path of the beam accept energy that causes them to vibrate in phase with the beam and act as secondary sources of radiation. Absorption of this type is proportional to the atomic number and density of the absorber. Further, absorption increases progressively with wavelength, as shown by the curve in figure 12.1. The second type of absorption shows abrupt changes at wavelengths characteristic of each element, represented by the vertical portions of the curve shown in figure 12.1. This absorption takes place at the critical energy at which an inner electron in an atom of the absorber is ejected. (Remember that wavelength and energy have an inverse relationship.) Thus, as energy impinging on an absorber increases and reaches a critical level, there is an abrupt increase in absorption. A further increase in energy (decrease in wavelength) produces progressively less absorption by electrons until the critical energy for another electron displacement is reached with a concomitant absorption jump, or **edge**.

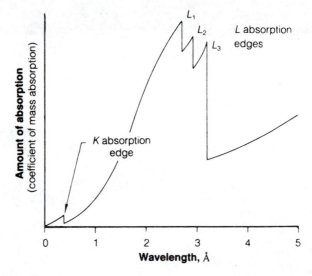

Figure 12.1
Absorption of a range of incident radiation by strontium. (Data from Jenkins and DeVries, 1969.)

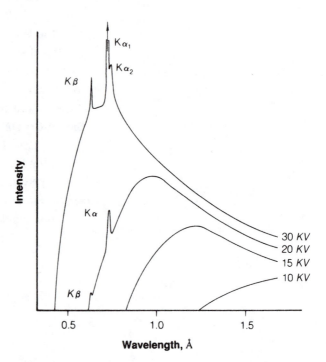

Figure 12.2
Generalized plot of the variation of x radiation with wavelength and tube voltage.

Most x rays used in laboratories are generated by the impact of high velocity electrons on a metal target in a vacuum tube. Electrons are given off from a heated tungsten filament, which serves as a cathode, and accelerated by a very high voltage to the metal target anode. The impact of the electrons generates x radiation with an energy distribution as shown in figure 12.2. The intensity increases, and both the wavelength of the maximum intensity and the minimum wavelength move toward shorter

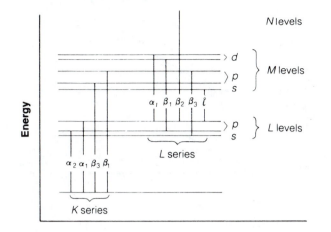

Figure 12.3
Electron transitions generating some characteristic x-ray lines.

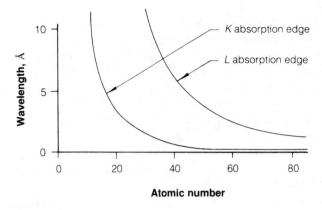

Figure 12.4
Change in wavelength of characteristic x radiation with atomic number.

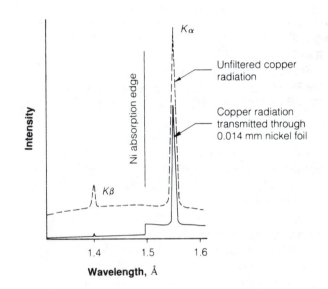

Figure 12.5
Effect of a nickel filter in suppressing copper $K\beta$ radiation.

wavelengths as the tube voltage is increased. At some specific critical voltage, depending on the target element, **characteristic radiation** appears as $K\alpha$ and $K\beta$ peaks of intensity above the bulge of continuous or "white" radiation. The radiation continuum represents the energy released by the cathode electrons as they interact with the target electrons.

The characteristic radiation coincides with the absorption edges previously mentioned. At the K absorption edge, an electron is ejected from the K shell, and electrons from higher energy levels drop into the vacated position, giving off the energy difference as characteristic radiation. A K electron may be replaced by an electron from an L shell or from an M shell, causing emission of $K\alpha$ or $K\beta$ radiation, respectively. Since the energy difference between K and M shells is greater than that between K and L shells (see figure 12.3), $K\beta$ radiation is more energetic (has a shorter wavelength) than $K\alpha$ radiation. On the other hand, the $K\alpha$ peak has greater intensity because of the larger number of L-K transitions. Both $K\alpha_1$ and $K\alpha_2$ wavelengths are emitted depending on which L sublevel shell supplied the electron, but the wavelengths are very close (e.g., 1.5401Å and 1.54433Å for copper) and generally an average ($K\alpha$) is used. A vacancy at an L or M level is filled by infall from higher levels with an additional energy release. The energy difference between energy levels diminishes, K-$L > L$-$M > M$-N . . . , and the emitted radiation above L is soft and seldom measured. The wavelength of characteristic radiation decreases with atomic number, as shown in figure 12.4.

Copper K radiation is commonly used for diffraction studies, but it consists of two wavelengths, $K\alpha$ plus $K\beta$, and monochromatic radiation is desired. Therefore, some method of eliminating the weaker $K\beta$ component is usually employed. The K absorption edge of nickel lies between the wavelengths of copper, $K\alpha$ and $K\beta$, as shown in figure 12.5. Thus, placing a nickel foil in the beam path essentially eliminates the $K\beta$ radiation relative to the $K\alpha$ radiation, which although diminished is affected to a

much lesser extent. The white (continuous) radiation is variously absorbed depending on the wavelengths of its components. The result is effectively monochromatic $K\alpha$ radiation. A second system for obtaining monochromatic radiation uses crystal monochromaters. Crystals are bent in an arc such that, when placed in a beam of polychromatic radiation, one preselected wavelength is brought into focus at a desired point. Such monochromaters are the best source of monochromatic radiation because one wavelength is selected, whereas with the use of filters one wavelength band is minimized.

Diffraction

Friedrich and Knipping, working under the direction of Max von Laue, conducted the first experiment demonstrating diffraction in 1912. Laue correctly interpreted and published the results of the experiment. In 1914 W. L. Bragg published an alternative, simplified explanation of

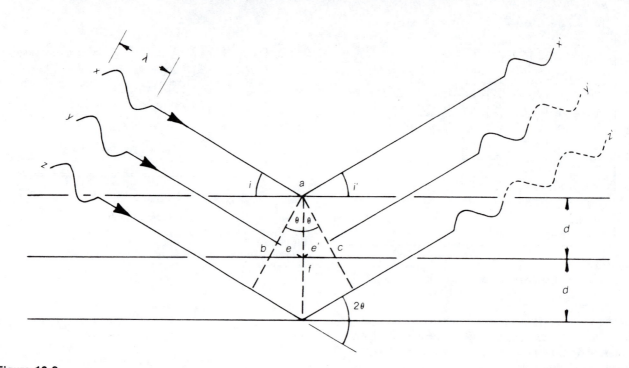

Figure 12.6
The derivation of the Bragg equation.

diffraction in terms of reflection, and Bragg's model has dominated the study of diffraction phenomena since. Keep in mind, however, that x-ray "reflections" are a convenient approximation and the actual mechanism of diffraction is not reflection.

A beam of monochromatic x rays falling on a crystal surface is diffracted in a manner partly analogous to the reflection of light from a surface. The beam is actually scattered by each atom encountered, but the scattered rays are reinforced and can be measured only in certain specific directions in which they are in phase. In figure 12.6, a beam of x rays of wavelength λ encounters a crystal surface with an angle of incidence i. Ray x impinges at point a, and the dashed lines ab and ac are normals to rays y and y' dropped from a. The distance between parallel, adjacent atom layers is d. The illustration shows that ray y–y' travels farther than the parallel ray x–x' by a distance equal to e plus e'. If the distance $e + e'$ is equal to zero or to a multiple of whole wavelengths, the rays x' and y' are in phase and are reinforced. The length of e is $d\sin\theta$, and since abf and acf are equal triangles,

$$e + e' = 2(d\sin\theta)$$

Therefore, if $n\lambda = 2d\sin\theta$, rays x' and y' are in phase and will reinforce when n equals any small whole number. Ray $z - z'$ is in phase by the same reasoning. Note that when the conditions of the Bragg equation are met, $i = i' = \theta$. Reinforcement of scattered rays will take place whenever the Bragg equation, $n\lambda = 2d\sin\theta$, is satisfied. Under these conditions, a beam of x rays will be diffracted from the

crystal. In these circumstances the angular difference in the direction of the incident and diffracted beams is equal to 2θ.

A more correct view of diffraction holds that each electron in the beam path is set vibrating at the frequency of the passing x rays. Each electron then acts as a source of x rays, and a crystal during diffraction is a three-dimensional framework containing uncounted x-ray sources vibrating at the frequency of the impinging beam. A row of atomic point sources is shown in figure 12.7 at the centers of circles. Lines drawn tangent to wave fronts indicate the direction of beams preserved and strengthened by constructive interference. Destructive interference eliminates beams in all other directions. Note that one wave front is parallel to the row of sources, and the others are inclined at increasing angles on either side. The beam propagation directions are drawn as open arrows normal to the wave fronts. These discrete, two-dimensional wave fronts can be transformed into a family of nested cones by rotating them about the row of sources, as shown in figure 12.8.

The diffracted beams from a single row of sources are confined to a series of opposite-facing nested cones, and in crystals three different intersecting source rows must be present. In this situation, the cones mutually interfere and eliminate all diffracted beams except those lying mutually in all three cones, as shown in figure 12.9. It is these highly selected few beams that escape destructive interference and that are recorded in diffraction experiments. The diffracted beam from any two or three atom layers is

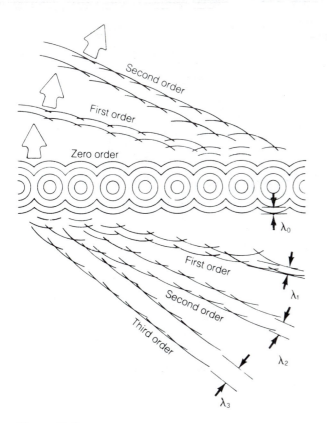

Figure 12.7
Waves generated by forced atomic vibrations due to impinging
x rays.

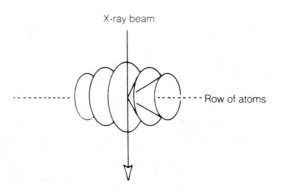

Figure 12.8
Cones of diffraction from a row of atoms.

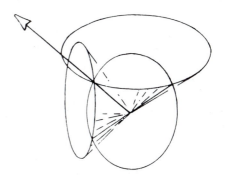

Figure 12.9
Three intersecting diffraction cones. Arrow is on the common
surface.

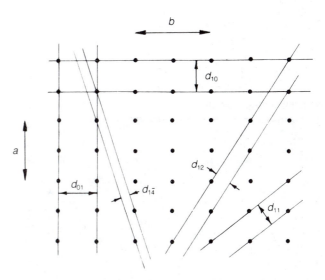

Figure 12.10
A few of the possible diffracting lines in a plane lattice.

too weak to be recorded, and reinforcing beams from perhaps 100 layers are required for a detectable beam. Diffraction becomes measurable only when a stack of atom
layers is in position to satisfy Bragg's law.

The Powder Method

There are many possible diffracted beams from a crystal,
as shown in figure 12.10. Because no two crystal structures are identical, no two structures will give all "reflections" at the same angles. Minerals can therefore be
identified by their pattern of diffraction peaks, as one might
use fingerprints to identify a person. Mineral identification by diffraction pattern is accomplished using either a
diffractometer or a powder camera. In either case, a very
finely powdered mineral is used as the sample, which en-
sures that all possible grain orientations are present and
that all possible diffraction peaks will be represented. Both
techniques allow the measurement of a series of 2θ angles
and intensity readings from which the mineral can be
identified.

Debye and Scherrer in Germany and Hull in the
United States in 1916 were the first to use fine-grained
crystalline aggregates to obtain characteristic x-ray patterns. The Bragg law states that when the angle θ between
the incident x-ray beam and a planar array of atoms in a
crystal satisfies the relation $n\lambda = 2d\sin\theta$, x rays are diffracted by the array. Thus, for each (hkl) plane in the
crystal, there is a discrete value of θ at which diffraction
occurs. Further, the angle depends only on the interplanar
spacing d_{hkl} and on the characteristic wavelength of the x
rays. X-ray diffraction patterns of powders can therefore
be used to identify minerals and to determine the size and
type of the unit cell of the crystal.

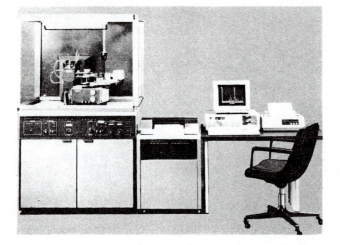

Figure 12.11
The D/Max RB x-ray diffraction system controlled by a personal computer. (Courtesy Rigaku/USA Inc., Danvers, MA.)

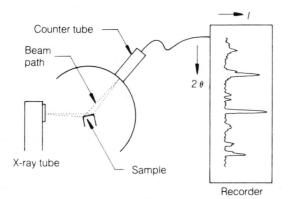

Figure 12.12.
Essentials of an x-ray diffractometer.

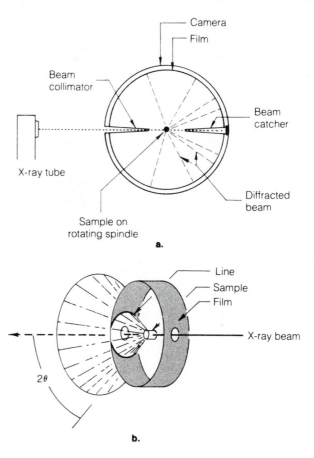

Figure 12.13
(a) Essentials of an x-ray powder camera. (b) Principles of a powder camera. The film intercepts the cone of diffraction from many particles in the sample yielding paired arcuate lines on the film.

Powder Diffractometry

A modern x-ray diffraction system is shown in figure 12.11 and its essential geometry in figure 12.12. An x-ray tube, usually having a copper target, produces x rays of known wavelength in a beam that falls on a flat-surfaced sample. The sample is mounted at the center of a circle with a detector approximately 20 cm away and rotating about the same center. The detector rotates at a rate of 2° for each degree of rotation of the sample. The detector is attached to a device that continuously records the 2θ angle and the intensity of the diffracted beam.

The general procedure in x-ray diffractometry is to scan the range of Bragg angles while recording the output of the detector on a strip chart. At each angle where the Bragg equation is satisfied, a peak appears, the height of which is a measure of the peak intensity. Peak positions are best measured not at the apex but at the center of the curve at about two-thirds of the peak height. Because of the small difference in wavelength for $CuK\alpha_1$ and $CuK\alpha_2$, a single diffraction can result in two peaks of slightly different 2θ. These are referred to as *doublets*. The positions of unresolved $\alpha_1 - \alpha_2$ doublets can be estimated by weighting their wavelengths according to the expression

$$\lambda = \frac{1}{3} \left(2\lambda_{\alpha_1} + \lambda_{\alpha_2}\right)$$

Powders of pure known materials can be added to the specimens to aid in the calibration of diffractograms. Quartz, fluorite, silicon metal, and α-alumina are commonly used as standards. The choice depends on the proximity of standard peaks to specimen peaks and a minimum number of peak interferences.

Powder Photography

A powder camera is a simple device, as illustrated in figure 12.13a. X radiation of known wavelength is admitted into a tubular cylindrical camera, usually with an internal diameter of 114.6 mm and depth of 35 mm. A cylindrical powder sample, approximately 0.5 mm in diameter and held together by some binder transparent to x rays (such as collodion) is mounted in the center of the camera and slowly rotated. A strip of 35 mm photographic film is wrapped around the inside of the camera and acts as the recorder.

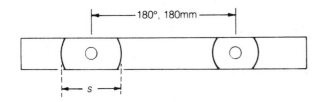

Figure 12.14
Arc *S* produced by the intersection of the cone of diffraction and the photographic film.

As the x-ray beam passes through the sample, it encounters many randomly oriented crystallites, among which there will be many oriented so that a particular set of (*hkl*) planes satisfies the Bragg law. At each value of θ where the Bragg law is satisfied, a diffracted beam (called a reflection) is given off by the powder. As discussed earlier, the individual reflections from like (*hkl*) planes of the randomly oriented crystals form a cone of half-opening 2θ (see figure 12.13b). This cone intersects the strip of film producing two nearly circular arcs, as shown in figure 12.14.

The distance between the entrance and exit holes is 180 mm and also 180°. The angle 2θ can therefore be measured directly on the film. The distance *S* between pairs of arcs on the film is proportional to the Bragg angle θ, because 4θ is the full-opening angle of the diffraction cone. Thus,

$$S = R \cdot 4\theta$$

or

$$\theta = \frac{S}{4R}$$

where *S* is the measured distance between a pair of arcs, θ is in radians, and *R* is the radius of the film. Once θ is known, d_{hkl} can be calculated from the Bragg law and the unit cell dimensions determined. The intensity of the lines is either estimated or can be measured with a densitometer.

In usual practice, the film is mounted in such a way that the position of the entrance and exit slits can be easily measured. After exposure to x rays for several hours, the film is developed and the line positions measured. Estimation or measurement of line intensities can be made at this time.

Figure 12.15 shows a sketch of a powder pattern for quartz. Each of the arcs produced corresponds to a particular θ value according to the Bragg equation. Each in turn then corresponds to the interplanar spacing d_{hkl} of the (*hkl*) planes that produced the reflection.

The distances *S* between the arcs of symmetrical pairs are determined by measuring the distances s_1, s_2, s_3, s_4,

. . . , where s_1 and s_2 are measurements for a low-angle pair and s_3 and s_4 are measurements for a high-angle pair. As a check of measurement accuracy, all the sums $s_1 + s_2$ should be equal, as should the sums of $s_3 + s_4$. Then, for a pair of arcs at $2\theta < 90°$

$$\frac{s_2 - s_1}{2B - 2A} = \frac{4\theta}{360}$$

and

$$\theta = (s_2 - s_1)\frac{90}{2B - 2A}$$

For reflections with $2\theta > 90°$,

$$\theta = (s_4 - s_3)\frac{90}{2B - 2A} + 90$$

Using the θ values thus determined, the spacings (*d*) can be calculated.

Identification of an Unknown

A common use of the powder method is in the identification of unknown specimens. This could be done by matching the pattern for the unknown with one for a standard substance. This technique, however, demands that a library of known patterns, either charts or films, be available. Also, it can be very cumbersome and time-consuming to visually inspect a number of charts or films. There exists, however, a reference file of diffraction data for nearly all substances, and this file is accessible by manual or computer searches.

The Joint Committee on Powder Diffraction Standards (JCPDS) provides an identification system consisting of two parts, the Powder Diffraction File and the Search Manuals. The data base is made up of a set of data cards (see figure 12.16) that are accessible in card files, on microfiche, in books, or in computer-readable form. Information on the data cards includes a complete listing of spacings and intensities, chemical and crystallographic data, the source of the material used, and the analytic technique. The Search Manuals are arranged according to one of the search techniques and provide reference numbers (5–0490 on figure 12.16) to the data file cards. Complete instructions for using the Powder Diffraction File and the Search Manuals are provided with each manual. An example of the procedure is, however, given here.

The oldest and most-used search procedure is the Hanawalt method, developed in 1936 by J. D. Hanawalt, an American chemist. In this system the three strongest lines of a diffraction pattern are tabulated in the Search

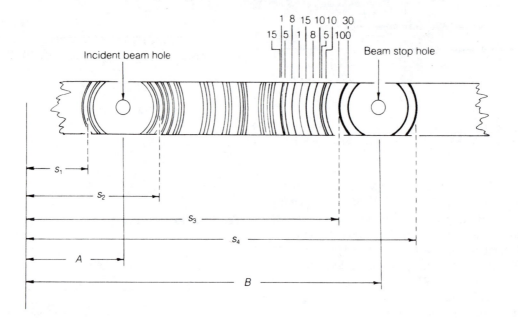

Figure 12.15

Sketch of a powder photograph of quartz showing relative intensities for some of the principal lines.

Manuals. The search procedure is straightforward. Say, for example, that a powder diffraction analysis of an unknown results in the pattern in figure 12.15. The spacings and intensities of the first eight lines are measured as follows:

Line	Relative Intensity	d
1	30	4.256
2	100	3.341
3	10	2.456
4	10	2.279
5	5	2.236
6	10	2.130
7	8	1.983
8	15	1.820

The three most intense peaks in order have d values of 3.341, 4.256, and 1.820. One then goes to the JCPDS Hanawalt Search Manual and turns to the section giving peaks of highest intensity in the range 3.34–3.30. The first column on these pages lists all peaks within the range 3.34–3.30 and is unordered. The second column lists the second most intense peaks and is ordered by decreasing value. In our example, the second most intense peak has a d of 4.256. The tables indicate four possibilities for the pattern using only the two most intense peaks: antimony sulfate, quartz, lead hydroxide carbonate hydrate, and mercury bromide. However, examination of the third column reveals that only one of these four, quartz, has a third peak with a d of 1.82. Each substance also has a pattern number listed, number 5–0490 for quartz, that can be located in the Powder Diffraction File. Reference to this data card (figure 12.16) allows confirmation of other peaks measured and their relative intensities.

Intensity is generally variable due to preferred orientation within the powder. Recognizing this, the com-

pilers of the index recorded the three strongest peaks three times in the listings: in 1–2–3 order; in 2–1–3 order; and in 3–1–2 order. If, however, the three dominant peaks determined are not the same three peaks commonly recognized, identification becomes arduous. The fact that peak positions vary in a solid-solution series causes some problems in identification but also permits identification of members within a series.

Determination of Cell Parameters

Structural analysis is well beyond the scope of this book, but the major factors and techniques involved in analyzing structures can be introduced. Although the patterns are complex, x-ray diffraction is the most powerful tool for determining total crystal structures, including the symmetry, the unit cell dimensions, and the coordinates of each atom position in the actual structure.

The pattern of diffraction peaks is not complex for simple structures, and it is possible to determine the indices of the planes responsible for each peak. This process of indexing a powder pattern is relatively simple for isometric minerals and increases in complexity with diminishing symmetry. If a primitive isometric cell is assumed, the interplanar spacing of atomic planes perpendicular to (100), d_{100}, equals a, which is the length of the unit cell. A general equation for any d in an isometric cell is

$$\frac{1}{d^2} = \frac{h^2 + k^2 + l^2}{a^2}$$

We can solve the general equation for all expected planes and determine the 2θ positions of their diffracted peaks or strong lines on film using various values of a. A plot of such data (see figure 12.17) can be used to index isometric crystals. Given a value for a, the index for each peak is found by scaling off the d of the diffraction peaks

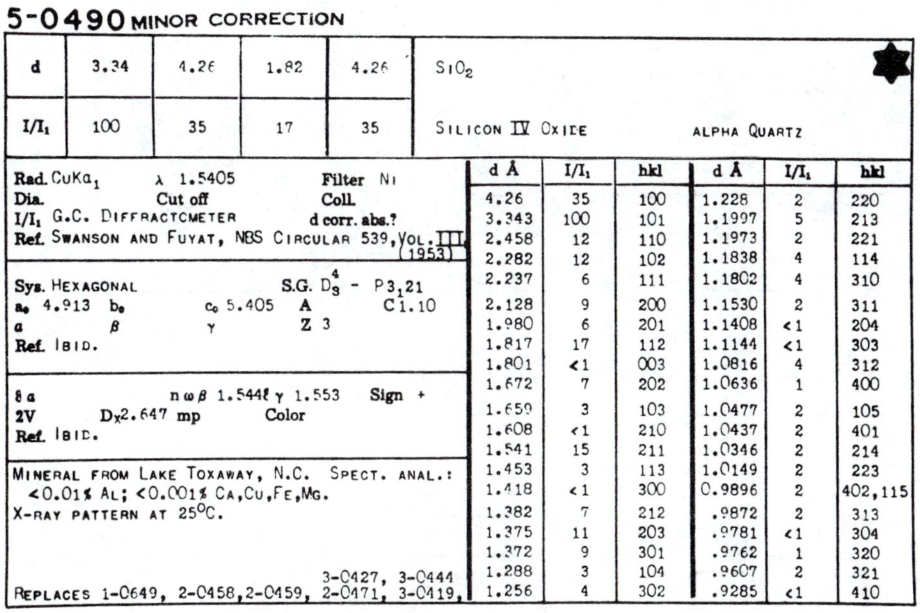

5-0490 MINOR CORRECTION

d	3.34	4.26	1.82	4.26	SiO_2
I/I_1	100	35	17	35	SILICON IV OXIDE ALPHA QUARTZ

Rad. $CuK\alpha_1$ λ 1.5405   Filter Ni
Dia.   Cut off   Coll.
I/I_1 G.C. DIFFRACTOMETER   d corr. abs.?
Ref. SWANSON AND FUYAT, NBS CIRCULAR 539, VOL. III. (1953)

Sys. HEXAGONAL   S.G. D_3^4 - $P3_121$
a_0 4.913 b_0 c_0 5.405   A   C 1.10
α β γ Z 3
Ref. IBID.

δa $n\omega\beta$ 1.544ℓ γ 1.553   Sign +
2V D_X 2.647 mp   Color
Ref. IBID.

MINERAL FROM LAKE TOXAWAY, N.C. SPECT. ANAL.:
<0.01% Al; <0.001% Ca,Cu,Fe,Mg.
X-RAY PATTERN AT 25°C.

REPLACES 1-0649, 2-0458, 2-0459, 2-0471, 3-0419, 3-0427, 3-0444

d Å	I/I_1	hkl	d Å	I/I_1	hkl
4.26	35	100	1.228	2	220
3.343	100	101	1.1997	5	213
2.458	12	110	1.1973	2	221
2.282	12	102	1.1838	4	114
2.237	6	111	1.1802	4	310
2.128	9	200	1.1530	2	311
1.980	6	201	1.1408	<1	204
1.817	17	112	1.1144	<1	303
1.801	<1	003	1.0816	4	312
1.672	7	202	1.0636	1	400
1.659	3	103	1.0477	2	105
1.608	<1	210	1.0437	2	401
1.541	15	211	1.0346	2	214
1.453	3	113	1.0149	2	223
1.418	<1	300	0.9896	2	402,115
1.382	7	212	.9872	2	313
1.375	11	203	.9781	<1	304
1.372	9	301	.9762	1	320
1.288	3	104	.9607	2	321
1.256	4	302	.9285	<1	410

Figure 12.16

Example of a data card from the JCPDS Powder Diffraction file. Card 5–0490 for quartz from the x-ray powder data file. (Courtesy Joint Committee on Powder Diffraction Standards, Swarthmore, PA.)

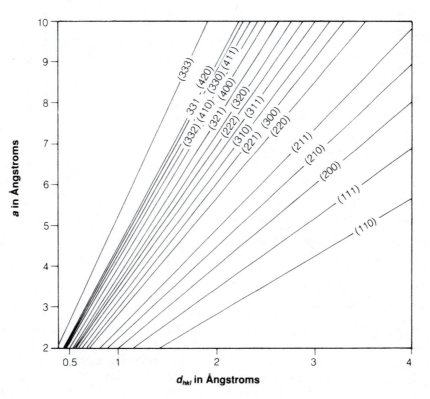

Figure 12.17

d versus cell edge length for isometric minerals.

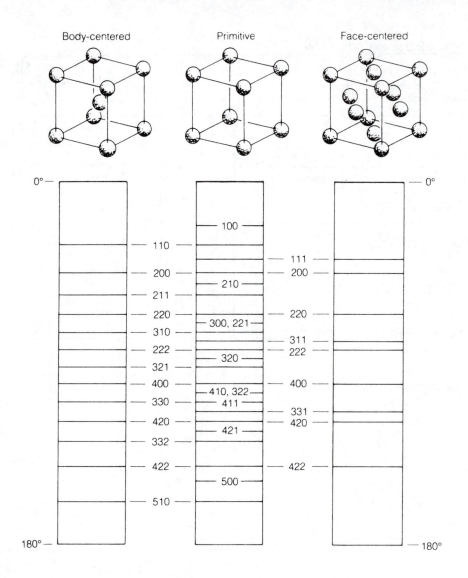

Figure 12.18
Patterns of diffraction peaks from cubic structures.

parallel to the abscissa. The peaks of an unknown isometric mineral can be indexed by the following steps:

1. Plot the peak positions as *d* at the edge of a strip of paper using the same scale as the abscissa of figure 12.17.
2. Slide the paper up the graph parallel to the abscissa to a position at which each peak matches some line. Not all possible peaks will be present.
3. Read the index of each peak from the graph and read the length of *a* from the ordinate. Some error is inherent in this process because there may be two or three apparent fits that must be checked against the expected peaks.

There are three cubic structures (see figure 12.18), each producing a different pattern of peaks. The primitive structure produces the greatest number of peaks, and the other structures exhibit a selected series of these peaks. The d_{100} peak is observed only from primitive isometric structures, the d_{200} and the d_{400} from all. This variation is explained by the fact that in nonprimitive isometric structures one or more atoms lie in a (100) plane at a spacing of $a/2$, and if the Bragg equation is satisfied for a, the beams diffracted from the $a/2$ spacing are $\lambda/2$ out of phase. The beams from a and $a/2$ cancel by destructive interference and no diffracted beam emerges from the crystal. If the Bragg equation is satisfied for (100) planes at a spacing of $a/2$, the beams reinforce and the peak is recorded as (200).

Some peak indices, for example (200) and (333), are multiples of simplest indices and therefore forbidden in morphological crystallography. To understand why, consider a primitive cubic lattice with $a = 8.0$Å for which the Bragg equation can be satisfied assuming that the retardation of adjacent rays equals 1.542Å (copper radiation):

$$1\lambda = 2d \sin\theta$$

$$1.542 = 2(8.0)\sin\theta$$

$$\sin\theta = 1.542/16 = 0.0965$$

$$\theta = 5.54°$$

giving a diffraction peak at $2\theta = 11.08°$.

If the angle of incidence of the x-ray beam is increased the Bragg equation is again satisfied by the same planes. For example, when n is 2,

$$2\lambda = 2d \sin\theta$$

$$3.084 = 2(8.0)\sin\theta$$

$$\sin\theta = 3.084/16 = 0.193$$

$$\theta = 11.13°$$

Thus, there will be a diffraction peak at a 2θ of $22.26°$.

In practice the Bragg equation is solved by assuming a retardation of unity, and the resulting d is indexed as d_{200}. Thus, d values of n times some index may represent real atom planes at a spacing of $d_{hkl/n}$ or a spacing of d_{hkl} at a retardation of $n\lambda$.

Graphs are available for the indexing of isometric, tetragonal, hexagonal, and orthorhombic patterns, but with the exception of the isometric they are complex. The equations for determining d in various sytems are as follows:

Tetragonal $\quad \dfrac{1}{d^2} = \dfrac{h^2 + k^2}{a^2} + \dfrac{l^2}{c^2}$

Hexagonal $\quad \dfrac{1}{d^2} = \dfrac{4}{3}\dfrac{h^2 + hk + k^2}{a^2} + \dfrac{l^2}{c^2}$

Orthorhombic $\quad \dfrac{1}{d^2} = \dfrac{h^2}{a^2} + \dfrac{k^2}{b^2} + \dfrac{l^2}{c^2}$

Monoclinic $\quad \dfrac{1}{d^2} = \dfrac{1}{\sin^2\beta}\dfrac{b^2}{a^2} + \dfrac{k^2\sin^2\beta}{b^2}$

$$+ \dfrac{l^2}{c^2} - \dfrac{2hl\cos\beta}{ac}$$

Triclinic $\quad \dfrac{1}{d^2} = \left[\dfrac{h^2}{a^2}\sin^2\alpha + \dfrac{k^2}{b^2}\sin^2\beta\right.$

$$+ \dfrac{l^2}{c^2}\sin^2\gamma$$

$$+ \dfrac{2hk}{ab}(\cos\alpha\cos\beta - \cos\gamma)$$

$$+ \dfrac{2kl}{bc}(\cos\beta\cos\gamma - \cos\alpha)$$

$$+ \left.\dfrac{2hl}{ac}(\cos\gamma\cos\alpha - \cos\beta\right]$$

$$[1 - \cos^2\alpha - \cos^2\beta - \cos^2\gamma$$

$$+ 2\cos\alpha - \cos\beta\cos\gamma].$$

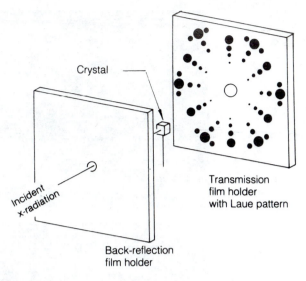

Figure 12.19
Geometry of a Laue camera with both transmission and back-reflection capabilities.

The indexing of powder patterns for minerals of low symmetry is quite difficult with powder-diffraction data alone, and today most patterns are indexed by iterative computer procedures. In most cases the indexing for non-cubic minerals given in the JCPDS file is adequate.

Single-Crystal Diffractometry

Single-crystal diffractometry techniques are commonly employed in determining the structure of minerals. Typically, single-crystal x-ray photographs exhibit resolved reflections from all planes. The planes can be indexed and their positions used to determine symmetry, cell parameters, and space groups. Their reflection intensities can also be used to predict atomic positions. Single-crystal techniques are also used if only very small amounts of the mineral are available.

Laue Patterns

The technique used in 1912 by Friedrich and Knipping to perform Laue's experiment is still in use today. Laue photographs of the diffraction by a single crystal are easy to make with rather simple apparatus. A single crystal is placed in the path of a beam of white x radiation from a source with a heavy-metal target such as tungsten. Photographic film can be placed behind the crystal to record transmission diffractometry, but back-reflection cameras in which the photographic film is placed in front of the crystal, are also used. Figure 12.19 shows schematically a modern Laue camera having both transmission and back-reflection capabilities.

A range of wavelengths is present as generated by the source tube. Thus, many planes will be in a position to satisfy the Bragg equation for some wavelength, and each set of planes will cause a spot or family of spots on the film. The arrangement of the spots on the film (see figure 12.20) is indicative of the symmetry along the crystal axis set parallel to the beam.

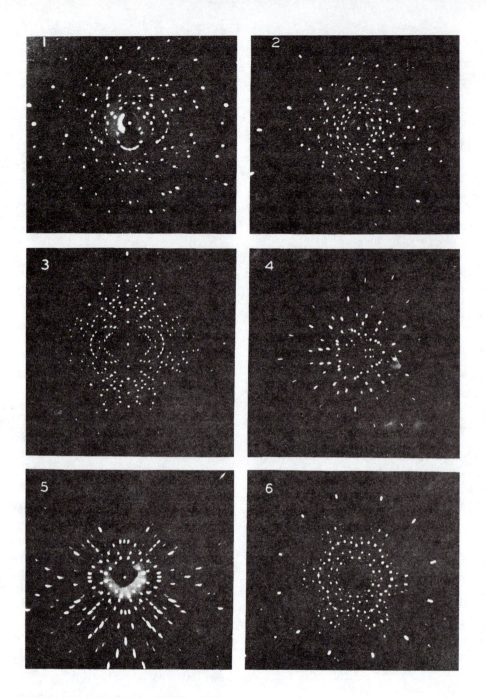

Figure 12.20
Examples of Laue photographs: (1) *m*, (2) 2, (3) 2*mm*, (4) 3*m*,
(5) 4*mm*, (6) 6*mm*. (Courtesy of The Natural History Museum,
London.)

The diffracted beams of the Laue technique have different wavelengths. These are selected from the incident beam of white radiation by the d's and θ values of the crystal planes that produce the diffraction. The position of any Laue spot will be unaltered by a change in the spacing d. The only effect of such a change will be to change the wavelength of the diffracted beam. Thus, two isostructural minerals with different cell parameters but the same orientation will produce identical Laue photographs. The Laue method, although good for revealing crystal symmetry, gives little other information and is little used today.

Rotation Methods

In the rotation method, an oriented single crystal is rotated about an axis parallel to the axis of a film cylinder and perpendicular to the collimated x-ray beam, as shown in figure 12.21. The film forms a cylinder around the axis of crystal rotation, and cones of diffracted rays with $n =$ 1, 2, 3, . . . intersect the film, producing spots in parallel rows. For example, if the crystal is rotated about its c-axis, the row including the direct beam represents all ($hk0$) planes, the next higher row includes all ($hk1$) planes, the next ($hk2$), and so on. Other cell edges can be determined by using each as the axis of rotation in another exposure.

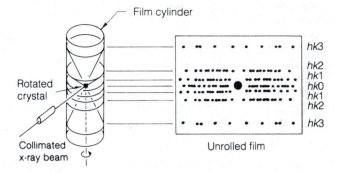

Figure 12.21
The rotating crystal technique.

If r is the radius of the cylindrical film and s is the distance of any diffraction row from the central row ($n = 0$), and if $(90 - \phi)$ is the exterior angle of the diffraction cone, then

$$\frac{s}{r} = \tan (90 - \phi)$$

and, according to the Bragg equation,

$$a = \frac{n\lambda}{\sin(90 - \phi)}$$

where a is the lattice period.

Scattering by Atoms and Unit Cells

Diffraction is actually energy scattering by individual electrons, and because the number and spatial distribution of electrons vary in different elements, the diffracting efficiency of elements also varies. The first proof of the face-centered structure of halite was based on this fact. Atom layers in (111) planes of halite are alternately Na and Cl (see figure 12.22a). Reinforcement by similar atoms occurs at a spacing of d_{111}, but interference by layers of atoms of the other element occurs at a spacing of $d_{111/2}$. Diffraction from chlorine atoms is more efficient than that from sodium, and the chlorine-diffracted beam is more intense. The d_{111} beam is not entirely canceled but has an amplitude or intensity equal to the difference in intensity of the chlorine and the sodium beams. At an angle such that $d_{111/2}$ satisfies the Bragg equation, the diffracted beams from chlorine and from sodium layers are in phase, and the resultant amplitude is the sum of the two. Therefore, halite exhibits weak diffraction peaks from (111) and (333) and strong peaks from (222) and (444), as shown in figure 12.22b.

Quantitative Analysis by Diffraction

The x-ray diffractometer can sometimes be a useful tool for compositional determinations in solid-solution series. The usual differences in size of the substituting ions in

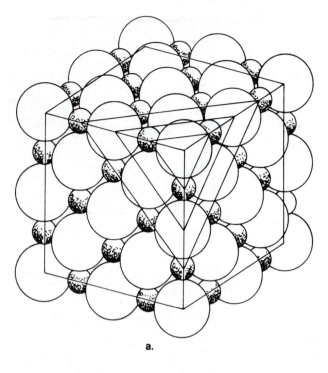

a.

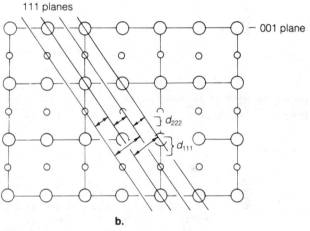

b.

Figure 12.22
(a) Halite structure showing (111) planes. (b) Vertical (110) section through halite cube corners. Size of ions reduced and unit cells increased to show the continuity of planes better.

such a series will cause slight changes in unit cell dimensions and volume. In many cases the dimensional changes are linearly related to the magnitude of the substitution. The cell dimension changes can be monitored by measuring shifts in the position of selected diffraction lines, lines that have simple relationships with the cell dimensions of the solid-solution mineral. Thus, if appropriate diffraction lines are selected, their positions will be seen to shift gradually with changes in composition.

Quantitative analysis by diffraction has found many uses in mineralogic studies. Among the solid-solution series that have been calibrated for composition are amphiboles, apatite, carbonates (both calcite and dolomite), chlorite,

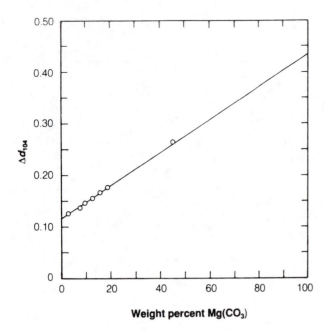

Figure 12.23
Variation of (104) spacing in Ca-Mg carbonates with substitution of Mg, referred to fluorite (111) as standard. (Modified from Goldsmith et al., 1955.)

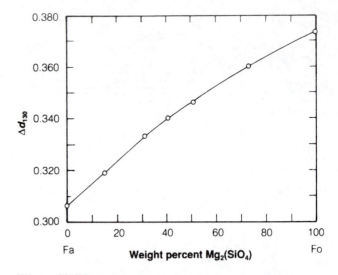

Figure 12.24
Variation of (130) spacing in olivine, referred to silicon (111) as a standard. (From data in Yoder and Sahama, 1957.)

feldspars (both alkali and plagioclase series), garnet, Fe-Ti oxides, micas, olivines, pyroxenes, and sphalerite. Following are some examples of the technique used with solid-solution series of some common minerals.

Magnesian Calcite

The magnesium content of calcite is of considerable geologic and mineralogic interest because it provides information about the environment in which this common mineral formed. Because of the large difference in size between Mg^{2+} and Ca^{2+} ions, respectively 0.80Å and 1.08Å, there is a large difference in cell-edge lengths of the isostructural minerals magnesite and calcite. Thus, the shrinkage in length of the c-axis of the calcite cell from 17.061Å toward that of magnesite, 15.016Å, with the substitution of Mg for Ca in calcite provides a means of determining the magnesium content.

If a series of Ca-Mg carbonate crystals of known Mg content is subjected to x-ray diffraction analysis, a calibration curve such as that shown in figure 12.23 will result. In this example, the (104) spacings in a series of specimens ranging from pure calcite through dolomite to pure magnesite were determined along with the (111) spacing for a pure fluorite as a standard ($d_{111} = 3.153$Å). Thus, for each specimen

$$\Delta d = [d_{111} \text{ (fluorite)} - d_{104}\text{(calcite)}]$$

and Δd is plotted as the ordinate versus the weight percent $Mg(CO_3)$. The $Mg(CO_3)$ content of an unknown Ca-Mg carbonate can be determined by carefully measuring $2\theta_{104}$

along with the $2\theta_{111}$ for an admixed fluorite standard, converting these figures to the respective d values, and calculating Δd.

Similar calibrations can be generated using other suitable standard materials, such as halite or silicon metal. Further, the calibration could be constructed in terms of 2θ rather than d, but the former will be dependent on the source wavelength employed.

Olivine

The compositions of members of the olivine series, $Mg_2(SiO_4)$-$Fe_2(SiO_4)$, can be determined by optical or density measurements. Because the difference in the ionic radii of Mg^{2+} and Fe^{2+}, 0.80Å and 0.83Å respectively, is quite small, a relationship between cell parameters and composition might not be expected. Indeed, the cell volumes of forsterite and fayalite, 290.8Å^3 and 308.1Å^3, are not very different. There are, however, variations in several interplanar spacings that are useful for compositional determinations.

Figure 12.24 shows the variation of d_{130} with composition for the olivine series. The curve was determined by powder diffractometry using the d_{130} value for olivine and the d_{111} value for silicon metal (3.1388Å) as a standard to give

$$\Delta d_{130} = [d_{111}(\text{silicon}) - d_{130}(\text{olivine})]$$

The d_{130} reflection for olivine is intense enough that rock powders containing as little as 10% olivine can be used for the determination.

The Plagioclase Series

The determination of feldspar composition presents a problem for which plagioclase can serve as an example. The plagioclase series, $Na(AlSi_3O_8)$-$Ca(Al_2Si_2O_8)$, would

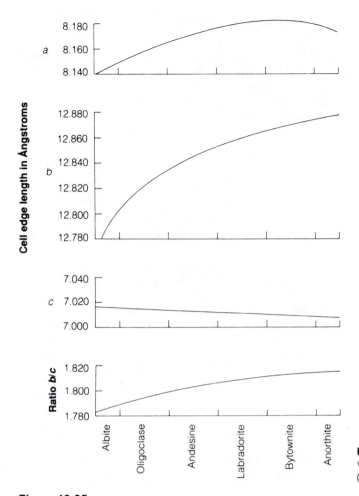

Figure 12.25

Changes in cell edge lengths with composition in the plagioclase series.

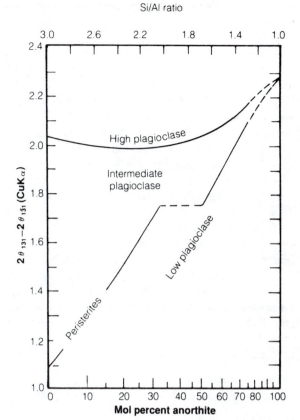

Figure 12.26

Variation of $2\theta_{131}-2\theta_{1\bar{3}1}$ (CuKα) with plagioclase composition. (Modified from Bambauer et al., 1967.)

not seem conducive to compositional discrimination because of the small size difference between Na^+ (1.24Å) and Ca^{2+} (1.20Å). The cell-edge length can be seen to vary with composition (figure 12.25), however, and it might be expected that absolute values or ratios of cell parameters could be used to determine plagioclase composition.

A difficulty arises, however, in that the structural state of the feldspars is controlled by the degree of Al-Si ordering (see chapter 21), which in turn is determined by the temperature of formation and the cooling rate. High-temperature plagioclases possess high Si-Al disorder, and low-temperature plagioclase and plagioclase that has cooled slowly possess high Si-Al ordering.

If x-ray diffractometry is performed on plagioclases with a range of composition and some measure of composition is discerned, for example, $2\theta_{131} - 2\theta_{1\bar{3}1}$ (CuKα), then an obvious relationship to Si-Al ordering can be seen. Figure 12.26 shows the results of such an analysis. It is evident that cell geometry bears as close a relationship to the Al/Si ratio as it does to the anorthite content. Figure 12.27 shows a determinative curve given in terms of two measures of cell parameters and plagioclase chemistry.

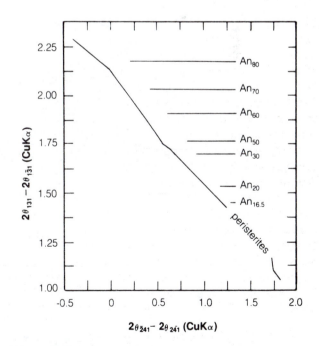

Figure 12.27

Relationship between the diffraction parameters ($2\theta_{131}-2\theta_{1\bar{3}1}$, $2\theta_{\bar{2}41}-2\theta_{2\bar{4}1}$) and plagioclase composition using CuKα radiation. (After Bambauer et al., 1967.)

However, the curve applies to ordered, low-temperature plagioclase alone and cannot be used for disordered specimens.

The determination of plagioclase composition by x-ray diffraction must be treated with caution. Reasonable estimates of composition can be made only if the degree of ordering is known. Geologic reasoning, however, can often provide clues for particular materials. Ordered low-temperature plagioclase is found in plutonic rocks where cooling has been relatively slow. On the other hand, the plagioclase of volcanic or subvolcanic rocks tends to be high-temperature and disordered. Volcanic plagioclases, however, will often have an intermediate structural state because of partial ordering during subsolidus cooling. Thus, it is often difficult to extract much useful compositional information from plagioclase diffraction data unless the structural state is known by independent means. If, however, the composition of the plagioclase is known by some other technique, then the diffraction data can provide much useful information as to its structural state.

Concluding Remarks

This chapter concludes part 2, in which we discussed aspects of practical mineral study. The various instrumental techniques described may seem exotic, but they are the stuff of modern mineralogic study. In practice, their application is much less complex than their description.

The student is urged to gain hands-on familiarity with minerals and their properties using the information provided in chapter 9. The ability to identify minerals is fundamental to geologic practice. Mineral identification should be coupled with an appreciation of the interrelationships of gross properties, symmetry, and structure. Knowledge of mineral chemistry and its variability within types is an important aspect of petrographic studies.

Suggested Readings

Azaroff, L. V., and Buerger, M. J. 1958. *The Powder Method in X-Ray Crystallography.* New York: McGraw-Hill. An older but excellent and detailed book on the fundamentals and practice of x-ray diffraction.

Cullity, B. D. 1978. *Elements of X-Ray Diffraction.* 2d ed. Reading, MA: Addison-Wesley. Written by a metallurgist, this book presents the principles of x-ray diffraction in a simple and clear manner.

Hutchison, C. S. 1974. *Laboratory Handbook of Petrographic Techniques.* New York: Wiley. A very useful book in which examples and procedures are given in a clear and concise manner.

Klug, H. W., and Alexander, L. E. 1974. *X-Ray Diffraction Procedures.* 2d ed. New York: Wiley. A standard textbook on x-ray diffraction.

Moore, D. M., and Reynolds, R. C. 1989. *X-Ray Diffraction and the Identification and Analysis of Clay Minerals.* Oxford: Oxford Univ. Press.

Part Three

Mineral Descriptions

Some 178 common, valuable, or theoretically important mineral species, series, or groups are described in the following chapters. These descriptions are intended for the use of beginning mineralogy students working principally with hand specimens and are therefore somewhat restricted in scope. Descriptions are arranged in the following order:

Chapter 13 Native Elements
Chapter 14 Sulfides and Sulfosalts
Chapter 15 Halides
Chapter 16 Oxides and Hydroxides
Chapter 17 Carbonates
Chapter 18 Borates
Chapter 19 Sulfates
Chapter 20 Phosphates, Arsenates, Vanadates, Chromates, Tungstates, and Molybdates
Chapter 21 Silicates

Each of the mineral classes is preceded by a tabulation of included mineral groups, series, and species, but only the more important of these are further described. General descriptions accompany each mineral class as well as the more important subdivisions. Physical and chemical properties noted in the descriptions are further discussed in chapter 9. Descriptions of individual minerals or mineral series are arranged by the following format. For supplementary details, students are referred to the sources given on page 222.

Mineral Description Format

NAME OF SPECIES OR SERIES
Formula

(discredited synonyms)
Pronunciation, etymology.

Crystallography, Structure, and Habit   Crystal system. Crystal Class. Cell size (in Ångstroms, Å) and angles. Space group. Crystal habit. Twinning.

Brief description of structure, including isostructural references.

Mineral habit.

Physical Properties   Manner of breaking: cleavage, parting, fracture, sectility. Hardness. Specific gravity. Luster. Diaphaneity. Color. Streak. Principal optical properties. Miscellaneous properties: fusibility (fusibility scale value), fluorescence, taste, etc.

Distinctive Properties and Tests   Summary of the most valuable determinative properties. Brief statement of any particularly useful confirmatory tests, including fusibility and solubility in common acids.

Association and Occurrence   Brief statement of the geological environment and rock types in which the mineral is found and a listing of other minerals often found with it.

Alteration   Minerals from or to which the mineral may readily change. Formulas are given for species not described elsewhere.

Confused with   Minerals similar in general appearance are noted.

Variants   Many minerals have variants that differ from the norm in physical properties or appearance. Long usage has established special terms for these variants, and the more important are given as form varieties. Most minerals also have minor variations in chemical content. The more important of these variations are included as chemical varieties. All minerals contain small amounts of nonformulary elements. Such contaminants are noted when significant.

Related Minerals   Mineral species related by solid solution and polymorphism are listed together with formulas. Mineral species not related by the above phenomena but having generally similar composition or structure are listed as *other similar species*.

Mineral Formulas

The customary organization of mineral formulas follows more or less closely the following rules.

1. The sequence of chemical symbols from left to right is cations, anion or anionic radical, interstitial components.
2. Cations are arranged in the order of decreasing ionic radius from left to right. If two or more cations occupy the same structural position, they are enclosed in parentheses, usually with the most abundant ion given first. Thus, because calcium is larger than magnesium, the formula for diopside is $CaMg(Si_2O_6)$. If some iron is substituting for magnesium, the formula becomes $Ca(Mg,Fe)(Si_2O_6)$, with the structurally identical Mg and Fe enclosed in parentheses (the more abundant first) and separated by a comma.
3. Minerals are broadly grouped into chemical classes based on their constituent anion or radical. Radicals are often enclosed in parentheses, especially if they are complex. For diopside, the anionic component is $(Si_2O_6)^{4-}$. In those rare instances when more than one radical is present in a mineral or when structurally interchangeable radicals exist, they are treated like cations with square brackets being used as needed.

Table III.1   **Formulas for Some Copper Minerals**

Mineral	Cations	Anions	Interstitial Components
Copper	Cu	—	—
Covellite	Cu	S	—
Chalcopyrite	CuFe	S_2	—
Crednerite	CuMn	O_2	—
Enargite	Cu	(AsS_4)	—
Malachite	Cu	(CO_3)	$(OH)_2$
Brochantite	Cu	(SO_4)	$(OH)_6$
Chalcanthite	Cu	(SO_4)	$5H_2O$
Chalconatronite	Na_2Cu	$(CO_3)_2$	$3H_2O$
Turquoise	$CuAl_6$	$(PO_4)_4$	$(OH)_8 \cdot 5H_2O$
Andrewsite	$(Cu,Fe^{2+})Fe^{3+}$	$(PO_4)_3$	$(OH)_2$
Chalcothallite	$Tl_2(Cu,Fe)_6$	(SbS_4)	—
Chalcophyllite	$Cu_{18}Al_2$	$(AsO_4)_3(SO_4)_3$	$(OH)_{27} \cdot 33H_2O$
Arsenosulvanite	Cu_3	$[(As,V)S_4]$	—
Cuprorivaite	CaCu	(Si_4O_{10})	—
Antimonpearcite	$(Ag,Cu)_{16}$	$[(Sb,As)_2S_{11}]$	—

4. Interstitial components in minerals are usually molecular water, H_2O, or hydroxyl, $(OH)^-$, or fluorine ions, F^-. In their more common sites, they are an integral part of a mineral but not an essential part of the mineral's structural skeleton, although $(OH)^-$, F^-, and related ions may also serve as essential anions. Which role they play can be determined by inspecting the formula; for example, the hydroxyl radical in kaolinite, $Al_2(Si_2O_5)(OH)_4$, is interstitial, and (Si_2O_5) is the radical. In contrast, the hydroxyl radical in gibbsite, $Al(OH)_3$, is essential.

5. Subscripts following the chemical symbol identify the relative numbers of ions present. Thus, in diopside there is an equal number of calcium and magnesium ions (or calcium and magnesium plus iron). The ratio of calcium to silicon is 1:2 and that of silicon to oxygen is 1:3. This latter ratio is particularly important in distinguishing the many silicate minerals based on complex silicate radicals (see page 306). Superscripts are sometimes employed, usually to give the valence state of the particular ion or radical, as in Fe^{2+}, Fe^{3+}, $(OH)^-$, $(CO_3)^{2-}$, and $(Si_2O_6)^{4-}$. Other superscript notation occasionally indicates an ion's coordination. For example, to show why the formula of magnetite is properly $FeFe_2O_4$ rather than Fe_3O_4, we can use valence and coordination notation. The formula is $^{VI}Fe^{2+} \, ^{IV}Fe^{3+}{}_2O_4$ (read "one ferrous iron ion in 6-fold coordination, two ferric iron ions in 4-fold coordination, four oxygen ions").

The formulas of some copper minerals are listed in table III.1 to provide an exercise in dismembering a mineral formula and obtaining the information that has been encoded. For example, using the table, what are the relative ionic sizes of copper and iron (chalcopyrite)? Which is the richer copper ore, covellite or enargite? For which minerals is there a possibility of physical change on modest heating (with H_2O evolved)? Which mineral may effervesce when treated with hydrochloric acid (have carbonate radical)? What is the nature of the silicate (Si_4O_{10}) skeleton in cuprorivaite? Which minerals will become magnetic on strong heating (have iron present)?

Mineral Structures

Mineral structures can be drawn in a number of ways to show different aspects of their makeup. The most common are (1) packing diagrams which bring out the organization of spherical ions or atoms of different size, (2) coordination polyhedra diagrams that show the arrangement of corner- or edge-sharing of coordination polyhedra, and (3) coordination, or ball-and-stick, figures emphasizing the bonding among ions. Examples of these types of illustrations for the mineral halite, NaCl, are shown in the figure on the next page.

Although these diagrams are very different in appearance, each incorporates, either directly or indirectly, the symmetry, coordination, and cell contents of the mineral.

In calculating the number of ions in a unit cell, realize that an interior ion counts as one, and an ion lying on a face is half in the cell and half in the adjacent cell and so counts as 1/2. Ions on edges are one-quarter in each of the four cells and count as 1/4, and corner ions count as 1/8 (see figure). The number of ions for NaCl, then, is as follows:

Sodium ions	Chlorine ions
8 corner ions × 1/8 = 1	12 edge ions × 1/4 = 3
6 face ions × 1/2 = 3	1 interior ion × 1 = 1
Total 4	Total 4

The cell contains four NaCl units or four formula units.

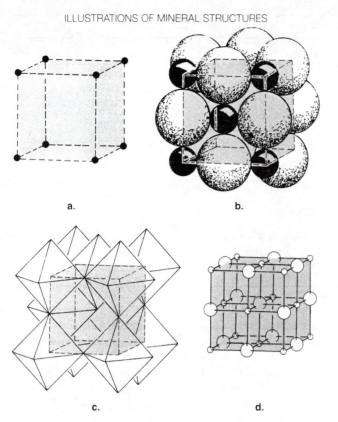

a.

b.

c.

d.

a. Cubic lattice serving as a reference for the halite, NaCl, structure
b. Packing diagram; larger ions are Cl, smaller Na
c. Polyhedral diagram; octahedra of Cl around Na
d. Coordination diagram; bonds between ions shown as solid lines

Suggested Readings

Dana, E. S., and Ford, W. E. 1932. *A Textbook of Mineralogy*. 4th ed. New York: Wiley.

Deer, W. A.; Howie, R. A.; and Zussman, J. 1962–1986. *Rock-forming Minerals*. Vol. 1A, *Orthosilicates;* Vol. 1B, *Disilicates and Ring Silicates;* Vol. 2, *Chain Silicates;* Vol. 2A, *Single-chain Silicates;* Vol. 3, *Sheet Silicates;* Vol. 4, *Framework Silicates;* Vol. 5, *Non-silicates.* London: Longmans.

———. 1966. *An Introduction to the Rock-forming Minerals.* London: Longmans.

Fleischer, M. 1991. *Glossary of Mineral Species.* 6th ed. Tucson, AZ.: The Mineralogic Record.

Hey, M. H. 1963. *Index of Mineral Species and Varieties Arranged Chemically.* London: British Museum.

Palache, C.; Berman, H.; and Frondel, C. 1944–1962. *The System of Mineralogy.* 7th ed. Vol. I, *Elements, Sulfides, Sulfosalts, Oxides;* Vol. II, *Halides, Nitrates, Borates, Carbonates, Sulfates, Phosphates, Arsenates, Tungstates, Molybdates,* etc.; Vol. III, *Silica Minerals.* New York: Wiley.

Phillips, W. R. and Griffen, D. T. 1981. *Optical Mineralogy, The Nonopaque Minerals.* San Francisco: Freeman.

Povarennykh, A. S. 1972. *Crystal Chemical Classification of Minerals.* Vols. 1 and 2. Translated by J. E. S. Bradley. New York: Plenum.

Strunz, H. 1977. *Mineralogische Tabellen.* 6th ed. Leipzig: Akademische Verlagsgesellschaft Geest & Porting K. G.

Zoltai, T. 1960. Classification of silicates and other minerals with tetrahedral structures. *Am. Mineral.* 45:960–73.

Native Elements

Native Elements

Metals
    Gold Group
        Gold, Au
        Silver, Ag
        Copper, Cu
    Platinum Group
        Platinum, Pt
    Iron Group
Semimetals
    Arsenic Group
        Arsenic, As
Nonmetals
    Tellurium Group
    Sulfur Group
        Sulfur, S
    Carbon Group
        Diamond, C
        Graphite, C

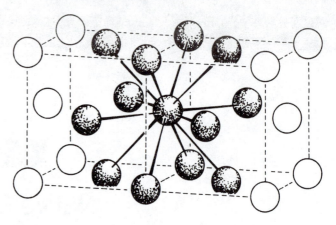

Figure 13.1
Coordination in the copper structure.

Table 13.1	**Packing in Some Copper-Zinc Compounds**		
Structure Type:	**Body-centered Cubic**	**Hexagonal Closest Packing**	**Cubic, 52 Atoms/Cell**
Example:	CuZn	CuZn$_3$	Cu$_5$Zn$_8$
Electrons, Cu (4s)	1 ⎫	1 ⎫	5 ⎫
	⎬ 3	⎬ 7	⎬ 21
Zn(4s)	2 ⎭	6 ⎭	16 ⎭
Atoms	2	4	13
Valence electrons/ atoms	3:2	7:4	21:13

About twenty elements are known to occur naturally in a pure or native form. These are about equally divided between metallic and nonmetallic minerals.

The metallic subdivision includes the valuable native metals gold, silver, copper, and platinum, which have closely similar structures and many physical similarities, such as metallic luster, high electrical conductivity, easy malleability, high specific gravity, softness, and (except for platinum) similar melting points.

The structure of these metals (figure 13.1) is based on a face-centered cubic cell containing four atoms (eight 1/8 atoms on the corners and six 1/2 atoms on the faces). In this structure, the stacking of face-centered cells causes each atom to have twelve identical neighbors. These neighbors lie on the corners of a regular cuboctahedron centered on the coordinating atom and may be outlined by connecting the coordinated atoms in figure 13.1. Identical coordination polyhedra exist around every atom in the structure. The polyhedra share faces when they surround atoms separated by a unit translation and are interpenetrating for adjacent corner and face atoms. This arrangement of atoms is called **cubic closest packing** and, together with **hexagonal closest packing,** represents the highest degree of space filling (74.1%) that can be attained by the packing of uniform spheres. Since the nature of the metallic bond does not dictate the spatial distribution of atoms, it is not unexpected that metals crystallize into the arrangements that fill space most efficiently.

The differences between cubic and hexagonal closest packing can be seen by stacking spheres (see Box 13.1). The first and second layers are identical, but in cubic closest packing the third layer has atoms over first-layer holes, whereas in hexagonal closest packing third-layer atoms are over first-layer atoms. Most metals have one of

these closest-packed arrangements, and those that do not have instead a cubic body-centered array:

> Cubic closest packing: Ag, Al, Au, α-Co, Cu, γ-Fe, Ir, α-Ni, Pb, Pd, Pt
>
> Hexagonal closest packing: Cd, β-Co, β-Cr, Mg, β-Ni, Os, α-Ti, Zn
>
> Body-centered cubic: α-Cr, α-Fe, Mo, Nb, Ta, β-Ti, V, W

In general, compound formation in metallic systems follows the **Hume-Rothery rule** that the structural type depends on the ratio of the number of valence electrons to the number of atoms in the compound. For example, consider the copper-zinc compounds in table 13.1.

The members of the gold group, together with such elements as mercury, lead, and palladium, are so similar in crystallochemical properties that they readily substitute for one another. In nature, for example, native gold is always silver-bearing and copper is commonly so, silver and mercury form a natural amalgam, and palladian gold is known. Sometimes these solid solutions or alloys are

Box 13.1

Close Packing of Atoms

Cubic close-packed (ccp) and hexagonal close-packed (hcp) structures (see figure 13.2) are a common basis for the packing of spheres of the same radius. The atoms in metals are uniform spheres, and metal structures nearly always exist in one of these two arrangements.

The geometry of packing may be understood by considering a layer-by-layer buildup of the two arrangements. The procedure can be modeled using marbles in a box. Layer 1 is composed of atoms centered on the vertices of a triangular mesh with each atom having six neighbors in the plane (see figure 13.2a). Layer 2 is identical but with its atoms located over the dimples of layer 1, that is, over the interatomic holes at the centers of the triangles of the layer 1 mesh (see figure 13.2b). Layer 3 can have its atoms located in the dimples of layer 2 at either position X or Y but not at both, because the atom radius (dashed circle) is too large. If the triangles of the layer 3 mesh are centered on position X and its equivalents—over the holes in layer 1—the packing is hcp. For triangles centered on position Y over atoms in layer 1, the packing is ccp. The layer repeat for hexagonal close-packing is thus 123123, whereas that for cubic close-packing is 121212.

Uniform spheres can also fill space in body-centered cubic and simple cubic arrays. These four arrangements are contrasted in table 13.2 and illustrated in figure 13.3.

Many minerals are composed of relatively large anions, such as sulfur or oxygen, in combination with smaller cations. Their structures often approximate a close-packing of the anions with cations nestled in the interstices.

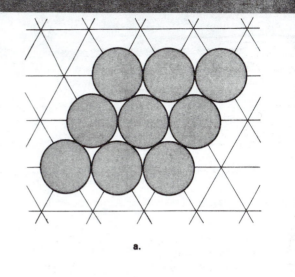

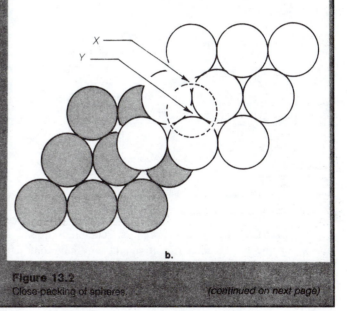

Figure 13.2
Close-packing of spheres.

(continued on next page)

Table 13.2	Packing of Uniform Spheres	
Arrangement	Coordination	Space-filling, %
Cubic close-packed, ccp	12	74.1
Hexagonal close-packed, hcp	12	74.1
Body-centered cubic, bcc	8	68.1
Simple cubic	6	52.4

stable only at higher temperatures and exsolve into ordered components on cooling. A solid solution of appropriate proportions of gold and copper, for example, will separate upon slow cooling into an arrangement with copper atoms on the cube corners and gold atoms on the cube faces. This compound, $AuCu_3$, is structurally distinct from a disordered distribution of the same elements and is also electrically and mechanically different.

Similar extensive and complex solid-solution relationships exist among the noble metals in the platinum group, and limited solid solution exists between it and the gold group.

Some elements from Groups IVB to VIB of the periodic table (see figure 4.4) crystallize into semimetallic and nonmetallic minerals with dominantly covalent or metallocovalent bonds. Those in the same column have similar structures because of their closely related electronic configurations, but structural diversity (including polymorphism) is much more common than for the native metals.

The **octet rule,** which states that each element may have as many nearest neighbors as there are missing electrons in its eight-electron outer shell ($s + p$ electrons), is a useful means of assessing a possible structure. Chlorine

Box 13.1 continued

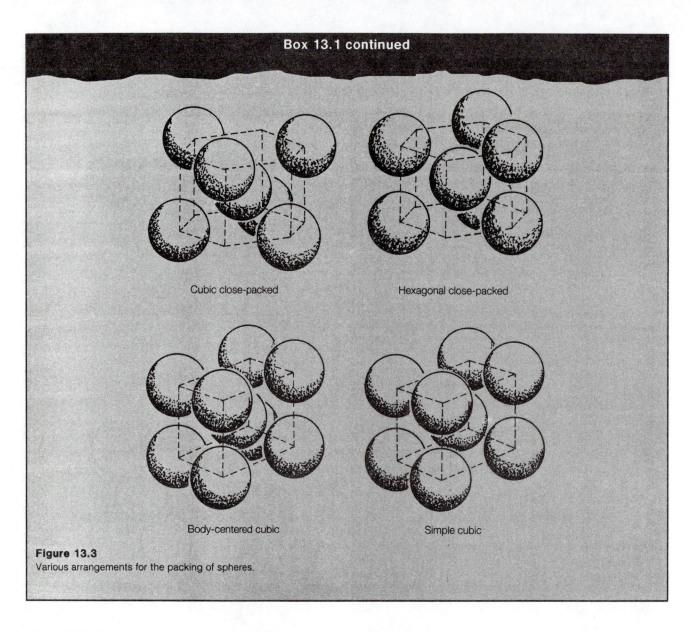

Cubic close-packed

Hexagonal close-packed

Body-centered cubic

Simple cubic

Figure 13.3
Various arrangements for the packing of spheres.

(Group VII*B*) lacks only one electron for a complete shell, and according to the octet rule it should only have one neighbor (8 − 7 = 1). One would predict (correctly) that chlorine "crystallizes" into a diatomic gas. A tellurium atom (Group VI*B*) may have two (8 − 6 = 2) tellurium neighbors, and since this is true of its tellurium neighbors also, crystals of tellurium may be anticipated to be based on chains or rings of atoms. Arsenic, Group V*B*, has three (8 − 5 = 3) neighbors and crystallizes as a layer structure, as do antimony and bismuth. Carbon (Group IV*B*) has four neighbors in the diamond network.

The strongly bonded chains or layers in these crystals are held together by relatively weak van der Waals forces, whose interatomic attraction arises from the dipolar nature of atoms. Electrons circulating around a positively charged nucleus give rise to oscillating electrical fields that provide for relatively long-range interatomic forces. Because of the presence of both strong and weak forces, minerals such as graphite are very easily deformed by the sliding of one

coherent layer of atoms over another, and sulfur loses its crystallinity but not its molecular makeup at slightly elevated temperatures.

Solid solution in these minerals may also be predicted through application of the octet rule, since those elements with similar electronic structures might be expected to proxy for formulary ions. Note that since the bonding is dominantly covalent, the effect of ionic radius is relatively unimportant. As predicted, arsenic and antimony show solid solution, as do sulfur and selenium, but no substitution of sulfur in graphite is found.

The economic importance of many of these native elements as mineral resources should be immediately apparent. Although the world supply of some comes from compounds that contain them (e.g., copper and silver in sulfides and sulfosalts), the main supply of such elements as gold, platinum, and sulfur and of the mineral diamond comes from their native forms.

Native gold is recovered from both ancient and modern sedimentary concentrations (placers), for example, at Witwatersrand, South Africa; in the Urals and Siberia of Russia; in the Yukon; and in Alaska. Gold is also produced from auriferous quartz veins in many localities, including the Canadian Shield, the Sierra Nevada, and Western Australia. Native platinum is found in placer deposits in the Russian Urals, but the largest production comes from magmatic concentrations in ultramafic rocks in South Africa. Native sulfur is formed both as a sublimate from volcanic gases and by bacterial reduction of sulfate minerals occurring in the caprock of salt domes, for example, in the Texas-Louisiana Gulf Coast region. Diamonds crystallize in ultramafic magmas at depths in excess of 100 km and are brought to the surface by explosive volcanism through pipe vents. The principal production is from placer deposits produced by the weathering of diamond-bearing igneous rocks. The bulk of world recovery of diamonds is on the African continent in South Africa, Zaire, Ghana, Angola, and the Congo. Other production is from Brazil, Venezuela, Australia, and Yakut, Siberia, west of the Lena River.

GOLD
Au

gold, from Anglo-Saxon *gold,* yellow; Latin *aurum.*

Crystallography, Structure, and Habit Isometric. $4/m\ \bar{3}\ 2/m.$ *a* 4.079. *Fm3m.* Crystals rough and distorted, usually octahedra and sometimes dodecahedra or cubes, often skeletal. Repeated twins on $\{111\}$ common.

Copper structure (see figure 13.1). Face-centered cubic cell containing four Au atoms.

Usually in rounded, irregular masses (nuggets) (see figure 13.4), grains, or scales (colors), or in arborescent forms.

Physical Properties No cleavage. Hackly fracture. Sectile, malleable, and ductile. Hardness 2.5–3. Specific gravity 19.23 pure, range 15.5–19.3. Splendid metallic luster. Opaque. Color and streak gold-yellow. Melting point 1,063° C.

Distinctive Properties and Tests Sectility, color, and specific gravity. Does not tarnish. Fuses easily (3) to a malleable globule. Insoluble in ordinary acids. Soluble in aqua regia.

Association and Occurrence Found with quartz, pyrite, chalcopyrite, galena, stibnite, sphalerite, arsenopyrite, tourmaline, and molybdenite. Widely distributed in small amounts. Usually in quartz veins related to both intrusive

Figure 13.4
Gold nugget, 17.5 × 9.5 × 4.5 cm, near Grenville, California. (Courtesy Royal Ontario Museum, Toronto.)

and extrusive silica-rich igneous rocks or concentrated by stream action into placer deposits.

Alteration Resistant to change.

Confused with Pyrite (fool's gold), chalcopyrite, and weathered biotite.

Variants Most natural gold contains alloyed Ag, Cu, or Fe; much contains some Pb, Ti, Al, Sb, Hg, V, Bi, Mn, Si, As, or Sn; and occasionally gold may contain Mg, Ni, Ca, Zn, Pd, Pt, or Tl.

The purity of gold, its **fineness,** is expressed in parts per thousand, and pure gold has a fineness of 1,000. Most native gold contains about 10% of other metals and has a fineness of about 900. The purity of gold is sometimes also measured in carats, with pure gold being 24.

Related Minerals *Solid solutions:* silver (electrum if more than 20% Ag), copper (auricupride, $\approx AuCu_3$), palladium (porpezite, 5–10% Pd), rhodium (rhodite 34–43% Rh), nickel. *Isotypes:* lead, Pb. *Other similar species:* maldonite, Au_2Bi.

SILVER
Ag

sil′vər, from Old English *seolfor,* original meaning lost; Latin *argentum,* silver.

Crystallography, Structure, and Habit Isometric. $4/m\ \bar{3}\ 2/m.$ *a* 4.086. *Fm3m.* Distorted crystals. Sometimes cubes, octahedra, or dodecahedra but more commonly in acicular forms. Twinning on $\{111\}$ common as pairs or cyclic groups.

Copper structure (see figure 13.1). Face-centered cubic cell containing four Ag atoms.

Usually in irregular masses, plates, or coatings, or in wirelike forms.

Physical Properties No cleavage. Hackly fracture. Sectile, malleable, and ductile. Hardness 2.5–3. Specific gravity 10.5 pure, range 9.6–12.0. Splendent metallic luster. Opaque. Color and streak silver-white and shining on fresh surface but tarnishing readily to gray or black. Melting point 961° C.

Distinctive Properties and Tests Sectility, color, and specific gravity. Fuses easily (2) to a bright, malleable globule. Readily soluble in HNO_3.

Association and Occurrence Found with sulfides and arsenides of lead, silver, copper, cobalt, and nickel. Commonly associated gangue minerals are calcite, quartz, barite, fluorite, and uraninite. Widely distributed in small amounts (more rare than native gold). Principal concentrations found in the oxidized zone of certain ore deposits: (1) hydrothermal veins containing silver and other sulfides; (2) hydrothermal veins containing cobalt and nickel sulfides, arsenides, and sulfarsenides; and (3) hydrothermal veins containing metallic sulfides and uraninite.

Alteration From silver halides, sulfides, and sulfosalts.

Confused with Platinum group metals.

Variants Aurian (küstelite), mercurian (amalgam, kongsbergite, arquerite), cuprian, arsenian, antimonian, and bismuthian. May also contain Pt or Fe.

Related Minerals *Solid solutions:* gold and amalgam, (Ag, Hg). *Polymorphs:* 2*H*, 4*H*. *Isotypes:* copper. *Other similar species:* dyscrasite, Ag_3Sb.

COPPER
Cu

kŏ′pər, from Greek *kyprios,* of Cyprus, the location of ancient copper mines; Latin *cuprum.*

Crystallography, Structure, and Habit Isometric. $4/m \bar{3} 2/m$. a 3.615. *Fm3m*. Cubes, dodecahedra, and octahedra. Crystals usually distorted and combined into branching groups. Dendritic. Contact, penetration, and cyclic twins are very common.

Copper structure (see figure 13.1). Face-centered cubic cell containing four Cu atoms.

Usually in irregular masses, plates, or scales; also flattened, elongate, and spearlike forms. Sometimes wirelike or arborescent.

Physical Properties No cleavage. Hackly fracture. Sectile, malleable, and ductile. Hardness 2.5–3. Specific gravity 8.9. Splendent metallic luster. Opaque. Color and streak copper-red on fresh surface darkening to brown on exposure. Melting point 1,083° C.

Distinctive Properties and Tests Sectility, color, and specific gravity. Readily soluble in HNO_3.

Association and Occurrence Found with native silver, chalcopyrite, bornite, calcite, chlorite, zeolites, cuprite, malachite, and azurite. Fills in cracks and amygdules and partially replaces the silicates of mafic lava flows or acts as a cement in associated conglomerates and sandstones. Also occurs in the oxidized zone above copper-sulfide deposits.

Alteration To cuprite, malachite, and azurite.

Confused with Red-gold.

Variants Arsenian. Often contains such elements as Ag, Fe, Bi, Sb, Hg, Ge, Sn, and Pb.

Related Minerals *Isotypes:* gold; silver; platinum; lead, Pb; iridium, Ir.

PLATINUM
Pt

plăt′n-əm, from Spanish *platina,* diminutive of *plata,* silver. The new metal found in large placer deposits during the sixteenth-century Spanish conquest of South America was called *platina del Pinto* after the Rio Pinto, Colombia.

Crystallography, Structure, and Habit Isometric. $4/m \bar{3} 2/m$. a 3.923. *Fm3m*. Rare and distorted cubes.

Copper structure (see figure 13.1). Face-centered cubic cell containing four Pt atoms.

Found in small grains and scales and, more rarely, in larger nuggets.

Physical Properties No cleavage. Hackly fracture. Sectile, malleable, and ductile. Hardness 4–4.5. Specific gravity 14–19, 21.5 when pure. Splendent metallic luster. Opaque. Color and streak steel-gray to dark gray. Infusible. Insoluble. May be weakly magnetic.

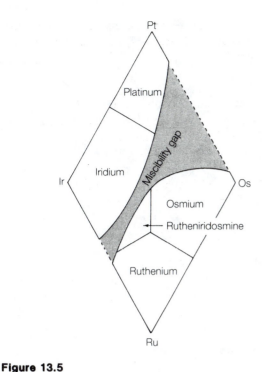

Figure 13.5
Classification of the platinum group elements (PGE). (After Harris and Cabri, 1991.)

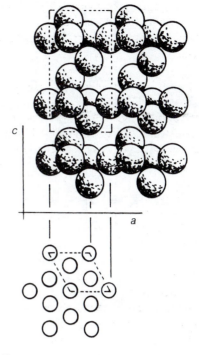

Figure 13.6
View along *b*-axis of the half cell of arsenic.

Distinctive Properties and Tests Sectility, color, and specific gravity. Infusible. Insoluble except in hot aqua regia.

Association and Occurrence Found with chromite, olivine, and spinels in mafic and ultramafic rocks. Also in placers.

Alteration Resistant to change.

Confused with Native silver.

Variants May alloy with up to 50% of other elements, especially Fe, and also often with Cu, Pd, Rh, Ir. Also may contain Os and, less commonly, Au or Ni.

Related Minerals *Isotypes:* copper, platinum group elements (PGE) (see figure 13.5).

ARSENIC
As

är'sə-nik, from Greek *arsenikon,* masculine, alluding to its potent properties.

Crystallography, Structure, and Habit Hexagonal. $\bar{3}\,2/m.$ *a* 3.760, *c* 10.555. $R\bar{3}m.$ Crystals rare, usually acicular, and sometimes pseudocubes. Twins on $\{01\bar{1}2\}$.

Covalently bonded corrugated sheets of As atoms perpendicular to the *c*-axis. The simple cell of arsenic is an acute rhombohedron containing two atoms, one at each lattice point and one at the center of the cell. These two atoms are structurally dissimilar. The unit cell of the arsenic structure is a multiple cell containing six As atoms. Figure 13.6 represents a half cell.

Massive, granular, and colloform.

Physical Properties Perfect basal $\{0001\}$ and imperfect $\{01\bar{1}2\}$ cleavage. Uneven fracture. Brittle. Hardness 3.5. Specific gravity 5.7. Splendid metallic luster. Opaque. Color tin-white on fresh surface tarnishing to dark gray. Streak gray.

Distinctive Properties and Tests Luster and habit.

Association and Occurrence Found with silver-cobalt-nickel ores in hydrothermal veins.

Alteration Oxidizes to form a black crust of arsenolite, As_2O_3.

Confused with Manganese oxides.

Variants Usually contains such contaminants as Sb, Fe, Ni, Ag, and S.

Related Minerals *Isostructures:* Antimony and bismuth. *Polymorphs:* arsenolamprite (orthorhombic). *Isotypes:* antimony, Sb, and bismuth, Bi. *Other similar species:* graphite and allemontite, a mixture of stibarsen, SbAs, and either As or Sb.

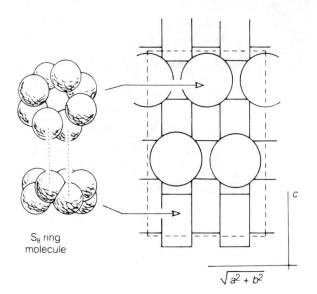

S₈ ring molecule

$\sqrt{a^2 + b^2}$

Figure 13.7
Arrangement of S₈ ring molecules in sulfur. View along [110].

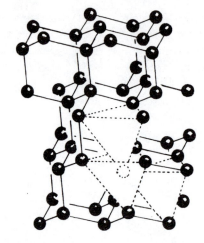

Figure 13.8
The diamond structure.

SULFUR
S

(sulphur, α-sulfur, brimstone)
sŭl′fər, from Middle English *sulphur,* brimstone.

Crystallography, Structure, and Habit Orthorhombic. $2/m\,2/m\,2/m$. a 10.467, b 12.866, c 24.486. *Fddd*. Crystals pyramidal, dipyramidal, thick tabular. Rare twinning on $\{101\}$.

The structure has 16 S₈ "puckered-ring" molecules per cell (see figure 13.7). The rings are stacked parallel to c and held together by residual bonds; hence its softness and low melting point.

Massive, colloform, encrusting, powdery.

Physical Properties Imperfect $\{001\}$, $\{110\}$, and $\{111\}$ cleavage. Conchoidal to uneven fracture. Brittle to subsectile. Hardness 1.5–2.5. Specific gravity 2.1. Resinous luster. Translucent to transparent. Color yellow, of various shades. Optical: $\alpha = 1.958$, $\beta = 2.038$, and $\gamma = 2.245$; positive; $2V = 70°$. Melting point 113° C. Negatively electrified by friction.

Distinctive Properties and Tests Hardness, specific gravity, and color. Melts and burns with a blue flame liberating acrid SO_2.

Association and Occurrence Found with celestine, gypsum, anhydrite, aragonite and calcite (1) in sedimentary rocks from reduction of sulfates (bacterial action), (2) as a volcanic sublimate, and (3) as a decomposition product of sulfides in metallic veins.

Alteration From sulfides. From and to sulfates.

Confused with Orpiment and sphalerite.

Variants Selenian. May contain small amounts of Te. Inclusions of clay and bitumen are common.

Related Minerals *Polymorphs:* sulfur is known in three forms, the common orthorhombic (α) modification and two rare monoclinic modifications known as β- and γ-sulfur (rosickyite).

DIAMOND
C

dī′ə-mənd, from Greek *adamas,* invincible. First known use by Manlius (A.D. 16) and Pliny (A.D. 100).

Crystallography, Structure, and Habit Isometric. $4/m\,\bar{3}\,2/m$. a 3.567. *Fd3m*. Usually in octahedra that are often distorted; also dodecahedra, cubes, tetrahedra. Twinning is common. Simple twins (spinel law) are often flattened and are called macles. Also multiple cyclic twins and cyclic groups.

Face-centered cubic unit cell containing eight C atoms. Each carbon atom is linked tetrahedrally by sp³ bonds to four neighbors (see figure 13.8).

Crystal grains, often with curved faces. Inclusions common.

Physical Properties Perfect {111} cleavage. Conchoidal fracture. Brittle. Hardness 10; varies slightly, hardest on {111}. Specific gravity 3.5. Adamantine luster. Transparent to translucent. Color blue-white, green, yellow, brown, black, colorless and, very rarely, pink to red. *Optical: n* = 2.4175. Triboelectric. Sometimes strongly fluorescent under ultraviolet light. Often fluoresces when rubbed on wood. Infusible. Wet by animal grease. Color may be changed from colorless to pale blue by irradiation.

Distinctive Properties and Tests Hardness and habit. Unattacked by acids and alkalis.

Association and Occurrence Found with pyrope garnet, kyanite, olivine, and zircon in altered ultramafic rocks or in placers. The stability of diamond is discussed in conjunction with figure 6.2.

Alteration Resistant to change.

Confused with Quartz and topaz.

Variants. *Form varieties:* Bort (bortz, boart), granular to cryptocrystalline, gray to brown, translucent to opaque; ballas, spherical masses arranged concentrically, very hard and tough; and carbonado, massive black or gray bort. All are widely used in industry.

May contain minute quantities of such elements as Si, Fe, Mg, Al, Ca, and Ti.

Diamond weights are measured in carats and points; one carat = 0.2 gram, one point = 0.01 carat.

Related Minerals *Polymorphs:* graphite, chaoite (hexagonal), and lonsdaleite (hexagonal).

GRAPHITE
C

(plumbago, black lead)
grǎf'ǐt, from Greek *graphein,* to write; named by German chemist and mineralogist A. G. Werner in 1789.

Crystallography, Structure, and Habit Hexagonal. $6/m\ 2/m\ 2/m$. *a* 2.461, *c* 6.708. $P3_6/mmc$. Hexagonal tablets. Glide twinning may produce striae on the base.

Structure is comprised of widely spaced planes of covalently bonded carbon atoms in hexagonal array. Structure somewhat similar to that of arsenic but the sheets are not corrugated (see figures 13.6 and 13.9). The unit cell contains four C atoms.

Embedded foliated masses, scales, columns, grains, earthy masses.

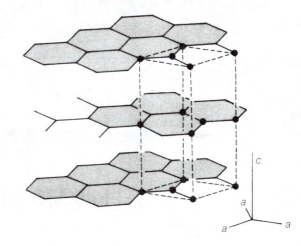

Figure 13.9
The graphite structure.

Physical Properties Perfect {0001} cleavage. Sectile. Hardness 1–2. Specific gravity 2.1–2.3, 2.25 pure. Submetallic to earthy luster. Opaque (transparent in extremely thin flakes). Color black to steel-gray. Streak black. Cleavage laminae are pliable but not elastic. Greasy feel. Thermoelectric.

Distinctive Properties and Tests Softness (greasy feel, soils fingers and marks paper) and cleavage. Infusible and unattacked by acids.

Association and Occurrence Found with calcite, quartz, orthoclase, pyroxene, garnet, spinel, and amphibole in metamorphic rocks, less commonly in quartz veins, and rarely in igneous rocks.

Alteration Resistant to change.

Confused with Molybdenite and pyrolusite.

Variants *Form varieties:* graphitite, cryptocrystalline, and shungite, vitreous. Often contains a small amount of Fe.

Related Minerals *Polymorphs:* diamond, "white carbon" chaoite (hexagonal), lonsdaleite (hexagonal), and a 3*R* polytype with every third rather than every second layer identical. *Other similar species:* arsenic.

Sulfides and Sulfosalts

Sulfides

Tetradymite Group	$Bi_{2-4}(Te,S)_{2-4}$
Copper Arsenide Group	$Cu_{3-6}(As,Sb,S)$
Argentite Group	$(Ag,Cu,Au)_{2-4}(S,Se,Te)_{1-2}$
Chalcocite Group	$(Cu,Ag)_2S$
Chalcocite	
Acanthite	
Bornite	
Galena Group	$(Pb,Mn,Ca)(S,Se,Te)$
Galena	
Sphalerite Group	$(Zn,Hg)(S,Se,Te)$
Sphalerite	
Chalcopyrite Group	$(Cu,Fe,Sn)_4S_4$
Chalcopyrite	
Wurtzite Group	$(Zn,Cd)S$
Nickeline Group	$(Fe,Ni,Cu,Ag)_{1-3}(S,As,Sb)_{1-3}$
Nickeline	
Pyrrhotite	
Pentlandite	
Covellite Group	$(Cu,Hg,As,Pt,Sn,Ag)(S,Se,Te)$
Covellite	
Cinnabar	
Realgar	
Orpiment	
Linnaeite Group	$(Co,Ni,Cu,Cr)_3S_4$
Stibnite Group	$(Sb,Bi)_2(S,Se)_3$
Stibnite	
Pyrite Group	$(Fe,Ni,Ru,Pt,Mn)(S,As,Se)$
Pyrite	
Cobaltite Group	$(Co,Ni)(As,Sb,S)$
Cobaltite	
Löllingite Group	$(Fe,Co,Ni)(As,S)_2$
Marcasite	
Arsenopyrite Group	$(Fe,Ni,Co,Cu)AsS$
Arsenopyrite	
Molybdenite Group	$(Mo,W)S_2$
Molybdenite	
Krennerite Group	$(Au,Ag,Ni)Te_2$
Skutterudite Series	$\cdot(Co,Ni)As_3$
Miscellaneous Species	

Sulfides are generally soft and fusible minerals. They are often heavy and opaque, and they have a metallic luster, dark color and streak. These family characteristics, which are macroscopic indications of their composition and bonding, are shared with the native elements and the sulfosalts.

About eighty sulfide mineral species are known, and only those in which sulfur is combined with iron, copper, or zinc are relatively abundant. Having a simple chemistry, they tend to be structurally simple. Cation coordination is almost exclusively octahedral or tetrahedral. Most structures are of a halite or nickeline type (6-fold coordination) or have a layered arrangement, as in molybdenite or covellite. Sulfides are often found as well-formed crystals, mostly in the hexagonal or cubic systems, but many distortional or substitutional derivatives are also known.

Chemically, sulfides are combinations of one or more metals with either sulfur or a chemically similar element, such as arsenic, selenium, or tellurium. A few chemical series exist, but most sulfide species exhibit a restricted amount of solid solution.

The sulfide minerals can be readily identified from their physical properties alone. The easily recognized properties of luster (metallic or nonmetallic), hardness, and cleavage, for example, can be used to separate the sulfide minerals into groups (see table 14.1). These groups can be further subdivided into individual species by color and miscellaneous properties (see table 14.2).

The identification of a sulfide can be confirmed by simple tests. For example, sulfides can be divided into groupings based on their fusibility and the magnetic properties of the resulting fusion. A magnetic residue gives positive evidence of the presence of iron in the mineral (see table 14.3).

Sulfur, in Group VI*B* of the periodic table, has two unpaired electrons (in $3p$) available for bonding and should be able to accommodate only two neighbors according to the octet rule (see page 225). When these neighbors are typical metals, however, one of the electrons may be raised to the $4s$ level, thus providing four electrons with unpaired spins to participate in covalent bonds. Electron states of the metals also undergo changes, and a degree of metallic bonding will usually be present. Generally, the interaction of sulfur and metal atoms results in a crystalline array in which the sulfur-to-metal proportions range from 2:1 to 1:2. The tendency of sulfur to form covalent bonds is offset by the marked preference of the metals for metallic bonding. Mixed bonding, even including a degree of ionic character, is not unusual.

The formulas of the sulfides do not, in general, provide a clue to the bonding mechanism and structure, as so often is the case in other mineral families. Galena, PbS, for example, is a cubic array of covalently bonded lead and sulfur atoms; sphalerite, ZnS, has mixed covalent-ionic bonding; and pyrrhotite, approximately FeS, is composed of a hexagonal closest packed network of sulfur atoms with interstitial iron. The iron is covalently bonded to the sulfur but also exhibits considerable Fe-Fe metallic interaction. Further, the semi-independent nature of iron in this compound allows some of it to be replaced by vacant sites, resulting in an iron-deficient structure.

Pyrite, FeS_2, contains discrete S_2 molecular groups that alternate with iron in the mineral. Chalcopyrite, $CuFeS_2$, on the other hand, is a three-dimensional alternation of Cu-S-Fe-S-Cu-S-Fe ad infinitum, and molybdenite, MoS_2, is made up of Mo-S sheets.

Sulfides are the source of many of the metals essential to modern civilization. Nearly all of the production of the base metals copper, lead, and zinc, as well as that of the ferroalloys nickel and molybdenum, comes from sulfide deposits of one or another kind.

Table 14.1 Physical Properties of Some Sulfide Minerals

Luster	Cleavage	Hardness	Minerals
Metallic	Present	>5.5	Arsenopyrite, skutterudite series
		2.5–5.5	Galena, cobaltite
		<2.5	Covellite, stibnite, molybdenite
	Absent	>5.5	Pyrite, marcasite
		2.5–5.5	Chalcocite, bornite, chalcopyrite, pyrrhotite, nickeline
		<2.5	Acanthite
Nonmetallic	Present	>5.5	
		2.5–5.5	Sphalerite, cinnabar
		<2.5	Realgar, orpiment

Table 14.2 Further Distinctions of Some Sulfide Minerals

Mineral	Fresh Color	Distinctive Characteristics
Chalcocite	Shining lead-gray	Subsectile
Bornite	Copper-red	Iridescent tarnish
Chalcopyrite	Brass-yellow	Greenish black streak
Pyrrhotite	Brown-bronze	Magnetic
Nickeline	Pale copper-red	Brown-black streak

Table 14.3 Simple Fusibility Tests for the Identification of Sulfide Minerals

Infusible	Fusible with a Magnetic Residue	Fusible with a Nonmagnetic Residue	Volatile
Sphalerite	Bornite	Acanthite	Cinnabar
Molybdenite	Chalcopyrite	Chalcocite	Realgar
	Pyrrhotite	Galena	Orpiment
	Pyrite	Nickeline	
	Marcasite	Covellite	
	Arsenopyrite	Stibnite	
		Cobaltite	
		Skutterudite series	

A major type of copper deposit is the large, low-grade occurrence of copper-bearing sulfides such as chalcopyrite and bornite disseminated in intrusive igneous rocks of intermediate composition. These "porphyry copper" deposits are formed by fractional melting in orogenic zones along continental margins. Deposits of this type are known along the entire length of the American cordillera from Alaska to Chile. Although having an ore grade of only 0.5–1.0% copper, the huge porphyry deposits can be worked profitably by using large-scale, open-pit methods. Further, the deposits are often enriched by weathering processes that reprecipitate chalcocite.

Molybdenum as molybdenite is typically a minor constituent of porphyry-copper ores and is produced as a by-product of the copper mining. Occasionally, however, the roles of copper and molybdenum are reversed, and porphyry molybdenum deposits are found.

The lead- and zinc-bearing sulfides, galena and sphalerite, can be found in hydrothermal vein deposits. Intermediate temperature deposits, such as that at Coeur d'Alene, Idaho, typically contain chalcopyrite, pyrite, tetrahedrite, and silver sulfides and sulfosalts. The principal lead-zinc deposits are, however, huge, blanketlike bodies formed by the replacement and open-space filling of carbonate rocks by metal-bearing groundwater. A typical location for such a deposit is along the margin of a large sedimentary basin. Major deposits in North America are in Missouri, in Wisconsin, along the western side of the Appalachian system, and at Pine Point in the Northwest Territories of Canada.

Some deposits of copper, lead, and zinc sulfides are formed on the ocean floor from rising hot waters related to the intrusion of mafic magma along zones of extension marked by midocean ridges. The sulfides are deposited from metal-laden waters around orifices marked by spectacular chimneylike structures called black smokers. Further, submarine volcanism of siliceous magmas can lead to large massive sulfide deposits, principally of pyrite but also containing significant amounts of chalcopyrite, sphalerite, and galena. Examples of this type of deposit include Kidd Creek, Ontario; Noranda, Quebec; and the Kuroko deposits of Japan.

The principal ore of nickel is pentlandite, $(Fe,Ni)_9S_8$, which occurs in mafic and ultramafic intrusive rocks in association with chalcopyrite and pyrrhotite. Major deposits are at Sudbury, Ontario; Lynn Lake, Manitoba; and Kalgoorlie, Western Australia.

CHALCOCITE
Cu₂S

(copper glance)
kăl′kə-sīt, from Greek *chalkos,* copper.

Crystallography, Structure, and Habit Monoclinic. $2/m$. a 15.22, b 11.88, c 13.48, β 116.35°. $P2_1/c$. Crystals rare, usually short prismatic to thick tabular. Twinning produces a stellate grouping of three individuals or a cruciform penetration twin.

The structure is based on hexagonal close-packed S with one-third of the Cu atoms in interstices of the sulfur layers. The remaining Cu atoms are in triangular sites between the sulfur hcp layers. The unit cell contains forty-eight formulary units.

Massive, compact, and impalpable.

Physical Properties Indistinct {110} cleavage. Conchoidal fracture. Subsectile. Hardness 2.5–3. Specific gravity 5.7–5.8. Metallic luster. Color and streak dark to lead-gray. Usually tarnished to a dull black. Sometimes sooty black. Easily fusible (2–2.5).

Distinctive Properties and Tests Hardness, color, and sectility. Soluble in HNO_3.

Association and Occurrence Found with covellite, cuprite, native copper, azurite, malachite, and bornite in secondarily enriched zone in copper-bearing sulfide deposits and in hydrothermal sulfide veins.

Alteration To native copper, covellite, malachite, and azurite.

Variants Often contains some Ag and Fe.

Related Minerals *Solid solutions:* berzelianite, Cu_2Se. *Isotypes:* stromeyerite, AgCuS. *Polymorphs:* hexagonal modification above 103° C. *Other similar species:* digenite, $Cu_{2-x}S$.

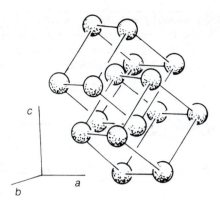

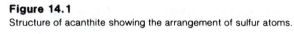

Figure 14.1
Structure of acanthite showing the arrangement of sulfur atoms.

ACANTHITE
Ag₂S

ə-kǎn′thīt, from Greek *acantha,* thorn, an allusion to its pointed crystals.

Crystallography, Structure, and Habit Monoclinic (pseudo-orthorhombic). $2/m$. a 4.228, b 6.928, c 7.862, β 99.6°. $P2_1/n$. Penetration twins.

Silver atoms are interstitial to a slightly distorted, body-centered cubic array of sulfur atoms (see figure 14.1). Half of the Ag atoms are in 2-fold coordination and half in 3-fold coordination. Those in 2-fold form zigzag chains perpendicular to (103) and those in 3-fold link the chains via AgS_3 groups. Four Ag_2S units per cell. Acanthite transforms from argentite below 173° C into a polycrystalline pseudomorph.

Usually massive, also groups of parallel individuals. Arborescent; coatings.

Physical Properties Poor cleavage. Subconchoidal fracture. Very sectile. Hardness 2–2.5. Specific gravity 7.3. Metallic luster. Opaque. Color dark lead-gray, darkening on exposure to light. Streak shining. Fuses readily (1.5) with intumescence.

Distinctive Properties and Tests Hardness, specific gravity, and sectility. Easy fusion that, if done on charcoal, produces a silver button.

Association and Occurrence Found with other silver minerals and cobalt-nickel sulfides in veins and as inclusions in argentiferous galena.

Alteration To native silver and silver sulfosalts.

Confused with Chalcocite and tetrahedrite.

Related Minerals *Polymorphs:* argentite, an isometric form of Ag_2S stable above 173° C. Most acanthite is pseudomorphous after argentite. *Other similar species:* hessite, Ag_2Te; petzite, Ag_3AuTe; fischesserite, Ag_3AuSe_2; naumannite, Ag_2Se; eucairite, $CuAgSe$; jalpaite, Ag_3CuS_2; and aguilarite, Ag_4SeS.

BORNITE
Cu_5FeS_4

(peacock ore)

bôr′nīt, after Ignaz von Born (1742–1791), Austrian mineralogist.

Crystallography, Structure, and Habit Tetragonal (pseudoisometric). $\bar{4}2m$. *a* 10.94, *c* 21.88. $P\bar{4}2_1c$. Rare crystals with rough and curved faces. Penetration twins common.

Spinel-type structure (see figure 16.8) with FeS_4 tetrahedra arranged as carbon atoms in diamond and Cu ions on edges of tetrahedra lacking Fe. Forty of forty-eight Cu sites occupied statistically, with some as Cu^{2+} for charge compensation. The unit cell contains sixteen formulary units.

Usually massive, occasionally granular or compact.

Physical Properties {111} cleavage in traces. Conchoidal to uneven fracture. Brittle to subsectile. Hardness 3. Specific gravity 5.1 ±. Metallic luster. Opaque. Color copper-red tarnishing to give a purplish, iridescent coating. Streak gray-black. Fusible at 2 yielding a black, magnetic globule.

Distinctive Properties and Tests Color and peacock tarnish. Soluble in HNO_3, giving a blue solution.

Association and Occurrence Widespread in copper deposits, usually with chalcopyrite and quartz. Also with chalcocite, enargite, covellite, pyrite, pyrrhotite, marcasite, and arsenopyrite. Found (1) in sulfide veins, (2) in zone of secondary enrichment in copper-bearing sulfide deposits, (3) possibly as a magmatic segregation, (4) as a minor mineral in pegmatites, and (5) in black shales.

Alteration To chalcocite, covellite, cuprite, chrysocolla, malachite, and azurite. From chalcopyrite.

Confused with Nickeline, pyrrhotite, chalcocite, and covellite.

Variants Forms solid solutions with chalcopyrite above 475° C that yield chalcopyrite lamellae on cooling. Often contains small amounts of Pb and, less commonly, Au and Ag.

Related Minerals *Polymorphs:* isometric with antifluorite structure above 228° C. *Other similar species:* pentlandite.

GALENA
PbS

(lead glance)

gə-lē′nə, from Greek *galene,* lead ore.

Crystallography, Structure, and Habit Isometric. $4/m\,\bar{3}\,2/m$. *a* 5.936. $Fm3m$. Commonly in cubes or cuboctahedra. Rarely octahedra. Penetration and contact twins common; also repeated lamellar twins.

Isostructural with halite, NaCl (see figure 15.2). Four PbS per unit cell.

Crystal aggregates, massive, fine granular to impalpable, and plumose.

Physical Properties Perfect {001} cleavage yielding cubic fragments. Parting on {111}. Brittle. Hardness 2.5. Specific gravity 7.6. Metallic luster. Opaque. Color and streak lead-gray. Melting point 1,115° C.

Distinctive Properties and Tests Cubic crystals and cleavage, and specific gravity. Fusible at 2 yielding a malleable button.

Association and Occurrence Widely distributed. Found with sphalerite, chalcopyrite, chalcocite, pyrite, quartz, silver ores, fluorite, and barite in sulfide veins and replacement deposits, usually in limestone.

Alteration To cerussite, anglesite, and pyromorphite.

Confused with Stibnite.

Variants Often contains Sb and As; may also contain Tl, Bi, Zn, Cd, Fe, Mn, and Cu. Native silver or acanthite may be included as an exsolved phase.

Related Minerals *Solid solution:* clausthalite, PbSe, and altaite, PbTe. *Isostructure:* halite, NaCl. *Isotypes:* alabandite, MnS; oldhamite, (Ca,Mn)S; niningerite, (Mg,Fe,Mn)S.

SPHALERITE
ZnS

(blende, zinc blende, blackjack)

sfǎl′ə-rīt, from Greek *sphaleros,* treacherous, an allusion to the ease with which dark varieties were mistaken for galena, but yielded no lead.

Crystallography, Structure, and Habit Isometric. $\bar{4}3m$. *a* 5.43. $F\bar{4}3m$. Tetrahedron, dodecahedron, and cubes as basic forms, but crystals are often highly modified, distorted, or rounded. Twins usually on {111} or {211} as simple or multiple contact twins or lamellar intergrowths, occasionally penetration twins.

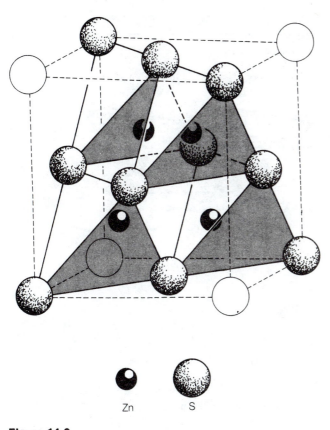

Figure 14.2
Coordination tetrahedra of zinc atoms in sphalerite. Unit cell dashed.

Zn   S

Isostructural with diamond (see figure 13.8); Zn and S atoms, held together with dominantly covalent bonds, occupy alternate C atom sites. Each S is coordinated by four Zn atoms at the corners of a regular tetrahedron, and Zn is surrounded by S in a similar way (see figure 14.2). The cell contains four ZnS.

Crystals have tetrahedral or dodecahedral habits and frequently exhibit curved faces.

Physical Properties Perfect {110} cleavage. Brittle. Hardness 3.5–4. Specific gravity 3.9–4.2, pure 4.1. Resinous, with adamantine to submetallic luster. Translucent to transparent. Color yellow, brown, or black. Streak white, pale yellow, pale brown. Optical: $n = 2.37 - 2.50$ (increasing with iron content). Pyroelectric. Occasionally triboluminescent and fluorescent. Infusible.

Distinctive Properties and Tests Cleavage and luster. Powdered mineral decomposed by warm HCl with the evolution of H_2S gas.

Association and Occurrence Found with galena, chalcopyrite, pyrite, barite, fluorite, dolomite, siderite, rhodochrosite, and quartz in low-temperature occurrences. Found with magnetite, garnet, rhodonite, fluorite, and apatite in high-temperature veins. Sphalerite is a common

mineral. It is found as replacements (1) in dolostone, limestone, and other sedimentary rocks, (2) in sulfide veins, and (3) in contact metamorphic deposits.

Alteration To goslarite, $ZnSO_4 \cdot 7H_2O$; hemimorphite; and smithsonite.

Confused with Siderite, ankerite, sulfur, and enargite.

Variants Cadmian, ferroan (marmatite), and christophite, (Zn,Fe)S, with up to 26% Fe. May also contain Ga, In, Tl, Hg, Mn, and Ge.

Related Minerals *Isotypes:* metacinnabar, (Hg,Fe, Zn)S; tiemannite, HgSe; coloradoite, HgTe; and stilleite, ZnSe. *Polymorphs:* wurtzite, stable form above 1,020° C; and matraite (hexagonal). *Other similar species:* hawleyite, CdS.

CHALCOPYRITE
CuFeS$_2$

(copper pyrites)
kăl′kə-pī′rīt, from Greek *chalkos,* copper, and *pyrites,* strike fire.

Crystallography, Structure, and Habit Tetragonal. $\overline{4}2m$. *a* 5.299, *c* 10.434. $I\overline{4}2d$. Usually in pseudotetrahedra with prominent, sphenoidal faces. Sphenoidal faces (112) are dull, oxidized, or striated, and the $(1\overline{1}2)$ faces are brilliant and unstriated. Penetration and lamellar twins.

Structure is a derivative of sphalerite (see figure 14.2). Cu and Fe atoms alternate in the sites occupied by Zn in sphalerite (see figure 14.3). The unit cell contains four formulary units.

The common form is a {112} disphenoid that, because $c/2 \approx a$, resembles a tetrahedron. Usually massive, and crystal aggregates are rare.

Physical Properties Indistinct cleavage on {011} and {111}. Hardness 3.5–4. Specific gravity 4.2. Metallic luster. Opaque. Color brass-yellow, sometimes showing an iridescent tarnish on exposure. Streak green-black. Readily fusible (2) yielding a magnetic globule.

Distinctive Properties and Tests Color and streak. Decomposed by HNO_3 with separation of sulfur and formation of a green solution. Ammonia added in excess generates a red ferric-hydroxide precipitate and a deep blue solution.

Association and Occurrence The principal ore of copper; widespread in occurrence and found in most rock types. Usually found with pyrite, sphalerite, bornite, galena,

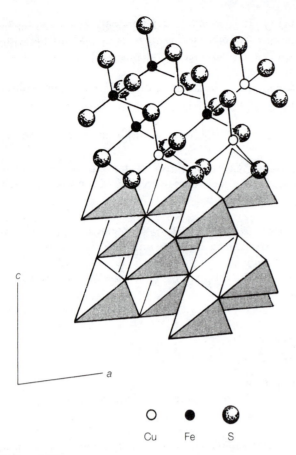

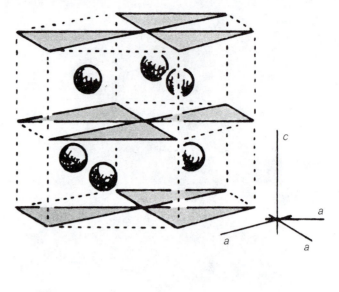

 As

Figure 14.4
Structure of nickeline. Nickel atoms on corners of triangles.

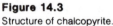

○   ●   ◍

Cu   Fe   S

Figure 14.3
Structure of chalcopyrite.

chalcocite, quartz, or calcite in (1) magmatic segregations with pentlandite and pyrrhotite; (2) vein deposits of wide temperature range; (3) pyrometasomatic deposits; and (4) black shales.

Alteration   To chalcocite, covellite, chrysocolla, malachite, and iron oxides.

Confused with   Pyrite, gold, bornite, and pyrrhotite.

Variants   Commonly contains small amounts of Ag, Au, and Zn. Many metals have been reported as contaminants.

Related Minerals Solid solutions: eskebornite, $CuFeSe_2$. *Isotypes:* stannite, Cu_2FeSnS_4; gallite, $CuGaS_2$; and roquesite, $CuInS_2$. *Other similar species:* talnakhite, $Cu_9(Fe,Ni)_8S_{16}$; mooihoekite, $Cu_9Fe_9S_{16}$; and haycockite, $Cu_4Fe_5S_8$.

NICKELINE
NiAs

(kupfernickel, niccolite)
nik'ə-lēn, from the German nickname for the ore miners called kupfernickel (copper nickel); after Old Nick and

his mischievous gnomes, because the ore seemed to contain copper but yielded none.

Crystallography, Structure, and Habit   Hexagonal. $6/m\,2/m\,2/m$. *a* 3.618, *c* 5.034. *C6/mmc*. Crystals rare; usually pyramidal. Cyclic twins (fourlings).

The structure has hexagonal closest-packed As atoms with interstitial Ni atoms (see figure 14.4). Covalent d^4sp hybrid bonds link the atoms into regular octahedra of As around Ni. There are two NiAs per unit cell.

Usually massive; sometimes colloform or columnar.

Physical Properties   No cleavage. Uneven fracture. Brittle. Hardness 5–5.5. Specific gravity 7.3–7.7. Splendent metallic luster. Opaque. Color pale copper-red. Streak brownish black. Fusible (2), yielding arsenical fumes and a magnetic residue.

Distinctive Properties and Tests   Color, association, and alteration. Green solution when dissolved in HNO_3.

Association and Occurrence   Found with skutterudite series minerals, native silver, silver sulfosalts, pyrrhotite, and chalcopyrite in veins with silver and cobalt minerals and in copper-iron-nickel sulfide deposits associated with mafic igneous rocks.

Alteration   To annabergite, $Ni_3As_2O_8 \cdot 8H_2O$, a green, earthy powder.

Confused with   Bornite and pyrrhotite.

Variants   Usually contains some Sb, and often Fe, Co, and S.

Figure 14.5
Platy crystals of pyrrhotite showing a pseudohexagonal symmetry. Trepca, Serbia. (Courtesy Royal Ontario Museum, Toronto.)

Related Minerals   *Solid solutions:* breithauptite, NiSb. *Isotypes:* pyrrhotite. *Isostructures:* freboldite, CoSe, and kotulskite, Pd(Te,Bi). *Other similar species:* millerite, NiS; pentlandite; imgreite, NiTe; langisite, (Co,Ni)As; sederholmite, β-NiSe; sobolevskite, PdBi; stumpflite, Pt(Sb,Bi); and sudburyite, (Pd,Ni)Sb.

PYRRHOTITE
Fe$_{1-x}$S

(magnetic pyrites)
pîr′ə-tīt, from Greek *pyrrhos,* flame-colored.

Crystallography, Structure, and Habit   Monoclinic. $2/m$. a 11.90, b 6.87, c 22.88, β 90.5°. $C2/c$. Rare tabular to platy crystals (see figure 14.5). Twinning on $\{10\bar{1}2\}$.

Distorted derivative of the NiAs structure. Hexagonal close-packed S atoms with interstitial Fe in 6-fold coordination between layers of sulfur atoms. Some of the Fe sites are vacant, and the Fe:S ratio varies from about 0.85 to 1.0. The unit cell contains two formula units.

Usually massive; also disseminated grains.

Physical Properties   No cleavage. Parting on $\{0001\}$ and $\{11\bar{2}0\}$. Uneven to subconchoidal fracture. Brittle. Metallic luster. Opaque. Color bronze-yellow. Streak black. Weak to intermediate magnetic susceptibility. Fusible (3).

Distinctive Properties and Tests   Color and magnetism. Yields H$_2$S when decomposed by HCl.

Association and Occurrence   Found with pyrite, chalcopyrite, pentlandite, galena, and magnetite in mafic igneous rocks, pegmatites, metamorphic rocks, and hydrothermal veins.

Alteration   To sulfates, oxides, or carbonates of iron.

Confused with   Bornite and pyrite.

Variants   Exsolution of pyrrhotite into Fe-rich and Fe-poor phases may occur. Ni, Co, Mn, and Cu are common impurities. Pyrrhotite exists in several closely related polymorphic forms showing different degrees of Fe-hole substitution and c-axis repetition of a nickeline (NiAs) substructure as given in table 14.4.

Related Minerals   *Isotypes:* nickeline; breithauptite, NiSb; mackinawite, (Fe,Ni)$_9$S$_8$; and troilite, FeS. *Polymorphs:* hexagonal modification above 254° C.

PENTLANDITE
(Ni,Fe)$_9$S$_8$

pĕnt′lən-dīt, after J. B. Pentland, who discovered the mineral at Sudbury, Ontario.

Crystallography, Structure, and Habit   Isometric. $4/m$ $\bar{3}$ $2/m$. a 10.04. $Fm3m$. Crystals rare.

Structure consisting of corner-sharing (Ni,Fe)S$_6$ octahedra with additional (Ni,Fe) atoms lying above alternate octahedral faces (see figure 14.6). Cell contains four (Ni,Fe)$_9$S$_8$. Structure similar to that of sphalerite with all sulfur positions occupied.

Usually massive or in granular aggregates. Sometimes in large pieces showing $\{111\}$ parting. Commonly as exsolution lamellae in pyrrhotite.

Physical Properties   No cleavage. Parting on $\{111\}$. Uneven to conchoidal fracture. Brittle. Hardness 3.5–4. Specific gravity 4.6–5.0. Metallic luster. Opaque. Color light bronze-yellow. Streak light bronze-brown. Readily fusible (1.5–2), with odor of sulfur dioxide, to a steel-gray magnetic bead.

Distinctive Properties and Tests   Color, association, and test for nickel. A sensitive and simple test for nickel is by wetting the specimen with a basic solution such as saliva and sprinkling on powdered dimethylgloxime. The solution will turn pink even if only a small amount of Ni is present.

Association and Occurrence   Pentlandite is the principal ore of nickel and is found in intimate association with pyrrhotite, chalcopyrite, and other nickel and iron sulfides

Table 14.4 Polymorphs of Pyrrhotite

Atomic Percent of Iron	Formula	Crystal System	c-Axis Repeat of "Basic" NiAs Substructure (5.48Å)	a-Axis Repeat of "Basic" NiAs Substructure (3.45Å)
50.0	FeS	Hexagonal	2	$\sqrt{3}$
47.8	$Fe_{11}S_{12}$	Hexagonal	6	2
47.6	$Fe_{10}S_{11}$	Orthorhombic	11	2
47.4	Fe_9S_{10}	Hexagonal	5	2
46.7	Fe_7S_8	Monoclinic	4	2

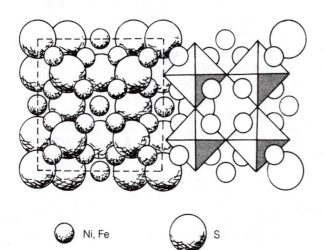

Ni, Fe   S

Figure 14.6
Structure of pentlandite.

as a late magmatic segregation in ultramafic to mafic igneous rocks. The largest deposits known are in the Sudbury District, Ontario.

Alteration   To violarite, Ni_2FeS_4.

Confused with   Pyrrhotite; distinguished by parting, magnetism, and test for nickel. Pyrrhotite, however, often contains small amounts of Ni.

Variants   Chemical varieties: cobaltian.

Related Minerals   *Solid solutions:* cobalt pentlandite, Co_9S_8. *Isostructures:* argentopentlandite, Ag $(Fe,Ni)_8S_8$; geffroyite, $(Ag,Cu,Fe)_9(Se,S)_8$; manganese shadlunite, $(Mn,Pb,Cd)(Cu,Fe)_8S_8$; and shadlunite, $(Pb,Cd)(Fe,Cu)_8 S_8$. *Other similar species:* bornite and nickeline.

COVELLITE
CuS

kō′və-līt, after N. Covelli (1790–1829), an Italian mineralogist.

Crystallography, Structure, and Habit   Hexagonal. $6/m\ 2/m\ 2/m$. a 3.792, c 16.34. $P6_3/mmc$. Crystals rare; hexagonal plates when found.

The structure has units with layers of CuS_3 triangles sandwiched between CuS_4 tetrahedral layers. The units are linked by covalent S-S bonds (see figure 14.7). The cell contains six CuS.

Usually massive or in tabular aggregates; often as a blue coating on other copper sulfides.

Physical Properties   Perfect $\{0001\}$ cleavage, flexible in very thin laminae. Hardness 1.5–2. Specific gravity 4.7. Submetallic luster. Opaque. Color indigo blue, sometimes with a purplish tarnish. Streak lead-gray to black. Turns vivid violet when wet. Fusible (2.5), giving sulfurous fumes.

Distinctive Properties and Tests   Color and association. Paper-thin laminae that may burn.

Association and Occurrence   Found with other copper minerals in veins and secondarily enriched zones. Often intergrown with pyrite or chalcopyrite.

Alteration   From chalcopyrite, chalcocite, enargite, and bornite.

Variants   Usually contains some iron.

Related Minerals   *Solid solutions:* klockmannite, CuSe.

CINNABAR
HgS

sin′ə-bär, from Persian *zinjifrah;* original meaning lost.

Crystallography, Structure, and Habit   Hexagonal. 32. a 4.149, c 9.495. $P3_121$. Crystals as rhombohedra, thick tablets, and prisms. Contact twins fairly common, also six-pointed stellate forms and penetration twins.

Structure composed of infinite Hg-S-Hg chains paralleling the c-axis (see figure 14.8). The unit cell contains three HgS.

Usually found as crystalline incrustations, granular; massive, or earthy.

Physical Properties   Perfect $\{10\bar{1}0\}$ cleavage. Subconchoidal to uneven fracture. Subsectile. Hardness 2–2.5.

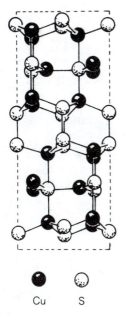

Cu   S

Figure 14.7
Structure of covellite. View along *a*-axis.

Specific gravity 8.1. Luster may be adamantine splendent in dark-colored crystalline varieties or earthy to dull in friable varieties. Color is a tint or shade of red. Streak red-brown to scarlet. Optical: $\omega = 2.905$, $\epsilon = 3.256$; positive.

Distinctive Properties and Tests   Color and streak, specific gravity, and tests for mercury. Wholly volatile before the blowpipe; *fumes are poisonous.*

Association and Occurrence   Found with pyrite, marcasite, realgar, calcite, stibnite, quartz, opal, fluorite, and barite (1) in veins, (2) disseminated, and (3) in masses in sedimentary and volcanic rocks. Sometimes with native mercury. The principal sources are Idria, Italy, and Almaden, Spain, where it has been mined for at least 2,000 years.

Confused with   Hematite, cuprite, and realgar.

Variants   Often admixed with clay, iron oxide, or bitumen.

Related Minerals   *Polymorphs:* metacinnabar (isometric), and hypercinnabar (hexagonal). *Other similar species:* coloradoite, HgTe, and tiemannite, HgSe.

REALGAR
AsS

rē-ăl′gär, from Arabic *rahj al-gahr,* powder of the mine.

Crystallography, Structure, and Habit   Monoclinic. $2/m$. *a* 9.29, *b* 13.53, *c* 6.57, β 106.6°. $P2_1/n$. Short pris-

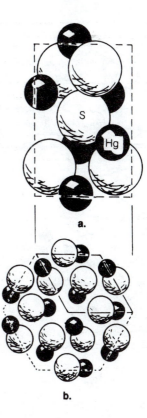

a.

b.

Figure 14.8
Structure of cinnabar. (a) View perpendicular to *c*-axis showing atomic packing. (b) View along *c*-axis.

matic crystals striated vertically. Contact twins with irregular composition surface.

Covalently bonded, 8-member puckered rings of alternating S and As (see figure 14.9). The As atoms lie alternately below and above the plane of S atoms and serve as a bridge between successive rings. The cell contains sixteen AsS units.

Found as crystalline aggregates or crusts, coarse to fine-grained and compact.

Physical Properties   Good {010} cleavage and less good {001}, {$\bar{1}$01}, and {120} cleavage. Conchoidal fracture. Sectile. Hardness 1.5–2. Specific gravity 3.5. Adamantine luster. Transparent (fresh) to translucent. Color and streak red to orange-yellow. Optical: $\alpha = 2.538$, $\beta = 2.684$, and $\gamma = 2.704$; negative; $2V = 46°$. Melting point approximately 310° C. AsS is unstable in the presence of light and alters to As_2S_3 and As_2O_3.

Distinctive Properties and Tests   Color and low melting point. Wholly volatile when heated on charcoal with garlic odor of arsine.

Association and Occurrence   Found with orpiment, arsenic minerals, and stibnite and with lead, gold, and silver ores (1) in hydrothermal veins, (2) in replacement deposits in sedimentary rocks, and (3) as a volcanic sublimate.

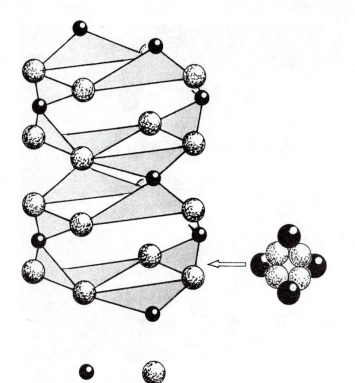

As   S

Figure 14.9
Structure of realgar.

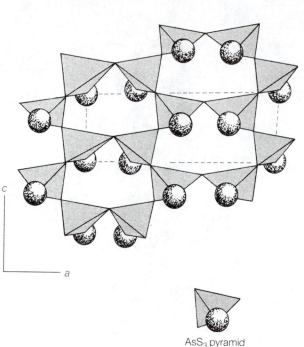

c

a

AsS$_3$ pyramid

Figure 14.10
Structure of orpiment. View along *b*-axis, unit cell dashed.

Alteration   Readily to a reddish yellow powder that is a mixture of orpiment and arsenolite, As$_2$O$_3$. From arsenical ores.

Confused with   Cinnabar, hematite, and cuprite.

Related Minerals   *Other similar species:* orpiment.

ORPIMENT
As$_2$S$_3$

ôr′pə-mənt, from Latin *aurum,* gold, and *pigmentum,* pigment, in allusion to its color.

Crystallography, Structure, and Habit   Monoclinic. 2/*m. a* 11.49, *b* 9.59, *c* 4.25, *β* 90.5°. *P*2$_1$/*n.* Small and poor, short tabular crystals.

Structure is crumpled sheets of flattened, corner-sharing, 6-membered rings of AsS$_3$ trigonal pyramids stacked parallel to (100) (see figure 14.10). Four formula units per cell.

Many habits; usually in foliated or columnar masses.

Physical Properties   Perfect {100} cleavage; cleavage laminae are flexible but not elastic. Sectile. Hardness 1.5–2. Specific gravity 3.5. Pearly and adamantine splendent luster. Transparent to translucent. Color yellow; streak yellow. Optical: *α* = 2.40, *β* = 2.81, and *γ* = 3.02; negative: 2*V* = 76°.

Distinctive Properties and Tests   Hardness and color. Orpiment is wholly volatile on heating.

Association and Occurrence   Found with stibnite, realgar, native arsenic, calcite, barite, and gypsum in low-temperature hydrothermal veins, hot springs, volcanic sublimates, and replacement deposits in sedimentary rocks.

Alteration   From realgar.

Confused with   Native sulfur.

Related Minerals   *Isotypes:* getchellite, AsSbS$_3$. *Other similar species:* realgar.

STIBNITE
Sb$_2$S$_3$

(antimonite, antimony glance)
stĭb′nĭt, mentioned by Dioscosides and Pliny as *stibi, stimmi,* and *platyophthalmon,* the last referring to the use of powdered stibnite (kohl) as an eye shadow.

Crystallography, Structure, and Habit   Orthorhombic. 2/*m* 2/*m* 2/*m. a* 11.229, *b* 11.310, *c* 3.839. *Pbnm.* Stout to slender prismatic crystals (see figure 14.11), often striated vertically and sometimes bent.

Structure has infinite bands of Sb$_2$S$_3$ parallel to the *c*-axis and approximately diagonal to the *a*- and *b*-axes (see figure 14.12). Four formulas per unit cell.

Usually found in aggregates of acicular to columnar crystals, also granular to impalpable.

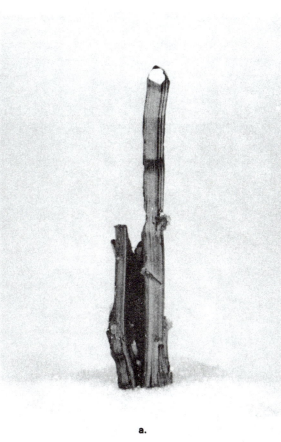

a.

b.

Figure 14.11
(a) Stibnite, Lengshwijiang, Hunan Province, People's Republic of China. (Courtesy University of Windsor. Photograph by D. L.

MacDonald.) (b) Stibiconite, pseudomorphous after stibnite. Mexico. (Courtesy Royal Ontario Museum, Toronto.)

Physical Properties   Perfect {010} cleavage and imperfect {100} and {110} cleavage. Subconchoidal fracture. Subsectile. Hardness 2. Specific gravity 4.6. Metallic luster. Opaque. Color and streak lead- to steel-gray, often with a black tarnish. Specimens are readily bent or twisted. Melting point 550° C. Transparent to infrared light.

Distinctive Properties and Tests   Hardness, habit, and cleavage. Slowly soluble in HCl. A drop of concentrated KOH produces a yellow coating and a brown spot on stibnite.

Association and Occurrence   Stibnite is widely distributed. It is found with realgar, orpiment, galena, marcasite, pyrite, cinnabar, calcite, barite, and chalcedonic quartz in low-temperature hydrothermal veins, replacement deposits, and hot-spring deposits.

Alteration   To kermesite, Sb_2S_2O; cervanite, $Sb_2O_3 \cdot Sb_2O_5$; senarmontite, Sb_2O_3; valentinite, Sb_2O_3; and stibiconite, $H_2Sb_2O_5$ (see figure 14.11).

Confused with   Jamesonite, enargite, and galena.

Variants   May be contaminated with traces of Fe, Pb, or Cu and, less often, with Zn, Co, Ag, or Au.

Related Minerals   *Isotypes:* bismuthinite, Bi_2S_3, and guanajuatite, Bi_2Se_3. *Polymorphs:* metastibnite (amorphous).

PYRITE
FeS₂

(iron pyrites, mundic)
pī'rīt, from Greek *pyr*, fire; an allusion to its property of yielding sparks when struck by steel.

Crystallography, Structure, and Habit   Isometric. $2/m \overline{3}$. a 5.418. *Pa*3. Crystals common, usually in cubes, less often in pyritohedra (see figure 14.13), and rarely in octahedra. Crystals commonly striated (see figure 14.14). Penetration twins called *iron cross twins* with a (011) twin plane and [011] twin axis are common; contact twins rare.

Structure composed of alternating Fe and S_2 pairs in cubic array analogous to halite (see figure 14.15). The unit cell contains four FeS_2.

Usually in crystals or crystal aggregates; also massive, granular.

Physical Properties   Indistinct {100} and {311} cleavage. Conchoidal to uneven fracture. Brittle. Hardness 6–6.5.

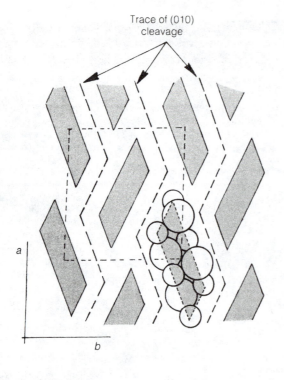

Trace of (010) cleavage

a

b

 Sb$_2$S$_3$ chains

Figure 14.12
Structure of stibnite showing arrangement of Sb$_2$S$_3$ chains.

Figure 14.13
Pyritohedral crystals of pyrite, FeS$_2$. (Courtesy J. S. Hudnall Mineralogy Museum, University of Kentucky. Photograph by T. A. Moore.)

Figure 14.14
Pyrite showing cubic {100} face striations and pyritohedral {210} modification. Kassandra Peninsula, Greece. (Courtesy Royal Ontario Museum, Toronto.)

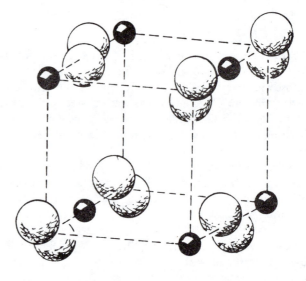

Fe   S

Figure 14.15
Structure of pyrite. Note S$_2$ pairs.

Specific gravity 4.8 to 5.0 with lower hardness and density values related to Co or Ni substitution. Splendent metallic luster. Opaque. Color pale brass-yellow, sometimes with an iridescent tarnish. Streak green-black to brown-black. Thermoelectric. Sparks when struck with steel. Fusible (2.5–3) to a magnetic globule.

Distinctive Properties and Tests Crystal form, hardness, color. Insoluble in HCl.

Association and Occurrence Pyrite is the most widespread and abundant of the sulfides. It is associated with most other minerals and is found over a wide range of environments. It is an accessory mineral in all igneous rocks, it is ubiquitous in metal-bearing hydrothermal veins, and it often occurs in sedimentary rocks such as black or green shales, limestones, and coals where reducing conditions exist. Pyrite may replace wood or shells or be replaced by iron oxides.

Alteration To sulfates and oxides of iron.

Confused with Chalcopyrite, marcasite.

Variants *Chemical varieties:* bravoite, $(Ni,Fe)S_2$; cobaltian; selenian. Often contains small amounts of V, Mo, Cr, W, Tl.

Related Minerals Some eighteen mineral species are in the pyrite group. *Solid solutions:* vaesite, NiS_2; cattierite, CoS_2. *Isotypes:* aurostibite, $AuSb_2$; laurite, RuS_2; sperrylite, $PtAs_2$; penroseite, $NiSe_2$; trogtalite, $CoSe_2$; erlichmanite, OsS_2; fukuchilite, Cu_3FeS_8; geversite, $PtSb_2$; insizwaite, $Pt(Bi,Sb)_2$; krutaite, CuS_2; maslovite, $PtBiTe$; michenerite, $(Pd,Pt)BiTe$; testibiopalladite, $Pd(Sb,Bi)Te$; villamaninite, $(Cu,Ni,Co,Fe)S_2$. *Isostructures:* hauerite, MnS_2. *Polymorphs:* marcasite. *Other similar species:* arsenoferrite, $FeAs_2$; mackinawite, $Fe_{1-x}S$; greigite, Fe_3S_4; smythite, $(Fe,Ni)_9S_{11}$.

COBALTITE
CoAsS

kō′bôl-tīt, from German *kobold,* goblin; early miners' reference to its troublesome properties.

Crystallography, Structure, and Habit Orthorhombic and pseudoisometric. $2mm$. $a \approx b \approx c = 5.58$. $Pca2_1$. In apparent cubes or pyritohedra with striated faces.

Isostructural with pyrite (see figure 14.15), with As and S pairs occupying the sites of sulfur pairs in pyrite. The unit cell contains four CoAsS.

Crystal aggregates, granular, and compact.

Physical Properties Imperfect {100} cleavage. Uneven fracture. Brittle. Hardness 5.5. Specific gravity 6.0–6.4.

Splendent metallic luster. Opaque. Color usually silver-white, sometimes tending toward reddish, violet-gray, or gray-black. Streak gray-black. Thermoelectric. Fusible (2–3).

Distinctive Properties and Tests Color, cleavage, and crystal habit. Powdered mineral imparts a pink color to warm HNO_3.

Association and Occurrence Found with cobalt and nickel sulfides and arsenides, pyrrhotite, chalcopyrite, galena, and magnetite in high-temperature deposits, either disseminated or in veins.

Alteration To erythrite, $Co_3(AsO_4)_2 \cdot 8H_2O$, a pink, powdery coating, or "bloom."

Confused with Skutterudite series minerals.

Variants *Chemical varieties:* ferrian. Nickel may be present as a minor contaminant.

Related Minerals *Solid solutions:* gersdorffite, NiAsS; ullmanite, NiSbS; willyamite, (Co,Ni)SbS; and intermediate Ni-Co or As-Sb solutions. *Isotypes:* pyrite (As-S for S-S pairs). *Polymorphs:* pyrite structure with complete As-S disorder above 850° C. *Other similar species:* hollingworthite, (Rh,Pt,Pd)AsS; irarsite, (Ir,Ru,Rh,Pt)AsS; platarsite, (Pt,Rh,Ru)AsS; tolovkite, IrSbS.

MARCASITE
FeS₂

(white iron pyrites)
mär′kə-sīt, from Assyrian *Markashitu,* referring to an old province of Persia, presumably an early source of the mineral. Marcasite was the term miners and mineralogists used for pyrite until about 1800.

Crystallography, Structure, and Habit Orthorhombic. $2/m\ 2/m\ 2/m$. a 4.443, b 5.423, c 3.388. $Pmnn$. Crystals tabular, frequently with curved faces. Twinning produces coxcomb arrays (see figure 14.16), spear shapes, and occasionally stellate fivelings.

Zigzag FeS_2 chains parallel to the c-axis (see figure 14.17). Neighboring chains in (100) layers have S pairs pointing in the same direction; the direction is reversed in adjacent layers. Two FeS_2 per unit cell.

Crystal aggregates; fine grained to impalpable; stalactitic, colloform, and concentric.

Physical Properties Poor {101} cleavage. Uneven fracture. Brittle. Hardness 6–6.5. Specific gravity 4.9. Splendent metallic luster. Opaque. Color tin-white on fresh

Figure 14.16
Marcasite with coxcomb habit. Ottawa County, Oklahoma. (Courtesy University of Windsor. Photograph by D. L. MacDonald.)

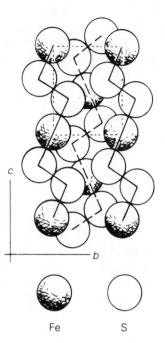

Fe   S

Figure 14.17
Structure of marcasite. Note FeS$_2$ chains.

surface deepening on exposure to pale bronze-yellow or dingy gray-green. Streak gray-black. Fusible (2.5–3) to magnetic globule.

Distinctive Properties and Tests   Coxcomb habit (see figure 14.16). Color. Distinguished from pyrite by treating finely ground mineral with cold HNO$_3$ and boil the solution. Marcasite is decomposed with the liberation of sulfur, yielding a milky solution. Pyrite treated in the same manner is completely dissolved.

Association and Occurrence   Found in lead and zinc ores. It is deposited at low temperatures from acid solutions in metalliferous veins and in sedimentary rocks, especially limestones, shales, and coals.

Alteration   To iron sulfates, iron oxides, and native sulfur.

Confused with   Pyrite.

Variants   May contain traces of Fe or Cu and also Tl.

Related Minerals   *Isotypes:* löllingite, FeAs$_2$; safflorite, (Co,Fe)As$_2$; and rammelsbergite, NiAs$_2$. *Isostructures:* hastite, CoSe$_2$; ferroselite, FeSe$_2$; frohbergite, FeTe$_2$; kullerudite, NiSe$_2$; and mattagamite, CoTe$_2$. *Polymorphs:* pyrite above 350° C.

ARSENOPYRITE
FeAsS

(mispickel)
ärs-ən-ō-pī′rīt, from Greek *arsenikon pyrites,* referring to its composition.

Crystallography, Structure, and Habit   Monoclinic. 2/*m*. *a* 5.760, *b* 5.690, *c* 5.785, *β* 112.2°. *P*2$_1$*c*. Prismatic and stout striated crystals. Penetration, contact, and cyclic twins.

Structure is a derivative of marcasite (see figure 14.17), with one half of the S replaced by As. Fe coordinates octahedra of three As and three S. Four FeAsS units per cell.

Found as disseminated crystals, crystal aggregates, and granular and compact masses.

Physical Properties   Distinct {101} cleavage. Uneven fracture. Brittle. Hardness 5.5–6. Specific gravity 6.1. Splendent metallic luster. Opaque. Color silver-white. Streak black. Readily fusible (2). May be thermoelectric.

Distinctive Properties and Tests   Color and crystal form. Sulfur remains when arsenopyrite is decomposed by HNO$_3$.

Association and Occurrence   Most abundant and widespread arsenic-bearing mineral. Found with pyrite, chalcopyrite, sphalerite, and ores of tin, cobalt, nickel, silver, gold, and lead (1) in high-temperature veins, (2) in pegmatites, (3) in contact metamorphic rocks, and (4) disseminated in marble or slate.

Alteration   To scorodite, FeAsO$_4$ · 2H$_2$O.

Confused with   Skutterudite series minerals, marcasite, and pyrite.

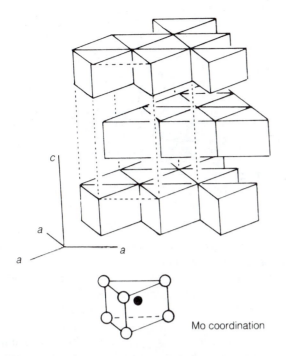

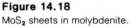
Mo coordination

Figure 14.18
MoS₂ sheets in molybdenite.

Variants *Chemical varieties:* cobaltian and bismuthian. Usually contains some Co and may contain traces of Ni, Sb, Pb, Zn, Au, or Ag.

Related Minerals *Solid solutions:* glaucodot, $(Co,Fe)AsS$. *Isotypes:* gudmundite, $FeSbS$. *Other similar species:* lautite, $CuAsS$; osarsite, $(Os,Ru)AsS$; and ruarsite, $RuAsS$.

MOLYBDENITE
MoS₂

mə-lĭb′də-nīt, from Greek *molybdos,* lead, from its early misidentification.

Crystallography, Structure, and Habit Hexagonal. $6/m\,2/m\,2/m$. a 3.160, c 12.295. $P6_3/mmc$. Crystals as hexagonal plates or short prisms.

Molybdenite structure (see figure 14.18). Layers perpendicular to c composed of sulfur ions on the corners of trigonal, face-sharing prisms, each having a central molybdenum ion. Successive layers alternate 180° in orientation and are separated by an unoccupied octahedral site. The unit cell contains two MoS₂.

Usually found in foliated masses or in discrete scales.

Physical Properties Perfect {0001} cleavage; laminae flexible but not elastic. Sectile. Hardness 1–1.5. Specific gravity 4.7. Metallic luster. Opaque. Color lead-gray. Streak grayish black to bluish black (greenish on glazed porcelain). Infusible.

Distinctive Properties and Tests Cleavage, hardness, and greenish streak on glazed porcelain. Soluble in aqua regia and decomposed by HNO_3.

Association and Occurrence Found with scheelite, wolframite, topaz, fluorite, chalcopyrite, cassiterite, and epidote in veins, pegmatites, granite, and contact metamorphic deposits.

Alteration To ferrimolybdite, $Fe_2(MoO_4)_3 \cdot 8H_2O$, a bright yellow powder; powellite, $Ca(MoO_4)$; molybdite, MoO_3; and ilsemannite, $Mo_3O_8 \cdot nH_2O$.

Variants May contain traces of Au, Ag, Re, and Se.

Related Minerals *Isotypes:* tungstenite, WS_2. *Polymorphs:* molybdenite-3R (hexagonal); jordisite (amorphous).

SKUTTERUDITE SERIES
(Co,Ni)As₃ ₋ ₓ

Skutterudite $CoAs_3$, skŭt′ə-roo̅-dīt, from its discovery at Skutterrud, Norway.
Smaltite $(Co,Ni)As_{3-x}$, smôl′tīt, from *smalt,* a coloring agent made of blue glass containing cobalt.
Chloanthite $(Ni,Co)As_{3-x}$, klō-an′thīt, from Greek *chlo,* young green vegetation, and *anthos,* flower.

Crystallography, Structure, and Habit Isometric. $2/m\,\bar{3}$. a 8.21 to 8.33. $Im3$. Cubes and cuboctahedra, frequently with curved faces. Cyclic twins and complex and distorted shapes.

Mutually perpendicular square AsS₄ rings are linked by (Co,Ni) in octahedral coordination (see figure 14.19). Occasional As atoms are missing. The unit cell contains eight formulary units.

Usually massive; dense to granular.

Physical Properties Distinct {011} and {111} cleavage, but this cleavage is variable and not characteristic. Variable cleavage may be the result of parting in these directions. Fracture conchoidal to uneven. Brittle. Hardness 5.5–6. Specific gravity 6.1–6.9. Splendent metallic luster. Opaque. Color tin-white to silver-gray. Streak black. Thermoelectric. Fusible (2–2.5).

Distinctive Properties and Tests Color. Fuses with garlic odor and a magnetic residue.

Association and Occurrence Usually found in veins with cobalt and nickel minerals, especially with cobaltite and nickeline, and also with arsenopyrite, native silver, silver sulfosalts, native bismuth, calcite, siderite, barite, and quartz.

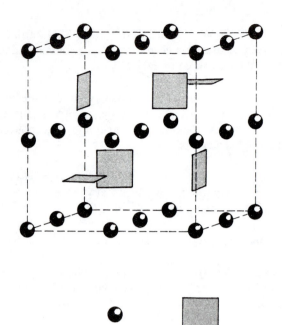

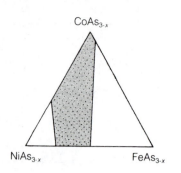

Figure 14.20
Solid solution in the skutterudite series.

Co,Ni   AsS$_4$

Figure 14.19
Structure of skutterudite. Unit cell dashed.

Alteration Cobalt-rich types alter to pink erythrite, Co$_3$As$_2$O$_8$ · 8H$_2$O. Nickel-rich types alter to green annabergite, Ni$_3$As$_2$O$_8$ · 8H$_2$O.

Confused with Arsenopyrite and cobaltite.

Variants *Chemical varieties:* ferrian and bismuthian. S commonly replaces a small amount of As.

Related Minerals The skutterudite series (see figure 14.20) includes the following solid-solution pairs: skutterudite, (Co,Ni)As$_{3-x}$ to nickel skutterudite, (Ni,Co)As$_{3-x}$, with x = 0 to 0.5; smaltite (Co,Ni)As$_{3-x}$ to chloanthite, (Ni,Co)As$_{3-x}$, with x = 0.5 to 1.0. *Other similar species:* linnaeite, Co$_3$S$_4$.

Sulfosalts

Polybasite Group	(Ag,Cu)$_{16}$(Sb,As)$_2$S$_{11}$
Argyrodite Series	Ag$_{5-8}$(Ge,Sn,Sb,As)S$_6$
Ruby Silver Group	(Ag,Cu)$_3$(Sb,As,Bi)S$_3$
Pyrargyrite	
Tetrahedrite Series	(Cu,Fe)$_{12}$(Sb,As)$_4$S$_{13}$
Sulvanite Group	Cu$_3$(V,Ge,Sn)S$_4$
Enargite Group	(Cu,Pb,Ag)$_{3-14}$(Sb,As,Bi)$_{1-7}$S$_{4-23}$
Enargite	
Bournonite Group	(Pb,Cu,Ag)$_{2-8}$(Sb,As,Bi)$_{1-5}$S$_{3-13}$
Bournonite	
Chalcostibite Group	(Cu,Tl,Pb)$_{1-8}$(Sb,As,Bi)$_{1-6}$S$_{2-14}$
Jamesonite	
Andorite Group	(Pb,Ag,Cu)$_{2-13}$(Sb,Bi,As)$_{3-8}$S$_{6-17}$
Plagionite Group	Pb$_{3-9}$Sb$_8$S$_{15-21}$
Miscellaneous Series	

The general physical characteristics of sulfosalts are the same as those of the sulfides, to which they are closely related. The principal difference between the two families is in sulfosalts chains and sheets of sulfur and arsenic, antimony, or bismuth occur, whereas in sulfides the sulfur (or arsenic, selenium, etc.) occurs singly or in pairs. Their complex structural geometry causes sulfosalts to have lower symmetry than the sulfides, usually as orthorhombic or monoclinic crystals, and to form mainly as aggregates rather than single well-formed crystals.

Sulfosalts are always composed of a limited number of elements. As-S, Bi-S, or Sb-S are coordinated by silver, copper, or lead. Only rarely are other elements present. A generalized formula for the sulfosalts might thus be written (Ag,Cu,Pb)(As,Bi,Sb)S. The limited number of elements found in sulfosalts is not an indication of their variability since various combinations in different proportions result in nearly eighty-five different species.

The various sulfosalts described in this chapter are best distinguished by the properties of cleavage, color, and streak, together with the mineral habit, as indicated in table 14.5.

The sulfosalt family may be distinguished from the sulfide family on the basis of fusibility. Only two sulfides, acanthite and stibnite, melt at less than two on the scale of fusibility, whereas all sulfosalts fuse at less than two.

Sulfosalts are occasionally of economic importance, particularly as ores of silver found in fissure-vein deposits formed by the passage of hydrothermal solutions through fractured rock. Much of the world's silver has come from a belt in the western Americas extending from Utah and Nevada southerly through Mexico to Honduras. The Coeur d'Alene district of Idaho and the Cobalt and Timmins districts of Ontario have also been important North American occurrences.

PYRARGYRITE
Ag$_3$SbS$_3$

(ruby silver)
pī-rär′jə-rīt, from Greek *pyr,* fire, and *argyros,* silver; an allusion to its deep red color.

<table>
<tr><th>Table 14.5</th><th colspan="6">Physical Properties of Some Sulfosalts</th></tr>
</table>

Mineral	Hardness	Cleavage	Color	Streak	Habit, etc.
Pyrargyrite	2.5	Yes	Red to black	Red	
Tetrahedrite	3–4.5	No	Gray-black	Brown to black	Brittle; tetrahedral crystals
Enargite	3	Yes	Black	Black	
Bournonite	2.5–3	No	Steel-gray to black	Steel-gray to black	Tubular crystals; cogwheel twins
Jamesonite	2–3	Yes	Steel-gray to black	Steel-gray to black	Feathery appearance

Crystallography, Structure, and Habit Hexagonal. $3m$. a 11.052, c 8.718. $R3c$. Striated prisms. Twins by growth into complex aggregations; also repeated and cyclic twins.

Structure composed of SbS_3 pyramidal groups, Sb at the apex and S on the base, located at the center and corners of a rhombohedral cell (see figure 14.21). The unit cell contains six formula units.

Crystal aggregates, massive and compact.

Physical Properties Distinct $\{10\bar{1}1\}$ and imperfect $\{10\bar{1}2\}$ cleavage. Conchoidal to uneven fracture. Brittle. Hardness 2.5. Specific gravity 5.8. Metallic luster. Translucent. Color deep red. Streak red. Melting point 486° C.

Distinctive Properties and Tests Color, diaphaneity, fusibility, and tests for silver and antimony. Decomposed by HNO_3 with the formation of S and Sb_2O_3.

Association and Occurrence Found with silver sulfosalts, native silver, galena, sphalerite, calcite, dolomite, barite, and quartz in low-temperature metalliferous veins and in secondarily enriched zones.

Alteration To acanthite, native silver, and rarely to chlorargyrite or stibnite. From argentite, native silver.

Confused with Proustite, Ag_3AsS_3, and cuprite.

Variants Some As may replace Sb.

Related Minerals *Solid solutions:* proustite, Ag_3AsS_3, above 300° C. *Polymorphs:* pyrostilpnite (monoclinic) and xanthoconite (monoclinic). *Isotypes:* seligmannite, $CuPbAsS_3$.

TETRAHEDRITE SERIES
$(Cu,Fe)_{12}(Sb,As)_4S_{13}$

(gray ore, fahlore, fahlerz, falkenhaynite, stylotypite)
Tetrahedrite $(Cu,Fe)_{12}Sb_4S_{13}$, tĕt′rə-hē′drīt, from Greek *tetra,* four, and *hedra,* base, describing its crystal form. *Tennantite* $(Cu,Fe)_{12}As_4S_{13}$, tĕn′ən-tīt, after the English chemist Smithson Tennant.

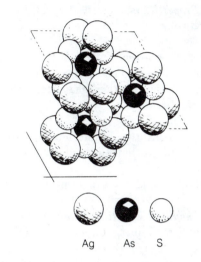

Figure 14.21
Structure of pyrargyrite. View along *c*-axis.

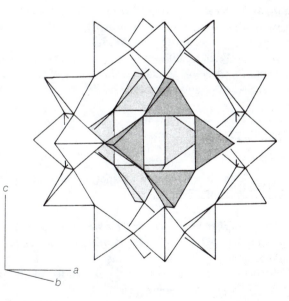

Figure 14.22
Structure of tetrahedrite. (After Johnson et al., 1988.)

Crystallography, Structure, and Habit Isometric. $\bar{4}3m$. a 10.190 (tennantite) to 10.327 (tetrahedrite). $I\bar{4}3m$. Tetrahedral crystals. Penetration twins; often repeated twins.

Structure (see figure 14.22) is analogous to that of sodalite, a framework of corner-sharing Cu-S tetrahedra surrounding cavities that contain As and Sb. The unit cell contains two formulary units.

Crystal aggregates, massive, compact.

Physical Properties No cleavage. Subconchoidal to uneven fracture. Brittle. Hardness 3–4.5, increasing with As content. Specific gravity 4.4–5.4, increasing with Sb content. Metallic luster. Opaque. Color gray to black. Streak black to brown to red. Fuses readily (1.5). Very thin splinters are cherry-red in transmitted light.

Distinctive Properties and Tests Crystal form, lack of cleavage, and color.

Association and Occurrence One of the more common sulfosalts with widespread occurrence and varied association. Found with chalcopyrite, sphalerite, galena, pyrite, quartz, siderite, barite, etc.

Alteration To azurite, malachite, antimony oxides, and occasionally to cuprite, limonite, and chrysocolla.

Confused with Chalcocite and acanthite.

Variants *Chemical varieties of tetrahedrite:* zincian, ferroan, argentian (freibergite), mercurian (schwatzite), plumbian, bismuthian, nickelian, and cobaltian. *Chemical varieties of tennantite:* zincian, ferroan, argentian, and bismuthian.

Related Minerals *Solid solutions:* The tetrahedrite series exhibits complete solid solution from tetrahedrite, $Cu_{12}Sb_4S_{13}$, to tennantite, $Cu_{12}As_4S_{13}$. Limited replacement of Cu by many different elements is very common. *Other similar species:* germanite, $Cu_3(Ge,Fe)S_4$; colusite, $Cu_3(As,Sn,Fe)S_4$; and sulvanite, Cu_3VS_4.

ENARGITE
Cu_3AsS_4

ĕn-är′jīt, from Greek *enarges,* distinct; in reference to its cleavage.

Crystallography, Structure, and Habit Orthorhombic. 2mm. a 6.426, b 7.422, c 6.144. $Pnm2_1$. Tabular or prismatic striated crystals. Cyclic twins.

Structure (see figure 14.23) closely related to that of wurtzite. Two formulary units per unit cell.

Crystal aggregates; massive, granular, and bladed.

Physical Properties Perfect {110} and distinct {100} and {010} cleavage. Uneven fracture. Brittle. Hardness 3.5. Specific gravity 4.4. Metallic luster. Opaque. Color and streak gray-black to iron-black. Readily fusible (1) after decrepitation, yielding a brittle black globule. Dull black tarnish on exposed surfaces. Insoluble.

Distinctive Properties and Tests Cleavage and color.

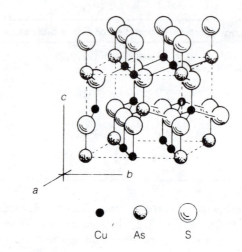

Figure 14.23
Structure of enargite.

Association and Occurrence Found with copper sulfides and pyrite in intermediate temperature veins and replacement deposits.

Alteration To tennantite, copper arsenates, and arsenic oxides.

Confused with Sphalerite, stibnite, bournonite, and jamesonite.

Variants *Chemical varieties:* antimonian. Commonly contains some Fe and, more rarely, Zn or Ge.

Related Minerals *Solid solutions:* famatinite, Cu_3SbS_4. *Polymorphs:* luzonite. *Other similar species:* tetrahedrite; sulvanite, Cu_3VS_4; germanite, Cu_3GeS_4; and colusite, $Cu_3(As,Sn,Fe)S_4$.

BOURNONITE
$PbCuSbS_3$

(cogwheel ore, endellionite)
bōor′nō-nīt, described by Count J. L. de Bournon in 1804.

Crystallography, Structure, and Habit Orthorhombic. 2/m 2/m 2/m. a 8.16, b 8.75, c 7.81. $Pn2_1m$. Short prisms and tablets. Striated. Twins to form cruciform or wheel-like aggregates on {110}.

Structure (see figure 14.24) composed of alternating CuS_4 tetrahedra and SbS_3 and PbS_3 flat pyramids. Four formula units per cell.

Subparallel crystal aggregates; massive, granular, and compact.

Physical Properties Imperfect {010} cleavage. Subconchoidal to uneven fracture. Brittle. Hardness 2.5–3. Specific gravity 5.8. Metallic luster. Opaque. Color and streak steel-gray to black. Easily fusible (1); decrepitates and forms a bubbly mass. Decomposed by HNO_3, yielding a pale blue-green solution.

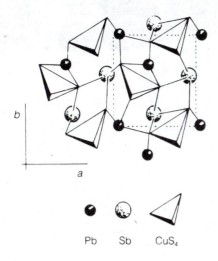

Figure 14.24
Structure of bournonite.

Distinctive Properties and Tests   Crystal form.

Association and Occurrence   One of the commonest sulfosalts. Found with galena, sphalerite, stibnite, chalcopyrite, siderite, chalcocite, quartz, dolomite, pyrite, etc., in hydrothermal veins formed at moderate temperatures.

Alteration   To lead antimonate (bindheimite), cerussite, malachite, and azurite.

Confused with   Enargite, stibnite, and jamesonite.

Variants   *Chemical varieties:* arsenian. Often contains small amounts of Ag, Zn, and Fe. *Form varieties:* endellionite, cogwheel twins from Wheal Boys (Boys Mine), Cornwall.

Related Minerals   *Solid solutions:* seligmannite, $PbCuAsS_3$. *Other similar species:* aikinite, $PbCuBiS_3$.

JAMESONITE
$Pb_4FeSb_6S_{14}$

(brittle feather ore)
jām′zŭ-nīt, after Robert Jameson (1774–1854), Scottish mineralogist.

Crystallography, Structure, and Habit   Monoclinic. $2/m$. a 15.57, b 18.98, c 4.03, β 91.8°. $P2_1/a$. Acicular to fibrous crystals. Striated.
   Structure illustrated in figure 14.25. Pyramidal SbS_3 groups are linked to form complex chains parallel to the c-axis. Fe in distorted octahedral coordination with S. PbS_7 and PbS_8 edge-sharing polyhedra are in complex chains. Two formulary units per unit cell.
   Needles and felted masses; massive, columnar, plumose.

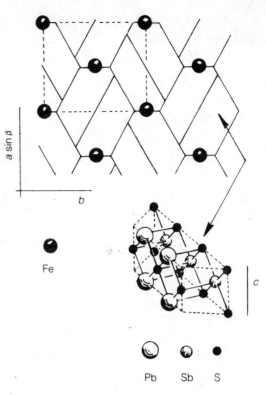

Figure 14.25
Structure of jamesonite showing arrangement of iron atoms and complex Pb-Sb-S units.

Physical Properties   Good {001} cleavage. Brittle. Hardness 2.5. Specific gravity 5.6. Metallic luster. Opaque. Color and streak gray-black; sometimes tarnishing iridescent. Easily fusible (1).

Distinctive Properties and Tests   Plumose (featherlike) habit. Dissolves in hot HCl with evolution of H_2S.

Association and Occurrence   Found in moderate- to low-temperature hydrothermal veins with lead sulfosalts, pyrite, sphalerite, galena, tetrahedrite, stibnite, quartz, siderite, dolomite, calcite, and rhodochrosite.

Alteration   To lead antimonate (bindheimite).

Confused with   Stibnite, enargite, and bournonite.

Variants   May contain significant amounts of Cu and Zn and traces of Ag and Bi.

Related Minerals   *Solid solutions:* benevidesite, $Pb_4(Mn,Fe)Sb_6S_{14}$. *Other similar species:* dufrenoysite, $Pb_2As_2S_5$; cosalite, $Pb_2Bi_2S_5$; zinkenite, $Pb_6Sb_{14}S_{27}$; boulangerite, $Pb_5Sb_4S_{11}$; meneghinite, $Pb_{13}Sb_7S_{23}$; plagionite, $Pb_5Sb_8S_{17}$; semseyite, $Pb_9Sb_8S_4$; and geocronite, $Pb_5(Sb,As)S_8$.

15

Halides

Halides

Halite Group	$(Na,K,Ag,NH_4)(Cl,F)$
Halite	
Chlorargyrite	
Nantokite Group	$(Cu,Ag)(Cl,I)$
Fluorite	
Lawrencite Group	$(Fe,Ni,Mg,Mn)Cl_2$
Matlockite Group	$(Pb,Bi)O(Cl,F)$
Atacamite Group	$(Cu,Mn)_2(OH)_3Cl$
Erythrosiderite Group	$(K,NH_4)_2FeCl_5 \cdot H_2O$
Hieratite Group	$(K,NH_4)_2SiF_6$
Malladrite Group	$(Na,NH_4)_2SiF_6$
Miscellaneous Species	

The halogen elements in Group VII of the periodic table are characterized by unit negative valences and poor shielding. These elements, therefore, tend to form strong ionic bonds with well-shielded alkali and alkali earth elements. The minerals that are formed are called halides, and only two of them, halite and fluorite, are common minerals. Bromine and iodine are geochemically rare, and so are minerals that contain them. Many multiple halides are known but none are common.

Halide structures are ordinarily quite simple, and most have high symmetry, usually isometric or hexagonal. No triclinic halides are known. Typical structures contain either one or two cations, usually Na, K, Mg, or Pb in chlorides and Ca, Na, or Al in fluorides. When two cations are present, they are usually a large ion–small ion pair. Halides as a family are characterized by low specific gravity, light colors, and softness, and many are soluble in water.

The halides are widely used in industry. Halite, in addition to its direct use as salt, is used in the manufacture of HCl and as a source of sodium and chlorine. Sylvite, KCl, provides an important source of potassium. Fluorite yields fluorine for numerous fluorine-bearing compounds and is used as a flux in the manufacturing of steel and aluminum. Fluorite, usually artificial, is also used as a component in spherically and chromatically corrected compound lenses and in spectrographic prisms that transmit ultraviolet or infrared light.

Halite is widely distributed as a rock-forming mineral and in solution as a principal component of saline waters. Extensive sedimentary evaporite deposits of rock salt have been formed worldwide by the evaporation of salty waters of both ancient and modern lakes and seas. Rock salt is typically associated with other evaporite rocks made from gypsum, anhydrite, and potash-bearing minerals. In North America, bedded salt deposits are extensively developed in New York, Michigan, Kansas, Texas, and New Mexico in the United States and in Nova Scotia, New Brunswick, Ontario, and the Prairie Provinces in Canada. The aggregate thickness of rock-salt beds in these various deposits can exceed 300 meters.

Rock salt is mechanically weak and will deform plastically when placed under differential pressure at somewhat elevated temperature. These conditions exist for deeply buried salt strata along the coast of the Gulf of Mexico from Mississippi to the Yucatan Peninsula, where the massive salt has been forced upward in numerous, pipelike intrusions called salt domes. These intrusions have brought the salt nearer to the surface, deformed the overlying rock, and modified groundwater flow, forming traps for petroleum accumulation and providing the environment for bacterial generation of native sulfur and precipitation of sulfide minerals.

Apart from bedded rock salt and salt domes, halite is also formed by the evaporation of saline surface or groundwater. For example, surface brines occur at the Great Salt Lake in Utah, and subsurface brines in the carboniferous rocks in Michigan, Ohio, Pennsylvania, New York, and West Virginia.

Sylvite and other potash-bearing evaporite minerals are much less common than halite. They are formed under the same geologic conditions, however, and are thus often intimately associated with halite. In North America, significant amounts of sylvite are recovered in Texas, New Mexico, and Saskatchewan.

The principal economic concentrations of fluorite are in vein and replacement deposits, usually in sedimentary rocks, which have been deposited from low to moderately high temperature groundwaters. Very commonly, the fluorite is associated with sulfide minerals (pyrite, galena, and sphalerite), calcite or dolomite, and barite. Extensive production has been from Mississippi Valley–type deposits in southern Illinois and western Kentucky. Presently, Mexico is the largest North American producer of fluorite.

HALITE
NaCl

(rock salt)
hā'līt, from Greek *hals*, salt.

Crystallography, Structure, and Habit   Isometric. $4/m$ $\bar{3}$ $2/m$. a 5.640. $Fm3m$. Crystals as cubes; sometimes "hopper" forms and, rarely, octahedra (see figure 15.1).

Halite structure of alternating Na and Cl ions in a face-centered cubic array (see figure 15.2). The ions are in cubic close-packing. Sodium ions are surrounded by regular edge-sharing octahedra of chlorine ions. Halite was the first crystal structure to be determined by x-ray diffraction methods, by Bragg in 1913. Four NaCl per unit cell.

Crystalline aggregates, massive, coarsely granular to compact.

Figure 15.1
Halite crystals with "hopper" form. Trona, California. (Courtesy University of Windsor. Photograph by D. L. MacDonald.)

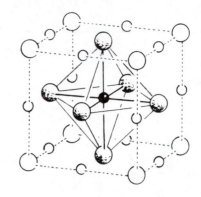

Figure 15.2
Octahedral coordination in halite. Unit cell dashed.

Related Minerals   *Isotypes:* bromoargyrite, AgBr; sylvite, KCl; villiaumite, NaF; and chlorargyrite. *Isostructures:* galena; monteponite, CdO; periclase, MgO; wüstite, FeO; bunsenite, NiO; manganosite, MnO; and altaite, PbTe. *Other similar species:* sal ammoniac, NH_4Cl.

CHLORARGYRITE
AgCl

(horn silver, cerargyrite)
klōr-är′jə-rīt, from Greek *chlor,* green (chlorine), and *argyros,* silver, describing its composition.

Crystallography, Structure, and Habit   Isometric. $4/m$ $\bar{3}\ 2/m$. a 5.549. *Fm3m.* Crystals rare; cubes.

Halite structure (see figure 15.2). Four AgCl per unit cell.

Usually massive and encrusting; sometimes columnar.

Physical Properties   No cleavage. Poor conchoidal fracture. Tough. Highly sectile. Hardness 2–3. Specific gravity 5.5. Resinous luster. Translucent to transparent. Color pearl-gray to gray-green to colorless. The color deepens with increasing amounts of iodine and also with exposure to light. Optical: $n = 2.07$. Resembles horn or wax. Melting point 445° C.

Distinctive Properties and Tests   Sectility and waxy appearance; readily fusible to globule of metallic silver.

Association and Occurrence   Chlorargyrite is a secondary silver mineral associated with native silver, cerussite, galena, limonite, calcite, barite, jarosite, and wulfenite.

Alteration   From other silver minerals.

Physical Properties   Perfect cubic cleavage. Conchoidal fracture. Brittle. Hardness 2. Specific gravity 2.2. Vitreous luster. Transparent to translucent. Colorless or white; also yellow, red, blue, or purple. Yellow, violet, and blue varieties owe their color to "F-centers"—structural sites where an electron is held in an anion vacancy. These colors disappear on heating. Optical: $n = 1.544$.

Distinctive Properties and Tests   Crystal form, cubic cleavage, and salty taste.

Association and Occurrence   Found (1) as extensive sedimentary beds with associated kainite, $KMg(SO_4)$ $Cl \cdot 3H_2O$; polyhalite, $K_2Ca_2Mg(SO_4)_4 \cdot 2H_2O$; carnallite, $KMgCl_3 \cdot 6H_2O$; gypsum; anhydrite; dolomite; and clay minerals; (2) as a precipitate from saline waters from oceans, springs, and desert efflorescence; and (3) as a volcanic sublimate. Halite precipitates from ocean waters evaporated to about one-tenth their original volume. It follows calcite, gypsum, and anhydrite and precedes soluble sulfates and chlorides of K and Mg.

Alteration   Readily dissolves in water, and casts of halite crystals are common.

Confused with   Sylvite, KCl. The bitter taste of sylvite is distinguishing.

Variants   *Chemical varieties:* argentian (huantajayite). Often mechanically mixed with KCl (sylvite), $CaSO_4$, $CaCl_2$, $MgCl_2$, or $MgSO_4$.

Figure 15.3
Cubic crystals of fluorite showing penetration twinning and lineage structure on (100) faces. The crystal at the front center of the mass exhibits {111} cleavage. Weardale, Durham, England. (Courtesy University of Windsor. Photograph by D. L. MacDonald.)

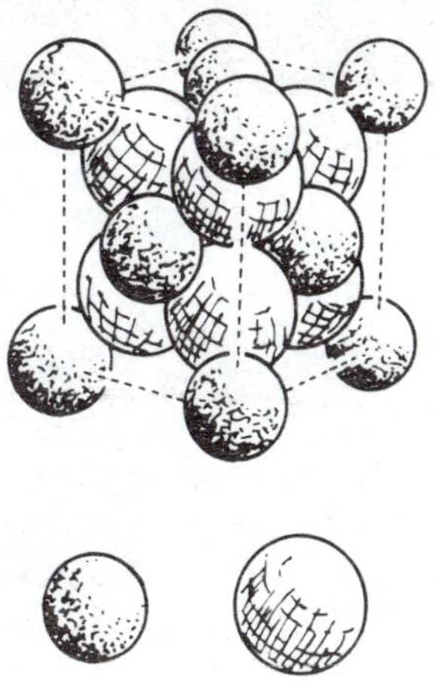

Figure 15.4
Structure of fluorite. Unit cell dashed.

Variants   *Chemical varieties:* bromian (embolite), iodian, and chlorian. May contain Hg or Fe.

Related Minerals   *Solid solutions:* bromargyrite, AgBr. *Isotypes:* halite; sylvite, KCl; and villiaumite, NaF. *Other similar species:* iodargyrite, AgI.

FLUORITE
CaF₂

(fluorspar, flusspat)
flŏŏ′rīt, from Latin *fluere,* to flow; from its ready fusibility and use as a flux in smelting.

Crystallography, Structure, and Habit   Isometric. 4/m 3̄ 2/m. *a* 5.464. *Fm3m.* Crystals usually cubes and rarely octahedra or dodecahedra (see figure 15.3). Penetration twinning common on {111}.

Fluorite structure (see figure 15.4) with four CaF₂ per unit cell. Each Ca is in the center of a cubic array of F ions, and each F is tetrahedrally surrounded by Ca ions.

Usually in crystals; also in cleavable, granular, and columnar masses.

Physical Properties   Perfect octahedral cleavage (see figure 9.4). Rare and indistinct parting on {011}. Flat conchoidal fracture; may be splintery for compact kinds. Brittle. Hardness 4. Specific gravity 3.2. Subvitreous luster. Transparent to translucent. Color commonly a shade of violet or green; also blue, white, yellow, pink, and colorless. The purple color is often due to irradiation and disappears on heating. Color is often in distinct bands or zones paralleling the cube faces. Optical: *n* = 1.433–1.44; increases with rare earth content. Frequently strongly fluorescent or phosphorescent when heated, scratched, or exposed to various kinds of electromagnetic radiation. Melting point 1,360° C.

Distinctive Properties and Tests   Crystal form, cleavage, and color. Loses color on heating. Soluble in hot, concentrated H₂SO₄ yielding *dangerous fumes.*

Association and Occurrence   Found (1) as a vein mineral associated with tourmaline, celestite, quartz, cassiterite, topaz, galena, sphalerite, calcite, and barite; (2) as a minor constituent of limestone and dolomite; (3) as a minor accessory in igneous rocks; (4) as a volcanic sublimate; and (5) in Mississippi Valley–type hydrothermal deposits. Sometimes coated with or containing inclusions of petroleum.

Alteration   Relatively soluble in carbonated water. Sometimes replaced by quartz or chalcedony.

Confused with   Calcite and quartz.

Variants   *Form varieties:* antozonite, dark purple fluorite, structurally damaged by radioactivity, that releases fluorine when crushed; chlorophane, thermoluminescent. *Chemical varieties:* yttrian and cerian (yttrocerite). Cl may replace F in minute quantities. May contain Sr, Ba, Fe, rare earths, and Na.

Related Minerals   *Isostructures:* cerianite-(Ce), (Ce, Th)O₂; thorianite, ThO₂; uraninite. *Other similar species:* sellaite, MgF₂, and frankdicksonite, BaF₂.

16

Oxides and Hydroxides

Oxides

Cuprite
Ice
Periclase Group   (Mg,Ni,Mn,Cd,Ca)O
Zincite Group   (Zn,Mn,Be,Cu,Hg,Pb)O
Hematite Group $(Al,Fe,Ti,Mg,Mn)_2O_3$
     Corundum
     Hematite
     Ilmenite Series
Arsenolite Series $(As,Sb,Mn)_2O_3$
Rutile Group $(Ti,Mn,Sn,Pb,Sb,Bi,V)O_2$
     Rutile
     Pyrolusite
     Cassiterite
Uraninite Group $(U,Th,Pb,Bi,Cu)O_{2-3} \pm H_2O$
     Uraninite
Spinel Group $(Mg,Fe,Zn,Mn,Ni)(Al,Fe,Cr,Mn)_2O_4$
     Spinel Series
     Magnetite Series
     Chromite
Hausmannite Group $(Mn,Zn,Be,Cu)(Mn,Al)_2O_4$
Columbite-Tantalite Series $(Fe,Mn)(Nb,Ta)_2O_6$
Pyrochlore-Microlite Series $(Na,Ca)_2(Nb,Ta)_2O_6(OH,F)$
Fergusonite Series $(Ce,La,Nd,Y)(Nb,Ta)O_4$
Stibiotantalite Series $Sb(Nb,Ta)O_4$
Tapiolite Series $(Fe,Mn)(Nb,Ta)_2O_6$
Euxenite-Polycrase Series $(Y,Ca,Ce,U,Th)(Nb,Ta,Ti)_2O_6$
Aeschynite Series $(Ce,Ca,Fe,Th)(Ti,Nb)_2(O,OH)_6$
Miscellaneous Species
     Romanechite

O xygen is the single most abundant element in the crust of the earth, comprising 46.4% of its weight and 94% of its volume. In view of this tremendous preponderance among the chemical elements, we can say that the outer portion of the earth is made up of closely packed oxygen ions with various metals in the interstices.

Oxygen is highly reactive and usually forms radicals or complex ions with a number of small, highly charged cations, e.g., C in $(CO_3)^{2-}$, S in $(SO_4)^{2-}$, P in $(PO_4)^{3-}$, and Si in silicate minerals of various complexity. In the absence of these or similar elements, or under conditions of high oxygen pressure, regular alternations of metal cations and oxygen can, however, form oxide minerals.

The oxides are, for the most part, simple in both composition and structure. Simple oxides are composed of one kind of metal and oxygen in ratios from M_2O to MO_2. Multiple oxides have two nonequivalent metal ions coordinating the oxygen ions. The metal-oxygen bond strengths are always about the same, and no discrete structural groups are found. Solid solution is common.

Oxide mineral structures are usually based on cubic or hexagonal closest packing of O or OH anions with cations in the resultant octahedral or, less often, tetrahedral sites. Structures are generally of high symmetry, with only a few oxide minerals crystallizing in the monoclinic system and a very rare few in the triclinic system.

Most of the metals in the common oxide minerals have ionic radii such that they coordinate six oxygen neighbors.

The resultant metal-oxygen octahedra are typically articulated into structures by the sharing of edges, as in periclase, MgO, and rutile, TiO_2. For simple oxides, each coordination octahedron surrounds the same metal, and for multiple oxides there is an alternation of the coordination octahedra of the two metals.

As a mineral class, oxides have widely different physical properties, as shown in table 16.1. Generally, however, they are harder than any other mineral class except the silicates, heavier than other classes except the sulfides, generally nonreactive to common acids, often opaque, and commonly have a high luster. The physical characteristics of greatest usefulness in distinguishing the various oxide minerals are hardness, color, and streak.

The oxide class is about equally divided into species whose hardness is less than, approximately equal to, and greater than that of a knife blade. Many species have a constant color, usually black or reddish, and streak colors are often distinctive.

As a class, oxides are common, widespread, and economically valuable minerals. Indeed, industrial civilization depends on the ready availability of iron ore, particularly hematite. Hematite is a very common weathering product of iron-bearing minerals and is responsible for the reddish color of many rocks and soils. At various times and places it has accumulated by chemical precipitation, been concentrated from iron-bearing rocks by the removal of soluble cations under conditions of intense weathering, or been deposited from mineralizing solutions in extensive beds or bodies. All these processes have led to deposits that can be exploited as iron ore.

Most of the world supply of iron ore comes from Precambrian iron formations, deposited when the oxygen waste produced by primitive anaerobic organisms two to three billion years ago was being scavenged from the oxygen-free atmosphere of the time by reaction with iron. Weathering of the siliceous iron formation under tropical conditions led to a leaching of silica, leaving high-grade hematite ore such as that found in the Lake Superior iron ranges, Ungava, and Labrador in Canada; at Mount Newman in Western Australia; at Cerro Bolivar in Venezuela; in the Carajas region and the Quadrilatero Ferrifero of Brazil; in major deposits in West Africa; in Bihar State, India; and in the Krivoy Rog in the Ukraine.

Another oxide of iron, the spinel called magnetite, $FeFe_2O_4$, is a very common accessory mineral in many igneous and metamorphic rocks. Because of its nearly unique magnetic properties (only pyrrhotite is also ferromagnetic and moderately common), it imparts a characteristic magnetic signature to the rocks in which it occurs. Magnetometers, either ground or airborne, are used to detect variations in the earth's field and provide data for geophysical mapping or prospecting. Magnetite in sufficient quantity is also an ore of iron having a very high iron content—72.4%, versus 70.0% for hematite.

Table 16.1 **Some Physical Properties of the Oxides**

Mineral	Hardness with Respect to Steel	Color	Streak	Miscellaneous
Cuprite	<	Red	Red-brown	Red crystals
Ice	<	Colorless	None	
Corundum	>	Various	None	Very hard
Hematite	< to =	Red-brown to black	Red	
Ilmenite	=	Black	Red to black	
Rutile	>	Red to black	Light brown	High luster
Pyrolusite	<	Steel gray	Black	Sooty
Cassiterite	>	Brown to black	None	Heavy
Uraninite	=	Black	Brown-black	Pitchy luster
Spinel	>	Various	None	Octahedral crystals
Magnetite	>	Black	Black	Magnetic
Chromite	=	Black to brown-black	Dark brown	
Columbite	>	Black	Red to black	High luster

The principal manganese deposits of the world were originally deposited as seafloor nodules by submarine exhalations, were chemically precipitated as sedimentary rocks, or were formed by the intense weathering of preexisting manganiferous rocks. The more important protoliths have been Mn garnet-bearing metamorphic rocks (gondites) and dolostones composed of manganoan dolomite.

Submarine exhalations have generated extensive deposits of manganese nodules on the floors of the present oceans, and several deposits in Cuba, for example, at Chario Redondo, are of this origin. Direct precipitation of manganese oxides in a shallow-water marine environment was responsible for the greatest manganese deposits in the world at Chiaturi, Georgia, and Nikopol, Ukraine. Intense weathering of gondite formed the major deposits at Amapa, Brazil; Nsuta, Ghana; and Madhya Pradesh, India. The ore at Postmasburg, South Africa, was formed on a manganiferous dolostone.

A number of other oxides are the ore minerals for various other metals. Ilmenite and rutile are sources of titanium; cassiterite and uraninite are the principal ore minerals of tin and uranium, respectively; chromite is the only economic source of chromium; and columbite-tantalite is the principal source of the rare metals niobium and tantalum.

Some oxides, particularly corundum and spinel, occur in crystals of gem quality such as sapphire, ruby, and many less valuable gemstones. Corundum (aluminum oxide, *not* Carborundum, silicon carbide) also finds wide use as an abrasive.

CUPRITE
Cu_2O

(ruby copper)

kōō'prīt, from Latin *cuprum,* copper, after *Cyprium,* Cyprus.

Crystallography, Structure, and Habit Isometric. $4/m$ $\bar{3} 2/m.$ *a* 4.270. *Pn3m.* Usually in octahedra; rarely dodecahedra, cubes, or capillary forms.

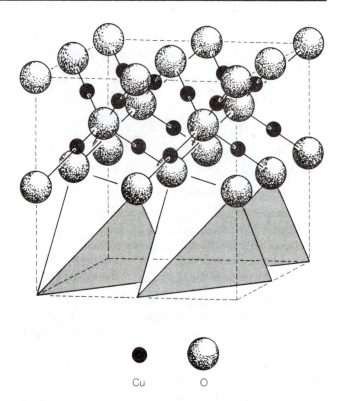

● Cu   ○ O

Figure 16.1
Structure of cuprite. Unit cell dashed.

Cuprite structure is shown in figure 16.1. Each Cu ion coordinates two O ions which are arranged in a body-centered cubic array. Each O ion is surrounded by four tetrahedrally disposed neighbors. Two unlinked and interpenetrating systems of Cu-O linkage are present and are responsible for the special property of electrical rectification shown by cuprite. Two Cu_2O per unit cell.

Usually in crystals or crystal aggregates; also massive, granular, or earthy.

Physical Properties Interrupted {111} cleavage. Conchoidal to uneven fracture. Brittle. Hardness 3.5–4. Specific gravity 6.1. Adamantine splendent to metallic to earthy luster. Subtransparent to subtranslucent. Color shades of red. Streak shining brown-red. Optical: $n =$ 2.849. Fusible (3).

Distinctive Properties and Tests Crystals, luster, and color. Soluble in HCl giving a blue solution.

Association and Occurrence Found in the oxidized portion of copper veins associated with limonite, native copper, malachite, azurite, and chrysocolla.

Alteration To malachite and native copper. From tetrahedrite and chalcopyrite.

Confused with Cassiterite, hematite, and cinnabar.

Variants *Form varieties:* capillary (chalcotrichite) and plushlike aggregates; earthy (tile ore). *Chemical variety:* may contain a little Fe.

Related Minerals *Similar species:* tenorite, CuO.

ICE
H_2O

īs, from old English *is,* cold, icy(?).

Crystallography, Structure, and Habit Hexagonal. *6mm. a* 4.521, *c* 7.367. *P6₃/mmc.* Crystals as lacy hexagonal plates (snowflakes), hexagonal prisms. Twinning with basal plane as twin plane.

In the structure of ice (see figure 16.2) water molecules form 6-membered puckered rings. The molecules occupy sites equivalent to the Si positions in trydimite. Four H_2O per unit cell.

Found in crystals in massive and granular forms.

Physical Properties No cleavage. Conchoidal fracture. Brittle, especially at low temperatures. Hardness 1.5. Specific gravity 0.92. Subvitreous luster. Transparent. Colorless to white. Optical: $\omega = 1.309$ and $\epsilon = 1.311$. Melting point 0° C.

Distinctive Properties and Tests Melting point, taste, and specific gravity.

Association and Occurrence Forms whenever temperature falls below the freezing point of water as precipitation (snow, hail) and as a coating over bodies of water. It is also a semipermanent feature of snowfields and glaciers. Some of the water absorbed on the surface of minute mineral grains, such as clay minerals, is probably a monolayer of ice.

Alteration Melts to water.

Related Minerals *Polymorphs:* Many polymorphic forms of ice, ices II through IX, are observed at increasingly higher pressures. Ices I (ordinary ice), III, V, VI, and VII have a disordered orientation of H_2O molecules, whereas ice II, VIII, and IX exhibit long-range ordering.

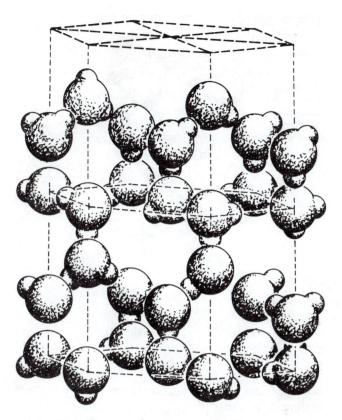

Figure 16.2
Arrangement of H_2O molecules in ice.

CORUNDUM
α-Al_2O_3

kə-rŭn'dəm, from Tamil *kuruntam,* the name for the rock in India from which it was first extracted.

Crystallography, Structure, and Habit Hexagonal. $\bar{3}\,2/m.\ a$ 4.759, *c* 12.989. $R\bar{3}c$. Crystals usually hexagonal prisms with tapered ends and flat terminations. Twinning on rhombohedral planes {10$\bar{1}$1} is very common, producing a lamellar structure and basal striae; contact and penetration twins also occur.

Corundum structure is shown in figure 16.3. Oxygen ions are in approximately hexagonal closest packing with interstitial Al ions in two-thirds of the available sites. Layers perpendicular to *c* comprised of six-sided rings of edge-sharing AlO_6 octahedra are linked by face- and corner-sharing octahedra. The unit cell contains six formulary units.

Found as crystals, massive, granular, and in rounded grains.

Physical Properties No cleavage. Good to perfect basal {0001} and rhombohedral {10$\bar{1}$1} parting. Uneven to conchoidal fracture. Brittle. Hardness 9. Specific gravity 4.0. Vitreous luster. Transparent to translucent. Color various; usually gray-blue or pink; sometimes brown, yellow, green, or colorless. Optical: $\omega = 1.760$ and $\epsilon = 1.768$; negative.

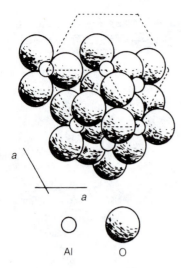

Figure 16.3
Structure of corundum. View along *c*-axis; hexagonal symmetry shown by dashed outline.

Al   O

Sometimes shows color zones. May change color on heating. Some varieties fluorescent and phosphorescent. Melting point 2,050° C.

Distinctive Properties and Tests Crystal form and hardness. Insoluble in acids.

Association and Occurrence Found (1) in Al-rich, Si-poor igneous and metamorphic rocks, (2) as rounded grains in placers, and (3) as a contact metamorphic mineral, especially around peridotites. Emery deposits are typically developed in contact zones of an igneous rock and a limestone. Gem corundum usually occurs as isolated crystals formed metasomatically in marble. The usual associates of corundum are chlorite, micas, olivine, serpentine, magnetite, spinel, kyanite, and diaspore.

Alteration Resistant, but may change to other aluminous minerals such as muscovite, zoisite, sillimanite, kyanite, and margarite.

Confused with Cordierite.

Variants *Form varieties:* ordinary—corundum; gems—sapphire (blue), ruby (red), oriental topaz (yellow), oriental emerald (green), oriental amethyst (purple), and several others; emery—granular black corundum with magnetite, hematite, or spinel. Asterism, arising from oriented rutile inclusions (e.g., star sapphire). Minor amounts of Fe, Ti, Cr, Ni, and Mn may replace Al.

Related Minerals *Isotypes:* hematite; ilmenite series; eskolaite, Cr_2O_3; and karelianite, V_2O_3. *Polymorphs:* synthetic β- and γ-alumina.

HEMATITE
α-Fe_2O_3

(haematite, oligiste)
hĕ′mə-tīt, from Greek *haema,* blood, referring, according to Theophrastus (315 B.C.), to its resemblance to dried blood.

Crystallography, Structure, and Habit Hexagonal. $\bar{3}\,2/m.$ $a\,5.033,$ $c\,13.749,$ $R\,\bar{3}c.$ Thick to thin tabular crystals. Penetration and lamellar twins.

Corundum structure with six Fe_2O_3 per unit cell (see figure 16.3).

Hematite is found in a wide variety of habits; as crystal groupings, often in parallel position or as rosettes; also columnar, granular, colloform, lamellar, compact, earthy, and micaceous.

Physical Properties No cleavage. Parting on $\{0001\}$ and $\{10\bar{1}1\}$ because of twinning. Subconchoidal to uneven fracture. Brittle in compact forms, elastic in thin plates, and soft and unctuous in loose, scaly kinds. Hardness 5–6.5. Specific gravity 5.3. Metallic to earthy luster. Translucent to opaque. Color reddish brown to black. Streak light to dark brick-red. Melting point 1,350° C. Becomes magnetic after heating.

Distinctive Properties and Tests Streak and tests for iron. Slowly soluble in concentrated HCl, giving a yellowish solution.

Association and Occurrence A very common and widespread mineral. Found (1) as extensive sedimentary beds, (2) in contact metamorphic deposits and metamorphosed, banded iron formation (jaspilite), (3) as a minor accessory in igneous rocks, (4) as coatings and nodules derived by alteration from other iron minerals, (5) in hydrothermal veins, and (6) in the oxidized zone above metalliferous veins.

Alteration From magnetite (called martite if pseudomorphous), siderite, pyrite, and marcasite.

Confused with Cinnabar.

Variants *Form varieties:* specularite—micaceous flakes with a splendent metallic luster from regional metamorphic rocks; compact columnar, fibrous, or colloform (as in kidney ore, see figure 16.4); ocherous—red and earthy; martite, pseudomorphous after magnetite.

Ti, Fe^{2+}, Al, and Mn may replace small amounts of Fe. Up to several percent water may be present in fibrous or ocherous varieties, as in turgite, $2Fe_2O_3 \cdot nH_2O$, and limonite, $Fe_2O_3 \cdot nH_2O$.

Related Minerals *Polymorphs:* maghemite, γ-Fe_2O_3 (isometric). *Isotypes:* corundum; eskolaite, Cr_2O_3; and

Figure 16.4
Colloform hematite (kidney ore). (Courtesy University of Windsor. Photograph by D. L. MacDonald.)

karelianite, α-V_2O_3. *Solid solutions:* ilmenite above 1,050° C. *Other similar species:* goethite, $FeO(OH)$; lepidocrocite, γ-$FeO(OH)$; akaganeite, β-$FeO(OH)$; feroxyhyte, $FeO(OH)$; bixbyite, $(Fe,Mn)_2O_3$; and heterogenite, $CoO(OH)$.

ILMENITE SERIES

(titanic iron ore, menaccanite, iserine)
Ilmenite $FeTiO_3$, ĭl′mə-nīt, from Ilmen Mountains, a chain of the Ural Mountains of Russia.
Geikielite $MgTiO_3$, gē′kə-līt, after Sir Archibald Geikie, a Scottish geologist.
Pyrophanite $MnTiO_3$, pī-rŏf′ə-nīt, from Greek *pyr*, fire, and *phainesthai*, to appear, referring to its red color.

Crystallography, Structure, and Habit Hexagonal. $\bar{3}$. *a* 5.05 to 5.16, *c* 13.90 to 14.18. $R\bar{3}$. Thick tabular crystals with prominent basal planes and small rhombohedral terminations. Lamellar twins on $\{10\bar{1}1\}$, simple twins on $\{0001\}$.

Corundum structure (see figure 16.3) with Fe, Mg, or Mn and Ti alternating in the Al sites of corundum, resulting in the drop in symmetry from $\bar{3}\,2/m$ to $\bar{3}$. Six formula units per cell.

Crystal aggregates, skeletal crystals, massive, granular, compact, scaly.

Physical Properties No cleavage for ilmenite, good to excellent cleavage for geikielite on $\{10\bar{1}1\}$ and pyrophanite on $\{02\bar{2}1\}$ and $\{10\bar{1}2\}$. Parting on $\{0001\}$ and $\{10\bar{1}1\}$ because of twinning. Conchoidal to subconchoidal fracture. Brittle. Hardness 5–6. Specific gravity 4.5–5.0. Submetallic luster. Opaque. Color black to brownish black (Mg-rich) or deep red (Mn-rich). Streak black to brownish red to yellow. Sometimes weakly magnetic because of the mechanical admixture or exsolution of magnetite.

Distinctive Properties and Tests Color and streak, and association. Very slowly soluble in hot HCl.

Association and Occurrence Ilmenite is found (1) as an accessory mineral in igneous rocks, especially mafic types, (2) as segregations of large extent from such igneous rocks, especially anorthosites, (3) in high-temperature hydrothermal veins, (4) in pegmatites, (5) as large dikelike masses, (6) in high-temperature, high-pressure metamorphic rocks, and (7) as a principal constituent of black beach sands. Ilmenite is often associated with magnetite in its primary occurrences. It is accompanied by magnetite, rutile, zircon, monazite, garnet, and quartz in the black sands.

Alteration Relatively stable. Sometimes changes to a yellowish white powdery material called **leucoxene,** which is a mixture of titanium oxides.

Confused with Magnetite, chromite, and hematite.

Variants *Chemical varieties:* ilmenite—ferrian (menaccanite), magnesian, and manganoan; geikielite—ferroan; pyrophanite—ferroan.

Related Minerals *Solid solutions:* complete between common ilmenite and rare geikielite and pyrophanite; with hematite above 1,050° C. *Isotypes:* hematite, corundum. *Other similar species:* senaite, $Pb(Ti,Fe,Mn)_{21}O_{38}$.

RUTILE
TiO_2

rōō′tēl, from Latin *rutilas*, golden red.

Crystallography, Structure, and Habit Tetragonal. $4/m\,2/m\,2/m$. *a* 4.594, *c* 2.962. $P4_2/mnm$. Prismatic crystals with pyramidal terminations; prism zone vertically striated. Commonly twinned on $\{011\}$ as contact "elbow" twins or cyclic twins; occasionally polysynthetic.

Rutile structure is shown in figure 16.5. Each Ti is surrounded by six O ions, and each O is bonded to three cations in triangular coordination. Slightly distorted TiO_6 octahedra sharing opposite edges form bands parallel to the *c*-axis. These bands are cross-linked by corner-sharing octahedra. Two TiO_2 per cell.

Common as well-formed tetragonal crystals; also compact massive.

Physical Properties Distinct $\{110\}$ and poor $\{100\}$ cleavage. Parting because of twin gliding on $\{102\}$ or $\{902\}$. Subconchoidal to uneven fracture. Brittle. Hardness 6–6.5. Specific gravity 4.2. Adamantine splendent to submetallic luster. Transparent to opaque. Color reddish-brown, red, black, and rarely violet, green, or yellowish.

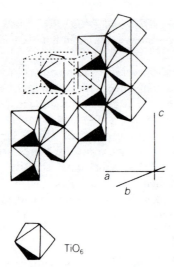

Figure 16.5
Arrangement of Ti-O coordination octahedra in rutile. Unit cell dashed.

Streak pale brown, yellowish, and rarely gray. Optical: $\omega = 2.621$, $\epsilon = 2.908$; positive. Turns black on heating and regains color on cooling. Infusible.

Distinctive Properties and Tests   Crystal form and twins. Luster, pale streak. Insoluble.

Association and Occurrence   Rutile is typically formed at high temperatures. It is found (1) in hornblende-rich igneous and metamorphic rocks, (2) as a vein mineral, (3) in pegmatites, and (4) as a constituent of black beach sands. The "blue quartz" found in high-temperature metamorphic rocks and granite owes its color to the inclusion of minute rutile needles.

Alteration   From other titanium minerals, such as titanite, ilmenite, and Ti-rich hornblende. To leucoxene, a yellowish-white earthy material.

Confused with   Cassiterite and titanite.

Variants   *Form varieties:* sagenite, crisscross pattern of acicular crystals often enclosed in other minerals, such as quartz or pyroxene; venus hairstone, smoky quartz penetrated with acicular rutile crystals. *Chemical varieties* (all black in color): ferrian, ferroan, tantalian (strüverite), and niobian (ilmenorutile). Small amounts of Sn, Cr, and V may be present.

Related Minerals   *Isotypes:* cassiterite; pyrolusite; plattnerite, PbO_2; paratellurite, TeO_2; and stishovite, SiO_2. *Polymorphs:* anatase (octahedrite), tetragonal, and brookite (orthorhombic). *Other similar species:* baddeleyite, ZrO_2.

PYROLUSITE
MnO_2

pī′rə-loo′sīt, Greek *pyr,* fire, and *louein,* to wash, referring to its use in decolorizing glass.

Crystallography, Structure, and Habit   Tetragonal. $4/m\,2/m\,2/m$. a 4.388, c 2.865. $P4_2/mnm$. Usually found as pseudomorphs after orthorhombic manganite. Rarely in prismatic crystals. Twinning rare as repeated twins.

Isostructural with rutile (see figure 16.5). Two MnO_2 per cell.

Usually in massive columnar, fibrous, or divergent forms. Also colloform, granular, and powdery. Often observed as dendritic growths on fracture surfaces in rocks (see figure 16.6).

Physical Properties   Perfect {110} cleavage (rarely observed). Uneven fracture. Brittle. Hardness 6–6.5 for crystals, and 2–6 for massive forms. Fibrous or pulverulent forms may be soft enough to soil the fingers. Specific gravity 5.0. Metallic luster. Opaque. Color light to dark steel-gray. Streak black. Infusible.

Distinctive Properties and Tests   Color, hardness, streak, habit, and low water content. Chlorine gas evolves when treated with HCl. Oxygen evolves when heated in the closed tube (as a test for oxygen, a glowing splinter of wood flares up when placed at the mouth of the tube).

Association and Occurrence   Pyrolusite is one of the more common manganese minerals. It is always formed under conditions of strong oxidation and is always a secondary mineral. It is found as coatings, nodules, and beds, probably resulting from the devitrification of manganese colloids transported by waters. Pyrolusite is usually accompanied by iron oxides and other manganese oxides.

Alteration   From other manganese minerals, such as manganite and rhodochrosite.

Confused with   Manganite, romanechite, and graphite.

Variants   *Form varieties:* ordinary; crystallized. Mixtures of manganese oxides are called **wad.** Pyrolusite may be relatively impure because of mechanical admixture of clay, silica, or iron oxides and because of adsorbed contaminants or proxy elements, including water, heavy metals, phosphorus, alkali metals, and alkaline earths, especially barium.

Related Minerals   *Isotypes:* rutile; cassiterite; and plattnerite, PbO_2. *Polymorphs:* nsutite, γ-MnO_2 (hexagonal); ramsdellite (orthorhombic); vernadite, δ-MnO_2 (hexagonal). *Other similar species:* manganite, $MnO(OH)$; romanechite; and birnessite, $Na_4Mn_{14}O_{27} \cdot 9H_2O$.

Figure 16.6
Dendritic growth of pyrolusite and romanechite on sandstone.
(Courtesy J. S. Hudnall Mineralogy Museum, University of Kentucky.
Photograph by T. A. Moore.)

CASSITERITE
SnO_2

(tinstone)
kə-sĭt′ə-rīt, from Greek *kassiteros,* tin.

Crystallography, Structure, and Habit Tetragonal. $4/m\ 2/m\ 2/m.$ a 4.738, c 3.188. $P4_2/mnm.$ Crystals as low pyramids or as prisms, sometimes striated. Contact and penetration twins on {101} and {011} are very common; also repeated twinning.

Isostructural with rutile (see figure 16.5). Two SnO_2 per unit cell.

Crystals; colloform with radiating fibers, massive, and granular.

Physical Properties Imperfect {100} cleavage. Distinct parting on {111} or {011}. Subconchoidal to uneven fracture. Brittle. Hardness 6–7. Specific gravity 6.8–7.1. Subadamantine to submetallic luster. Transparent to opaque. Color usually brown or black, rarely gray, red, white, or yellow. Streak white to brown. Optical: $\omega = 2.006$ and $\epsilon = 2.097$; positive. Infusible.

Distinctive Properties and Tests Crystal form, specific gravity, luster, and association. Insoluble.

Association and Occurrence Characteristically found in high-temperature hydrothermal veins or contact meta-morphic deposits genetically related to quartzose igneous rocks. Also noted in pegmatites, as an accessory mineral in highly siliceous plutonic igneous rocks, and in the zone of oxidation over tin deposits. The resistance of cassiterite to weathering and its high specific gravity often cause it to be concentrated in stream and beach placers. Associated with B- and F-bearing minerals such as topaz, tourmaline, fluorite, muscovite, and lepidolite in the mineral aggregation called **greisen.** Also associated with wolframite, quartz, arsenopyrite, and molybdenite.

Alteration Resistant to change.

Confused with Rutile.

Variants *Form varieties:* ordinary—in crystals or massive; stream tin—rounded pebbles or sand in placers; wood tin—colloform structure of radiating acicular crystal layers with intergrown needles of quartz, hematite, or topaz; brown in color, looking like dry wood. Wood tin is formed by colloidal precipitation at low temperatures. *Chemical varieties:* ferrian and tantalian. Small amounts of Nb, W, Sc, and Mn may be present.

Related Minerals *Isotypes:* rutile; pyrolusite; plattnerite, PbO_2; and paratellurite, TeO_2.

URANINITE
UO$_2$

(pitchblende, nasturan)

yŏŏ-rā'nə-nīt, after Uranus. Uranium was discovered by M. H. Klaproch in 1789 and named in honor of the 1781 discovery of the planet Uranus.

Crystallography, Structure, and Habit Isometric. $4/m$ $\bar{3}$ $2/m$. a 5.47 decreasing with oxidation to about 5.36. $Fm3m$. Rare octahedral, cubic, or cuboctahedral crystals. Rarely twinned.

Isostructural with fluorite (see figure 15.4). Face-centered isometric cell with four UO$_2$.

Usually massive; often colloform.

Physical Properties No cleavage. Uneven to conchoidal fracture. Brittle. Hardness 5–6. Specific gravity 10.8 but decreasing as U^{4+} is oxidized to U^{6+}. Luster submetallic to pitchlike and dull. Opaque. Color black. Streak brownish black. Highly radioactive. Infusible.

Distinctive Properties and Tests Habit, association, and high specific gravity. Radioactive. Easily soluble in HNO$_3$ or aqua regia.

Association and Occurrence Found (1) replacing wood or other reducing precipitants in sedimentary rocks, (2) in pegmatites with zircon, tourmaline, monazite, and carbonaceous material, (3) in high-temperature hydrothermal tin veins with cassiterite, pyrite, chalcopyrite, arsenopyrite, galena, and Co-Ni arsenide minerals, (4) in hydrothermal veins formed at moderate temperatures with Co-Ni-Bi-Ag-As minerals and pyrite, chalcopyrite, galena, various carbonates, barite, or fluorite, and (5) as in (4) but without the Co-Ni minerals.

Alteration Alters readily to a variety of highly colored secondary uranium compounds (gummite, UO$_3$ · nH$_2$O) of which carnotite is the most important; usually yellow, orange, or green. Auto-oxidation produces UO$_3$, so the mineral formula is U$^{4+}_{1-x}$ U$^{6+}_x$ O$_{2+x}$.

Confused with Columbite.

Variants *Form varieties:* crystallized. Structure is always more or less damaged by radioactive bombardment. Massive (pitchblende). *Chemical varieties:* thorian (bröggerite), cerian, and yttrian. Products of radioactive decay are always present, namely, Pb, He, and Ra. Other common contaminating elements are N, Ar, Fe, Ca, Zr, rare earths, and water.

Related Minerals *Solid solutions:* thorianite, ThO$_2$. *Isostructures:* fluorite. *Other similar species:* baddeleyite, ZrO$_2$; and cerianite, (Ce,Th)O$_2$.

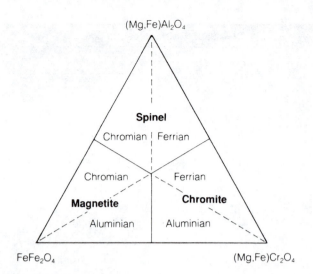

Figure 16.7
Compositional variations in the spinel group.

Table 16.2	Structural Arrangements in Spinels		
Spinel Type	***A***	***B***	**Example**
Normal	8A^{2+}	16B^{3+}	Hercynite, FeAl$_2$O$_4$
Inverse	8A^{3+}	8B^{2+}+8B^{3+}	Magnetite, FeFe$_2$O$_4$ or Fe^{3+}(Fe^{2+}Fe^{3+})O$_4$
Random	Intermediate between normal and inverse		

SPINEL GROUP

The spinel group is an extensive collection of species and series of multiple oxides with a common structural pattern based on alternating metal-oxygen octahedra and tetrahedra. About twenty mineral species and a large number of artificial compounds are known in this group. The cubic unit cell of a normal spinel contains eight formulary units comprising eight divalent metal ions, sixteen trivalent metal ions, and thirty-two oxygen ions.

The structure of spinels is a particularly efficient space-filling scheme of alternating face-sharing octahedra and tetrahedra, as shown in figure 16.7. The anion is usually oxygen, but spinel structures involving sulfur and selenium are also known. The octahedral sites in the many oxide minerals are designated as A sites and occupied by divalent cations whose radii are in the range 0.65–0.80Å. Tetrahedral sites, designated B, contain smaller trivalent cations with radii of 0.50–0.65Å. The cation-anion bond strength (see page 62) is the same for both positions, $3/6 = 0.5$ for octahedral sites and $2/4 = 0.5$ for tetrahedral sites. Structurally, (111) planes of closest packed oxygens alternate with layers of cations in 6-fold coordination and layers with both 4- and 6-fold coordination.

The type formula for a spinel is AB_2O_4, where A is Mg, Fe^{2+}, Mn^{2+}, Zn or similar divalent cations and B is Al, Fe^{3+}, Cr, or V. Two structural arrangements that differ

Table 16.3 **Chemical Variation of the Spinels**

	A^{2+}	B_2^{3+}	Solid Solution[1]
Spinel Series			
Spinel	Mg	Al	**ferroan** (pleonaste), **zincian, chromian**
Hercynite	Fe	Al	**magnesian** (pleonaste), ferrian, **chromian** (picotite)
Gahnite	Zn	Al	**magnesian**, ferroan
Galaxite	(Mn,Fe,Mg)	(Al,Fe)	ferroan, ferrian
Magnetite Series			
Magnetite[2]	Fe	Fe	**magnesian, titanian**
Magnesioferrite	Mg	Fe	**ferroan**
Franklinite	(Zn,Mn,Fe)	(Fe,Mn)	manganoan
Jacobsite	(Mn,Fe,Mg)	(Fe,Mn)	manganian, manganoan
Trevorite	Ni	Fe	magnesian, ferroan
Brunogeirite	(Ge,Fe)	Fe	
Cuprospinel	(Cu,Fe)	Fe	
Chromite Series			
Chromite	Fe	Cr	**magnesian**, ferrian, zincian
Magnesiochromite	Mg	Cr	**ferroan**, aluminian, ferrian
Cochromite	(Co,Ni,Fe)	(Cr,Al)	
Nichromite	(Ni,Co,Fe)	(Cr,Fe,Al)	
Miscellaneous Spinels			
Coulsonite	Fe	V	
Ringwoodite[2]	Si	(Mg,Fe)	
Ulvöspinel[2]	Ti	Fe^{2+}	
Vuorelainenite	(Mn,Fe)	(V,Cr)	

[1]Boldfaced varieties can have complete solid solution.
[2]Inverse spinels.

in the distribution of cations in the *A* and *B* positions are recognized and are called normal and inverse spinels (see table 16.2).

Substitution of *A*-site cations occur readily, and most spinels show mixed occupations of this site. On the other hand, solid solution involving the *B* site is limited, and the spinels can be divided into aluminum, ferric iron, chrome, and miscellaneous spinels on this basis. Figure 16.7 illustrates this division for the three most common spinels together with their more common *B*-site compositional variation. Table 16.3 lists the species and chemical variants of these complex minerals.

SPINEL
$MgAl_2O_4$

spī-nĕl′, from Latin *spina,* thorn; a reference to the sharp angles of its crystals.

Crystallography, Structure, and Habit   Isometric. $4/m$ $\overline{3}\,2/m$. *a* 8.050. $Fd3m$. Octahedral crystals. The common twinning of crystals in class $4/m\,\overline{3}\,2/m$ with $\{111\}$ as the twin plane or $[111]$ as the twin axis is called **spinel twinning** (see figure 5.38).

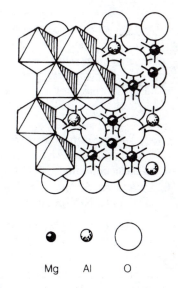

Mg   Al   O

Figure 16.8
Packing and coordination in spinel.

Spinel structure is shown in figure 16.8. Closely-packed edge-sharing MgO_6 octahedra and AlO_4 tetrahedra. Eight formula units per cell.

Usually as crystals; also massive, granular, and compact.

Physical Properties No cleavage. Indistinct {111} parting. Conchoidal to splintery fracture. Brittle. Hardness 7.5–8. Specific gravity 3.6. Vitreous luster. Transparent. Variously colored. White streak. Optical: $n = 1.719$–1.780, increasing with substitution of Mg and Al by other elements. Melting point 2,135° C.

Distinctive Properties and Tests Crystals and hardness. Red spinel turns brown or black and opaque when heated, and on cooling it changes color from black to green to colorless to red.

Association and Occurrence Spinel is a high-temperature mineral found (1) as an accessory in mafic igneous rocks, (2) in alumina-rich schists and gneisses, (3) in contact metamorphosed limestone, and (4) in placer deposits. The usual mineral associates are chondrodite, diopside, chalcopyrite, and pyrite; magnetite, garnet, and vesuvianite; sillimanite, andalusite, and cordierite; forsterite, chondrodite, and scapolite; or plagioclase, corundum, and zircon. Hercynite is mostly found in iron- and aluminum-rich metamorphic rocks and granitic granulites. Gahnite is chiefly found in granitic pegmatites. Galaxite occurs in manganiferous vein deposits with rhodonite, spessartine, and tephroite.

Alteration Resistant to change but may alter to talc, mica, or serpentine.

Confused with Rutile, garnet, zircon, and corundum (variant ruby).

Variants *Form varieties:* Ordinary: gems—ruby spinel, balas ruby (red), rubicelle (yellow to orange-red), almandine spinel (violet), and many blue varieties. *Chemical varieties:* ferroan spinel (pleonaste), ferroan-chromian spinel, chromian hercynite (picotite), manganoan gahnite; also zincian and ferrian.

Related Minerals *Solid solutions:* hercynite, $FeAl_2O_4$; gahnite, $ZnAl_2O_4$; galaxite, $MnAl_2O_4$; zincochromite, $ZnCr_2O_4$; and magnesiochromite, $MgCrO_4$. *Isotypes:* other spinels. *Isostructures:* bornhardite, $CoCo_2Se_4$; linnaeite, $CoCo_2S_4$; polydymite, $NiNi_2S_4$; indite, $FeIn_2S_4$; and greigite, $FeFe_2S_4$.

MAGNETITE
$FeFe_2O_4$

(lodestone, loadstone)
măg'ni-tīt. Natural magnets (lodestones) were known to the ancient Greeks, such as Thales of Miletus (c. 625–547 B.C.) who called this mineral *magnet* because it was found in the lands of the Magnetes (Magnesia) in Thessaly.

Figure 16.9
Magnetite (var. lodestone), Utah, showing its diagnostic magnetic properties. (Courtesy Royal Ontario Museum, Toronto.)

Crystallography, Structure, and Habit Isometric. $4/m \bar{3} 2/m$. *a* 8.394. *Fd3m*. Crystals usually octahedra. Contact and lamellar twins on {111}.

Isostructural with spinel (see figure 16.8). Fe^{2+} occupies the *A* sites in the spinel structure and Fe^{3+} occupies the *B* sites. Eight formula units per cell.

Massive, granular, and as crystals.

Physical Properties No cleavage. Good {111} parting. Subconchoidal to uneven fracture. Brittle. Hardness 5.5–6.5. Specific gravity 5.2. Submetallic luster, splendent to dull. Opaque. Color and streak black. Melting point 1,591° C. Strongly magnetic, sometimes with polarity, as in lodestone (see figure 16.9).

Distinctive Properties and Tests Strongly magnetic; black. Slowly soluble in HCl.

Association and Occurrence Magnetite is one of the more abundant and widespread of oxide minerals. It is found (1) as magmatic segregations in igneous rocks with apatite and pyroxene, (2) as an accessory mineral in all igneous rocks, (3) in metamorphic rocks, especially marble, with garnet, diopside, olivine, pyrite, and chalcopyrite, (4) in high-temperature sulfide veins, and (5) in placers. Magnesioferrite is formed by fumarolic action on magnesian rocks. Franklinite is found with metasomatic zinc ores as at its type locality in Franklin, New Jersey.

Alteration   To hematite (martite) and goethite. From other iron-bearing minerals.

Confused with   Ilmenite and chromite.

Variants   Chemical varieties: magnesian, manganoan, nickelian, aluminian, chromian, titanian, and vanadian.

Related Minerals   Solid solutions: jacobsite, $(Mn,Fe,Mg)(Fe,Mn)_2O_4$; and magnesioferrite, $MgFe_2O_4$; possibly ulvöspinel, $FeTiO_4$; partial with maghemite, $\gamma\text{-}Fe_2O_3$. *Isotypes:* other spinels. *Other similar species:* hausmannite, $MnMn_2O_4$.

CHROMITE
$FeCr_2O_4$

(chromic iron)
krō'mīt, from Greek *chromo,* color, referring to the brilliant red, yellow, or green colors of chromium compounds.

Crystallography, Structure, and Habit   Isometric. $4/m$ $\bar{3}$ $2/m$. *a* 8.36. *Fd3m.* Crystals rare; usually as octahedra.
   Isostructural with spinel (see figure 16.8). Fe in 6-fold and Cr in 4-fold coordination. Eight formula units per cell.
   Usually occurs as discrete grains or granular to compact masses.

Physical Properties   No cleavage. Uneven fracture. Brittle. Hardness 5.5. Specific gravity 4.6. Submetallic luster. Opaque. Color black. Streak brown. Sometimes weakly magnetic.

Distinctive Properties and Tests   Streak and association.

Association and Occurrence   Found with olivine, pyroxene, spinel, chlorite, magnetite, pyrrhotite, or nickeline in ultramafic rocks and their derived serpentines (as an accessory mineral and in segregations) and in placers.

Alteration   To goethite, stichite, $Mg_6Cr_2(OH)_{16}CO_3 \cdot 4H_2O$; maghemite, $\gamma\text{-}Fe_2O_3$; and chlorite.

Confused with   Magnetite and ilmenite.

Variants   Chemical varieties: magnesian (magnesiochromite) ferrian, aluminian, and zincian.

Related Minerals   Solid solutions: magnesiochromite, $MgCr_2O_4$; hercynite, $FeAl_2O_4$. *Isotypes:* other spinels; hercynite above 510° C. *Polymorphs:* donathite.

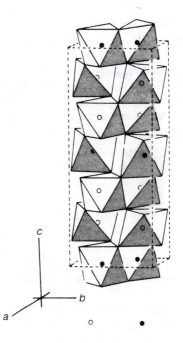

Fe or Mn   Nb or Ta

Figure 16.10
Structure of columbite.

COLUMBITE-TANTALITE SERIES
$(Fe,Mn)(Nb,Ta)_2O_6$

kə-lŭm'bīt, from *columbium,* after Christopher Columbus, an early visitor to the Americas. Columbium is an earlier name for *niobium,* after Niobe, daughter of the ancient Greek king Tantalus, because of the close association of niobium with tantalum.

tăn'tə-līt, from *tantalum,* after the Greek myth of Tantalus.

Crystallography, Structure, and Habit   Orthorhombic. $2/m$ $2/m$ $2/m$. Columbite: *a* 5.74, *b* 14.27, *c* 5.09; tantalite: *a* 5.73, *b* 14.24, *c* 5.08. *Pcan.* Short prismatic or equant crystals. Striated on {100} parallel to *b.* Contact twins common; also penetration twins and stellate forms.
   Structure (see figure 16.10) based on complex edge-sharing chains of $(Fe,Mn)O_6$ and $(Nb,Ta)O_6$ octahedra. Four formula units per cell.
   Crystal aggregates; massive.

Physical Properties   Distinct {010} and rare {100} cleavage. Subconchoidal to uneven fracture. Brittle. Hardness 6–6.5. Specific gravity 5.2 (columbite) to 8.2 (tantalite), increasing linearly with an increase in Ta_2O_5. Submetallic luster. Translucent to transparent. Color black to brownish black, often with an iridescent tarnish. Streak dark red to black. Infusible.

Distinctive Properties and Tests Luster, streak, specific gravity, and association. Insoluble in acids.

Association and Occurrence Found (1) in granitic pegmatites with albite, lithium silicates (such as lepidolite and spodumene), and phosphates (2) in carbonatites, and (3) in placers.

Alteration Resistant to change.

Confused with Uraninite and wolframite.

Variants *Chemical varieties:* ferroan (Fe:Mn > 3:1), manganoan (Fe:Mn < 3:1), magnesian, niobian (ilmenorutile), tantalian (strüverite), and uranian. Small amounts of Sn (for Fe or Mn) and W (for Nb or Ta) may be present, and also Ti, Y, and rare earths.

Related Minerals *Solid solutions:* complete and independent solid solution of Fe-Mn and Nb-Ta in this series:

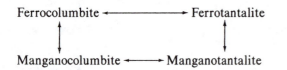

Niobium-rich types can contain significant amounts of Mg. *Isostructures:* fersmite, $(Ca,x)(Nb,Ti)_2(O,OH)_6$ where x = rare earths. *Polymorphs:* tapiolite, $(Fe,Mn)(Ta,Nb)_2O_6$ (tetragonal). *Other similar species:* brookite, TiO_2.

ROMANECHITE
$(Ba,H_2O)(Mn^{4+},Mn^{3+})_5O_{10}$

(psilomelane)
rō-măn′ə-kīt, from its type locality in Romaneche, France.

Crystallography, Structure, and Habit Monoclinic. $2/m$. a 9.56, b 2.88, c 13.85, β 92.5°. $A2/m$. Never shows distinct crystals.

Structure (see figure 16.11) is a grid of chains of edge-sharing and distorted Mn-O tetrahedra. Ba and H_2O occupy holes in the grid and alternate parallel to the b-axis in a double row:

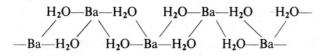

Massive, in colloform crusts, and earthy.

Physical Properties Hardness 5–6. Specific gravity 4.7. Submetallic to dull luster. Opaque. Color black to gray. Shining black streak. Infusible.

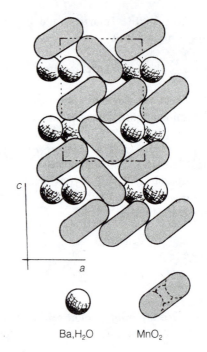

Ba,H₂O   MnO₂

Figure 16.11
Structure of romanechite. Unit cell dashed.

Distinctive Properties and Tests Color, luster, and habit. Chlorine gas evolves when dissolved in HCl.

Association and Occurrence Romanechite is a secondary mineral formed by weathering of other manganese minerals; it is often associated with pyrolusite, goethite, and calcite.

Alteration From manganese carbonates and silicates.

Confused with Pyrolusite.

Variants A number of contaminants, either admixed, adsorbed, or replacing Ba or Mn, may be present. As, V, and W are often present, and small amounts of Co, Cu, Ni, Mg, Ca, and alkali metals are usually found.

Related Minerals *Other similar species:* pyrolusite; manganite, $MnO(OH)$; wad, a mixture of manganese oxides and hydroxides; hollandite, $Ba(Mn^{4+},Mn^{2+})_8O_{16}$; and cryptomelane, $K_2Mn_8O_{16}$.

Hydroxides

Brucite Group	$(Mg,Mn,Ca,Fe,Ni)(OH)_2$
Brucite	
Hydrous Oxides	
Lepidocrocite Group	$(Fe,Al)O(OH)$
Diaspore	
Goethite	
Hydrotalcite and	$(Mg,Ni)_6(Al,Cr,Fe,Mn)_2(CO_3)(OH)_{16}$
Sjögrenite Group	
Gibbsite	

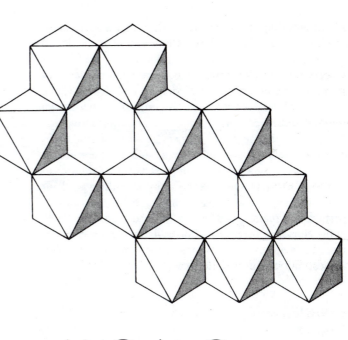

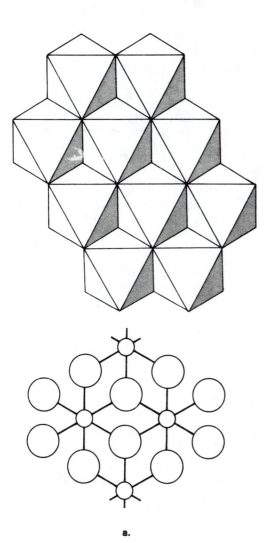

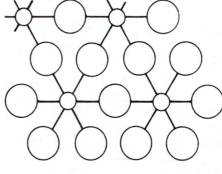

a.

b.

Figure 16.12
Dioctahedral and trioctahedral coordination. (a) Brucite, Mg(OH)$_2$, dioctahedral. (b) Gibbsite, Al(OH)$_3$, trioctahedral. Packing of cation-

OH octahedra above and bond distribution below. Small circles are cations.

The true hydroxides are differentiated from other hydroxyl-bearing minerals by the fact that the hydroxyl radical is a necessary component of the basic structure in which it alternates with cations. On the other hand, in hydroxyl-bearing minerals such as kaolinite, Al$_2$(Si$_2$O$_5$)(OH)$_4$, the hydroxyl groups occupy interstices in the sheets of linked Si-O tetrahedra and are required by the structure only for charge balance.

Several members of the hydroxide family are very common because they represent stages in the weathering of common primary minerals. Goethite and the closely related hydrous iron oxides that are called **limonite** are found nearly everywhere in consequence of the breakdown of iron-bearing minerals. These iron hydroxides may occur in amounts ranging from surface films on rocks up to deposits minable as iron ore. Goethite is also used as a natural yellow or brown pigment described variously as ocher, umber, and sienna.

The oxides and hydroxides of manganese tend to transform one into the other and are sometimes found as complex mixtures. They often contain significant amounts of nonformulary metals adsorbed from the waters in which they form. For example, large amounts of copper and cobalt are found in seafloor manganese nodules. This scavenging action occurs because of the flocculent nature of the precipitating minerals which provides a very large surface area for adsorption.

The surface area of very fine-grained aggregates is impressive, ranging from tens to hundreds of square meters per gram of such materials as chalk, clay, or carbon black.

The structures of brucite and gibbsite are basically layer structures made up of sheets of metal-hydroxyl octahedra, but because of the different cation charges on Mg and Al, the brucite sheet is complete whereas the gibbsite sheet contains holes, as shown in figure 16.12.

Economically, the most important of the hydroxides is gibbsite, which in mixtures with diaspore and colloidal aluminum hydrates constitutes **bauxite,** the ore of aluminum. Gibbsite is formed from any aluminous rock subjected to intense tropical weathering, and deposits take the form of residual blankets, interstratified deposits on unconformities, pockets in limestone, and transported and redeposited sediments. Briefly stated, the conditions for bauxite formation are a mean annual temperature of 25° C, alternating dry and wet seasons with a total rainfall of 2,000 mm, and a topography providing good internal drainage. These conditions are met in present and past tropical regions, and bauxite is found in quantity today in the Northern Territory of Australia, in Indonesia, in the Caribbean and Amazon basins of the Americas, in West Africa, and in southern Europe.

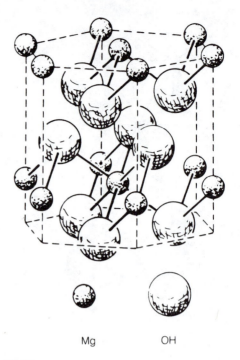

Mg   OH

Figure 16.13
Structure of brucite.

BRUCITE
Mg(OH)₂

(nemalite)
brōō'sīt, named for A. Bruce (1777–1818), a New York mineralogist.

Crystallography, Structure, and Habit Hexagonal. $\bar{3}$ $2/m$. a 3.147, c 4.769, $C\bar{3}m$. Platy crystals.

Brucite structure is shown in figure 16.13. Each layer of octahedra consists of two closest-packed layers of $(OH)^-$ with Mg^{2+} between them in octahedral coordination. The $(OH)^-$ ions are polarized and the octahedral plates are held together by hydrogen and Van der Waal's bonding. One $Mg(OH)_2$ per unit cell.

Found as platy crystals, foliated masses, and fibrous aggregates.

Physical Properties Perfect {0001} cleavage, yielding flexible folia. Sectile. Hardness 2.5. Specific gravity 2.4. Pearly luster on cleavage faces, vitreous elsewhere. Transparent. Color white to pale green, gray, brown, or blue. Streak white. Optical: $\omega = 1.560$ and $\epsilon = 1.580$; positive. Infusible. Water lost at 410° C.

Distinctive Properties and Tests Perfect cleavage and flexible folia. Softness. Soluble in acids. Association.

Association and Occurrence Found in low-temperature veins in Mg-rich metamorphic rocks, such as serpentine, chlorite, and dolomite schists associated with calcite, aragonite, talc, and magnesite.

Alteration From periclase, MgO. To hydromagnesite, $Mg_5(CO_3)_4(OH)_2 \cdot 4H_2O$; serpentine.

Confused with Gypsum and gibbsite.

Variants *Form varieties:* massive, foliated, and fibrous. *Chemical varieties:* ferroan and manganoan. Zn may substitute in minor amounts for Mg.

Related Minerals *Isotypes:* pyrochroite, $Mn(OH)_2$; amakinite, $(Fe,Mg)(OH)_2$; and theophrastite, $Ni(OH)_2$; and portlandite, $Ca(OH)_2$.

DIASPORE
α-AlO(OH)

dī'ə-spōr, from Greek *diaspora*, scattering, referring to the mineral's violent decrepitation on strong heating.

Crystallography, Structure, and Habit Orthorhombic. $2/m$ $2/m$ $2/m$. a 4.401, b 9.421, c 2.845. *Pbnm.* Tiny, platy, bladed, or acicular crystals. Twinning is rare.

Diaspore structure (see figure 16.14) has two octahedra-wide bands of edge-sharing AlO(OH) octahedra parallel to the c-axis. The bands alternate in position along the a direction with successive bands sharing octahedral corners. Four formula units per cell.

Usually in fine-grained aggregates of plates or scales; may be foliated massive, scaly, or stalactitic.

Physical Properties Perfect {010} and fair {110}, {210} or {100} cleavages. Conchoidal fracture. Very brittle.

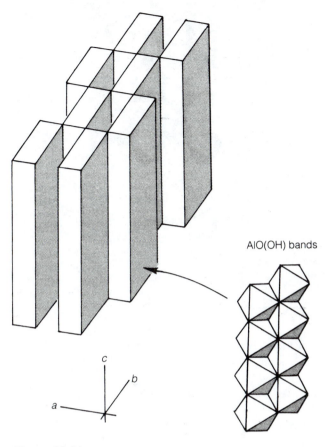

AlO(OH) bands

c
b
a

Figure 16.14
Structure of diaspore.

Hardness 6.5–7. Specific gravity 3.2–3.5. Brilliant to vitreous luster. Transparent. Color white or colorless; sometimes variously tinted. Optical: α = 1.685–1.706, β = 1.70–1.725, and γ = 1.730–1.752; positive; $2V$ = 85°. Infusible; decrepitates violently on strong heating.

Distinctive Properties and Tests   Brittleness, luster, and association. Insoluble in acids.

Association and Occurrence   Found (1) in emery deposits with corundum, magnetite, spinel, and chlorite; (2) in bauxite deposits with gibbsite and böhmite, in lateritic soils, and with aluminous clays; and (3) as a late hydrothermal mineral in alkali pegmatites.

Alteration   From corundum, nepheline. To kaolinite by silicification or to corundum by dehydration.

Confused with   Brucite.

Variants   *Chemical varieties:* ferrian, color brown; manganian, color red. Admixed phosphorus or silica may be present.

Related Minerals   *Isotypes:* goethite. *Polymorphs:* böhmite (orthorhombic). *Other similar species:* bauxite, a mechanical mixture of gibbsite, böhmite, AlO(OH), and diaspore.

GOETHITE
α-FeO(OH)

(bog iron ore, xanthosiderite)
gœ'tĩt, after Johann Wolfgang Goethe (1749–1832), German writer and scientist.

Crystallography, Structure, and Habit   Orthorhombic. $2/m\ 2/m\ 2/m$. a 4.576, b 9.957, c 3.021. *Pbnm*. Striated prisms, tablets, or scales. Acicular to filiform crystals. Isostructural with diaspore (see figure 16.14). Four FeO(OH) per cell. Usually massive, colloform, or stalactitic, showing concentric banding and a radial fibrous structure parallel to c.

Physical Properties   Perfect {010} and fair {100} cleavages. Uneven fracture. Brittle. Hardness 5–5.5. Specific gravity 4.3 (crystals) or 3.3–4.3 (massive material). Submetallic to dull luster; sometimes silky on fibrous kinds and sometimes adamantine in crystals. Translucent to transparent. Color a shade of brown, often color banded perpendicular to the fibers in radiating colloform masses. Streak brownish yellow. Optical: negative.

Distinctive Properties and Tests   Habit and streak. Soluble in HCl, giving a yellowish solution. Evolves water in the closed tube.

Association and Occurrence   One of the most common minerals. Occurs as the oxidation product of Fe-bearing minerals and is found concentrated in sedimentary beds, gossans, and lateritic soils. It is commonly associated with hematite, pyrolusite, romanechite, calcite, quartz, clay minerals, and limonite.

Alteration   From Fe-minerals, especially siderite, pyrite, and magnetite.

Confused with   Hematite.

Variants   *Form varieties:* bog iron ore; limonite (ocherous and poorly crystallized); and onegite, as yellow tufts or needles in quartz. Usually contains Mn and has adsorbed water, as in turgite, $2Fe_2O_3 \cdot nH_2O$.

Related Minerals   *Isotypes:* diaspore. *Polymorphs:* lepidocrocite, γ-FeO(OH)(orthorhombic); akaganeite (tetragonal); and feroxyhyte (hexagonal). *Other similar species:* manganite, MnO(OH); heterogenite CoO(OH); and montroseite, (V,Fe)O(OH).

GIBBSITE
γ-Al(OH)₃

(hydrargillite)
gĭb'zīt, after Colonel George Gibbs (1777–1834), American mineral collector.

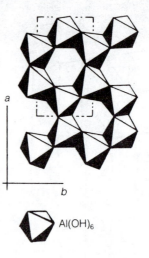

Figure 16.15
Structure of gibbsite. Unit cell dashed.

Crystallography, Structure, and Habit Monoclinic and pseudohexagonal. $2/m$. a 9.719, b 5.071, c 8.641, β 94.6°. $P2_1/n$. Rare tabular and prismatic crystals. Often nodular. Contact twins common on (001), also parallel twins on (130).

In gibbsite structure (see figure 16.15) each sheet is made up of two close-packed (OH)⁻ layers with two-thirds of the interstices occupied by Al^{3+} in 6-fold coordination. Eight formula units per cell.

Usually tiny crystals with pseudohexagonal outline or crystal aggregates; also colloform with radiating structure, encrusting, earthy, compact, and enamel-like.

Physical Properties Perfect {001} cleavage. Tough. Hardness 2.5–3.5. Specific gravity 2.4. Pearly luster on cleavage surfaces, otherwise vitreous. Transparent. Color white but often colored by impurities. Optical: $\alpha = 1.56$, $\beta = 1.56$, and $\gamma = 1.58$; positive; $2V = 0\text{-}40°$ (usually $< 20°$). Strong, earthy smell when breathed on. Infusible. Not readily soluble in acids.

Distinctive Properties and Tests Cleavage, earthy smell, and yields water in the closed tube; often in nodules.

Association and Occurrence Found with diaspore and böhmite in bauxite deposits and with corundum in talc schists.

Alteration From aluminous minerals. To kaolinite by silicification.

Confused with Kaolin and brucite.

Variants May contain a little Fe^{3+}.

Related Minerals *Polymorphs:* bayerite (monoclinic), doyleite, and nordstrandite (triclinic). *Other similar species:* bauxite, a mechanical mixture of gibbsite, böhmite, and diaspore.

17

Carbonates

Carbonates

Calcite Group $(Ca,Mg,Fe,Co,Zn,Cd,Mn)(CO_3)$
   Calcite
   Magnesite
   Siderite
   Rhodochrosite
   Smithsonite
Aragonite Group $(Ca,Ba,Sr,Pb)(CO_3)$
   Aragonite
   Witherite
   Strontianite
   Cerussite
Dolomite Group $Ca(Mg,Fe,Mn)(CO_3)_2$
   Dolomite
   Ankerite
Carbonates with Hydroxyl or
   Halogen
     Malachite
     Azurite
Miscellaneous Species

Table 17.1 Distinctive Tests for the Carbonate Minerals

Mineral	Color	Reaction with Cold, Dilute HCl	Flame Color
Calcite	Various	Effervesces	Brick-red
Magnesite	White	No action	None
Siderite	Brown	No action	None
Rhodochrosite	Pink	No action	None
Smithsonite	White	Effervesces	Blue-green streaks
Aragonite	White	Effervesces	Brick-red
Witherite	White	Effervesces	Yellowish green
Strontianite	White	Effervesces	Crimson
Cerussite	White	No action	Pale bluish white
Dolomite	White	No action	Brick-red
Malachite	Green	Effervesces	Green
Azurite	Blue	Effervesces	Green

The carbonate minerals are easily distinguished from other mineral classes by their softness, distinctive colors, and effervescence in warm, dilute HCl. About seventy species are known, with calcite, dolomite, and siderite being by far the most abundant.

Members of the calcite group are usually well crystallized, have a blocky habit, and cleave into rhomb-shaped fragments (with the exception of smithsonite, which is usually colloform). Members of the aragonite group have an elongated or massive habit. The hydrous carbonates have distinctive colors and a colloform habit.

Simple distinguishing tests for the various carbonate minerals are given in table 17.1. The colors of siderite, rhodochrosite, malachite, and azurite are essentially constant. Colors of the other carbonates are usually white, but they may take on a wide range of tints. Reactions with cold HCl are quite distinctive if the sample is coarse-grained. Flame colors are distinctive.

Most natural carbonate salts are isostructural with either calcite or aragonite. In general, the calcite arrangement is used by small cations such as Li, Na, Mg, Ca, Fe, Zn, Co, or Cd, whereas the aragonite structure better accommodates large cations such as K, Ba, Sr, and Pb. Polymorphs using the two structures are found when the cation-oxygen radius ratio is near the critical value of 0.73, as in polymorphs of $Ca(CO_3)$.

All the members of the calcite group are isotypic and show considerable solid solution, as shown in figure 17.1. Minerals in the aragonite group are also all isotypic, but solid solution between species is limited.

Many of the carbonates are minor ores of various metals. Others are used for various special purposes, such as the manufacture of magnesia brick from magnesite or red pyrotechnics from strontianite.

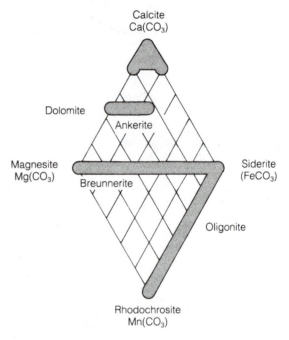

Figure 17.1
Solid solution relations (shaded) in the calcite group.

The sedimentary rocks limestone and dolostone and their metamorphosed equivalent, marble, are composed principally of the rock-forming minerals calcite and dolomite. These calcareous rocks, found in nearly every nation of the world, represent a major industrial mineral resource that is exploited for a variety of end uses, including agricultural applications, construction uses, and chemical and industrial uses in metallurgy, chemistry, sanitation, pulp and paper processing, ceramics, foods, and petroleum technology. The dominant uses are in quicklime (in mortar and plaster) and as a principal ingredient in the manufacture of portland cement.

Quicklime has been known since ancient times; the Egyptian pyramids, for example, were plastered with quicklime and medieval castles were mortared with it.

Quicklime is produced by heating (burning) a calcareous rock in the absence of oxygen in a lime kiln. At about 900° C, calcite dissociates into CaO (quicklime) and CO_2 (gas). Addition of water to quicklime causes it to hydrate or slake (see page 97) into a cementitious substance that will set hard on drying.

Apart from its applications in plasters and mortars, hydrated limes are the feedstock in many chemical processes and are widely used for both soil conditioning and stabilization. Calcium ions have the ability to displace H^+, Na^+, and K^+ from clay particle surfaces and in the presence of CO_2 (abundant in many soils due to bacterial fermentation) to form a noncrystalline calcium silicate cement, thus stabilizing the soil for engineering purposes.

The largest single use of limestone is in the manufacture of portland cement. Like plaster and mortar, cement has been known since ancient times. For example, it was widely used by Roman builders. Before the nineteenth century, however, cement as the binding material for concrete was derived from fortuitous mixtures of rock materials, such as those found as volcanic ash in Pozzuoli, Italy. In England in 1824, Joseph Aspdin succeeded in formulating a mixture whose quality could be controlled. He termed this portland cement from the similarity of its appearance to portland stone, a limestone widely used for building in the British Isles. As used in concrete, portland cement is the binder for the fine aggregate that fills the interstices of the coarse aggregate.

CALCITE
Ca(CO₃)

(calc spar, Iceland spar)
kăl'sīt, from Latin *calx,* lime.

Crystallography, Structure, and Habit Hexagonal. $\bar{3}\ 2/m.$ *a* 4.990, *c* 17.064. $R\bar{3}c.$ Calcite crystals show a greater variety of forms than any other mineral; at least 328 and possibly over 600 have been described. The four most common, however, are (1) prismatic with various terminations; (2) scalenohedral, often complex (e.g., dogtooth spar); (3) rhombohedral, both flat to very acute; and (4) tabular with a prominent basal face (see figures 17.2, 17.3, and 17.4). Repeated parallel twinning on $\{01\bar{1}2\}$ (see figure 5.34) is common; twinning also on $\{0001\}$.

Calcite structure resembles that of halite with strongly covalent planar $(CO_3)^{2-}$ groups taking the place of Cl ions. The CO_3 groups are arranged in parallel position normal to the 3-fold axis, [0001], which corresponds to a cube diagonal (see figure 17.5). Six $Ca(CO_3)$ per cell.

Crystals and crystal aggregates, coarse granular to impalpable, stalactitic, nodular, encrusting; also a component of shells.

Physical Properties Perfect rhombohedral cleavage on $\{10\bar{1}1\}$ (see figure 9.4). Parting on $\{01\bar{1}2\}$ and sometimes

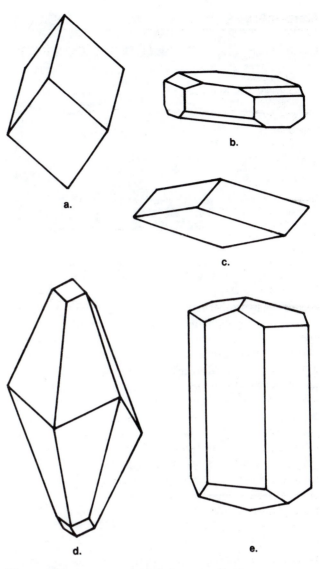

Figure 17.2
Common forms of calcite. (a) Acute rhombohedral. (b) Tabular. (c) Flat rhombohedral. (d) Scalenohedral (dog-tooth spar). (e) Prismatic.

on $\{0001\}$. Conchoidal fracture. Brittle. Hardness 3, approximately 2.7 on (0001) and 3.5 on $(10\bar{1}1)$. Specific gravity 2.7. Vitreous luster. Transparent. Colorless or white but may assume various colors; Mn varieties tend to be pink, Fe varieties brown or yellow, and Co varieties rosyred. White streak. Optical: $\omega = 1.658$, $\epsilon = 1.486$; negative. Sometimes fluorescent or phosphorescent under electromagnetic radiation (x rays, ultraviolet, sunlight). Often thermoluminescent. Infusible.

Distinctive Properties and Tests Cleavage, crystals, hardness. Soluble in cold dilute HCl with effervescence.

Association and Occurrence One of the more common and widely distributed of all minerals. It is the principal constituent of certain sedimentary rocks (limestone and chalk) and their metamorphic equivalents (marble). Also

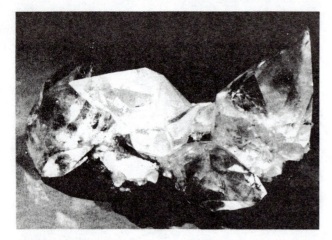

Figure 17.3
Twinned scalenohedra of calcite. Elmwood Mine, Carthage, Tennessee. (Courtesy Royal Ontario Museum, Toronto.)

occurs as a hydrothermal vein mineral with sulfides, barite, quartz, fluorite, dolomite, and siderite; and as a primary constituent of alkali-rich, silica-poor igneous rocks along with nepheline, orthoclase, and Nb-Ta minerals. Calcite is also formed by carbonation during weathering of most rocks, and is used by organisms in the production of shells.

Alteration   To solution. From calcium-bearing minerals.

Confused with   Dolomite, aragonite.

Next to quartz, calcite is the most common mineral in surficial rocks. The name was applied in its present usage by W. K. Haidinger in 1845, but the mineral had long been known and its study forms a significant body of mineralogical knowledge. Calcite was mentioned by Pliny around A.D. 77. Erasmus Bartholinus described its perfect cleavage and strong double refraction in 1669. Christian Huygens led a study of calcite to discover the laws of double refraction (1678). R. J. Hauy in 1781–1801 developed a theory of crystal structure by examining its cleavage. E. L. Malus discovered the polarization of light using Iceland spar in 1808.

Variants   *Form varieties:* ordinary, fibrous, lamellar, concretionary, oolitic, chalky, fetid (contains H_2S which is liberated when broken), Iceland spar (clear crystalline), nail-head spar (flat rhomb and prism), dogtooth spar (steep rhomb and prism), and numerous others. *Chemical varieties:* manganoan, ferroan, zincian, cobaltian, plumbian, barian, strontian, and magnesian.

Related Minerals   *Solid solutions:* extensive solid solution is known between calcite and rhodocrosite; limited solid solution with other members of the calcite group (see figure 17.1). *Isotypes:* magnesite; siderite; sphaerocobaltite, $Co(CO_3)$; smithsonite; nitratine (soda niter), $Na(NO_3)$; dolomite; gaspéite, $(Ni,Mg,Fe)(CO_3)$; and otavite, $Cd(CO_3)$. *Polymorphs:* aragonite, vaterite.

Figure 17.4
Prismatic crystals of calcite. Sweetwater Mine, Reynolds County, Missouri. (Courtesy University of Windsor. Photograph by D. L. MacDonald.)

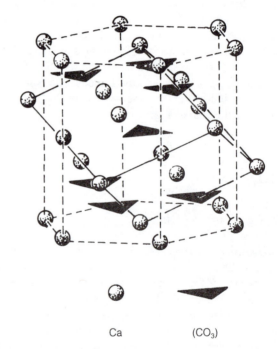

Ca (CO_3)

Figure 17.5
Structure of calcite showing relation between hexagonal (dashed lines) and rhombohedral (solid lines) symmetry.

MAGNESITE
$Mg(CO_3)$

măg′nə-sīt, from Greek *magnesia,* a kind of ore from Magnesia, an ancient city in Asia Minor.

Crystallography, Structure, and Habit   Hexagonal. $\bar{3}\ 2/m.\ a$ 4.633, c 15.016. $R\bar{3}c$. Crystals rare.
Isostructural with calcite. Six $Mg(CO_3)$ per cell.

Usually massive; sometimes finely granular, compact, earthy, and fibrous. Often porcelainlike because of admixed opal.

Physical Properties Perfect rhombohedral cleavage $\{10\bar{1}1\}$. Conchoidal fracture. Brittle. Hardness 4. Specific gravity 3.0. Vitreous to earthy luster. Translucent to transparent. Color white or colorless. Streak white. Optical: $\omega = 1.700$ and $\epsilon = 1.509$; negative. Sometimes shows green or blue fluorescence or phosphorescence in ultraviolet light. Triboluminescent. Infusible.

Distinctive Properties and Tests Chalky or porcelainlike appearance. Effervesces in warm HCl.

Association and Occurrence Massive magnesite forms by alteration of Mg-rich rocks during metamorphism. Limestone and dolomite may be replaced by $Mg(CO_3)$. Rarely found as a hydrothermal vein mineral, as an accessory in igneous rocks, and filling the cavities of lava.

Alteration From Mg-rich rocks such as serpentine, dunite, or peridotite by carbonation. Replaces calcite and dolomite.

Confused with Calcite, dolomite, and kaolin.

Variants *Chemical varieties:* ferroan (breunnerite), manganoan, calcian, nickeloan (hoshiite). May contain minor amounts of Zn.

Related Minerals *Solid solutions:* siderite and gaspéite, $(Ni,Mg,Fe)(CO_3)$. *Isotypes:* see calcite. *Other similar species:* hydromagnesite, $Mg_5(CO_3)_4(OH)_2 \cdot 4H_2O$.

SIDERITE
$Fe(CO_3)$

(spathic iron ore, chalybdite)
sid′ə-rīt, from Greek *sideros,* iron.

Crystallography, Structure, and Habit Hexagonal. $\bar{3}\ 2/m$. a 4.688, c 15.373. $R\bar{3}c$. Usually rhombohedral crystals, also tablets, prisms, and scalenohedra. Faces often curved or composite. Parallel repeated twinning on $\{01\bar{1}2\}$.

Isostructural with calcite. Six $Fe(CO_3)$ per cell.

Fine- to coarse-grained anhedral aggregates; rhombohedral crystals, often with curved faces; colloform, fibrous, oolitic, and earthy.

Physical Properties Perfect rhombohedral cleavage $\{10\bar{1}1\}$. Uneven to conchoidal fracture. Brittle. Hardness 4–4.5. Specific gravity 3.9. Subadamantine luster. Translucent to transparent. Color yellow-brown or grayish brown to brown and red-brown. Streak white. Optical: $\omega = 1.875$ and $\epsilon = 1.635$; negative. Fusible (4–4.5).

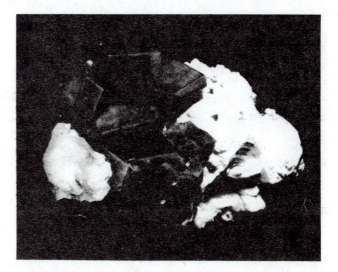

Figure 17.6
Rhombohedral crystals of rhodocrosite. Mount St.-Hilaire, Quebec. (Courtesy Royal Ontario Museum, Toronto.)

Distinctive Properties and Tests Color and cleavage. Specific gravity. Blackens and becomes magnetic when heated in a closed tube. Effervesces in warm HCl.

Association and Occurrence Siderite is a common mineral found usually (1) in bedded deposits with clay, or shale, coal, and quartz; (2) as a constituent of some hydrothermal veins; (3) in bogs and less commonly in lava cavities, carbonatites, pegmatites, and metamorphosed sedimentary rocks; and (4) as a replacement of limestone.

Alteration To goethite.

Confused with Sphalerite.

Variants *Form varieties:* spherosiderite (spherules in "clay ironstone"). *Chemical varieties:* calcian, magnesian, manganoan, oligonite, zincian, and cobaltian.

Related Minerals *Solid solutions:* rhodochrosite and magnesite. *Isotypes:* see calcite.

RHODOCHROSITE
$Mn(CO_3)$

(dialogite)
rō′də-krō′sīt, from Greek, *rhodochros,* rose-colored.

Crystallography, Structure, and Habit Hexagonal. $\bar{3}\ 2/m$. a 4.777, c 15.664. $R\bar{3}c$. Crystals are rare, usually rhombohedral (see figure 17.6). Rare lamellar twins on $\{01\bar{1}2\}$.

Isostructural with calcite. Six $Mn(CO_3)$ per cell.

Usually massive; also granular, compact, botryoidal, and columnar.

Physical Properties Perfect rhombohedral cleavage $\{10\bar{1}1\}$. Parting parallel to $\{10\bar{1}2\}$. Uneven to conchoidal fracture. Brittle. Hardness 3.5–4. Specific gravity 3.3–3.7. Vitreous to pearly luster. Subtranslucent to transparent. Color is a shade of pink, brown, or brownish yellow. Streak white. Optical: $\omega = 1.816$, $\epsilon = 1.597$; negative. Infusible, but dissociates at 300° C.

Distinctive Properties and Tests Color. Soluble with effervescence in warm HCl. Alteration to black or brown manganese oxides.

Association and Occurrence Found (1) as a primary gangue mineral in sulfide veins and replacement bodies, where it is also associated with such metalliferous minerals as tetrahedrite and Ag sulfosalts and the gangue minerals calcite, siderite, dolomite, fluorite, barite, quartz, and manganese oxides; (2) in high-temperature metamorphic deposits with rhodonite, garnet, and manganese oxides; and (3) as a secondary mineral in residual deposits of iron or manganese oxides.

Alteration To manganese oxides, such as pyrolusite, which changes the color to brown or black in the early stages of alteration.

Confused with Rhodonite.

Variants *Chemical varieties:* calcian, ferroan, ferroanzincian, magnesian, zincian, cobaltian, and cadmian.

Related Minerals *Solid solutions:* siderite, calcite, and kutnohorite, $CaMn(CO_3)_2$. *Isotypes:* see calcite.

SMITHSONITE
$Zn(CO_3)$

(dry-bone ore)
smith′sə-nīt, after James Smithson (1765–1829), benefactor of the Smithsonian Institution.

Crystallography, Structure, and Habit Hexagonal. $\bar{3}\ 2/m$. a 4.653, c 15.025. $R\bar{3}c$. Crystals rare; rhombohedral.

Isostructural with calcite. Six $Zn(CO_3)$ per cell.

Usually colloform; also stalactitic and earthy, often as secondary encrustations.

Physical Properties Good $\{10\bar{1}1\}$ cleavage. Uneven to subconchoidal fracture. Brittle. Hardness 4–4.5. Specific gravity 4.4. Subadamantine luster. Translucent. Color varies: grayish white, gray, greenish white, brownish white, green, blue-green, yellow, brown, or white. Streak white. Optical: $\omega = 1.850$, $\epsilon = 1.625$; negative. Infusible.

Distinctive Properties and Tests Habit, often color, and association. Soluble in cold HCl with effervescence.

Figure 17.7
Prismatic crystals of aragonite. Cyclic twins on $\{110\}$ result in a pseudohexagonal cross-section. Morocco. (Courtesy University of Windsor. Photograph by D. L. MacDonald.)

Association and Occurrence Found as a secondary mineral in the oxidized zone of ore deposits containing sphalerite or replacing adjacent calcareous rocks. Usually associated with hemimorphite, cerussite, malachite, azurite, or anglesite.

Alteration From sphalerite, to hemimorphite.

Confused with Hemimorphite.

Variants *Form varieties:* dry-bone ore (encrusting or honeycombed masses) and turkey-fat ore (yellow, cadmian). *Chemical varieties:* ferroan, calcian, cobaltian (pink), cuprian (green), cadmian (yellow), magnesian, and manganoan. May also contain small amounts of Ge or Pb.

Related Minerals *Solid solutions:* otavite, $Cd(CO_3)$. *Isotypes:* see calcite.

ARAGONITE
$Ca(CO_3)$

ə-rǎg′ə-nīt, after Aragon, the province in Spain where it was discovered as reddish twinned crystals.

Crystallography, Structure, and Habit Orthorhombic. $2/m\ 2/m\ 2/m$. a 5.741, b 7.968, c 4.959. *Pmcn*. Crystals acicular, prismatic, or tabular (see figure 17.7). Twinning common as contact or cyclic twins on $\{110\}$ sometimes giving a pseudohexagonal appearance.

Aragonite structure (see figure 17.8) composed of CaO_9 edge-sharing polyhedra and CO_3 triangular radicals. Each CaO_9 polyhedron is surrounded by six CO_3 groups, three of which share edges and three corners with the polyhedron. Each polyhedron shares six edges with adjoining polyhedra. Four $Ca(CO_3)$ per cell.

Found as columnar and acicular crystals and crystal aggregates; also colloform, fibrous, and pisolitic.

Physical Properties Imperfect $\{010\}$ and $\{110\}$ cleavage. Subconchoidal fracture. Brittle. Hardness 3.5–4. Specific

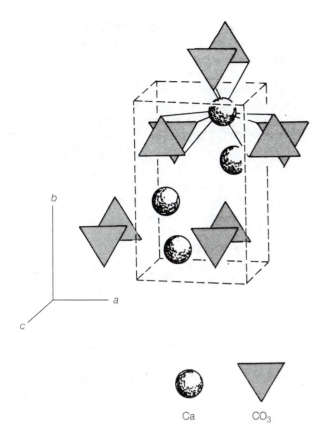

Ca   CO₃

Figure 17.8
Structure of aragonite. Unit cell dashed. 9-fold coordination of one
calcium ion shown. (After Smyth and Bish, 1988.)

gravity 2.9. Vitreous to resinous luster. Transparent to
translucent. Colorless to white. Streak white. Optical:
$\alpha = 1.530$, $\beta = 1.680$, and $\gamma = 1.685$; negative; $2V =$
18°. Sometimes fluorescent under x rays or ultraviolet
light. Sometimes thermoluminescent. Infusible, but
decrepitates at red heat as it transforms to calcite.

Distinctive Properties and Tests Crystals. Effervesces
with cold dilute HCl. Stained lilac by hot cobalt-nitrate
solution.

Association and Occurrence Found (1) as crystals, pi-
solites, and sinter from hot springs or on recent sea bot-
toms; (2) as disseminated crystals and masses in gypsum
or clay beds; (3) with sedimentary limonite and siderite;
(4) in the oxidized zone of ore deposits with limonite, calcite, malachite, azurite, smithsonite, or cerussite; (5) as
the constituent of certain shells; (6) in cave deposits; and
(7) in deep-seated, low-temperature metamorphic rocks
with lawsonite, glaucophane, and jadeite.

Alteration To calcite.

Confused with Strontianite and calcite. Aragonite can
be distinguished from calcite by the following stain tests.
Boil fragments in cobalt-nitrate solution and wash. Aragonite assumes a lilac-rose tint more rapidly than does

calcite. Wash with ammonium-sulfide solution. Aragonite
turns black, calcite gray. These tests are not reliable for
very fine-grained material.

Variants *Form varieties:* flosferri, with intertwined
slender branches. *Chemical varieties:* plumbian, strontian, and zincian.

Related Minerals *Isostructures:* strontianite, witherite, cerussite, and niter, KNO₃. *Isotypes:* alstonite,
CaBa(CO₃)₂. *Polymorphs:* calcite, vaterite.

WITHERITE
Ba(CO₃)

with′ə-rīt, after William Withering, English physician and
naturalist, who in 1784 showed this mineral to be chemically distinct from barite.

Crystallography, Structure, and Habit Orthorhombic.
$2/m\ 2/m\ 2/m$. a 6.430, b 8.904, c 5.314. *Pmcn*. Crystals
always twinned on {110}, forming pseudohexagonal pyramids. Deep, horizontal striations sometimes make the
crystals look like a stack of pyramids.
   Isostructural with aragonite. Four Ba(CO₃) per cell.
As crystals, globular, colloform, columnar, or granular.

Physical Properties Distinct {010} and imperfect {110}
cleavage. Uneven fracture. Hardness 3–3.5. Specific
gravity 4.3. Vitreous to resinous luster. Transparent to
translucent. Colorless to white, often with tints of yellow,
brown, lavender, or green. Streak white. Optical: $\alpha =$
1.529, $\beta = 1.676$, and $\gamma = 1.677$; negative; $2V = 16°$.
May fluoresce or phosphoresce in x rays or ultraviolet light.
May be thermoluminescent. Fusible (2.5–3).

Distinctive Properties and Tests Specific gravity and
crystals. Soluble with effervescence in cold dilute HCl.
Flame test for barium. Even very cold dilute solutions give
a precipitate of barium sulfate when sulfuric acid is added.

Association and Occurrence Found in low-temperature
hydrothermal veins associated with barite and galena.

Alteration To and from barite.

Confused with Barite.

Variants Usually contains some Ca, Sr, or Mg.

Related Minerals *Isotypes:* alstonite, CaBa(CO₃)₂.
Polymorphs: transforms at 811° C to a hexagonal form
and at 982° C to an isometric form.

STRONTIANITE
Sr(CO₃)

stron'chē-ə-nīt, after Strontian, a village in west central Scotland where it has been known since at least 1764; not recognized as a mineral species until 1787 when William Cruikshank first detected the element strontium in the mineral.

Crystallography, Structure, and Habit   Orthorhombic. $2/m\ 2/m\ 2/m$. a 6.029, b 8.414, c 5.107. *Pmcn*. Short to long prismatic crystals; often acicular. Commonly twinned on {110} into pseudohexagonal or lamellar forms.

Isostructural with aragonite. Four $Sr(CO_3)$ per cell.

Found as pseudohexagonal prismatic crystals; massive and columnar to fibrous.

Physical Properties   Perfect {110} cleavage. Uneven to subconchoidal fracture. Brittle. Hardness 3.5. Specific gravity 3.8. Vitreous to resinous luster. Transparent to translucent. Colorless to gray with tints of yellow, green, or red. Streak white. Optical: $\alpha = 1.519$, $\beta = 1.665$, and $\gamma = 1.667$; negative; $2V = 8°$. May fluoresce or phosphoresce under x rays or ultraviolet light. Sometimes thermoluminescent. Fuses at 1,497° C.

Distinctive Properties and Tests   Crystals. Flame test for strontium. Swells and sprouts when strongly heated. Effervesces in cold dilute HCl. From aragonite by {110} cleavage.

Association and Occurrence   Strontianite is a low-temperature hydrothermal mineral associated with barite, celestine, and calcite (1) in veins in limestone and marl; (2) in sulfide veins, and; (3) in geodes or as concretionary masses in limestone or clay beds.

Alteration   To and from celestine.

Confused with   Aragonite.

Variants   *Chemical varieties:* calcian and barian. May also contain Pb.

Related Minerals   *Isotypes:* see aragonite. *Polymorphs:* transforms to a hexagonal form at 929° C.

CERUSSITE
Pb(CO₃)

sə-roōs'īt, from Latin *cerussa*, white lead.

Crystallography, Structure, and Habit   Orthorhombic. $2/m\ 2/m\ 2/m$. a 6.152, b 8.436, c 5.195. *Pmcn*. Crystals varied but often tabular. Almost always twinned on {110} into pseudohexagonal forms.

Isostructural with aragonite. Four $Pb(CO_3)$ per cell.

Found as crystal clusters; massive, granular to compact, and earthy.

Physical Properties   Distinct {110} and fair {021} cleavage. Conchoidal fracture. Very brittle. Hardness 3–3.5. Specific gravity 6.6. Subadamantine; resinous or pearly luster. Often chatoyant. Transparent to translucent. Colorless to white, gray, blue, or green. Streak white. Optical: $\alpha = 1.303$, $\beta = 2.074$, and $\gamma = 2.076$; negative; $2V = 8°$. May be fluorescent (yellow) under x rays or ultraviolet light. Readily fusible (1.5) and yields a lead globule when heated on charcoal.

Distinctive Properties and Tests   Specific gravity, luster, and ready fusion. Soluble with effervescence in dilute HNO_3; slowly soluble in HCl.

Association and Occurrence   Cerussite is a common secondary lead mineral found in the oxidized portion of ore deposits, where it is associated with anglesite or iron oxides.

Alteration   From galena and anglesite.

Confused with   Anglesite.

Variants   May contain trace amounts of Sr, Ag, or Zn.

Related Minerals   *Isotypes:* see aragonite. *Other similar species:* hydrocerussite, $Pb_3(CO_3)_2(OH)_2$.

DOLOMITE
CaMg(CO₃)₂

(pearl spar)
dō'lə-mīt, in honor of Deodat de Dolomieu (1750–1801), French geologist.

Crystallography, Structure, and Habit   Hexagonal. $\bar{3}$. a 4.808, c 16.010. $R\bar{3}$. Crystals usually rhombohedra or prisms, often with curved faces. Twinning on {0001} and {10$\bar{1}$1}. Glide twinning on {02$\bar{2}$1}.

Isostructural with calcite. The Ca and Mg atoms of dolomite alternate in the Ca sites of calcite. Three formula units per cell.

Fine- to coarse-grained anhedral cleavage aggregates and as rhombohedral crystals.

Physical Properties   Perfect rhombohedral cleavage {10$\bar{1}$1}. Parting parallel to {02$\bar{2}$1}. Subconchoidal fracture. Brittle. Hardness 3.5–4. Specific gravity 2.9, Vitreous to pearly luster. Translucent. Colorless to white, brown, or pink. Streak white. Optical: $\omega = 1.679$ and $\epsilon = 1.500$; negative. May fluoresce under ultraviolet light. May be thermoluminescent. Infusible.

Distinctive Properties and Tests Cleavage. Effervesces in hot dilute HCl. Poorly soluble in cold dilute HCl unless freshly powdered.

Association and Occurrence Found (1) as massive sedimentary rock, often mixed with calcite; (2) in hydrothermal veins with fluorite, barite, calcite, siderite, quartz, and metallic ores; (3) in cavities in limestone with calcite, celestine, gypsum, and quartz; and (4) as grains in serpentine and talcose rocks.

Alteration From calcite by action of Mg-rich solutions. Under thermal metamorphism, dolomite may break down into calcite and periclase, MgO, or combine with silica to form diopside or other Ca-Mg silicates.

Confused with Calcite.

Variants *Chemical varieties:* cobaltian (reddish), plumbian, manganoan, ferroan (brown), calcian, magnesian, zincian, and cerian.

Related Minerals *Solid solutions:* ankerite, $Ca(Fe,Mg)(CO_3)_2$; kutnohorite, $(CaMn)(CO_3)_2$; minrecordite, $CaZn(CO_3)_2$; and norsethite, $BaMg(CO_3)_2$. *Isostructures:* nordenskiöldine, $CaSnB_2O_6$. *Other similar species:* calcite; alstonite, $BaCa(CO_3)_2$; paralstonite, $BaCa(CO_3)_2$; barytocalcite, $BaCa(CO_3)_2$; and huntite, $CaMg_3(CO_3)_4$.

MALACHITE
$Cu_2(CO_3)(OH)_2$

(green copper carbonate)
măl′ə-kīt, from Greek *moloche,* a mallow, for its resemblance in color to mallow leaves.

Crystallography, Structure, and Habit Monoclinic. $2/m$. a 9.502, b 11.974, c 3.204, β 98.8°. $P\,2_1/a$. Crystals rare; prismatic acicular or rounded. Commonly contact twinned on {001}.

In the structure (see figure 17.9) each CO_3 triangle is surrounded by three $Cu(O,OH)_6$ octahedral pairs to which it contributes the oxygen. Four formula units per cell.

Massive, encrusting, and colloform (see figure 17.10).

Physical Properties Perfect {$\overline{2}$01} and distinct {010} cleavages. Subconchoidal to uneven fracture. Hardness 3.5–4. Specific gravity 4. Subadamantine luster. Translucent to opaque. Bright green in color. Streak pale green. Optical: $\alpha = 1.655$, $\beta = 1.875$, and $\gamma = 1.909$; negative; $2V = 43°$. Fusible (3). Loses water at 315° C.

Distinctive Properties and Tests Habit and color. Soluble with effervescence in cold dilute HCl. Evolves much water in the closed tube.

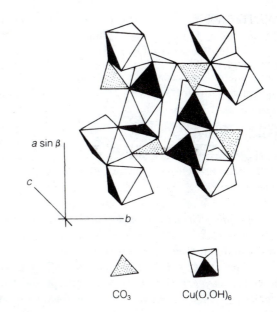

Figure 17.9
Structure of malachite.

Figure 17.10
Colloform malachite. Shaba Province, Zaire. (Courtesy University of Windsor. Photograph by D. L. MacDonald.)

Association and Occurrence Malachite is a secondary copper mineral formed by oxidation and carbonation. It is usually associated with azurite, cuprite, limonite, calcite, chalcedony, and chrysocolla.

Alteration From azurite, cuprite.

Confused with Garnierite, melanterite, chrysocolla, and other secondary copper minerals.

Variants Zincian, cobaltian.

Related Minerals *Similar species:* azurite; phosgenite, $Pb_2(CO_3)Cl_2$; hydrozincite, $Zn_5(CO_3)_2(OH)_6$; aurichalcite, $(Zn,Cu)_5(CO_3)_2(OH)_6$; glaukosphaerite, $(Co,Ni)_2$

$(CO_3)(OH)_2$; kolwezite, $(Cu,Co)_2(CO_3)(OH)_2$; mcguinessite, $(Mg,Cu)_2(CO_3)(OH)_2$; nullaginite, $Ni_2(CO_3)(OH)_2$; rosasite, $(Cu,Zn)_2(CO_3)(OH)_2$; and zincrosasite, $(Zn,Cu)_2(CO_3)(OH)_2$.

AZURITE
$Cu_3(CO_3)_2(OH)_2$

(blue copper carbonate, chessylite)
azh'ə-rīt, from French *azur,* azure, referring to its sky-blue color.

Crystallography, Structure, and Habit Monoclinic. $2/m.$ a 5.008, b 5.844, c 10.336, β 92.5°. $P\,2_1/c.$ Crystals varied in habit and often distorted. Twinning is rare.

In the structure (see figure 17.11), each Cu ion coordinates two oxygens of adjacent CO_3 groups and two OH radicals in a square planar array. Two formula units per cell.

Massive, encrusting, and earthy. Tabular and prismatic crystals.

Physical Properties Perfect but interrupted {011} and {100} cleavage. Conchoidal fracture. Brittle. Hardness 3.5–4. Specific gravity 3.8. Vitreous luster. Transparent to opaque. Light to dark azure blue. Streak blue. Optical: $\alpha = 1.730$, $\beta = 1.758$, and $\gamma = 1.838$; positive; $2V = 68°$. Fusible (3). Loses water at 410°.

Distinctive Properties and Tests Color and habit. Association with malachite. Soluble with effervescence in cold dilute HCl.

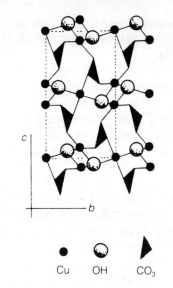

Figure 17.11
Structure of azurite. Unit cell dashed.

Cu   OH CO_3

Association and Occurrence Azurite, like malachite, is a secondary copper mineral formed by oxidation and carbonation of copper oxides and sulfides.

Alteration To malachite.

Related Minerals Similar species: hydrocerussite, $Pb_3(CO_3)_2(OH)_2$; hydromagnesite, $Mg_5(CO_3)_4(OH)_2 \cdot 4H_2O$; and malachite.

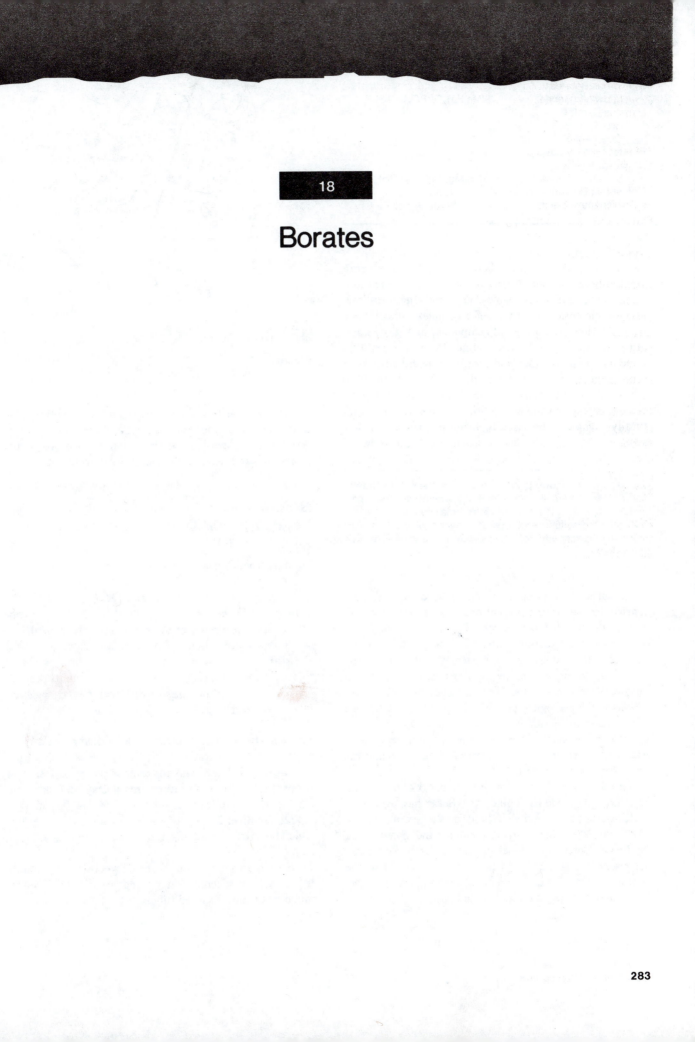

18

Borates

Borates

Anhydrous Borates
    Ludwigite Series   (Mg,Fe)$_2$Fe(BO$_5$)
Hydrous Borates
    Borax
    Colemanite
Borates with Hyrodyxl or
    Halogen
        Sussexite Series   (Mn,Mg,Zn)(BO$_2$)(OH)
Compound Borates
Miscellaneous Species

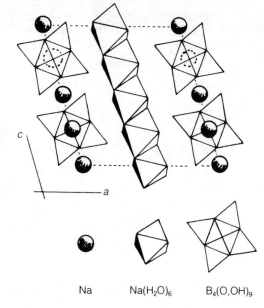

Na   Na(H$_2$O)$_6$ B$_4$(O,OH)$_9$

Figure 18.1
Structure of borax. Unit cell dashed.

The very small, trivalent boron cation has a mineralogical role analogous to that of chemically related aluminum in that boron is capable of entering structures either as a simple cation or as part of a complex radical. As a simple cation, it is a typical component of high-temperature minerals such as tourmaline. In low-temperature minerals, boron is capable of coordinating either three or four oxygen anions into triangular (BO$_3$)$^{3-}$, tetrahedral (BO$_4$)$^{5-}$, or more complex radicals. Because of these coordination and sharing possibilities, the borates are structurally complex and usually form in crystals of low symmetry. About forty-five borate minerals are known.

An elegant and showy test for borates can be conducted as follows. Place a small amount of the powdered mineral on a watch glass, soak the powder with a small amount of concentrated sulfuric acid, flow on a layer of alcohol, and light. The presence of boron is indicated by a yellowish-green flame.

Hydrous borates have a volcanic association. The extensive deposits in the present desert areas of the world, such as the southwestern United States, Tibet, and Chile, are reconcentrated from boron-containing volcanic emanations by the action of groundwater. Waters rich in boron are drawn to the surface by capillary action and deposit borates upon evaporation. The existence of similar conditions in the geological past is indicated by the presence of borates in many saline deposits that developed by evaporation in volcanic regions, such as the salt deposits at Stassfurt, Germany, the salt domes of Louisiana, and the brine springs of northern Italy.

Borates, usually transformed into borax or boric acid, H$_3$(BO$_3$), find wide use in industry. Borax has a very low melting point and is extensively used as a flux in such applications as welding or soldering and in the manufacture of tile and porcelain enamels for coating iron. Borax is a solvent for casein, and this property is exploited in the production of plaster, paint, and calcimine and in the manufacture of plywood and coated paper.

Borax is easily soluble in water, and the resulting boric acid solution is mildly antiseptic. Wide use of this property is made in medicine, in the manufacture of soaps, disinfectants, and deodorants, and in food preservation.

BORAX
Na$_2$(B$_4$O$_7$) · 10H$_2$O

(tincal)
bōr′ăks, from Persian *burah,* white.

Crystallography, Structure, and Habit Monoclinic. 2/*m*. *a* 11.858, *b* 10.674, *c* 12.197, *β* 106.7°. *C* 2/*c*. Tabular to short prismatic crystals, often poorly formed.

Edge-sharing chains of Na(H$_2$O)$_6$ octahedra parallel the *c*-axis, alternating with rows of Na ions and B$_4$(O, OH)$_9$ groups (see figure 18.1). Four formula units per cell.

Crystals as stubby prisms with rectangular or octagonal cross section; granular aggregates.

Physical Properties Perfect {100} cleavage, less good {110}. Conchoidal fracture. Brittle. Hardness 2–2.5. Specific gravity 1.7. Vitreous to resinous luster. Translucent to opaque. Colorless to white or tints of gray, blue, or green. Streak white. Optical: α = 1.447, β = 1.470, and γ = 1.472; negative; 2V = 40°. Sweetish alkaline taste. Fusible (1–1.5) with swelling and frothing.

Distinctive Properties and Tests Sweetish taste, soluble in water, low specific gravity, and association. Evolves much water in the closed tube.

Association and Occurrence Borax is the most wide-spread of the borate minerals. It is found (1) in evaporite deposits and muds of saline lakes and playas with halite, other borates, gypsum, and calcite; (2) as an efflorescence on soils in arid regions; and (3) as a deposit from hot springs.

Alteration Dehydrates readily to tincalconite. Readily dissolves.

Confused with Other borates.

Related Minerals *Similar species:* colemanite; kernite, $Na_2(B_4O_6)(OH)_2 \cdot 3H_2O$; tincalconite, $Na_2(B_4O_5)(OH)_4 \cdot 3H_2O$; and ulexite, $NaCa(B_5O_6)(OH)_6 \cdot 5H_2O$.

COLEMANITE
$Ca_2(B_6O_{11}) \cdot 5H_2O$

kōl'mə-nīt, after W. T. Coleman (1824–1893), mine owner and a founder of the California borax industry.

Crystallography, Structure, and Habit Monoclinic. $2/m$. a 8.743, b 11.264, c 6.102, β 110.1°. $P2_1/a$.

Short prismatic crystals with numerous faces.

Rings of BO_4 tetrahedra and BO_3 triangles are cross-linked into endless crumpled sheets parallel to (010) (see figure 18.2). Four formula units per cell.

Massive, granular, compact.

Physical Properties Perfect {010} and distinct {001} cleavage. Uneven to subconchoidal fracture. Hardness 4–4.5. Specific gravity 2.4. Vitreous to adamantine luster. Transparent to translucent. Colorless to white or with tints of yellow or gray. Streak white. Optical: $\alpha = 1.586$, $\beta = 1.592$, and $\gamma = 1.614$; positive; $2V = 56°$. Fusible (1.5) with exfoliation and crumbling.

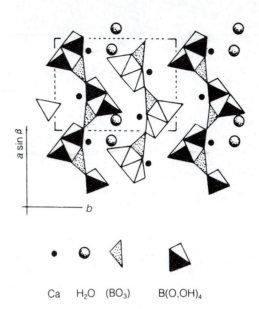

Ca   H₂O   (BO₃)   B(O,OH)₄

Figure 18.2
Structure of colemanite. Unit cell dashed.

Distinctive Properties and Tests Cleavage. Evolves water in the closed tube. Soluble in hot HCl.

Association and Occurrence Found in stratified deposits in ancient lake beds with other borate minerals.

Alteration From other borates.

Confused with Other borates.

Variants Minor amounts of Mg^{2+} or alkali metals may replace calcium.

Related Minerals *Similar species:* inyoite, $Ca_2(B_6O_{11}) \cdot 13H_2O$; meyerhofferite, $Ca_2(B_6O_{11}) \cdot 7H_2O$; veatchite, $Sr_2(B_{11}O_{16})(OH)_5 \cdot H_2O$; and kurnakovite, $Mg_2(B_6O_{11}) \cdot 15H_2O$.

19

Sulfates

Sulfates

Mascagnite Group	$(K,NH_4)_2(SO_4)$
Barite Group	$(Ba,Sr,Pb)(SO_4)$
Barite	
Celestine	
Anglesite	
Anhydrite	
Langbeinite Group	$K_2(Mn,Mg)_2(SO_4)_3$
Blödite Group	$(Na,K)_2Mg(SO_4)_2 \cdot 4H_2O$
Picromerite Group	$(K,NH_4)_2(Mg,Cu)(SO_4)_2 \cdot 6H_2O$
Tamarugite Group	$Na(Al,Fe)(SO_4)_2 \cdot 6H_2O$
Mendozite Group	$(Na,K)Al(SO_4)_2 \cdot 11H_2O$
Alum Group	$(K,Na,NH_4)Al(SO_4)_2 \cdot 12H_2O$
Kieserite Group	$(Mg,Fe,Mn)(SO_4) \cdot H_2O$
Gypsum	
Chalcanthite Group	$(Cu,Fe,Mg)(SO_4) \cdot 5H_2O$
Hexahydrite Group	$(Mg,Mn,Zn)(SO_4) \cdot 6H_2O$
Melanterite Group	$(Fe,Cu,Co,Mn,Mg)(SO_4) \cdot 7H_2O$
Melanterite	
Epsomite Group	$(Mg,Zn,Ni,Fe)(SO_4) \cdot 7H_2O$
Epsomite	
Halotrichite Group	$(Mg,Fe,Mn,Zn)(Al,Fe)_2(SO_4)_4 \cdot 22H_2O$
Alunite Group	$(K,Na,NH_4,Ag,H_2O,Pb)(Al,Fe)_3$
	$(SO_4)_2(OH)_6$
Jarosite	
Alunite	
Copiapite Group	$(Fe,Mg,Cu)Fe_4(SO_4)_6(OH)_2 \cdot 20H_2O$
Miscellaneous Species	

Sulfur is found in nature in a reduced state as a native element and in metalloidal compounds. In its oxidized state, it is the central cation of a regular SO_4 tetrahedron in sulfate minerals, and only in very rare instances do these radicals share oxygens. Structurally, the sulfates are built up of alternating sulfate radicals and cations that are usually divalent, and when two cations are present, they are usually a large ion-small ion pair. The cation is commonly an alkaline earth or one of the metals Pb, Cu, or Fe. Water in the hydrous sulfates is very loosely bound and may be readily removed by low heating or dessication.

Nearly 130 sulfates are known and, as a class, they are characterized by softness and light color; and some are readily soluble in water or have high specific gravity. Sulfates are often confused with carbonates, but they can be distinguished from them by the ready effervescence of the latter in warm HCl. An excellent test for sulfates is to add a few drops of barium hydroxide to a dilute acid solution of the mineral; a dense white precipitate of barium sulfate will develop. The flame colors of the cations distinguish the different sulfate species.

Gypsum and anhydrite are rock-forming sulfates. Gypsum is the first mineral to precipitate in large amounts when seawater is evaporated; it is followed by anhydrite. Limestones and rock salt are usually found with the gyprock formed in this environment.

The most useful sulfate is gypsum, used principally in the manufacture of plaster and plaster products but also in statuary (alabaster). It is also an additive to portland cement, an ingredient in the manufacture of chalks and crayons, a filling and filtering medium, a component of land plaster, and the starting material for the manufacture of ammonium sulfate and sulfuric acid.

When heated to 128° C, the water content of gypsum is reduced from 2 to 0.5 moles. This hemihydrate is plaster of paris, named for deposits at Montmartre near Paris, France. Addition of water to plaster of paris causes gypsum to reform with an increase in volume and evolution of heat. In the manufacture of gypsum wallboard, a laminate of gypsum and cardboard is held together by the newly formed needlelike gypsum crystals to form a strongly bonded, fireproof material that can be sawed or nailed.

Barite, because of its low cost and high density, is widely used in the petroleum industry for the compounding of drilling muds. These muds not only lubricate the drilling tools but also float rock chips to the surface and stop fluids in the rocks from entering the drill hole.

Anglesite and some copper sulfates are minor ores of lead and copper. Celestine is used in pyrotechnics (red) and in the processing of beet sugar.

BARITE
$Ba(SO_4)$

(heavy spar, barytes)
bar'īt, from Greek *barys,* heavy.

Crystallography, Structure, and Habit Orthorhombic, $2/m\ 2/m\ 2/m$. a 8.878, b 5.450, c 7.152. *Pnma.* Thin to thick tabular crystals (see figure 19.1).

Barite structure (see figure 19.2) exhibits an alternation of Ba ions and SO_4 groups in a manner similar to Na and Cl in halite. Each Ba^{2+} ion is surrounded by six $(SO_4)^{2-}$ radicals and coordinates twelve oxygens. Four $Ba(SO_4)$ per unit cell.

Crystals, rosettes (see figure 19.3), or concretions.

Physical Properties Perfect $\{001\}$ and good $\{210\}$ cleavages. Uneven fracture. Brittle. Hardness 3–3.5. Specific gravity 4.5. Vitreous to resinous luster. Transparent to translucent. Colorless to white with tints of yellow, brown, red, gray, green, or blue; the blue color is probably due to irradiation by Ra^{2+}. Streak white. Optical: $\alpha = 1.636$, $\beta = 1.638$, and $\gamma = 1.648$; positive; $2V = 37°$. May be fluorescent, phosphorescent, or thermoluminescent. Melting point 1,580° C; decrepitates.

Distinctive Properties and Tests Crystal form, cleavage, and specific gravity. Insoluble in acids. Flame test for Ba.

Figure 19.1
Thick tabular crystals of barite (white) with stibnite (gray). (Courtesy J. S. Hudnall Mineralogy Museum, University of Kentucky. Photograph by T. A. Moore.)

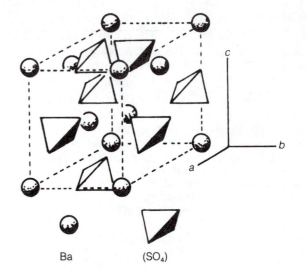

Figure 19.2
Structure of barite. Unit cell dashed.

Ba   (SO₄)

Figure 19.3
Barite rosette. Norman, Oklahoma. (Courtesy Department of Geology, University of Windsor. Photograph by D. L. MacDonald.)

Association and Occurrence   Barite is the most common barium mineral. It is found (1) as a gangue mineral in hydrothermal veins, especially those of low to moderate temperature where it is associated with fluorite, calcite, siderite, dolomite, quartz, and galena; (2) disseminated or as lenses in limestones and as veins, lenses, fillings, and replacements in other sedimentary rocks; (3) in sedimentary iron and manganese deposits; (4) filling cavities in lava; and (5) as a hot-spring deposit.

Alteration   Quite resistant; may change to witherite.

Confused with   Calcite and dolomite.

Variants   Chemical varieties: strontian, calcian, plumbian (hokutolite). Minor amounts of Hg^{2+}, Co^{2+}, and Ra^{2+} may substitute for Ba^{2+}; and $(CrO_4)^{2-}$, $(SeO_4)^{2-}$, and $(MnO_4)^{2-}$ radicals may substitute for $(SO_4)^{2-}$.

Related Minerals   Solid solutions: celestine and anglesite, although usually very limited in natural barite. *Isotypes:* anglesite. *Isostructures:* avogadrite, (K,Cs)(BF₄). *Polymorphs:* transforms at 1,149° C to an apparent monoclinic form.

CELESTINE
Sr(SO₄)

(celestite)

sə-lĕs'tīn, from Latin *caelestis,* celestial, referring to the sky-blue color of some specimens.

Crystallography, Structure, and Habit   Orthorhombic, $2/m\ 2/m\ 2/m$. *a* 8.539, *b* 5.352, *c* 6.866. *Pnma.*

Crystals as thin to thick tablets, laths, and equant forms. Isostructural with barite (see figure 19.2). Four Sr(SO₄) per cell.

Tabular crystals or crystal groups; fibrous, nodular, granular, and banded.

Physical Properties Perfect {001} and good {210} cleavage. Uneven fracture. Brittle. Hardness 3–3.5. Specific gravity 4.0. Vitreous to pearly luster. Transparent to translucent. Colorless to pale blue or white; occasionally reddish, greenish, or brownish. Optical: $\alpha = 1.621$, $\beta = 1.623$, and $\gamma = 1.630$; positive; $2V = 50°$. Occasionally fluorescent and thermoluminescent. Melting point 1,605° C; decrepitates and fuses to a white pearl. Insoluble in acids.

Distinctive Properties and Tests Crystal form, cleavage, and specific gravity.

Association and Occurrence Found chiefly in sedimentary rocks (1) with bedded gypsum, anhydrite, or halite deposits often associated with native sulfur; (2) as cavity fillings and veins or disseminated in limestone and dolostone with strontianite, gypsum, calcite, dolomite, and fluorite; (3) disseminated in shale, marl, and sandstone; (4) in evaporite deposits with potassium salts and borates. It is also found as a gangue mineral in hydrothermal veins.

Alteration To strontianite.

Confused with Barite, anhydrite.

Variants *Chemical varieties:* barian, plumbian, and calcian.

Related Minerals *Solid solutions:* barite, limited in natural materials. *Isotypes:* anglesite. *Isostructures:* avogadrite, $(K,Cs)(BF_4)$. *Polymorphs:* transforms at 1,152° C to hexagonal α-$Sr(SO_4)$.

ANGLESITE
Pb(SO₄)

(lead vitriol)
ăng′glĭ-sīt, after Anglesey, an island of Wales where it was discovered.

Crystallography, Structure, and Habit Orthorhombic. $2/m\ 2/m\ 2/m$. *a* 8.480, *b* 5.398, *c* 6.958. *Pnma*. Crystal forms various; usually thin to thick tabular, prismatic, or equant.

Isostructural with barite (see figure 19.2). Four formula units per cell.

Usually massive; also granular to compact, nodular, or banded. Crystals prismatic, tabular, or bipyramidal.

Physical Properties Good {001} and {210} cleavage. Conchoidal fracture. Brittle. Hardness 2.5–3. Specific gravity 6.4. Subadamantine to resinous luster. Transparent to opaque. Colorless to white, or with tints of gray, yellow, green, or blue. Optical: $\alpha = 1.878$, $\beta = 1.883$, and $\gamma = 1.895$; positive; $2V = 70°$. Often fluorescent. Fusible (1.5) and yields a lead globule.

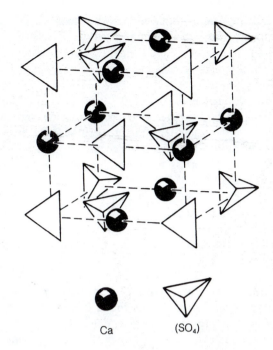

Ca   (SO₄)

Figure 19.4
Structure of anhydrite.

Distinctive Properties and Tests Specific gravity and fusibility. Acquires a white coating ($PbCO_3$) in an ammonium carbonate solution and a black lustrous coating (PbS) in an ammonium sulfide solution. Soluble with difficulty in HNO_3.

Association and Occurrence A common secondary mineral after galena. Associated with cerussite, pyromorphite, native sulfur, wulfenite, gypsum, and cerargyrite.

Alteration From galena, to cerussite.

Confused with Cerussite.

Variants *Chemical varieties:* barian.

Related Minerals *Solid solutions:* barite. *Isotypes:* celestine. *Isostructures:* avogadrite, $(K,Cs)(BF_4)$. *Polymorphs:* transforms at 864° C to a monoclinic form.

ANHYDRITE
Ca(SO₄)

ăn-hī′drīt, from Greek *anhydros,* waterless, in contrast to gypsum.

Crystallography, Structure, and Habit Orthorhombic. $2/m\ 2/m\ 2/m$. *a* 6.991, *b* 6.996, *c* 6.238. *Amma.* Crystals rare; usually equant or thick tablets. Contact or repeated twins on {011}.

In anhydrite structure (see figure 19.4) edge-sharing SO_4 tetrahedra and CaO_8 polyhedra alternate in three mutually perpendicular directions with chainlike Ca-SO₄ repetition most prominent parallel to the *c*-axis. Structure is similar to that of zircon. Four Ca(SO₄) per unit cell.

Usually massive; also fine granular or fibrous.

Physical Properties Perfect {010}, nearly perfect {100}, and good {001} cleavages yielding cubic or oblong fragments. Uneven to splintery fracture. Brittle. Hardness 3.5. Specific gravity 3.0. Pearly, vitreous, or glassy luster. Transparent. Colorless to blue-white, reddish, purplish, brown, or gray. Optical: $\alpha = 1.569$, $\beta = 1.574$, and $\gamma = 1.609$; positive; $2V = 43°$. Melting point about 1,450° C; fuses to a white enamel.

Distinctive Properties and Tests Pseudocubic cleavage. Soluble in HCl.

Association and Occurrence Anhydrite is a rock-forming mineral and is often found in bedded deposits alone or with gypsum, limestone, dolostone, or halite. It is also found (1) as an accessory mineral in sedimentary rocks with celestine, dolomite, calcite, gypsum, and quartz; (2) as a gangue mineral with pyrite in hydrothermal veins; (3) in cavities in lava with prehnite and zeolites; (4) as a component of caliche; (5) as a deposit around fumaroles; and (6) in the cap rocks of salt domes.

Alteration Changes readily from and to gypsum.

Confused with Gypsum and celestine.

Variants *Chemical varieties:* strontian. May also contain small amounts of Ba and H₂O.

Related Minerals *Polymorphs:* transforms at 1,193° C to α-Ca(SO₄). *Other similar species:* gypsum; apthitalite (glaserite), (K,Na)₃Na(SO₄)₂; and glauberite, Na₂Ca(SO₄)₂.

GYPSUM
Ca(SO₄) · 2H₂O

(gips, yeso)
jip′səm, probably from Arabic *jibs,* plaster.

Crystallography, Structure, and Habit Monoclinic. 2/*m*. *a* 5.68, *b* 15.18, *c* 6.29, *β* 113.8°. *A*2/*n*. Thick to thin tabular crystals, usually elongated; also prismatic and acicular forms (see figure 19.5). Often twinned on {100}, yielding contact "swallowtail" twins (see figure 19.6).

Gypsum structure is shown in figure 19.7. Each Ca²⁺ is surrounded by four (SO₄)²⁻ radicals and two H₂O molecules. Layers of water molecules alternate with layers of calcium ions and sulfate radicals parallel to (001). Four Ca(SO₄) per cell.

Massive, foliated, and as crystal aggregates or rosettes.

Figure 19.5
Acicular crystals of gypsum. Crystal Falls, Michigan. (Courtesy Royal Ontario Museum, Toronto.)

Figure 19.6
Gypsum showing "swallow-tail" twinning. Mexico. The largest crystal is 12 × 8 × 4 cm. (Courtesy Royal Ontario Museum, Toronto. Photograph by J. A. Chambers.)

Physical Properties Eminent {010}, distinct {100}, and good {011} cleavage, yielding flattened rhombic fragments. Cleavage folia may be bent but are not elastic. Hardness 2. Specific gravity 2.3. Vitreous luster. Transparent to translucent. Colorless, white, gray, yellow, brown, orange, or black. Optical: $\alpha = 1.519$, $\beta = 1.523$, and $\gamma = 1.529$; positive; $2V = 58°$. Fusible (3).

Distinctive Properties and Tests Softness and cleavage. Yields copious water in the closed tube. Soluble in hot dilute HCl.

Association and Occurrence Gypsum is the most common sulfate mineral. It is found (1) in bedded sedimentary deposits associated with limestone, red shale, sandstones, marl, and clay; (2) around fumaroles with native sulfur; (3) as an efflorescence from soils; (4) in gossans; and (5) in salt pans and dry lake beds.

Alteration Changes readily to and from anhydrite.

Confused with Anhydrite and calcite.

Variants *Form varieties:* selenite, crystallized (see figure 19.8); satin spar, fibrous; alabaster, massive; Kopi gypsum, earthy. Gypsum is one of the few minerals with no significant chemical variation. Traces of barium and strontium are sometimes present, and admixed clay, sand, or bitumens are common.

Related Minerals *Isostructures:* brushite, $CaH(PO_4) \cdot 2H_2O$; pharmacolite, $CaH(AsO_4) \cdot 2H_2O$; churchite-(Y) (weinshenkite), $Y(PO_4) \cdot 2H_2O$; and ardealite, $Ca_2H(SO_4)(PO_4) \cdot 4H_2O$. *Other similar species:* anhydrite; mirabilite, $Na_2(SO_4) \cdot 10H_2O$; and bassanite, $Ca(SO_4) \cdot \frac{1}{2}H_2O$, the hemihydrate of commerce called plaster of paris.

MELANTERITE
$Fe(SO_4) \cdot 7H_2O$

(copperas, green vitriol)
mə-lăn′tə-rīt, from Greek *melanteria,* black metallic dye.

Crystallography, Structure, and Habit Monoclinic. $2/m$. a 14.072, b 6.503, c 11.041, β 105.6°. $P2_1/c$. Equant to short prismatic crystals.

Some water molecules are coordinated with iron in $Fe(H_2O)_6$ octahedra; others are uncoordinated (see figure 19.9). Four formula units per cell.

Usually stalactitic or concretionary; also fibrous, encrusting, and pulverulent.

Physical Properties Perfect {001} and distinct {110} cleavage. Conchoidal fracture. Brittle. Hardness 2. Specific gravity 1.9. Vitreous luster. Translucent. Color green, green-blue, greenish white, or blue. Colored streak. Optical: $\alpha = 1.470$, $\beta = 1.479$, and $\gamma = 1.486$; positive; $2V = 80°$. Sweetish astringent and metallic taste. Turns yellowish white and opaque on exposure to air.

Distinctive Properties and Tests Taste and color. Yields copious water in the closed tube.

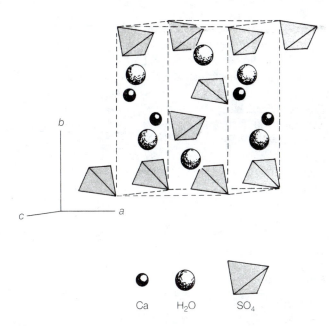

Figure 19.7
Structure of gypsum. Unit cell dashed.

Figure 19.8
Gypsum, variety selenite. Kingdon Mine, Galetta, Ontario. (Courtesy Royal Ontario Museum, Toronto.)

Association and Occurrence Melanterite is a secondary mineral formed by alteration of iron-bearing sulfides. It is found typically as efflorescence in mines or pyritic deposits and in the oxidized zones of ore deposits, as well as in coals and lignites.

Alteration From iron sulfides, such as pyrite and marcasite. From and to other ferrous and ferric sulfates. Dehydrates readily to siderotil.

Variants *Chemical varieties:* cuprian, magnesian, zincian, and manganoan. May also contain small amounts of Co or Ni.

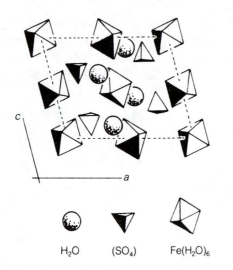

H₂O   (SO₄)   Fe(H₂O)₆

Figure 19.9
Structure of melanterite. Unit cell dashed.

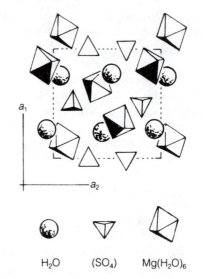

H₂O   (SO₄)   Mg(H₂O)₆

Figure 19.10
Structure of epsomite. Unit cell dashed.

Related Minerals *Solid solutions:* boothite, $Cu(SO_4) \cdot 7H_2O$. *Isotypes:* mallardite, $Mn(SO_4) \cdot 7H_2O$; and bieberite, $Co(SO_4) \cdot 7H_2O$. *Other similar species:* siderotil, $Fe(SO_4) \cdot 5H_2O$; chalcanthite, $Cu(SO_4) \cdot 5H_2O$; coquimbite, $Fe_2(SO_4)_3 \cdot 9H_2O$; and rozenite, $Fe(SO_4) \cdot 4H_2O$.

EPSOMITE
$Mg(SO_4) \cdot 7H_2O$

(epsom salt)
ep′sə-mīt, after Epsom, England, where epsom salts were originally prepared from mineral waters.

Crystallography, Structure, and Habit Orthorhombic. 222. *a* 11.799, *b* 12.050, *c* 6.822. *P222*. Crystals rare.

Some water molecules are coordinated with magnesium in $Mg(H_2O)_6$ octahedra; others are uncoordinated (see figure 19.10). Four formula units per cell.

Usually fibrous, woolly, or colloform.

Physical Properties Perfect {010} and distinct {011} cleavage. Conchoidal fracture. Hardness 2–2.5. Specific gravity 1.7. Vitreous to silky luster. Transparent to translucent. Colorless to white or tints of pink or green. Optical: $\alpha = 1.433$, $\beta = 1.455$, and $\gamma = 1.461$; negative; $2V = 50°$. Bitter salt taste.

Distinctive Properties and Tests Habit, low specific gravity, and taste. Melts in its own water of crystallization when heated in the closed tube.

Association and Occurrence Found as (1) a crust or efflorescence in coal or metal mines, in limestone caves, and in sheltered spots on outcrops of limestone, dolostone, gypsum, and magnesia-rich igneous rocks; (2) in the oxidized zone of pyritic sulfide deposits in arid regions; and (3) in mineral spring waters. Epsomite is usually associated with melanterite and other iron sulfates or with gypsum.

Alteration From kieserite, $Mg(SO_4) \cdot H_2O$; serpentine; talc; and magnesite. From and to hexahydrite, $Mg(SO_4) \cdot 6H_2O$.

Confused with Other hydrous sulfates.

Variants *Chemical varieties:* nickelian (green), manganoan (pink), and zincian.

Related Minerals *Isomorphs:* goslarite, $Zn(SO_4) \cdot 7H_2O$; and morenosite, $Ni(SO_4) \cdot 7H_2O$.

JAROSITE
$KFe_3(SO_4)_2(OH)_6$

jär′ə-sīt, from Barranco Jaroso, the site of its discovery, in the Sierra Almagrera, Spain.

Crystallography, Structure, and Habit Hexagonal. 3*m*. *a* 7.30, *c* 17.21. *R3m*. Crystals minute and indistinct, in the form of pseudocubes or tablets.

Structure (see figure 19.11) made up of layers perpendicular to *c* of isolated SO_4 tetrahedra alternating and sharing corners with $Fe(OH)_6$ octahedra. Three formula units per cell.

Found as crusts and coatings; also granular, fibrous, and nodular.

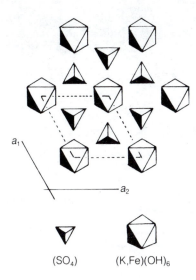

Figure 19.11
Structure of jarosite. Unit cell dashed.

(SO_4) $(K,Fe)(OH)_6$

Physical Properties   Distinct $\{0001\}$ cleavage. Uneven to conchoidal fracture. Brittle. Hardness 2.5–3.5. Specific gravity 2.9–3.3. Adamantine to vitreous luster. Translucent. Color ocherous yellow to brown. Streak pale yellow. Optical: $\omega = 1.820$ and $\epsilon = 1.715$; negative. Strongly pyroelectric.

Distinctive Properties and Tests   Color, habit, and occurrence. Soluble in HCl. Evolves sulfur dioxide at red heat.

Association and Occurrence   Jarosite is a widespread secondary mineral found as crusts and coatings on iron ores and in veins in secondary rocks. It is commonly associated with limonite, barite, quartz, pyrite, and hematite.

Alteration   Tarnishes quickly on exposure. From iron sulfides. To iron oxides.

Confused with   Limonite and hematite.

Variants   May contain minor amounts of Ag or Pb.

Related Minerals   *Solid solutions:* natrojarosite, $NaFe_3(SO_4)_2(OH)_6$. *Isotypes:* The formula for jarosite can be generalized to $XY_3(SO_4)(OH)_6$, where X can be K, Na, Pb, NH_4, H_3O, or Ag and Y can be either Fe (jarosites) or Al (alunites). The following species are recognized: alunite; ammoniojarosite, $(NH)_4Fe_3(SO_4)_2$ $(OH)_6$; argentojarosite, $AgFe_3(SO_4)_2(OH)_6$; beaverite, $Pb(Cu,Fe,Al)_3(SO_4)_2(OH)_6$; hydronium jarosite, $(H_3O)Fe_3(SO_4)_2(OH)_6$; minamiite, $(Na,Ca,K)Al_3$

$(SO_4)_2(OH)_6$; natroalunite, $NaAl_3(SO_4)_2(OH)_6$; osarizawaite, $PbCuAl_2(SO_4)_2(OH)_6$; and plumbojarosite, $PbFe_6(SO_4)_4(OH)_{12}$. *Isostructures:* svanbergite, $SrAl_3(PO_4)(SO_4)(OH)_6$; woodhousite, $CaAl_3(PO_4)(SO_4)$ $(OH)_6$; and hidalgoite, $PbAl_3(AsO_4)(SO_4)(OH)_6$.

ALUNITE
$KAl_3(SO_4)_2(OH)_6$

(alumstone)
ăl′yə-nīt, from Latin *alumen*, alum. Aluminum was named by Sir Humphrey Davy in 1812 for its presence in this mineral.

Crystallography, Structure, and Habit   Hexagonal. $3m$. a 6.97, b 17.38. $R3m$. Crystals rare but may be in $\{01\bar{1}2\}$ rhombohedrons which are pseudocubic in appearance.
  Isostructural with jarosite, the Fe^{3+} analog (see figure 19.11). Three formula units per cell.
  Found as crusts and coatings; usually massive, also fibrous to columnar, sometimes granular.

Physical Properties   Imperfect $\{0001\}$ and poor $\{01\bar{1}2\}$ cleavages. No twinning. Hardness 3.5–4. Specific gravity 2.6–2.9. Transparent to translucent. Vitreous to pearly luster. Color white to gray, sometimes reddish or yellowish. Streak white. Optical: $\omega = 1.568$ and $\epsilon = 1.590$; positive. Infusible; decrepitates on heating. Slowly soluble in sulfuric acid.

Distinctive Properties and Tests   Association. Turns blue when heated in cobalt nitrate solution. Yields acidic water when heated in a closed tube.

Association and Occurrence   Formed by sulfurous gases and sulfate-bearing hydrothermal solutions acting on potassium-bearing rocks. Commonly found in felsic and intermediate volcanic flows and pyroclastics. May form where pyrite is oxidized in the presence of water to yield sulfuric acid. Usually associated with fine-grained quartz, kaolinite, allophane, and diaspore. Often occurs as a gangue mineral in epithermal gold deposits.

Alteration   From potassic feldspars.

Confused with   Brucite, anhydrite, magnesite, clay minerals, limestone.

Variants   *Solid solutions:* natroalunite, $NaAl(SO_4)_3$ $(OH)_6$. Minor substitution of Fe^{3+} for Al and P for S.

Related Minerals   See jarosite.

20

Phosphates, Arsenates, Vanadates, Chromates, Tungstates, and Molybdates

Phosphates, Arsenates, Vanadates, Chromates

Phosphates

Triphylite Group	$(Li,Na)(Fe,Mn,Ca)(PO_4)$
Sicklerite Series	$(Li,Fe,Mn)(PO_4)$
Alluaudite Series	$(Na,Ca)_4(Fe,Mn,Mg)_8Fe_4(PO_4)_{12}$
Heterosite Series	$(Fe,Mn)(PO_4)$
Monazite	
Fairfieldite Group	$Ca_2(Mn,Fe,Mg)(PO_4)_2 \cdot 2H_2O$
Reddingite Series	$(Mn,Fe)_3(PO_4)_2 \cdot 3H_2O$
Herderite Series	$CaBe(PO_4)(F,OH)$
Amblygonite Series	
Frondelite Series	$(Mn,Fe)Fe_4(PO_4)_3(OH)_5$
Apatite Series	
Childrenite Series	$(Fe,Mn)Al(PO_4)(OH)_2 \cdot H_2O$
Turquoise Group	$Cu(Al,Fe)_6(PO_4)_4(OH)_8 \cdot 5H_2O$
Turquoise	
Lazulite Series	$(Mg,Fe)Al_2(PO_4)_2(OH)_2$
Autunite Group	$(Cu,Ca,Ba,Mg)(UO_2)_2(PO_4)_2 \cdot 10-12H_2O$

Combined Phosphates and Arsenates

Brushite Group	$CaH[(PO_4), (AsO_4)] \cdot 2H_2O$
Rösslerite Group	$MgH[(PO_4), (AsO_4)] \cdot 7H_2O$
Vivianite Group	$(Fe,Co,Ni,Zn)_3[(PO_4), (AsO_4)]_2 \cdot 8H_2O$
Variscite Group	$(Al,Fe)[(PO_4), (AsO_4)] \cdot 2H_2O$
Plumbogummite Group	$(Pb,Ba,Sr,Ca,Ce)(Al,Fe)_3[(PO_4), (AsO_4)]_2(OH)_5 \cdot H_2O$
Triploidite Group	$(Mn,Fe)_2[(PO_4), (AsO_4)](OH)$
Olivenite Group	$(Cu,Zn)_2[(PO_4), (AsO_4)](OH)$
Pyromorphite Series	
Beudantite Group	$(Pb,Ca,Sr)(Fe,Al)_3[(AsO_4), (PO_4)](SO_4)(OH)_6$

Arsenates

Roselite Group	$Ca_2(Co,Mg,Mn)(AsO_4)_2 \cdot 2H_2O$
Berzeliite Series	$(Ca,Na)_3(Mn,Mg)_2(AsO_4)_3$
Chlorophoenicite Group	$(Mn,Mg)_3Zn_2(AsO_4)(OH,O)_6$
Adelite Group	$(Ca,Pb)(Mg,Cu,Zn)(AsO_4)(OH)$

Vanadates

Descloizite Group	$(Zn,Cu,Mn)(Pb,Ca)(VO_4)(OH)$
Carnotite	

Chromates
Crocoite

The various radicals can be distinguished by simple chemical tests as follows:

Phosphates A canary-yellow precipitate results when a few milliliters of a nitric acid solution of the mineral are added to an ammonium molybdate solution.

Arsenates Fuse the mineral with about six volumes of $NaCO_3$. Boil the fused mass for about one minute in water, filter, and acidify the filtrate with HCl. Add an excess of NH_4OH and a few drops of magnesium sulfate. The presence of arsenic is indicated by a white precipitate of NH_4MgAsO_4.

Vanadates A red-brown color is obtained when a few drops of H_2O_2 are added to an acidic solution of a vanadate.

Chromates If chromium is present, the borax bead will turn yellow-green in an oxidizing flame and emerald-green in a reducing flame.

Tungstates Fuse the mineral with about six volumes of $NaCO_3$. Pulverize the fused mass and boil the powder in a test tube with 5–10 ml H_2O, filter, and acidify the filtrate with HCl. A white precipitate that turns yellow on boiling suggests the presence of tungsten. Add a few grains of metallic Sn or Zn and boil. Tungsten is present if the solution turns first blue and then brown. (Solutions with niobium remain blue with continued boiling.)

Molybdates Add a small quantity of the powdered mineral to about 3 ml HCl in a test tube and boil until the acid is nearly all evaporated. Cool, add about 5 ml H_2O and a fragment of metallic Sn. A deep blue color indicates the presence of molybdenum.

The phosphates, arsenates, vanadates, and chromates make out a diverse group of minerals encompassing over 250 species. Although this group is very large, most of its members are so rare that they are seldom seen. Only apatite and possibly monazite can be considered common because of their wide distribution as accessory minerals in igneous rocks and their relatively high concentrations in some detrital sediments.

Insofar as is known, the minerals in these families do not share oxygen ions between their ZO_4 radicals. Their structures, therefore, must be an alternation of cations and these tetrahedra. Partial solid solution is possible between PO_4 and AsO_4, between AsO_4 and VO_4, and between CrO_4 and VO_4. None, however, is known between PO_4 and VO_4 nor between PO_4 and CrO_4.

Apatite is the principal source of phosphorus used in fertilizers as well as in a large number of other applications. Some crystalline apatite from igneous rocks is exploited, but most production is from collophanite, a massive, cryptocrystalline variety. Apatite-rich sedimentary rocks, called **phosphorites,** formed under marine conditions have been deposited as remarkably uniform beds over large areas. They are found in the southeastern United States, mainly Florida, and in the northern Rocky Mountain states; in North Africa; and in Russia west of the Urals.

Apatite is a principal component of human teeth and bones, and as such it plays a critical role in human health. For example, the hydroxylian-carbonatian apatite of normal teeth is more subject to dental caries than is fluorapatite. A solid-solution transformation can be made by adding fluorine to the tooth surface from toothpaste or from body fluids with fluoridated water.

Another phosphate of economic importance is monazite, which is the principal source of many of the rare earth metals and an important source of thorium, which may be present in amounts up to 12% ThO_2. Workable concentrations of monazite have usually been found in beach placers along with other resistant minerals such as zircon, magnetite, rutile, ilmenite, and garnet. These "black sands" are widely distributed, but most of the economically valuable deposits are in the southern hemisphere in Australia, China, India, Malaysia, Sri Lanka, Thailand, and South Africa.

Carnotite, together with a number of closely related uranium- and vanadium-bearing minerals, is widely distributed and occasionally concentrated in sedimentary rocks. Under oxidizing conditions, the uranyl radical, $(UO_2)^{2+}$, is soluble in groundwater. Under reducing conditions, however, because of the presence of carbonaceous materials or sulphate-reducing bacteria, uranium is reduced from U^{6+} to U^{4+} as uraninite, UO_2, or to some other UO_2-containing mineral. Coal, black shales, and near-surface sulfate-bearing groundwater regimes, such as those found in sandstones or salt-dome cap rocks, are thus potential locations of mineralization.

MONAZITE
(Ce,La,Nd,Th)(PO₄)

mŏn′ə-zīt, from Greek *monazein,* to live alone, an allusion to its rare occurrence in its first known localities.

Crystallography, Structure, and Habit Monoclinic. $2/m.$ a 6.78, b 7.00, c 6.45, β 104.4°. $P2_1/n.$ Crystals usually small with elongate, prismatic, and equant forms. Contact twins common on {100}.

The monazite structure is shown in figure 20.1. Slightly distorted PO_4 tetrahedra alternate with cations in 9-fold coordination with oxygen. Four formula units per cell.

Crystals, masses, and rounded grains.

Physical Properties Distinct {100} and imperfect {010} cleavage. Parting often on {001}, rarely on {$\bar{1}$11}. Conchoidal to uneven fracture. Brittle. Hardness 5–5.5. Specific gravity 5.0–5.3. Luster variable; may be adamantine, resinous, waxy, or vitreous. Translucent. Color yellow or red-brown, brown, green, or white. Streak white. Optical: $\alpha = 1.770–1.800, \beta = 1.777–1.801,$ and $\gamma = 1.825–1.850;$ positive; $2V = 3–19°.$ Infusible.

Distinctive Properties and Tests Color, habit, and occurrence. Usually radioactive. Turns gray when strongly heated. Bluish-green flame color obtained when the powdered mineral is moistened with H_2SO_4.

Association and Occurrence Monazite is widely disseminated as an accessory mineral in granitic or syenitic igneous rocks, pegmatites, and gneisses; it may also be concentrated in beach or stream placers. It is associated with zircon, other rare earth minerals, magnetite, apatite, columbite, ilmenite, and garnet.

Alteration Resistant to change but usually acquires a yellow or red-brown opaque coating on exposure to air.

Confused with Zircon.

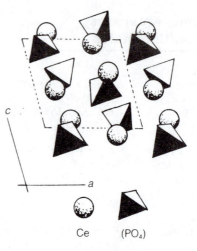

Figure 20.1
Structure of monazite. Unit cell dashed.

Variants *Form varieties:* metamict, structurally damaged by recoil of fission fragments from decay of included radioelements. *Chemical varieties:* All of the rare earths may be present but Ce, La, and Nd are the more abundant. The names monazite-(Ce), monazite-(La), and monazite-(Nd) are used according to the dominance of Ce, La, and Nd. Coupled substitution of $Ca^{2+} + Th^{4+}$ for two trivalent rare earth ions is common. $(SO_4)^{2-}$ or $(ScO_4)^{4-}$ commonly replaces some phosphatic groups.

Related Minerals *Solid solutions:* huttonite, $ThSiO_4$. *Isotypes:* crocoite. *Isostructures:* brabantite, $CaTh(PO_4)$; cheralite, $(Ca,Ce,Th)[(P,Si)O_4]$; and rooseveltite, $Bi(AsO_4)$. *Other similar species:* xenotime-(Y), $Y(PO_4)$; and pucherite, $Bi(VO_4)$; gasparite-(Ce), $(Ce,La,Nd)AsO_4$.

AMBLYGONITE SERIES
Amblygonite LiAl(PO₄)F

ăm-blig′-ə-nīt, from Greek *amblygonios,* obtuse-angled, a reference to its cleavage.

Montebrasite LiAl(PO₄)(OH)

mŏn-tə-brā′sīt, from its discovery at Montebra, Creuse, France.

Crystallography, Structure, and Habit Tricilinic. $\bar{1}.$ a 5.18, b 7.11, c 5.03, α 112.1°, β 97.8°, γ 68.1°. $P\bar{1}.$ Large, coarse, and poorly formed crystals. Repeated parallel twins on {$\bar{1}\bar{1}$1} and parallel to {111} are common.

Structure comprised of alternating PO_4 tetrahedra and $AlO_5(F,OH)$ octahedra linked by Li ions. Two formula units per cell.

Usually columnar to compact massive; often in irregular segregations.

Physical Properties Perfect {001}, good {100}, and sometimes distinct {0$\bar{2}$1} cleavage. Uneven to subconchoidal fracture. Brittle. Hardness 5.5–6. Specific gravity 2.9–3.2. Vitreous to greasy luster; pearly on (100). Transparent to translucent. Color white with tints of yellow, pink, green, blue, or gray; also colorless. Optical: $\alpha = 1.58–1.60$, $\beta = 1.59–1.62$, and $\gamma = 1.60–1.63$; negative; $2V = 50–90°$. Readily fusible (2) with bubbling.

Distinctive Properties and Tests Crystals, cleavage, fusibility, and association. Soluble in H_2SO_4 but not readily soluble in HCl.

Association and Occurrence Found only in granite pegmatites rich in lithium and phosphorus associated with spodumene, apatite, lepidolite, and tourmaline and in high-temperature tin veins and greisen with cassiterite, topaz, and mica.

Alteration To kaolin; mica; turquoise; wavellite, $Al_3(PO_4)_2(OH,F)_3 \cdot 5H_2O$.

Confused with Spodumene.

Variants *Chemical varieties:* sodian.

Related Minerals *Solid solutions:* amblygonite and montebrasite are completely miscible. *Other similar species:* herderite, $CaBe(PO_4)(F,OH)$.

APATITE SERIES
$Ca_5(PO_4)_3(F,Cl,OH)$

(tavistockite)
ăp′ə-tīt, from Greek *apate,* deceit, because it is often mistaken for other minerals.

Crystallography, Structure, and Habit Hexagonal. $6/m$. Fluorapatite: a 9.368, c 6.884; chlorapatite: a 9.629, c 6.778; hydroxylapatite: a 9.418, c 6.883; carbonate-apatite: a 9.436, c 6.883. $P6_3/m$. Short to long prismatic crystals. Twinning on {11$\bar{2}$1} and {10$\bar{1}$3}.

Apatite structure is shown in figure 20.2. Ca^{2+} ions on a 3-fold axis are bonded to oxygen neighbors of adjacent PO_4 tetrahedra to form Ca-PO_4 chains parallel to the c-axis. The remaining Ca^{2+} ions surround channels parallel to the c-axis in which fluorine ions are located. Two formula units per cell.

Usually in crystals; also massive, granular, or colloform.

Physical Properties Imperfect {0001} cleavage. Conchoidal to uneven fracture. Brittle. Hardness 5. Specific gravity 3.1–3.4. Vitreous luster. Transparent to opaque. Color strongly dependent on amount and oxidation state

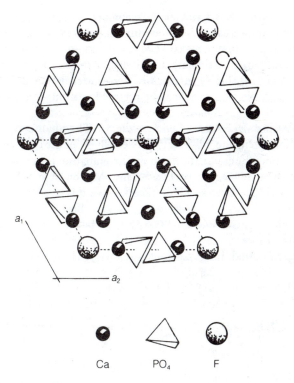

Figure 20.2
Structure of apatite. Unit cell dashed.

of manganese and iron contaminants; usually shades of green or red—yellow-green, blue-green, violet-blue, blue, pale green, dark red-brown—but also colorless and white. Streak white. Optical: $\omega = 1.633$ and $\epsilon = 1.630$; negative. Most specimens fluoresce or phosphoresce. Thermoluminescent. Fuses with difficulty (4–4.5); purplish kinds lose color on heating.

Distinctive Properties and Tests Crystal habit, color, hardness, and lack of good cleavage. Soluble in HCl or HNO_3. A fragment moistened with H_2SO_4 and then heated colors the flame pale bluish green (phosphoric acid).

Association and Occurrence Although normally occurring in only minor amounts, apatite is among the most widespread of minerals. It is found (1) as an accessory mineral in all igneous rocks; (2) as segregations from alkalic igneous rocks; (3) in pegmatites; (4) in magnetite deposits; (5) in hydrothermal veins; (6) in metamorphic rocks, especially marble with titanite, zircon, pyroxenes, amphiboles, spinel, vesuvianite, and phlogopite; (7) in talc and chlorite schists; (8) in bedded marine sediments; (9) as a replacement of limestone by guano; and (10) as a component of modern and fossil bones, shells, and teeth.

Alteration Resistant to change.

Confused with Beryl and epidote.

Variants *Form varieties:* collophane, a massive cryptocrystalline carbonatian apatite that makes up sedimentary phosphate rocks. *Chemical varieties:* chlorian, hydroxylian, carbonatian (dahllite, francolite), manganoan (dark blue, in pegmatites only), and strontian. S, Mg, Fe, As, Na, Y, and Si may be present in varying amounts.

Related Minerals *Solid solutions:* The apatite series shows complete miscibility between fluorapatite (the most common), chlorapatite, and hydroxylapatite. Further, Ca may be replaced in part by Sr, Mn, Mg, or Fe and (PO_4) by (AsO_4). *Isostructures:* The following isotypous phosphate species are recognized: alforsite, $Ba_5(PO_4)_3Cl$; belovite, $(Sr,Ce,Na,Ca)_5(PO_4)_3OH$; carbonate-fluorapatite (francolite), $Ca_5[(PO_4)(CO_3)]_3F$; carbonate-hydroxyapatite (dahllite), $Ca_5[(PO_4)(CO_3)]_3OH$; chlorapatite, $Ca_5(PO_4)_3Cl$; fermorite, $(Ca,Sr)_5[(AsO_4)(PO_4)]_3OH$; fluorapatite, $Ca_5(PO_4)_3F$; hydroxylapatite (hydroxyapatite), $Ca_5(PO_4)_3(OH)$; morelandite, $(Ba,Ca,Pb)_5[(AsO_4)(PO_4)]_3Cl$; pyromorphite; and strontium-apatite, $(Sr,Ca)_5(PO_4)_3(OH,F)$. A number of arsenates, vanadates, silicates, and sulfates are also isostructural with the apatite group; hedyphane, $Ca_2Pb_3(AsO_4)_3Cl$; johnbaumite, $Ca_5(AsO_4)_3$ OH; mimetite; svabite, $Ca_5(AsO_4)_3F$; turneaureite, $Ca_5[(As,P)O_4]_3Cl$; vanadinite; britholite-(Ce), $(Ce,Ca)_5[(SiO_4)(PO_4)]_3(OH,F)$; britholite-(Y) (abukumalite), $(Y,Ca)_5[(SiO_4)(PO_4)]_3(OH,F)$; fluorellestadite, $Ca_5[(SiO_4)(PO_4)(SO_4)]_3(F,OH,Cl)$; hydroxylellestadite, $Ca_{10}(SiO_4)_3(SO_4)_3(OH,Cl,F)_2$; and cesanite, $Na_3Ca_2(SO_4)_3OH$.

TURQUOISE
$CuAl_6(PO_4)_4(OH)_8 \cdot 4H_2O$

(turquois, henwoodite)
tûr′koiz, from French, *turquois*, Turkish. Gem turquoise has been mined in Iran (Persia) near Nishapur in Khorasan for many centuries, but the finest stones came to Europe via Turkey.

Crystallography, Structure, and Habit Triclinic. $\bar{1}$. *a* 7.42, *b* 9.95, *c* 7.69, α 111.65°, β 115.38°, γ 69.43°, $P\bar{1}$.

Structure (see figure 20.3) composed of isolated PO_4 tetrahedra linked at their corners with single and paired AlO_6 octahedra. Holes in the structure contain Cu in octahedral coordination with 4(OH) and $2H_2O$. One formula unit per cell.

Typically cryptocrystalline, rarely as minute crystals. Massive, compact, colloform.

Physical Properties Perfect {001} and good {010} cleavages, but rarely seen. Conchoidal to smooth fracture in massive material. Hardness 5–6. Specific gravity 2.6–2.8.

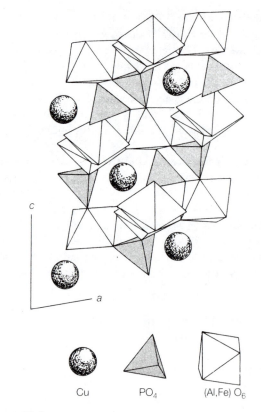

c

a

Cu   PO₄   (Al,Fe) O₆

Figure 20.3
Structure of turquoise.

Waxy luster. Color sky-blue to apple-green. Streak white to pale green. Translucent on thin edges. Optical: $\alpha = 1.61$, $\beta = 1.62$, and $\gamma = 1.65$; positive; $2V = 40°$. Infusible.

Distinctive Properties and Tests Color. Infusible but turns brown on heating. Soluble in HCl after ignition. Tests for copper.

Association and Occurrence Turquoise is a secondary mineral deposited from groundwater in decomposed alumina-rich volcanic rocks as small stringers, veins, and incrustations. Associated with kaolinite, limonite, and chalcedony.

Alteration May crumble if dehydrated. Color may fade with drying or exposure to sunlight.

Confused with Chrysocolla. Ivory, fossil bones, or teeth colored blue by natural or artificial copper-bearing solutions is called bone or fossil turquoise (odontolite).

Variants Zincian.

Related Minerals *Solid solutions:* chalcosiderite, $CuFe_6(PO_4)_4(OH)_8 \cdot 4H_2O$. *Isotypes:* faustite, $(Zn,Cu)Al_6(PO_4)\cdot 4H_2O$.

PYROMORPHITE SERIES
Pyromorphite $Pb_5(PO_4)_3Cl$

pī′rə-môr′fīt, from Greek *pyr,* fire, and *morphe,* form, because a fused fragment assumes a faceted form on solidifying.

Mimetite $Pb_5(AsO_4)_3Cl$

mī′mə-tīt, from Greek *mimos,* imitator, because of its close relation to pyromorphite.

Vanadinite $Pb_5(VO_4)_3Cl$

və-nād′n-īt, from its vanadium content.

Crystallography, Structure, and Habit Hexagonal. $6/m$. Pyromorphite: a 9.99, c 7.33; mimetite: a 10.26, c 7.44; vanadinite: a 10.32, c 7.35. $P6_3/m$. Prismatic and equant crystals; often rounded and barrel-shaped; sometimes cavernous or hollow prisms (see figure 20.4).

    Isostructural with apatite. Two formula units per cell. Crystals, granular, or colloform; encrusting.

Physical Properties No distinct cleavage. Uneven to subconchoidal fracture. Brittle. Hardness 3–4. Specific gravity 6.5–7.1 (pyromorphite), 7–7.2 (mimetite), and 6.9 (vanadinite). Resinous to subadamantine luster. Translucent. Color usually green, also yellow, brown, or reddish for pyromorphite and mimetite; red, orange, brown, or yellow for vanadinite.

Optical:

	Pyromorphite	Mimetite	Vanadinite
ω	2.058	2.144	2.146
ϵ	2.048	2.129	2.350
	negative	negative	negative

Piezoelectric. Readily fusible (1.5), yielding a lead globule that assumes a polyhedral form on cooling.

Distinctive Properties and Tests Crystal form, specific gravity, luster, color, and lead globule when fused. Soluble in HNO_3.

Association and Occurrence Secondary products in the oxidized zone of lead-ore deposits with cerussite, limonite, smithsonite, hemimorphite, and wulfenite.

Alteration From galena and cerussite.

Confused with Apatite.

Figure 20.4
Prismatic crystals of mimetite. Tsumeb, Namibia. (Courtesy Royal Ontario Museum, Toronto. Photograph by Arthur Williams.)

Variants *Chemical varieties:* calcian, (brown, green, or white), vanadian (black), chromian (red or orange), arsenatian (endlichite, green to white), and phosphorian.

Related Minerals *Isostructures:* apatite group.

CARNOTITE
$K_2(UO_2)_2(VO_4)_2 \cdot 3H_2O$

kär′nə-tīt, after Marie-Adolph Carnot (1839–1920), inspector general of mines in France.

Crystallography, Structure, and Habit Monoclinic. $2/m$. a 10.47, b 8.41, c 6.91, β 103.7°. $P2_1/n$. Tiny platy crystals with six sides.

    Structure (see figure 20.5) based on sheets of edge-sharing uranium and vanadium polyhedra parallel to (001) that are cross-linked by interlayer potassium ions and H_2O. Vanadium coordinates five oxygen ions in a square pyramid and uranium has seven neighbors—two in the linear uranyl radical and five in the plane normal to the uranyl axis. Two formula units per cell.

    Crystalline powder or pulverulent masses.

Physical Properties Perfect {001} cleavage. Hardness $\approx$ 4. Specific gravity 4.9. Earthy to resinous luster. Color bright yellow. Optical: $\alpha = 1.76, \beta = 1.90,$ and $\gamma = 1.92$; negative; $2V = 38–44°$. Strongly radioactive.

Distinctive Properties and Tests Color and habit. Radioactivity.

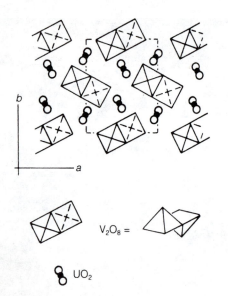

$V_2O_8 =$

UO_2

Figure 20.5
Structure of carnotite. Unit cell dashed.

Association and Occurrence   Disseminated or localized in sandstones or conglomerates, especially near carbonized vegetable matter with other secondary uranium minerals.

Confused with   Limonite and hydrated uranium oxides.

Related Minerals   Similar species: tyuyamunite, $Ca(UO_2)_2(VO_4)_2 \cdot nH_2O$; torbernite, $Cu(UO_2)_2(PO_4)_2 \cdot nH_2O$; and autunite, $Ca(UO_2)_2(PO_4)_2 \cdot nH_2O$.

CROCOITE
PbCrO₄

krō'kō-it, from Greek, *krokos,* saffron, referring to its orange color.

Crystallography, Structure, and Habit   Monoclinic. $2/m$. a 7.11, b 7.41, c 6.81, β 102.55°. $P2_1/n$. Slender prismatic crystals striated along c.

Monazite structure (see figure 20.1) with slightly distorted CrO_4 tetrahedra alternating with Pb in 9-fold coordination. Four formula units per cell.

Crystals and in columnar aggregates; also massive and granular.

Physical Properties   Imperfect {110} cleavage. Sectile. Hardness 2.5–3. Specific gravity 6.1. Vitreous to adamantine luster. Transparent to translucent. Color hyacinth red, orange-red, orange. Streak orange-yellow. Optical: $\alpha = 2.29$, $\beta = 2.36$, and $\gamma = 2.66$; positive; $2V = 57°$. Readily fusible (1.5).

Distinctive Properties and Tests   Color, luster, specific gravity. Test for chromium. Fusion with $NaCO_3$ on charcoal leaves a lead globule.

Association and Occurrence   As a rare secondary mineral in oxidized lead ores where it is associated with cerussite, pyromorphite, and wulfenite.

Alteration   From lead minerals.

Confused with   Wulfenite, vanadinite; distinguishable by its brighter colors.

Related Minerals   Isotypes: monazite. *Isostructures:* huttonite, $ThSiO_4$; brabantite, $CaTh(PO_4)$; cheralite, $(Ca,Ce,Th)[(P,Si)O_4]$; monazite-(La), $(La,Ce,Nd)(PO_4)$; rooseveltite, $Bi(AsO_4)$. *Other similar species:* xenotime, $Y(PO_4)$; pucherite, $Bi(VO_4)$.

Tungstates and Molybdates

Wolframite Series	
Scheelite Group	$Ca[(WO_4),(MoO_4)]$
Scheelite	
Wulfenite Group	$Pb[(WO_4),(MoO_4)]$
Wulfenite	

Tungsten and molybdenum are poorly shielded elements that coordinate four oxygen ions into a distorted ZO_4 tetrahedron. Tungstate and molybdate structures are made up of alternating cations and tetrahedra. No instances of the formation of complex ions are known.

Both tungsten and molybdenum impart valuable and important properties to steel, with which they are alloyed. Tungsten is a steel-hardening metal, and steels so alloyed keep their temper at high temperatures. Such steels are especially desirable for use in high-speed cutting tools. The extremely high melting point of tungsten (3,370° C) also makes this metal valuable for the manufacture of filaments for incandescent lamps, electrical contacts, spark plugs, and crucibles. Another use is in the manufacture of tungsten carbide, the hardest known cutting agent next to diamond. More than twenty tungsten-bearing minerals are known, but the only economically important ones are scheelite and the wolframite series.

Molybdenum metal is principally obtained from the sulfide molybdenite, but the lead molybdate wulfenite is a minor ore of both lead and molybdenum. Molybdenum is used in both ferrous and nonferrous alloys intended for use in furnace windings, permanent magnets, rustless steels, and tough steels for railroad forgings and agricultural machinery.

Tungstates and molybdates are relatively rare minerals. Both wolframite and scheelite are formed at high temperatures in close association with felsic igneous rocks. Deposits are typically found in contact zones, as pegmatites or high-temperature quartz veins, and as stockworks. Placer deposits are also worked. Scheelite is often closely associated with quartz, with which it is easily confused because of the very similar physical appearance of these two minerals. Scheelite, however, always fluoresces a characteristic electric blue to yellow color when irradiated with ultraviolet light, which greatly facilitates both prospecting for scheelite and the evaluation of scheelite deposits.

WOLFRAMITE SERIES

wool′frə-mīt, from German *wolfram*, possibly from *wolf*, wolf, and *rahm*, soot.

Hübnerite Mn(WO₄)

$Mn(WO_4)$

hœb′nə-rīt, for Adolph Hübner, metallurgist at Freiberg, Germany.

Ferberite Fe(WO₄)

$Fe(WO_4)$

fûr′bə-rīt, for chemist Rudolph Ferber of Gena, Germany.

Crystallography, Structure, and Habit Monoclinic. $2/m$. Hübnerite: a 4.834, b 5.758, c 4.999, β 91.2°; Ferberite: a, 4.732, b 5.708, c 4.965, β 90.0°. $P2/c$. Crystals as short to long prisms or tablets, usually striated vertically. Contact twins on {100} or {203}.

Structure (see figure 20.6) has layers of discrete, distorted WO_4 tetrahedra alternating with planes of (Mn,Fe). Two formula units per cell.

Subparallel crystal groups; granular.

Physical Properties Perfect {010} cleavage. Parting on {100} and {102}. Uneven fracture. Brittle. Hardness 4–4.5. Specific gravity 7.1 (hübnerite) to 7.5 (ferberite). Submetallic luster. Transparent to opaque. Color black, brownish black, or yellowish black. Streak brown-black to black. Optical: $\alpha = 2.17$–2.31, $\beta = 2.22$–2.40, and $\gamma = 2.30$–2.46; positive; $2V = 73$–79°; indices and $2V$ increase with Fe content. Occasionally has an iridescent tarnish. Sometimes weakly magnetic. Fusible (2.5–3) to a magnetic globule with a polyhedral surface.

Distinctive Properties and Tests Color, cleavage, luster, and streak. Partially dissolved in concentrated sulfuric acid; metallic tin added to the solution causes it to turn intense blue.

Association and Occurrence Found (1) in greisen, quartz-rich veins, or pegmatites immediately associated with granitic rocks; (2) in high-temperature hydrothermal veins with sulfides such as pyrrhotite, pyrite, chalcopyrite and arsenopyrite, scheelite, native bismuth, molybdenite, hematite, magnetite, tourmaline, topaz, albite, and apatite; (3) in mesothermal veins with sulfides, scheelite, native bismuth, quartz, siderite, barite, fluorite, and tourmaline; and (4) in placer deposits.

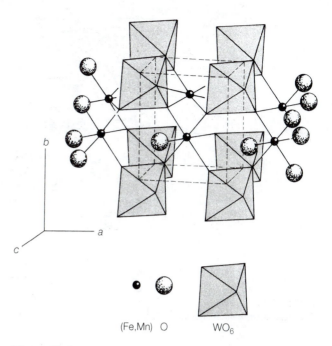

Figure 20.6
Structure of Wolframite. Unit cell dashed.

Alteration Resistant to change but may alter to iron and manganese oxides.

Confused with Hornblende.

Variants Calcian. May contain small amounts of Mg, Sc, Nb, or Ta.

Related Minerals *Solid solutions:* A complete solid-solution series exists between hübnerite and ferberite. Most specimens have an intermediate composition. *Isotypes:* sanmartinite, $(Zn,Fe)WO_4$. *Other similar species:* raspite, $Pb(WO_4)$.

SCHEELITE
Ca(WO₄)

$Ca(WO_4)$

shē′līt, after Karl Wilhelm Scheele (1742–1786), Swedish chemist.

Crystallography, Structure, and Habit Tetragonal. $4/m$. a 5.242, b 11.372. $I4/a$. Pseudo-octahedral and tabular crystals. Contact and penetration twins on {110}.

Scheelite structure (see figure 20.7) is a distortion of the structure of zircon. Four formula units per cell.

Found as crystals; massive, granular, and columnar.

Physical Properties Distinct {101} and interrupted {112} cleavage. Uneven to subconchoidal fracture. Brittle. Hardness 4.5–5. Specific gravity 6.1. Subadamantine luster. Transparent to translucent. Colorless to white, or tints of yellow, brown, green, gray, or red. Streak white.

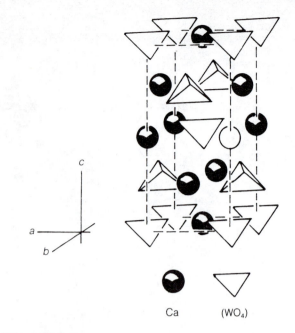

Ca   (WO₄)

Figure 20.7
Structure of scheelite.

Optical: $\omega = 1.920$ and $\epsilon = 1.937$; positive. Fluoresces bluish white to yellow in x rays and shortwave ultraviolet light, depending on the amount of Mo present. Thermoluminescent. Fusible with difficulty (5).

Distinctive Properties and Tests Association and fluorescence. Decomposed by boiling HCl or HNO₃, yielding a yellow powder that is soluble in NH₄OH. An HCl solution boiled with tin assumes a blue color, which changes later to brown.

Association and Occurrence Scheelite is typically a high-temperature mineral. It is found (1) in limestones metamorphosed at granite contacts with garnet, diopside, tremolite, hornblende, epidote, wolframite, vesuvianite, titanite, molybdenite, fluorite, pyrite, or chalcopyrite; (2) in quartzose hydrothermal veins and greisen; and (3) in pegmatites.

Alteration To tungstite, WO₃ · H₂O, or hydrotungstite, H₂(WO₄) · H₂O, a yellow or yellow-green powder.

Confused with Quartz and feldspar.

Variants *Chemical varieties:* molybdian, cuprian, and manganian. May also contain Sc.

Related Minerals *Solid solutions:* powellite, Ca(MoO₄). *Isotypes:* cuprotungstite, Cu₂(WO₄)(OH)₂; stolzite, Pb(WO₄); and pucherite, Bi(VO₄).

WULFENITE
Pb(MoO₄)

wool′fə-nīt, after Franz X. von Wulfen (1728–1805), Austrian mineralogist.

Crystallography, Structure, and Habit Tetragonal. $4/m$. a 5.435, b 12.110. $I4_1/a$. Crystals in square tablets. Rare twinning on {001}.
     Isostructural with scheelite. Four Pb(MoO₄) per cell. Usually in crystals; also granular or compact.

Physical Properties Distinct {101} cleavage. Subconchoidal to uneven fracture. Not very brittle. Hardness 3. Specific gravity 6.5–7.0. Resinous to adamantine luster. Transparent. Color orange to yellow. Streak white. Optical: $\omega = 2.404$ and $\epsilon = 2.283$; negative. Melting point 1,065° C. Easily fusible (2) to a lead globule.

Distinctive Properties and Tests Crystals and color. Decomposed on evaporation with HCl, yielding a residue of lead chloride and molybdic acid. This residue has an intense blue color when moistened with water in the presence of metallic zinc.

Association and Occurrence Wulfenite is a secondary mineral found in the oxidized zone of lead deposits associated with pyromorphite, cerussite, limonite, calcite, galena, and manganese oxides.

Confused with Native sulfur.

Variants *Chemical varieties:* calcian, vanadian, and tungstate. May also contain trace amounts of chromium and arsenic.

Related Minerals *Solid solutions:* stolzite, Pb(WO₄). *Other similar species:* raspite, Pb(WO₄).

21

Silicates

Silicates

Tectosilicates (Networks)
Silica Group
    α-Quartz SiO_2
    Opal $SiO_2 \cdot nH_2O$
Feldspar Group
    Alkali Feldspars $(K,Na)(AlSi_3O_8)$
      Monoclinic
        Sanidine-monalbite
        Series
        Orthoclase Series
      Triclinic
        Microcline Series
    Plagioclase Series $(Na,Ca)(Si,Al)_4O_8$
Feldspathoid Group
    Leucite $K(AlSi_2O_6)$
    Nepheline $Na(AlSiO_4)$
    Sodalite $Na_4(AlSiO_4)_3Cl$
    Cancrinite $Na_3Ca(AlSiO_4)_3(CO_3)$
Zeolite Group
    Stilbite $NaCa_2(Al_5Si_{13}O_{36}) \cdot 14H_2O$
    Chabazite $Ca(Al_2Si_4O_{12}) \cdot 6H_2O$
    Natrolite $Na_2(Al_2Si_3O_{10}) \cdot 2H_2O$
Miscellaneous Tectosilicates
    Scapolite Group $(Na,Ca)_4[(Al,Si)_4O_8]_3(Cl,CO_3)$
    Cordierite $Mg_2(Al_4Si_5O_{18})$
    Analcime $Na(AlSi_2O_6) \cdot H_2O$

Phyllosilicates (Sheets)
Clay Family
    Kaolinite Group
      Kaolinite $Al_2(Si_2O_5)(OH)_4$
    Smectite Group
    Illite Group
    Palygorskite Group
Talc Group
    Pyrophyllite $Al_2(Si_4O_{10})(OH)_2$
    Talc $Mg_3(Si_4O_{10})(OH)_2$
Chlorite Group $X_{5-6}(Z_4O_{10})(OH)_8$
                      $X = Al,Fe^{2+},Fe^{3+},Mg,Mn^{3+},Ni$
                      $Z = Al,Fe^{3+},Si$

Vermiculite Group
Brittle Mica Group
    Margarite $CaAl_2(Al_2Si_2O_{10})(OH)_2$
Elastic Mica Group
    Muscovite Series
      Muscovite $KAl_2(AlSi_3O_{10})(OH,F)_2$
    Biotite Series
      Phlogopite $KMg_3(AlSi_3O_{10})(OH,F)_2$
      Biotite $K(Mg,Fe)_3(AlSi_3O_{10})(OH,F)_2$
    Lepidolite Series $K(Li,Fe,Al)_3[(Si,Al)_4O_{10}](OH,F)_2$
Miscellaneous Phyllosilicates
    Serpentine Group $(Mg,Fe)_3(Si_2O_5)(OH)_4$
    Glauconite $(K,Na)(Fe,Mg,Al)_2[(Al,Si)_4O_{10}](OH)_2$
    Prehnite $Ca_2Al(AlSi_3O_{10})(OH)_2$

Inosilicates (Chains) and Cyclosilicates (Rings)
Double Chains
    Amphibole Group
      Anthophyllite Series $(Mg,Fe,Al)_7[(Al,Si)_8O_{22}](OH)_2$
      Cummingtonite Series $(Mg,Fe,Al)_7[(Al,Si)_8O_{22}](OH)_2$
      Actinolite Series $Ca_2(Mg,Fe)_5(Si_8O_{22})(OH)_2$
      Hornblende Series $Ca_2(Fe,Mg)_4Al(AlSi_7O_{22})(OH,F)_2$
Single Chains
    Pyroxene Group
      Enstatite Series $(Mg,Fe)_2(Si_2O_6)$
      Diopside Series $Ca(Mg,Fe,Al)[(Al,Si)_2O_6]$
      Augite $XY(Z_2O_6)$
      Spodumene $LiAl(Si_2O_6)$

    Pyroxenoids
      Rhodonite $Mn(SiO_3)$
      Chrysocolla $(Cu,Al)(SiO_3) \cdot nH_2O$
      Wollastonite-1T $Ca(SiO_3)$
      Pectolite $NaCa_2H(Si_3O_9)$

Cyclosilicates (Rings)
Tourmaline Group $WX_3Y_6(BO_3)_3(Si_6O_{18})(OH,F)_4$
Beryl $Be_3Al_2(Si_6O_{18})$

Sorosilicates (Pairs)
    Melilite Group $Ca_2(Mg,Al)[(Al,Si)_2O_7]$
    Hemimorphite $Zn_4(Si_2O_7)(OH)_2 \cdot H_2O$
Combined Pairs and Units
    Vesuvianite $Ca_{10}(Mg,Fe)_2Al_4(Si_2O_7)_2(SiO_4)_5$
                          $(OH,F)_4$
    Epidote Series $Ca_2(Al,Fe)_3(SiO_4)(Si_2O_7)(O,OH)$

Nesosilicates (Units)
Olivine Group $(Mg,Fe,Mn)_2(SiO_4)$
Garnet Group
    Garnet $X_3Y_2(SiO_4)_3$
    Ugrandite $Ca_3(Cr,Al,Fe)_2(SiO_4)_3$
    Pyralspite $(Mg,Fe,Mn)_3Al_2(SiO_4)_3$
Zircon $Zr(SiO_4)$
Humite Group $(Mg,Fe)_x(SiO_4)_y(F,OH)_2$
Topaz $Al_2(SiO_4)(F,OH)_2$
Chloritoid $(Fe^{2+},Mg,Mn)_2(Al,Fe^{3+})Al_3O_2$
                        $(SiO_4)_2(OH)_2$

Subsaturates
Aluminosilicate Group
    Andalusite $AlAl(SiO_4)O$
    Sillimanite $AlAl(SiO_4)O$
    Kyanite $AlAl(SiO_4)O$
Titanite $CaTi(SiO_4)O$
Staurolite $(Fe,Mg,Zn)_2Al_9[(Si,Al)_4O_{16}]O_6$
                        $(OH)_2$

The prevalence of silicon and oxygen in the earth's crust, coupled with their ability to join into a stable radical, controls the nature of the most abundant minerals, the silicates. The silicate minerals are a very large and diverse class containing between a quarter and a third of the known mineral species and making up at least 90% of the earth's crust.

The basic building block for the silicate minerals is a tetrahedral radical having an Si^{4+} ion surrounded by four O^{2-} ions. Diagrammatically, this SiO_4 unit can be represented as a tetrahedron whose corners represent the centers of the oxygen ions and whose edges equal two oxygen radii. A silicon ion is centered within the tetrahedron.

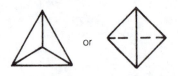

The bonding of the SiO_4 tetrahedral unit can be represented as in figure 21.1. Each oxygen contributes one-half of its valence of 2$-$ to the silicon ion of valence 4$+$.

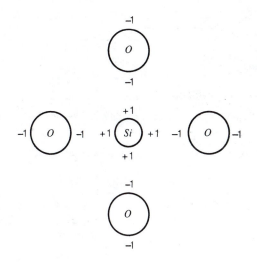

Figure 21.1
Electrostatic balance for (SiO₄)⁴⁻.

Figure 21.2
Electrostatic balance for (Si₂O₇)⁶⁻.

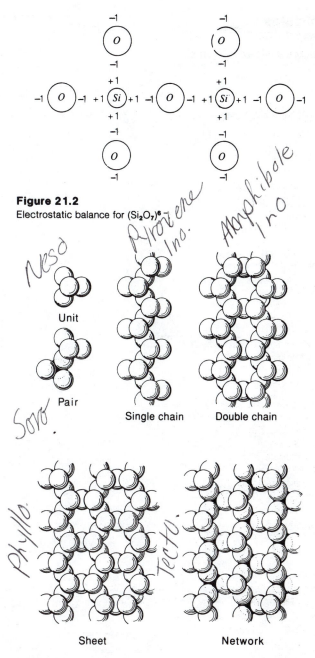

Figure 21.3
Oxygen sharing in silicates.

The valence requirements resulting from the unused charge of 1− per oxygen in the fundamental unit can be satisfied in a number of ways. Cations may be used to neutralize the unit and join it to adjacent units. These types of compounds are called **nesosilicates,** or units, and the mineral forsterite, $Mg_2(SiO_4)$, is a good example.

Tetrahedral units can share one or more oxygens with adjacent units. This is an important method of charge neutralization and is the basis for the subdivision of the silicate minerals into structural groups. A simple example of oxygen sharing is for two tetrahedra to have a common oxygen (see figure 21.2), such that the net charge on the two-tetrahedral unit, Si_2O_7, is reduced from 8− to 6−. If two oxygens on each tetrahedron are shared as in the single-**chain** structures called **inosilicates,** or in the **ring** structures called **cyclosilicates,** the Si:O ratio equals 1:3, and the net charge on the SiO_3 unit is 2−. A variation of the chain structure occurs when double chains are produced by alternate sharing of two or three oxygens. This generates a basic unit of Si_4O_{11} with a net charge of 6−.

If three oxygens in each tetrahedron are shared by neighboring tetrahedra to form an infinite **sheet** structure, or **phyllosilicate,** the basic unit becomes Si_2O_5 with a net charge of 2−. Finally, if all oxygens are shared with adjacent tetrahedra, **networks,** or **tectosilicates,** are formed with a basic unit of SiO_2 and a net charge of 0.

The Si-O complexes that act as the skeletons of the silicate minerals are illustrated in figure 21.3. Typically, these structures are held together by positive ions lying in and between them. Thus, many silicate structures have an alternation of complex radical and cation layers. Table 21.1 gives a summary of silicate structural types with examples.

Electrical neutrality is also attained by the addition of other constituents in the interstices of the basic tetrahedral structure. These may be anions or cations, hydroxyl groups, or even additional oxygens. The number

and types of these additions are numerous but are controlled by the size (coordination) and charge on the position. The relative abundances of the elements and the physical conditions of mineral formation are also factors in predicting the chemistry of silicates. Of special importance is Al^{3+}, the second most abundant cation in the crust.

The aluminum ion can reside in both 4-fold and 6-fold coordination positions. Its substitution for Si^{4+} in a tetrahedral site imposes an electric imbalance. The tectosilicate quartz, SiO_2, is nominally balanced electrically. On the other hand, in the feldspar albite there are one Al^{3+} and three Si^{4+} for each set of four tetrahedra. Thus, the

Table 21.1 Summary of Silicate Structures

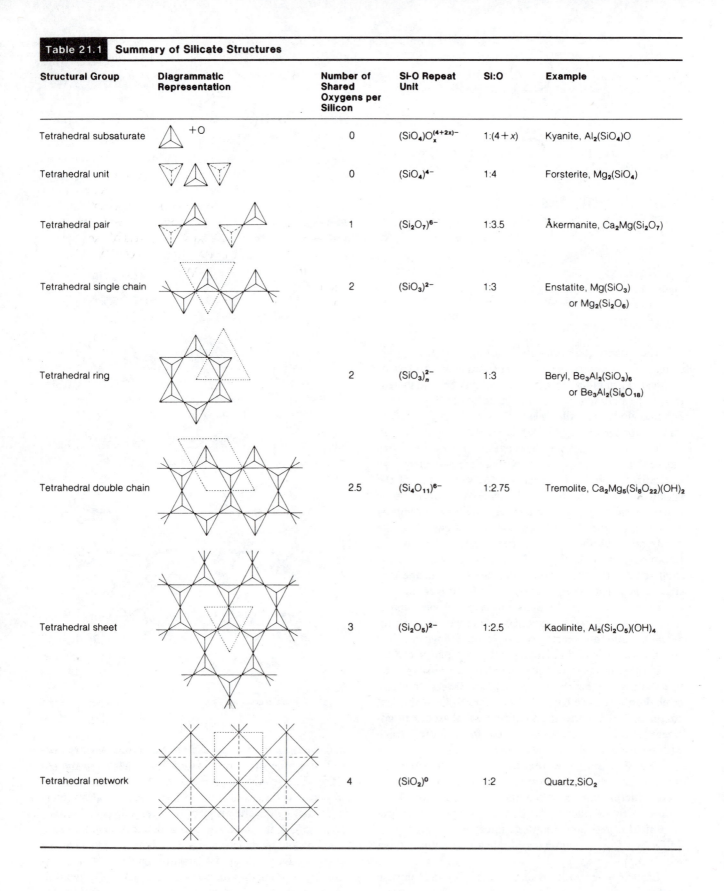

Structural Group	Diagrammatic Representation	Number of Shared Oxygens per Silicon	Si-O Repeat Unit	Si:O	Example
Tetrahedral subsaturate		0	$(SiO_4)O_x^{(4+2x)-}$	$1:(4+x)$	Kyanite, $Al_2(SiO_4)O$
Tetrahedral unit		0	$(SiO_4)^{4-}$	1:4	Forsterite, $Mg_2(SiO_4)$
Tetrahedral pair		1	$(Si_2O_7)^{6-}$	1:3.5	Åkermanite, $Ca_2Mg(Si_2O_7)$
Tetrahedral single chain		2	$(SiO_3)^{2-}$	1:3	Enstatite, $Mg(SiO_3)$ or $Mg_2(Si_2O_6)$
Tetrahedral ring		2	$(SiO_3)_n^{2-}$	1:3	Beryl, $Be_3Al_2(SiO_3)_6$ or $Be_3Al_2(Si_6O_{18})$
Tetrahedral double chain		2.5	$(Si_4O_{11})^{6-}$	1:2.75	Tremolite, $Ca_2Mg_5(Si_8O_{22})(OH)_2$
Tetrahedral sheet		3	$(Si_2O_5)^{2-}$	1:2.5	Kaolinite, $Al_2(Si_2O_5)(OH)_4$
Tetrahedral network		4	$(SiO_2)^0$	1:2	Quartz, SiO_2

radical ($Al^{3+}Si_3^{4+}O_8$) has a net charge of $1+$, which is made up by the addition of a sodium ion, Na^+, or $Na(AlSi_3O_8)$.

Silicates typically crystallize with relatively low symmetry, most commonly in the monoclinic or orthorhombic systems. Isometric silicates can form only if the mineral is a tectosilicate or nesosilicate. The structures and chemistries of the silicates are complex, and certain conventions are used to help write and interpret mineral formulas. In this book, the general formula for the silicates includes the symbols W, X, Y, and Z, which represent cationic groups based on valence and coordination. The principal cations found in silicates arranged according to their structural positions are as follows:

W cations with 8-fold coordination (Ca^{2+}, Fe^{2+}, Mn^{2+}, Na^+, Zr^{4+}) to 12-fold coordination (K^+, Ba^{2+})

X mainly divalent cations in octahedral (6-fold) coordination (Ca^{2+}, Fe^{2+}, Li^+, Mg^{2+}, Mn^{2+})

Y trivalent and quadrivalent ions in 6-fold coordination (Al^{3+}, Cr^{3+}, Fe^{3+}, Ti^{4+}, Zr^{4+})

Z mainly trivalent or quadrivalent cations in tetrahedral or 4-fold coordination (Al^{3+}, B^{3+}, Be^{2+}, Fe^{3+}, Si^{4+}, Ti^{4+})

A further convention is used here in regard to the common substitution of more than one cationic type into a structural position. In these cases, the dominant element in that position is given first. For example, the X position in an almandine garnet is nominally Fe-rich; Mg and Mn will normally be present also but in smaller amounts. Thus, the garnet formula can be given as $(Fe,Mg,Mn)_3$ $(Al_2Si_3O_{12})$. Similarly, the Z position in a phyllosilicate may contain variable amounts of Al and Si. Thus, a general phyllosilicate formula can be written as $WXY[(Si,Al)_4O_{10}](OH)_2$. In contrast, if the Al:Si ratio in the Z position is fixed, as in the feldspars, the formula would read $K(AlSi_3O_8)$.

The number of silicate minerals is extremely large, and the various subclasses are described in separate sections in this chapter. No generalizations hold for all silicates, but the majority are translucent to transparent, of moderate specific gravity, have a vitreous luster, and are chemically inert. The major diagnostic tests for the silicate minerals are summarized in table 21.2.

Tectosilicates (Networks)

Silicate network (or framework) minerals include the most abundant minerals of the earth's crust, the feldspars and the widely distributed and ubiquitous quartz. They are characterized by the sharing of all oxygen ions among their constituent Si-O or (Al,Si)-O tetrahedra.

Both because of their ready availability and useful properties, quartz and feldspar are widely used in the manufacture of glass and ceramics. Aside from its use in the manufacture of glass, quartz is widely employed as an abrasive and molding sand, and both clear and colored varieties serve as semiprecious gems.

SILICA GROUP

SiO_4 tetrahedra sharing all four oxygens with adjacent tetrahedra may be arranged in at least four distinctively different ways, each of which may have several minor modifications. These possibilities give rise to an extensive group of SiO_2 polymorphs. Figure 5.16 illustrates the structures of the various polymorphs under conditions of normal pressure. High-pressure members of the group—coesite, stishovite, and keatite—are also known; these have structures analogous to feldspar, which are different from those illustrated. Transformations (inversions) between forms of the same name, such as high quartz (β-quartz) and low quartz (α-quartz), are displacive transformations that are reversible and take place instantaneously at the transformation temperature. Transformations between different mineral species, such as quartz and tridymite, are reconstructive transformations which are nonreversible and very sluggish.

The minerals of the silica group are structurally related to a number of other mineral species through interchange of Si^{4+} with geologically common Al^{3+}. These minerals admit an alkali metal of appropriate size to maintain electrical neutrality. Examples of minerals isotypic with the silica group are

High cristobalite, SiO_2	High carnegieite, $Na(AlSiO_4)$
High tridymite, SiO_2	Kaliophilite, $K(AlSiO_4)$ Nepheline, $Na(AlSiO_4)$
Low quartz, SiO_2	Eucryptite, $Li(AlSiO_4)$

In each case, alternate Si ions in the basic structure are replaced by Al ions, thus enlarging the unit cell. The unit charge deficiency $(1-)$ resulting from each interchange is made up by the addition of one alkali metal ion $(1+)$. The alkali is located in the interstices of the basic structure, and the size of this hole largely controls the kind of alkali metal that may enter.

Although many isotypes of the silica group are known, solid solution in quartz is very limited. There are perhaps a few hundred contaminant atoms per million silicon atoms at most.

The only common representative of the silica group is low quartz. This species, however, is among the most ubiquitous of minerals, being formed and found in a very wide range of geologic environments. Quartz crystallizes as a stable phase from any natural silica-bearing solution, such as groundwater, hydrothermal solutions, or magma, at temperatures between 80 and 570° C. It is an essential constituent of many intrusive and extrusive igneous rocks,

Table 21.2 Some Diagnostic Tests for the Silicates

Mineral	Reaction with HCl	Yields Water in the Closed Tube	Fusibility	Miscellaneous
Analcime	Soluble[1]	Yes	3.5	
Andalusite	Insoluble	No	Infusible	
Anthophyllite	Insoluble	Yes	5*	
Augite	Insoluble	No	4–4.5	
Beryl	Insoluble	No	5.5	
Biotite	Insoluble	Yes	5*	Decomposes in boiling H_2SO_4, giving a milky solution
Cancrinite	Gel[2]	Yes	2	Effervesces
Chabazite	Soluble	Yes	3	
Chlorite	Insoluble	Yes	5.5*	As biotite
Chondrodite	Gel	Yes	Infusible	
Chrysocolla	Soluble	Yes	Infusible	
Cordierite	Insoluble	No	5.5	
Diopside	Insoluble	No	4	
Enstatite series				
Enstatite	Insoluble	No	Infusible	Solubility and fusibility increases with increasing iron
Ferrosilite	Soluble	No	5*	
Epidote	Insoluble	Yes	3–4(*)	Gelatinizes if previously ignited
Garnet	Insoluble	No	3–3.5(*)	As epidote
Glauconite	Soluble	Yes	Easy*	
Hemimorphite	Gel	Yes	6	
Hornblende	Insoluble	Yes	4*	
Kaolinite	Insoluble	Yes	Infusible	
Kyanite	Insoluble	No	Infusible	
Lepidolite	Insoluble	Yes	2	
Leucite	Soluble	No	Infusible	
Margarite	Insoluble	Yes	4–4.5	Slowly decomposes
Melilite	Gel	No	3	
Montmorillonite	Insoluble	Yes	Infusible	
Muscovite	Insoluble	Yes	5	
Natrolite	Gel	Yes	2.5	

[1]Yields free silica when boiled in HCl.
[2]Dissolves in boiling HCl and yields first a jelly and then granular silica on evaporation.
*An asterisk indicates the fused mineral is magnetic. An asterisk in parentheses indicates the fused mineral is sometimes magnetic.

arenaceous sediments and sedimentary rocks, and various metamorphic rocks. For some rocks, such as quartzose sandstone, chert, novaculite, and quartz veins, it is the dominant, and possibly the sole, constituent.

Opal is a water-bearing silica mineral deposited by near-surface groundwater in arid areas such as Australia or Mexico. The distinctive play of colors exhibited by many varieties arises from its unique structure composed of submicroscopic spheres of opaline matter arranged in more or less regular closely packed arrays. These diffract light in the manner of a diffraction grating, and the colors are analogous to the diffracted beams of x rays impinging on atoms in a crystal structure.

α-Quartz
SiO₂

kwôrts, from German, *quarz,* of uncertain derivation; perhaps from West Slav *kwardy,* hard, or Old English *querklufterz,* cross-vein ore.

Crystallography, Structure, and Habit Hexagonal. 32. *a* 4.914, *c* 5.405. $P3_121$ (right). Hexagonal prisms with pyramidal termination (see figures 7.24 and 21.4). Left- and right-handed forms. Prism faces often horizontally striated. Crystals sometimes distorted. Penetration and contact twins common, usually with {10$\bar{1}$0} twin plane, as in Dauphine, with {11$\bar{2}$0} twin plane, as in Brazil, or combined.

Table 21.2 *Continued*

Mineral	Reaction with HCl	Yields Water in the Closed Tube	Fusibility	Miscellaneous
Nepheline	Gel	No	4	
Olivine	Gel	No	Infusible	Slow reaction
Opal	Insoluble	Yes	Infusible	
Orthoclase	Insoluble	No	5	
Pectolite	Soluble	Yes	2.5–3	
Phlogopite	Insoluble	Yes	4.5–5	As biotite
Plagioclase series				
Albite	Insoluble	No	4–4.5	
Anorthite	Soluble	No	5	
Prehnite	Insoluble	Yes	2.5	Gelatinizes after being fused
Pyrophyllite	Insoluble	Yes	Infusible	
Quartz	Insoluble	No	Infusible	
Rhodonite	Insoluble	No	3	
Scapolite	Soluble	No	3	Imperfectly decomposed
Serpentine	Soluble	Yes	Infusible	
Sillimanite	Insoluble	No	Infusible	
Sodalite	Gel	No	3.5–4	
Spodumene	Insoluble	No	3.5	
Staurolite	Insoluble	Yes	Infusible	
Stilbite	Soluble	Yes	3	
Talc	Insoluble	Yes	5	
Titanite	Insoluble	No	4	
Topaz	Insoluble	No	Infusible	
Tourmaline	Insoluble	No	Varies with composition from 3 to infusible	
Tremolite	Insoluble	Yes	3–4	
Vesuvianite	Insoluble	Yes	3	Gelatinizes after being fused
Wollastonite	Soluble	No	Infusible	
Zircon	Insoluble	No	Infusible	

Figure 21.4
Quartz. Collier Creek, Arkansas. (Courtesy Royal Ontario Museum, Toronto. Photograph by Brian Boyle.)

Quartz structure (see figure 21.5) has interlocking spirals of shared silica tetrahedra which may be either clockwise or counterclockwise parallel to the c-axis. Three SiO_2 per cell.

Found massive, as crystals; rarely in other habits.

Physical Properties   No cleavage. No parting. Irregular to conchoidal fracture (see figure 9.1). Brittle. Hardness 7. Specific gravity 2.65. Vitreous luster. Transparent to translucent. Colorless, white, and less commonly, purple, pink, yellow, or black. Optical: $\omega = 1.544$ and $\epsilon = 1.553$; positive. Strong piezoelectric and pyroelectric properties. May be colored by radiation. Some varieties luminescent. Melting point 1,410° C.

Distinctive Properties and Tests   Hardness, lack of cleavage, and luster. Infusible and insoluble except in HF.

Association and Occurrence   Quartz is among the most common of minerals and is found almost everywhere. It is an important constituent in many igneous, metamorphic, and sedimentary rocks and the major component of some, such as sandstone, quartzite, and novaculite. It is also the most common mineral in veins, either as a gangue or sole constituent, and an almost universal residual and detrital mineral found in soils, sediments, and on beaches.

Alteration   Resistant to change. Chalcedonic variety commonly replaces wood, shell, and bones.

Confused with   Beryl, calcite, cordierite, scheelite, and feldspars.

Variants   Several form varieties. *Coarsely crystalline varieties:* amethyst (purple), rose quartz (pink), smoky quartz (yellow-brown to black), cairngorm (smoky yellow or brown), morion (nearly black), milky quartz, rock crystal (clear), and citrine (yellow). Enclosing rutile (blue, rutilated), tourmaline, hematite scales, or mica (aventurine). Replacing asbestos (tiger eye, cat's eye). *Fibrous cryptocrystalline varieties:* chalcedony, including carnelian (red), chrysoprase (apple-green), sard (brownish), plasma (white-flecked green), bloodstone or heliotrope (red-flecked green), agate (banded and variegated), onyx (straight white and gray or brown banding), sardonyx (blue-white and reddish bands), moss agate, or mocha stone (with dark wispy growths). *Granular cryptocrystalline varieties:* flint (dull to dark in color), chert (light in color), jasper (red), and prase (dull green). *Chemical varieties:* May contain minute amounts of Li, Na, Al, Ti, and OH.

Related Minerals   *Isostructures:* eucryptite, $Li(AlSiO_4)$, and berlinite, $Al(PO_4)$. *Polymorphs:* β-quartz (high quartz); high, middle, and low tridymite; high and

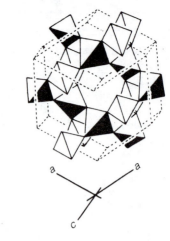

Figure 21.5
Structure of quartz.

low cristobalite; lechatelierite (silica glass); coesite; melanophlogite; keatite; and stishovite. *Other similar species:* opal.

OPAL
$SiO_2 \cdot nH_2O$

o′pəl, from Sanskrit *upala,* precious stone.

Crystallography, Structure, and Habit   Amorphous. Never in crystals. Composed of an aggregate of submicroscopic silica spheroids containing crystalline patches with a cristobalite structure. The spheroids are in cubic or hexagonal close packing.

Massive; often stalactitic or encrusting. Commonly banded.

Physical Properties   No cleavage. Conchoidal fracture. Brittle. Hardness 5–6. Specific gravity 1.9–2.2. Vitreous to resinous luster. Transparent to translucent. Colorless, white, and various tints. Optical: $n = 1.435–1.455$, decreasing with increasing water content. Often milky or opalescent. May display a brilliant play of colors, which results from light interference caused by closely packed silica globules.

Distinctive Properties and Tests   Luster and habit. Yields water in the closed tube. Attacked by hot concentrated HCl and alkaline solutions. Readily soluble in HF.

Association and Occurrence   Found (1) lining cavities in rocks, (2) replacing wood or other organic matter, (3) as accumulations of diatom or radiolarian tests (diatomaceous and radiolarian earth), and (4) as siliceous sinter or geyserite.

Confused with Chalcedony, lechatelierite (silica glass formed by lightning), volcanic glass.

Variants *Form varieties:* common opal (without internal reflections), precious opal (with internal play of colors, in white, black, and fire opal), hyalite (colorless globular form), geyserite (hot-spring deposit), wood opal (wood replaced by opal). *Chemical varieties:* may contain traces of Fe and Al.

Related Minerals *Similar species:* quartz, tridymite, and cristobalite.

FELDSPAR GROUP

fĕld'spär, from German *feld,* field, and *spat,* spar or crystal.

The feldspars constitute nearly 60% of the igneous and metamorphic rocks that make up over 90% of the earth's crust. The importance of the feldspars to the geologist is evident from these figures, and the complex mineralogical relations within the family make understanding them a challenge for the mineralogist.

Chemically, feldspars are alumino-silicates of potassium, sodium, calcium, or, rarely, barium with a considerable degree of solid solution. The formulas of the end members may be written as follows:

Alkali feldspars
$$\begin{cases} K(AlSi_3O_8) \\ Na(AlSi_3O_8) \end{cases}$$

Plagioclase feldspars
$$\begin{cases} Na(AlSi_3O_8) \\ Ca(Al_2Si_2O_8) \end{cases}$$

Celsian $\qquad Ba(Al_2Si_2O_8)$

The distinction between alkali and plagioclase feldspars is not always clear-cut because of solid solution, but the alkali feldspars are monoclinic, or if triclinic, have more K than Ca, whereas the plagioclases are triclinic with less K than Na or Ca. A diagram showing the general nature of solid solution in the feldspars is given in figure 21.6.

Virtually continuous solid solution exists at high temperatures between potassic and sodic feldspars and within the plagioclase feldspars. Both series, however, show more or less exsolution on cooling. The sanidine-monalbite series, $K(AlSi_3O_8)$-$Na(AlSi_3O_8)$, exsolves into K- and Na-rich domains in the $AlSi_3O_8$ network below about 660° C (see figure 21.7). Growth of these domains over time and under favorable temperature conditions leads to truly exsolved regions of potassic and sodic feldspar as distinct phases. With falling temperature, randomly distributed Ca, Na, and K first separate as **cryptoperthite** without causing structural discontinuity, although the network is strained. At lower temperatures, a structural and chemical discontinuity appears between domains accompanied by a rearrangement of Si and Al. This two-phase material is called perthite and appears as suboriented threads or lenses of sodic feldspar of slightly different color or luster in the

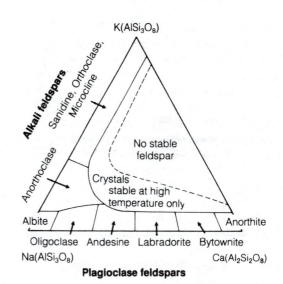

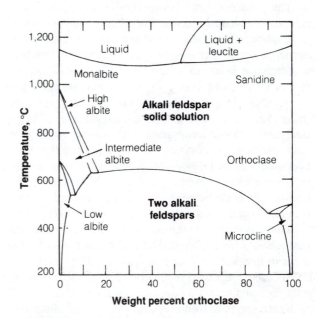

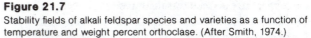

Figure 21.6
Solid solution, approximate stability fields, and nomenclature of the feldspars.

Figure 21.7
Stability fields of alkali feldspar species and varieties as a function of temperature and weight percent orthoclase. (After Smith, 1974.)

potassic feldspar matrix (see figure 5.8). The reciprocal configuration with potassic lamellae in a sodic matrix is called **antiperthite.**

The phase relations in the plagioclase series are shown in figure 21.8. Essentially complete solid solution exists at high temperatures, but at lower temperatures immiscibility gaps occur at An_2 to An_{15} (peristerite), An_{47} to An_{58} (Bøggild), and An_{60} to An_{85} (Huttenlocher). The optical evidence for the gaps is a characteristic schiller resulting from internal reflections. No perthitic exsolution textures are observed, possibly because the rearrangement of Si and Al is more difficult than in alkali feldspar.

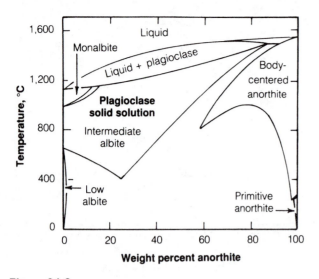

Figure 21.8
Stability fields of plagioclase as a function of temperature and weight percent anorthite. (After Smith, 1974.)

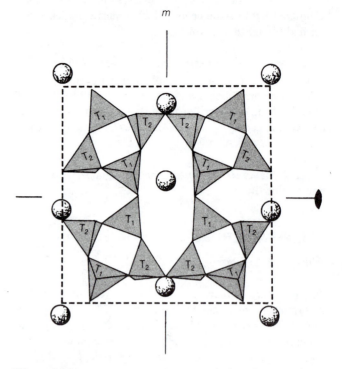

Figure 21.9
Articulation of SiO_4 tetrahedra in sanidine. View along a-axis projected onto ($20\overline{1}$). Potassium atoms, 2-fold axis, and mirror (m) shown. T_1 and T_2 identify crystallographically distinct tetrahedra. Unit cell dashed. (After Taylor, 1933.)

The structure of all feldspars is very similar. These minerals are networks in which one-quarter to one-half of the tetrahedrally coordinated silicon ions are replaced by aluminum ions. The charge discrepancy resulting from this substitution is balanced by the entry into the structure of Na^+, K^+, Ca^{2+}, and, rarely, by other alkali metals or alkaline earths. In orthoclase, $K(AlSi_3O_8)$, the potassium ion just fits the cation site, fixing the angle between the a- and b-axes and the cleavages at 90°. In the plagioclase series, this site is occupied by varying proportions of sodium and calcium, which, being smaller than potassium, allow a small slumping of the network of linked tetrahedra. The angle between the a- and b-axes and the two directions of cleavages is then close to but not exactly 90°.

The structure of the simplest feldspar, sanidine (monoclinic, $2/m$), is shown in a ($20\overline{1}$) projection along the a-axis in figure 21.9. The principal structural units are 4-membered rings of crystallographically distinct tetrahedra T_1 and T_2.

There are four formula units per cell in a feldspar. Thus, there are sixteen tetrahedral sites per cell, and four Al^{3+} and twelve Si^{4+} to fill them. The bulk chemistry requires that there be one Al^{3+} in the average 4-membered ring. In sanidine there are only two symmetrically nonequivalent sites, T_1 and T_2. If t_1 represents the Al content of the T_1 site, then $2t_1 + 2t_2 = 1.0$, and if Al and Si are distributed randomly, the structure is disordered and $t_1 = t_2 = 0.25$. This will be the case for the high-sanidine–monalbite series formed at high temperatures.

At lower temperatures (and under conditions of slower crystallization), the Al^{3+} migrates preferentially to the T_1 site. This satisfies the requirement of local electrostatic charge balance because the oxygens coordinating T_1 are more closely bonded to the large alkali ion than those coordinating T_2. In the orthoclase series, also monoclinic, the predicted T_1 site occupancy for Al^{3+} is $0.74 < 2t_1 < 1.0$.

Remember that in monoclinic feldspars there are only two nonequivalent tetrahedral sites, T_1 and T_2. Thus, t_1 cannot exceed 0.5. If further ordering is to occur, the T_1 sites and the T_2 sites in the 4-membered rings must be further differentiated by destruction of the 2-fold axis parallel to b and the (010) mirror. This produces four nonequivalent tetrahedral sites traditionally designated T_1O, T_1m, T_2O, and T_2m. If Al^{3+} concentrates in T_1O at the expense of T_1m, the sites are no longer equivalent and the symmetry degenerates to $\overline{1}$. The Al and Si distribution is completely ordered when $t_1o = 1.0$ and $t_1m = t_2o = t_2m = 0.0$. This is characteristic of low albite or low (maximum) microcline. Intermediate degrees of Al-Si ordering are evident in the microcline-albite series, where $t_1o > t_1m > t_2o = t_2m$.

Twinning is a very common phenomenon in feldspars, occurring usually with simple or repeated twins. The more common types are shown in figure 21.10.

Hand-specimen distinction between orthoclase and microcline is usually impossible unless a variety with special characteristics is being examined. It is probable that much of the potassic feldspar commonly classed as orthoclase is really microcline. Because the distinction often cannot be made in hand specimens, however, it is common practice to combine the two series under one of the terms *potassium feldspar, potash feldspar,* or *K-feldspar.* Similarly, the general term *alkali feldspar* is useful when the Na:K ratio is unknown.

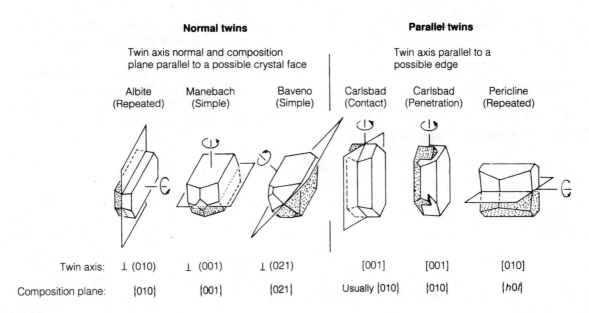

Normal twins

Twin axis normal and composition plane parallel to a possible crystal face

Parallel twins

Twin axis parallel to a possible edge

	Albite (Repeated)	Manebach (Simple)	Baveno (Simple)	Carlsbad (Contact)	Carlsbad (Penetration)	Pericline (Repeated)
Twin axis:	$\perp$ (010)	$\perp$ (001)	$\perp$ (021)	[001]	[001]	[010]
Composition plane:	{010}	{001}	{021}	Usually {010}	{010}	{h0l}

Figure 21.10
Feldspar twins.

Plagioclase minerals belong to a solid-solution series whose end members are albite, $Na(AlSi_3O_8)$, and anorthite, $Ca(Al_2Si_2O_8)$. This series can be divided arbitrarily into species:

Albite		Andesine		Bytownite	
	Oligoclase		Labradorite		Anorthite

$Na(AlSi_3O_8)$ ———————————————— $Ca(Al_2Si_2O_8)$

These species are also shown in Figure 21.6. The individual species are difficult to distinguish without petrographic, x-ray, chemical, or similar studies, and it is preferable when making hand-specimen determinations to use the series name *plagioclase*.

Certain of the plagioclase species can sometimes be distinguished on the basis of width of the polysynthetic twin lamellae, play of colors, and association. Plagioclase can usually be distinguished from the potash feldspars by twinning.

ORTHOCLASE SERIES
K(AlSi₃O₈)

(orthose, potash feldspar, potassium feldspar, K-spar) ôr′thə-klās, from Greek *orthos,* a right angle, and *klasis,* to break, referring to its orthogonal cleavage.

Crystallography, Structure, and Habit Monoclinic. $2/m$. a 2.562, b 12.996, c 7.193, β 116.0°. $C2/m$. Crystals as stubby to elongate prisms, often flattened and doubly terminated. Twinning common as simple penetration or contact twins (see figure 21.11).

Orthoclase structure. 4-membered tetrahedral rings parallel to (001) grouped by fours to surround holes that are occupied by alkali metals. Four formula units per cell.

Figure 21.11
Orthoclase with carlsbad twinning. Both crystals are from the Tertiary volcanics at Crested Butte, Colorado. The smaller crystal is 2 cm in the longest direction. (Courtesy University of Windsor. Photograph by D. L. MacDonald.)

Commonly found as more or less well-developed crystals; also massive and in formless grains.

Physical Properties Perfect {001} and {010} cleavages, yielding square pseudoprismatic fragments. Parting parallel to {100} and also at a small angle to (100), giving rise to schiller. Conchoidal to uneven fracture. Brittle. Hardness 6–6.5. Specific gravity 2.6. Vitreous luster. Transparent to translucent. Colorless, white, or flesh-red in color for most specimens. Streak white. Optical: $\alpha = 1.518$, $\beta = 1.522$, and $\gamma = 1.523$; negative; $2V = 60\text{--}85°$. Fusible with difficulty (5).

Distinctive Properties and Tests   Cleavage, crystal habit, twinning, hardness, association, and color. Soluble in HF.

Association and Occurrence   Orthoclase is among the more common minerals. It is (1) a principal constituent of igneous rocks, such as granite and syenite, (2) a major constituent of the sedimentary rock arkose, (3) common in many kinds of metamorphic rocks, and (4) found in pegmatites and in veins. It is usually associated with quartz and muscovite.

Alteration   To muscovite and kaolin.

Confused with   Microcline, calcite, corundum, and plagioclase.

Variants   *Form varieties:* moonstone, usually as transparent crystals in cavities in schists and veins; perthite; and adularia. *Chemical varieties:* sodian and barian (hyalophane). May contain small amounts of Cs, Rb, Li, or Ca. Also small amounts of Ti, Fe, Mg, Sr, Mn, Tl, and Pb.

Related Minerals   *Solid solutions:* sodian orthoclase, $(K,Na)(AlSi_3O_8)$; anorthoclase, $(Na,K)(AlSi_3O_8)$; and celsian, $Ba(Al_2Si_2O_8)$. *Polymorphs:* sanidine (complete Al-Si disorder, monoclinic) and microcline (complete Al-Si order). *Other similar species:* kaliophilite, $K(AlSiO_4)$; leucite; buddingtonite, $NH_4(AlSi_3O_8) \cdot \frac{1}{2}H_2O$; and banalsite, $BaNa_2(Al_4Si_4O_{16})$.

MICROCLINE SERIES
K(AlSi₃O₈)

mī′krō-klīn, from Greek *mikros,* small, and *klinein,* to lean, an allusion to the small deviation of cleavage angles from 90°.

Crystallography, Structure, and Habit   Triclinic. $\bar{1}$. a 8.577, b 12.967, c 7.223, α 90.65°, β 115.93°, γ 87.78°. $C\bar{1}$. Crystals as stubby to elongate prisms, often doubly terminated. Twinning common as simple penetration or contact twins. Polysynthetic twinning by both albite and pericline laws leads to the characteristic grating or grid twins, also called "tartan twinning," seen in thin section.

Structure is similar to sanidine (see figure 21.9), but microcline is triclinic and thus less symmetrical.

Commonly found as more or less well developed crystals, also massive and as formless grains.

Physical Properties   Perfect $\{001\}$ and good $\{010\}$ cleavage at an angle of 89.5°. Conchoidal to uneven fracture. Hardness 6. Specific gravity 2.54–2.57. Vitreous luster. Transparent to translucent. Colorless, white, yellowish, reddish, or green. Streak white. Optical: $\alpha = 1.518$, $\beta = 1.522$, and $\gamma = 1.525$; negative; $2V = 80°$. Fusible with difficulty (5).

Distinctive Properties and Tests   Cleavage, crystal habit, twinning, hardness, association, and color. Soluble in HF.

Association and Occurrence   Microcline is a very common alkali feldspar. It is usually found in rocks formed at lower temperatures than those that contain orthoclase. It is found in (1) granite and pegmatite, where it is the principal alkali feldspar, (2) veins, (3) metamorphic rocks, and (4) arkosic sedimentary rocks. Microcline is commonly associated with quartz and muscovite.

Alteration   To muscovite and kaolin.

Confused with   Orthoclase, calcite, corundum, and plagioclase.

Variants   *Chemical varieties:* sodian, barian (hyalophane), and rubidian (green, amazonite).

Related Minerals   *Solid solutions:* albite, $Na(AlSi_3O_8)$. *Polymorphs:* sanidine and orthoclase (monoclinic).

PLAGIOCLASE SERIES
Albite Na(AlSi₃O₈),
Anorthite Ca(Al₂Si₂O₈)

plā′jē-ə-klās, from Greek *plagios,* oblique, and *klasis,* breaking, an allusion to the obliquity of the angle between its $\{001\}$ and $\{010\}$ cleavages.

Crystallography, Structure, and Habit   Triclinic. $\bar{1}$. Albite: a 8.139, b 12.788, c 7.160, α 94.3°, β 116.6°, γ 87.7°. Anorthite: a 8.177, b 12.877, c 14.169, α 93.2°, β 115.85°, γ 91.2°. $P\bar{1}$(anorthite)–$C\bar{1}$(albite). Crystals usually tabular to bladed. Frequently twinned; often both simple and repeated twins.

Structure essentially that of orthoclase with a small distortion. Four formula units per cell for albite, and eight for anorthite.

Rarely in distinct crystals; usually found as cleavable masses or as irregular embedded grains.

Physical Properties   Perfect $\{001\}$ and distinct $\{010\}$ cleavage intersecting at about 86°. Uneven to conchoidal fracture. Brittle. Hardness 6–6.5. Specific gravity 2.6

(albite)–2.75 (anorthite). Vitreous luster. Transparent to translucent. Color white, gray, and occasionally reddish. Albitic varieties are lighter colored. Anorthitic varieties are often medium to dark gray and sometimes bluish gray. Streak white. Optical as follows:

Species	α	β	γ	Sign	2V
Albite (An$_{0-10}$)	1.527–1.533	1.531–1.537	1.538–1.582	+	77–82°
Oligoclase (An$_{10-30}$)	1.533–1.543	1.537–1.548	1.542–1.551	±	82–90°
Andesine (An$_{30-50}$)	1.543–1.544	1.548–1.558	1.551–1.562	±	76–90°
Labradorite (An$_{50-70}$)	1.554–1.564	1.558–1.569	1.562–1.572	+	76–90°
Bytownite (An$_{70-90}$)	1.564–1.573	1.569–1.579	1.573–1.585	−	79–88°
Anorthite (An$_{90-100}$)	1.573–1.577	1.579–1.585	1.585–1.590	−	77–79°

Melting point 1,100° C (albite) to 1,550° C (anorthite). A play of colors is often observed in labradorite (Ab$_{40}$An$_{60}$) and less often in andesine and albite.

Distinctive Properties and Tests Cleavage, polysynthetic twinning, hardness, and association. The polysynthetic twins tend to be wider in calcic varieties (see figure 10.20a). Albite is insoluble in acids (except in HF), but anorthite is decomposed with HCl and yields gelatinous silica.

Association and Occurrence Plagioclase is the most abundant mineral in the earth's crust. It is the principal constituent of many igneous rocks, and intermediate members are the most common. Albitic plagioclase is found in igneous rocks, such as granite, syenite, rhyolite, and trachyte, associated with orthoclase and quartz, and in pegmatites. Intermediate members of the plagioclase series are found principally in igneous rocks. The more calcic members are found in less siliceous rock types. Anorthite is a relatively rare plagioclase found in very mafic igneous rocks, in contact-metamorphosed marbles, and in meteorites.

Alteration To muscovite, kaolin, and calcite. Calcic plagioclase plus water yields zoisite or epidote (saussuritization).

Confused with Potash feldspars, quartz, and calcite.

Variants *Form varieties:* clevelandite (albite as curved lamellar masses in pegmatites). *Chemical varieties:* may contain Rb, Ba, Ti, Fe, Mg, Sr, and Mn.

Related Minerals *Solid solutions:* The plagioclase series is a continuous solid-solution series at elevated temperatures. The low-temperature series is discontinuous with breaks at 2, 47, and 60% anorthite component. Members of the plagioclase series are named as shown in figure

21.6 and the bar diagram on page 313. *Isostructures:* slawsonite, $(Sr,Ca)(Al_2Si_2O_8)$. *Polymorphs:* high albite (complete Al-Si disorder if monoclinic). *Isotypes:* other feldspars. *Other similar species:* nepheline; reedmergnerite, $Na(BSi_3O_8)$; eudidymite, $Na[BeSi_3O_7(OH)]$; danburite, $Ca(B_2Si_2O_8)$; and hurlbutite, $Ca(Be_2P_2O_8)$.

FELDSPATHOID GROUP

Feldspathoids, as implied by their name, are feldsparlike in appearance and general composition. They are interesting compositionally in that several of the species contain chlorine, sulfur, or the carbonate radical, all of which are unusual constituents in silicate minerals.

Feldspathoids serve as useful geologic indicators since they crystallize from silica-poor magmas in lieu of feldspars and hence are never found in primary association with quartz.

The minerals of the feldspathoid group usually fuse readily and either decompose or gelatinize when treated with HCl. The more common species are

Leucite	$K(AlSi_2O_6)$
Nepheline	$Na(AlSiO_4)$
Sodalite	$Na_4(AlSiO_4)_3Cl$
Cancrinite	$(Na,Ca)_4(AlSiO_4)_3(CO_3) \cdot nH_2O$

LEUCITE
K(AlSi₂O₆)

(amphigene)
$\overline{loo}$'sīt, from Greek *leukos,* white, referring to its typical color.

Crystallography, Structure, and Habit Tetragonal and pseudoisometric. $4/m$. a 13.074, c 13.738. $I4_1/a$. Commonly in pseudotrapezohedra. Faces often show fine striations as a result of repeated twinning.

Network structure (see figure 21.12) containing 4-, 6-, and 8-membered loops of tetrahedra. There are two different sites for K^+, only one of which is occupied, resulting in an open, low-density structure. Sixteen formula units per cell.

Usually in well-formed trapezohedral crystals; rarely in disseminated grains or massive.

Physical Properties Poor {110} cleavage. Conchoidal fracture. Brittle. Hardness 5.5–6. Specific gravity 2.6. Vitreous to dull luster. Translucent to opaque. Color white or gray. Streak uncolored. Optical: $\omega = 1.508$ and $\epsilon = 1.509$; positive. Infusible.

Distinctive Properties and Tests Crystal form, color, and association. Decomposed by HCl without gelatinization.

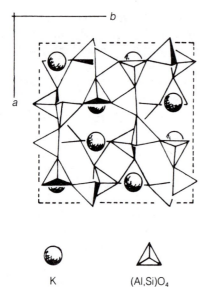

Figure 21.12
Structure of leucite. Unit cell dashed.

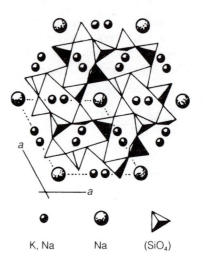

K, Na   Na   (SiO₄)

Figure 21.13
Structure of nepheline. View along *c*-axis; unit cell dashed.

Association and Occurrence   Found only in silica-poor and potassium-rich volcanic rocks, and especially in recent lavas.

Alteration   To analcime and kaolin, and pseudoleucite, an intergrowth of nepheline and K-feldspar.

Confused with   Analcime, garnet, and melilite.

Variants   *Chemical varieties:* sodian. Traces of Fe, Na, Li, Rb, and Cs may be present.

Related Minerals   *Polymorphs:* isometric above 605° C. *Other similar species:* pollucite, $(CsNa)_2$ $(Al_2Si_4O)_{12} \cdot H_2O$; orthoclase; kaliophilite, $K(AlSiO_4)$; and analcime.

NEPHELINE
Na(AlSiO₄)

(nephelite)
nĕf′ə-lēn, from Greek *nephele,* cloud, because clear crystals turn cloudy when immersed in hydrochloric acid.

Crystallography, Structure, and Habit   Hexagonal. 6. *a* 9.986, *c* 8.330. *P6₃*. Short prismatic to tabular crystals.
Isostructural with tridymite (see figure 5.16). Alternate Si ions of tridymite are replaced with Al ions (see figure 21.13). Na (or K) occupies the large holes in the tridymite structure. Eight formula units per cell.
Massive, compact, or embedded grains.

Physical Properties   Distinct {10$\bar{1}$0} and poor {0001} cleavage. Subconchoidal fracture. Brittle. Hardness 5–6.

Specific gravity 2.6. Vitreous luster; greasy on cleaved surfaces. Transparent to opaque. Usually colorless, white, or yellowish; also greenish, grayish, or reddish. Optical: $\omega = 1.530–1.545$ and $\epsilon = 1.527–1.541$; negative. Fusible.

Distinctive Properties and Tests   Greasy luster and association. Decomposed by HCl with separation of gelatinous silica.

Association and Occurrence   Typically found in silica-poor igneous rocks, such as syenites or, rarely, basalts. Associated with feldspars, cancrinite, biotite, sodalite, corundum, and zircon. Never in primary association with quartz.

Alteration   To a micaceous aggregate including zeolites, sodalite, cancrinite, and kaolinite.

Confused with   Feldspar, quartz, and apatite.

Variants   *Form varieties:* glassy, massive with a greasy luster. *Chemical varieties:* potassian (all natural nepheline contains about one K for every three Na ions). May also contain Li, Ga, and an excess of Si or Al.

Related Minerals   *Solid solutions:* kalsilite, $K(AlSiO_4)$. *Isotypes:* tridymite, SiO_2. *Polymorphs:* high nepheline above 900° C. *Other similar species:* albite and kaliophilite, $K(AlSiO_4)$.

SODALITE
Na₄(AlSiO₄)₃Cl

sōd′l-līt, from Middle English *soda,* saltwort.

Crystallography, Structure, and Habit   Isometric. $\bar{4}$3*m*. *a* 8.9., *P*$\bar{4}$3*n*. Rare dodecahedral crystals. Contact twins.

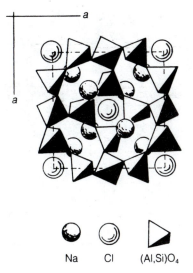

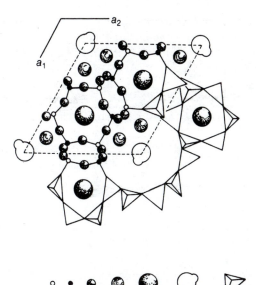

Na   Cl   (Al,Si)O₄

Figure 21.14
Structure of sodalite. Unit cell dashed.

Al Si O Na [Na H₂O]⁺ H₂O (Al,Si)O₄

Figure 21.15
Structure of cancrinite. View along *c*-axis; unit cell dashed. (Adapted from Hassan and Grundy, 1991.)

Rings of four $(Al,Si)O_4$ tetrahedra are centered on each cube face and are linked to 6-membered tetrahedral rings on the cell corners (see figure 21.14). The large holes accommodate Cl, S, and SO_4. One formula unit per cell.

Compact, cleavable, nodular, or disseminated masses.

Physical Properties Distinct dodecahedral $\{011\}$ cleavage. Uneven to conchoidal fracture. Hardness 5.5–6. Specific gravity 2.1–2.4 increasing with Ca and SO_4. Subvitreous luster on faces and greasy on cleavages. Transparent to opaque. Usually blue in color; also white, green, reddish, or gray. Streak colorless. Optical: $n = 1.484$. Colored varieties turn white on heating. Fusible (3.5) with intumescence to a colorless glass.

Distinctive Properties and Tests Color, association, and color loss on heating. Gelatinizes with HCl.

Association and Occurrence Associated with nepheline, cancrinite, leucite, feldspar, and zircon in silica-poor, alkali-rich igneous rocks.

Alteration To fibrous zeolites, clays, or cancrinite.

Confused with Leucite, analcite.

Variants Chemical varieties: sulfurian (hackmanite), potassian, calcian, and hydroxylian.

Related Minerals Isotypes: haüyne, $(Na,Ca)_{4-6}$ $[Al_6Si_6(O,S)_{24}](SO_4,Cl)_{1-2}$; nosean, $Na_8(AlSiO_4)_6(SO_4)$; and lazurite (lapis lazuli), $(Na,Ca)_{7-8}[(Al,Si)_{12}(O,S)_{24}]$ $[(SO_4),Cl,(OH)_2]$; hackmanite, $Na_8(Al_6Si_6O_{24})(Cl,S)$.

CANCRINITE
Na₆Ca₂(Al₆Si₆O₂₄)(CO₃)₂

kăn′kri-nīt, after Count Georg Cancrin (1774–1845), Russian Minister of Finance.

Crystallography, Structure, and Habit Hexagonal. $6/m\ 2/m\ 2/m$. *a* 12.58–12.76, *c* 5.11–5.20. $C6_3$. Rare prismatic crystals.

Corner-sharing AlO_4 and SiO_4 tetrahedra form a framework enclosing chains of cages and channels parallel to the a_3-axis. The cages are bounded by 6-membered rings of alternating AlO_4 and SiO_4 tetrahedra, and the channels by puckered 12-member rings of $(Si,Al)O_4$ tetrahedra (see figure 21.15). Other ions lie in the large channels. One formula unit per cell.

In compact, lamellar, or disseminated masses.

Physical Properties Perfect $\{10\bar{1}0\}$ and poor $\{0001\}$ cleavage. Uneven fracture. Hardness 5–6. Specific gravity 2.5. Subvitreous to greasy luster. Transparent to translucent. Color usually yellow, also white, gray, green, blue, or reddish. Streak colorless. Optical: $\omega = 1.507–1.528$ and $\epsilon = 1.495–1.503$; negative. Fusible (2) with intumescence to a white blebby glass. Colored varieties lose color on heating.

Distinctive Properties and Tests Color, association, and loss of color on heating. Evolves water in the closed tube. Effervesces with HCl and forms a jelly on heating.

Association and Occurrence Found in silica-poor igneous rocks, such as nepheline syenite. Associated with nepheline, sodalite, biotite, feldspar, titanite, and apatite.

Alteration From nepheline.

Confused with Other feldspathoid minerals.

Variants Chemical varieties: sodian-sulfatic (vishnevite), chlorian (davyne), and potassian-chlorian (microsommite).

Related Minerals Isotypes: wenkite, $Ba_4Ca_6(Si,Al)_{20}$ $O_{39}(OH)_2(SO_4)_3 \cdot nH_2O$; afghanite, $(Na,Ca,K)_8[(Si,Al)_{12}$ $O_{24}][(SO_4),Cl,(CO_3)]_3 \cdot H_2O$. *Other similar species:* sodalite; franzinite, $(Na,Ca)_7[(Si,Al)_{12}O_{24}]$ $[(SO_4),$ $(CO_3),(OH),Cl]_3 \cdot H_2O$; giuseppettite, $(Na,K, Ca)_{7-8}$ $[(Si,Al)_{12}O_{24}][(SO_4),Cl]_{1-2}$; liottite, $(Ca,Na,K)_8$ $[(Si,Al)_{12}O_{24}][(SO_4),(CO_3),Cl,(OH)]_4 \cdot H_2O$; microsommite, $(Na,Ca,K)_{7-8}[(Si,Al)_{12}O_{24}]$ $[Cl,(SO_4),$ $(CO_3)]_{2-3}$; and sacrofanite, $(Na,Ca,K)_9[(Si,Al)_{12}O_{24}]$ $[(OH)_2,(SO_4),(CO_3),Cl]_3 \cdot nH_2O$.

ZEOLITE GROUP

The zeolites include over forty species of hydrous aluminum silicates with sodium and calcium as the more important cations. Structurally, zeolites are built from loops of SiO_4 or $AlSiO_4$ tetrahedra arranged into interpenetrating chains. These structures have large interstitial holes and channels (see figure 21.16) that are filled with water. This water is not structurally bound and is continuously driven off as the mineral is heated.

Three different patterns of tetrahedral linkages divide the zeolites into fibrous, layered or foliated, and equidimensional or blocky types. Tetrahedra of the fibrous types, such as natrolite or scolecite (see figure 21.17), are in chains with some cross-linkages parallel to the *c*-axis; these zeolites have tetragonal or pseudotetragonal symmetry. Layered zeolites, such as stilbite, are built up from sheets of tetrahedra with some cross-linkage parallel to (010) and display prominent {010} cleavage. Blocky zeolites, such as chabazite, are based on tetrahedral frameworks. A listing of the species in this extensive and structurally complex group of network silicates is given in table 21.3.

Most zeolites assume either an equidimensional, often polyhedral habit or a tabular, sometimes sheaflike habit. The members of this group are commonly found in marine sediments and as fillings and coatings of cavities in mafic lavas such as basalt. They commonly occur in well-formed crystals. All zeolites are quite soft with hardnesses less than that of a knife, have low specific gravity (2–2.4), and are generally colorless or of pale color. They are readily

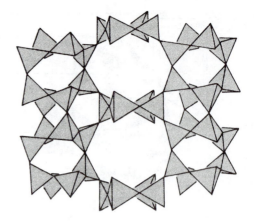

Figure 21.16
Interconnected $(Al,Si)O_4$ loops of tetrahedra in an open structure typical of the zeolites.

Figure 21.17
Scolecite in basalt vug. Nasik, Bombay State, India. (Courtesy Royal Ontario Museum, Toronto. Photograph by Leighton Warren.)

decomposed by hot HCl and usually gelatinize when the solution is evaporated. Zeolites' characteristic intumescence in response to heating is reflected in the name of the group, which derives from the Greek word *zein,* meaning "to boil."

Water readily enters and leaves the large interstices in zeolite structures so its content is largely fixed by temperature and humidity. Dehydrated zeolites are structurally intact, and depending on the particular structure and the size of the molecules involved, the channelways may be selectively refilled with such vapors as nitrogen, ammonia, iodine, or mercury. Such molecular sieves have wide use in industry including, for example, the production of oxygen from air.

Table 21.3 The Zeolites

	Common Cations			Other Cations	Al:Si	Crystal System	Number of (Al,Si)O$_4$ Tetrahedra per Ring
	Na	Ca	K				
Amicite	X		X		1:1	Mc	
Analcime	X				1:2	Isom	
Ashcroftine-Y	X	X	X	Y	1:2	Tet	
Barrerite	X	X	X		1:3.5	Ortho	
Brewsterite		X		Sr,Ba	1:3	Mc	
Chabazite		X			1:2	Hex	4,6,8
Clinoptilolite	X	X	X		Varies	Mc	
Cowlesite		X			1:1.5	Ortho	
Dachiardite	X	X	X		1:3.8	Mc	
Edingtonite				Ba	1:1.5	Ortho	
Epistilbite		X			1:3	Mc	
Erionite	X	X	X		1:7	Hex	4,6,8
Faujasite	X	X			1:2	Isom	4,6,12
Ferrierite	X		X	Mg	1:5	Ortho	
Garronite	X	X			1:1.7	Ortho	
Gismondine		X			1.1	Mc	
Gmelinite	X	X			1:2	Hex	4,6,8,12
Gonnardite	X	X			1:1.5	Ortho	
Goosecreekite		X			1:3	Mc	
Harmotome			X	Ba	Varies	Mc	4,8
Herschelite	X	X	X		1:2	Hex	
Heulandite	X	X			1:3.5	Mc	5,6,8
Laumontite		X			1:2	Mc	
Levyne	X	X	X		1:2	Hex	4,6,8
Mazzite		X	X	Mg	Varies	Hex	
Merlinoite	X	X	X	Ba	1:2.6	Ortho	
Mesolite	X	X			1:1.5	Mc	
Mordenite	X	X	X		1:5	Ortho	
Natrolite	X				1:1.5	Ortho	
Offretite		X	X		1:2.6	Hex	
Paranatrolite	X				1:1.5	Mc(?)	
Paulingite	X	X	X	Ba	1:3.2	Isom	
Phillipsite	X	X	X		Varies	Mc	4,8
Pollucite	X			Cs	1:2	Isom	
Scolecite		X			1:1.5	Mc	
Stellerite		X			1:3.5	Ortho	
Stilbite	X	X			1:2.6	Mc	
Thomsonite	X	X			1:1	Ortho	
Wairakite		X			1:2	Mc	
Wellsite		X	X	Ba	1:3	Mc	
Yugawaralite		X			1:3	Mc	

Water moving freely into and out of the zeolite structure may carry dissolved ions that can be exchanged for structural cations depending on ionic properties of the ions and their concentration gradients. By this means, hard water containing Ca^{2+} ions can be exchanged for sodium ions in a Na-zeolite, thus softening the water. (These ions also combine with oleates from soap to form calcium oleate, better known as the bathtub ring.) The Na-zeolite can be regenerated by passing a concentrated NaCl brine through the material. This cation or base exchange process is the basis for water softening systems, although more often resins having this property are used instead of zeolites.

STILBITE
$NaCa_2(Al_5Si_{13}O_{36}) \cdot 14H_2O$

(desmine)
stĭl'bīt, from Greek *stilbein,* to shine, describing its pearly luster.

Crystallography, Structure, and Habit Monoclinic. $2/m$. a 13.63, b 18.17, c 11.31, β 129.2°. $C2/m_1$. Simple crystals are unknown; all crystals are cruciform penetration twins on {001}.

Structure is based on chains of 6-membered tetrahedral rings parallel to the c-axis with channels along [100] and [001]. Four formula units per cell.

Typically in sheaflike aggregates (see figure 21.18).

Physical Properties Perfect {010} and poor {100} cleavage. Uneven fracture. Brittle. Hardness 3.5–4. Specific gravity 2.2. Vitreous luster; pearly on cleavage. Transparent to translucent. Color white; also yellowish, brownish, or reddish. Optical: $\alpha = 1.484–1.500$, $\beta = 1.492–1.507$, $\gamma = 1.494–1.513$; negative; $2V = 30–50°$. Fuses easily (2–2.5) with swelling and exfoliation to a white enamel.

Distinctive Properties and Tests Sheaflike habit and manner of fusion. Decomposed by HCl without gelatinization. Evolves water in the closed tube.

Association and Occurrence Typically found in cavities in basalt and related rocks associated with other zeolites and calcite. Rarely found in hydrothermal veins or in granite or gneiss.

Confused with Other fibrous or tabular zeolites.

Variants *Chemical varieties:* sodian, potassian, and calcian (epistilbite); Si-poor (hypostilbite) and Si-rich (stellarite). Adsorbed ions are always present.

Related Minerals *Isotypes:* heulandite, $(Na,Ca)_{2-3}$ $Al_3[(Al,Si)_2Si_{13}O_{36}] \cdot 12H_2O$. *Other similar species:* other zeolites.

CHABAZITE
$Ca(Al_2Si_4O_{12}) \cdot 6H_2O$

chăb'ə-zīt, from Greek *chabazios*, from *chalaza*, hail.

Crystallography, Structure, and Habit Hexagonal and pseudoisometric. $\bar{3}\ 2/m$. a 13.78, c 14.97. $R\bar{3}m$. Crystals usually simple rhombohedra close to cubes in angle. Penetration twins on {0001} common.

Basketwork of (Al,Si)-O tetrahedral chains sharing oxygens to form a network (see figure 21.19). Three formula units per cell.

Found as crystals and crystal crusts.

Physical Properties Distinct {10$\bar{1}$1} cleavage. Uneven fracture. Brittle. Hardness 4–5. Specific gravity 2.1. Vitreous luster. Transparent to translucent. Color flesh-red to white. Streak colorless. Optical: $\omega = 1.484$ and $\epsilon = 1.481$; negative. Readily fusible (3) with swelling to a nearly opaque blebby glass.

Distinctive Properties and Tests Crystal form, color, and association. Decomposed by HCl with the separation of slimy silica but without gelatinization. Evolves abundant water in the closed tube.

Figure 21.18
Sheaflike aggregates of stilbite crystals. Bombay, India. (Courtesy of University of Windsor.)

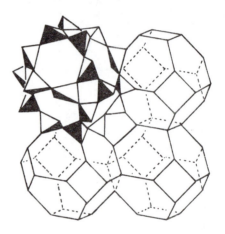

Figure 21.19
Structure of chabazite. (Al,Si)O$_4$ tetrahedra are centered on the vertices of multifaceted polyhedra.

Association and Occurrence Typically found in cavities in lavas of basaltic composition and rarely in cavities in other rocks. Associated with other zeolites and calcite. Deposited in thermal springs and hydrothermal veins.

Confused with Calcite.

Variants The composition is variable and adsorbed ions are always present.

Related Minerals *Solid solutions:* Chabazite shows continuous variation from Ca-rich to (Na,K)-rich forms and (Na,Ca)-rich forms (gmelinite, levyne). *Similar species:* other zeolites.

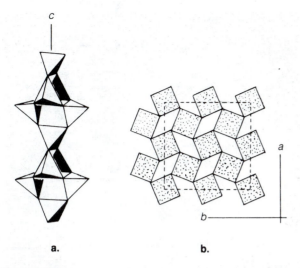

Figure 21.20
(a) Tetrahedral chains of the kind found in natrolite and other fibrous zeolites. (b) Their packing viewed along the *c*-axis.

NATROLITE
$Na_2(Al_2Si_3O_{10}) \cdot 2H_2O$

nā′trə-līt, from Greek *natron,* soda.

Crystallography, Structure, and Habit Orthorhombic and pseudotetragonal. *2mm. a* 18.30, *b* 18.63, *c* 6.60. *Fdd*2. Crystals with a square cross section as thin prisms or acicular; often as radial aggregates. Vertically striated. Cruciform twins.

Chains of (Al,Si)O_4 tetrahedra that are cross-linked by oxygen-sharing run parallel to the *c*-axis (see figure 21.20). Lengthwise cavities between the chains accommodate Na and H_2O. Eight formula units per cell.

Usually in radiating crystal groups; also massive and granular.

Physical Properties Perfect {110} cleavage. Uneven fracture. Hardness 5–5.5. Specific gravity 2.2–2.4. Subvitreous to silky or pearly luster. Transparent to translucent. Colorless or white; also greenish, yellowish, or reddish. Optical: $\alpha = 1.479$, $\beta = 1.482$, $\gamma = 1.491$; positive; $2V = 60°$. Easily fusible (2) to a colorless glass.

Distinctive Properties and Tests Habit, cleavage, and association. Evolves water, whitens, and becomes opaque in the closed tube. Soluble in HCl and yields gelatinous silica on evaporation.

Association and Occurrence A secondary mineral found filling cavities in basaltic lavas. It is associated with other zeolites and calcite.

Alteration From nepheline, sodalite, and plagioclase.

Confused with Aragonite and pectolite.

Variants Always contains minor amounts of K or Ca. Adsorbed ions are always present.

Related Minerals *Similar species:* other zeolites, especially mesolite, scolecite, thomsonite, gonnardite, and edingtonite. *Polymorphs:* tetranatrolite (tetragonal).

MISCELLANEOUS TECTOSILICATES

A number of tectosilicates have affinities for other families or groups but for various reasons should be considered separately. As with most other networks, they are all alumino-silicates, usually of Ca or Na, and are based on complex linkages of multimembered tetrahedral rings. Scapolite contains halogens, CO_3, and SO_4, as found also in some feldspathoids. Cordierite, although traditionally considered a ring structure, is here grouped with the networks based on the tetrahedral coordination of inter-ring Al^{3+}. Analcime has many of the characteristics of a feldspathoid but is often classified with the zeolites.

SCAPOLITE GROUP

(wernerite)
skăp′ə-līt, from Greek *skapos,* stalk, alluding to the shape of its crystals.

MARIALITE
$Na_4(Al_3Si_9O_{24})Cl$

mə′-rē-ə-lit, after Maria Rosa, wife of Gerhard vom Rath (1830–1888), a professor of mineralogy at the University of Bonn, Germany.

MEIONITE
$Ca_4(Al_6Si_6O_{24})[(CO_3), (SO_4)]$

mī′ə-nīt, from Greek *meion,* less, in reference to the less acute pyramidal terminations of its crystals compared to vesuvianite.

Crystallography, Structure, and Habit Tetragonal. *4/m.* Marialite: *a* 12.064, *c* 7.514. Meionite: *a* 12.174, *c* 7.652. *I* 4/*m.* Coarse prismatic crystals, often woody looking.

Rings of 4-membered (Al,Si)-O tetrahedra (see figure 21.21) are linked to form columns parallel to the *c*-axis. The smaller holes are occupied by Na or Ca cations and the larger by Cl or (CO_3) anions. Two formula units per cell.

In crystals, massive, and granular.

Physical Properties Distinct but interrupted {100} and {110} cleavage. Subconchoidal fracture. Brittle. Hardness 5–6.5. Specific gravity 2.5–2.8. Vitreous to pearly luster.

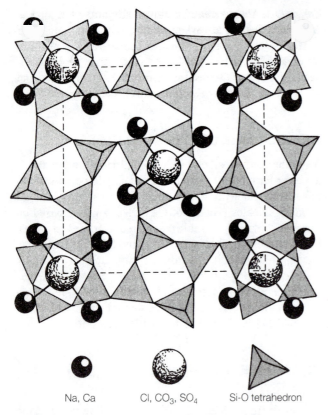

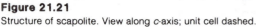

Na, Ca   Cl, CO₃, SO₄   Si-O tetrahedron

Figure 21.21
Structure of scapolite. View along c-axis; unit cell dashed.

Transparent to translucent. Color white or gray; also greenish, bluish, or reddish. Streak white. Optical: ω = 1.545–1.610 and ϵ = 1.540–1.570 increasing from marialite to meionite; negative. Fuses (3) with intumescence to a white blebby glass. Sometimes luminescent.

Distinctive Properties and Tests   Square prismatic crystals with a woody surface, cleavages intersecting at 45°, and association. Imperfectly decomposed by HCl and yields granular silica but not a jelly.

Association and Occurrence   Typically found in metamorphic rocks, especially those rich in calcium, such as marble. Associated with pyroxenes, amphiboles, garnet, apatite, titanite, zircon, and biotite.

Alteration   Readily to mica, epidote, talc, kaolin, and zeolites. From plagioclase.

Confused with   Feldspar and pyroxene.

Variants   K may replace Na, and (OH), F, or (SO₄) may replace Cl and (CO₃).

Related Minerals   Solid solutions: Marialite and meionite form a complete solid solution and the pure end-members are not found in nature.

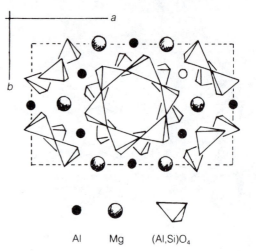

Al   Mg   (Al,Si)O₄

Figure 21.22
Structure of cordierite. View along c-axis; unit cell dashed.

CORDIERITE
Mg₂(Al₄Si₅O₁₈)

-1861),
al.

hombic
17.062,
winned

-mem-
ed into
e figure

d {001}
racture.
reous to
hades of
blue (fresh) to shades of green (altered). Optical: α = 1.54, β = 1.55, γ = 1.56; negative; $2V$ = 60–90°. Fusible with difficulty (5–5.5).

Distinctive Properties and Tests   Cleavage, color, and association. Insoluble except in HF.

Association and Occurrence   Common in Mg- and Al-rich metamorphic rocks, especially those in contact zones. Rare as an accessory mineral in granite, gabbro, pegmatite, and felsic lavas. Associated with anthophyllite, garnet, mica, quartz, andalusite, sillimanite, staurolite, and spinel.

Alteration   Alters readily, first developing a basal parting and a gray-green color and later changing, to an aggregate of chlorite and mica (pinite).

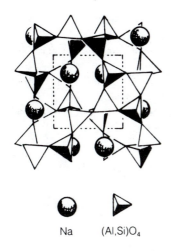

Na $(Al,Si)O_4$

Figure 21.23
Structure of analcime. Unit cell dashed.

Confused with   Quartz, corundum, and plagioclase.

Variants   *Chemical varieties:* ferroan (sekaninaite); all natural cordierite is ferroan but the iron content is small. Also ferrian, manganoan, calcian, beryllian, and hydoxylian. K and Na usually present.

Related Minerals   *Solid solutions:* sekaninaite, $(Fe,Mg)_2(Al_4Si_5O_{18})$. *Isotypes:* beryl. *Polymorphs:* complete range of ordering in Si-Al tetrahedra; high and low indialite (hexagonal). *Other similar species:* osumilite, $(K,Na)(Fe,Mg)_2(Al,Fe)_3[(Si,Al)_{12}O_{30}] \cdot H_2O$.

ANALCIME
$Na(AlSi_2O_6) \cdot H_2O$

(analcite)
ă-năl′sēm, from Greek *analkimos,* weak, referring to the weak electrostatic charge that develops when the mineral is heated (pyroelectricity) or rubbed.

Crystallography, Structure, and Habit   Isometric. $4/m$ $\bar{3}2/m$. a 13.733. $Ia3d$. Usually in trapezohedra and sometimes in cubes.

Network structure (see figure 21.23) containing 4-, 6-, and 8-membered loops of tetrahedra. Sixteen formula units per cell.

As well-formed crystals and crystal crusts; also granular and compact.

Physical Properties   No cleavage. Subconchoidal fracture. Brittle. Hardness 5–5.5. Specific gravity 2.3. Vitreous luster. Transparent to nearly opaque. Colorless to white and occasionally gray, green, yellowish, or reddish. Optical: $n = 1.48$. Fuses easily (2.5) to a colorless glass.

Distinctive Properties and Tests   Crystal form and association. Gelatinizes when treated with HCl and evaporated. Yields water in the closed tube. Readily fusible.

Association and Occurrence   Found in cavities and as a primary mineral in the groundmass of mafic igneous rocks and especially lavas; also found in lake sediments. Associated with zeolites and calcite.

Confused with   Garnet and leucite.

Variants   Usually contains a little K and Ca. Adsorbed ions are always present. Si:Al somewhat variable.

Related Minerals   *Solid Solutions:* pollucite, $(Cs,Na)_2$ $(AlSi_2O_6)_2 \cdot H_2O$. *Isotypes:* wairakite, $Ca(Al_2Si_4O_{12}) \cdot 2H_2O$. *Similar species:* leucite; viseite, $NaCa_5Al_{10}$ $(SiO_4)_3(PO_4)_5(OH)_{14} \cdot nH_2O$; and zeolite group.

Phyllosilicates (Sheets)

The structure of a large number of silicate minerals is based on sheets of Si-O tetrahedra stacked one above the other. The individual sheets are made up of tetrahedra sharing all their basal oxygens and with apical oxygen ions pointed in the same direction. Each sheet thus contains a basal oxygen layer that is electrically neutral (except when Al proxy for Si occurs) and an apical oxygen layer composed of oxygen ions with unsatisfied charges. Hydroxyl ions are also located in this layer, centered in the ring of apical oxygens.

In most sheet structures, small cations such as Mg^{2+}, Fe^{2+}, or Al^{3+} lie just above the apical oxygen layer and coordinate two oxygens and one hydroxyl from this layer with a similar combination in another sheet facing in the opposite direction. The sheets or layers of octahedra are complete and the $(OH)^-$ is shared among three octahedra, **trioctahedral,** for a divalent cation. The sheets are incomplete with each $(OH)^-$ shared by two octahedra, **dioctahedral,** for trivalent cations. Figure 21.24 shows the more usual kinds of stacking. Figure 21.25 illustrates the building of tetrahedral-octahedral-tetrahedral layers and shows how successive silica sheets are shifted with respect to each other.

Strongly bonded sandwiches of tetrahedral "bread" and octahedral "pastrami" are stacked one above the other and have only weak intersheet bonds, which accounts for the excellent basal cleavage of all the minerals with this structure. In some species, van der Waals forces account for the intersheet bonds; in others, hydrogen bonds associated with interlayer water are present; and in still others, large cations such as K^+ neutralize the excess negative charge arising from the Al^{3+} proxy for Si^{4+}. These large cations are located in two facing rings of basal oxygens, which gives them a regular 12-fold coordination.

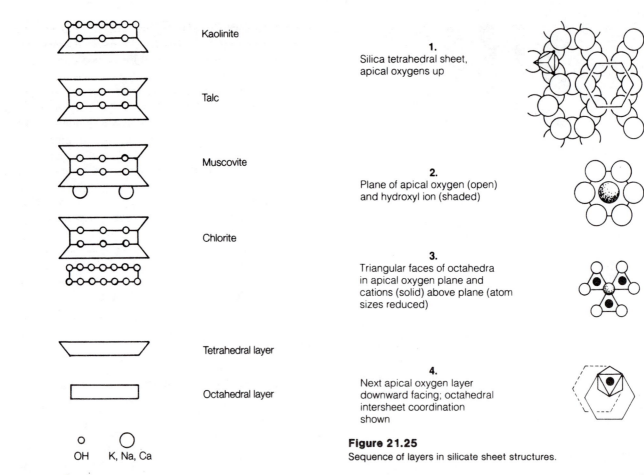

Figure 21.24
Common stacking sequences in phyllosilicates.

Kaolinite

Talc

Muscovite

Chlorite

Tetrahedral layer

Octahedral layer

o   O
OH   K, Na, Ca

1.
Silica tetrahedral sheet, apical oxygens up

2.
Plane of apical oxygen (open) and hydroxyl ion (shaded)

3.
Triangular faces of octahedra in apical oxygen plane and cations (solid) above plane (atom sizes reduced)

4.
Next apical oxygen layer downward facing; octahedral intersheet coordination shown

Figure 21.25
Sequence of layers in silicate sheet structures.

The need to shift equivalent locations from one sheet to another in the double-layer sandwich, coupled with the hexagonal plane symmetry of basal oxygen layers, leads to interesting and important consequences in the buildup of sheet structures by the stacking of sandwiches. An analogy is making a pile of real sandwiches with the crusts alternating in orientation.

The shift within a given double layer can be indicated by a vector (see figure 5.22). If successive layers have the same orientation, successive vectors are aligned and a monoclinic crystal with a one-layer repeat results ($1M$). Since, however, successive sandwiches can be rotated by multiples of 60° and maintain correlation of the basal oxygen layers, many regular stacking sequences are possible. For example, a 60° rotation of each successive layer, either right- or left-handed, leads to spirals that repeat after six layers and have hexagonal symmetry ($6H$). Other possibilities are spirals based on (1) rotations of 120° with a three-layer repeat yielding trigonal symmetry ($3T$), (2) 180° alternations that repeat every other layer and give orthorhombic symmetry ($2O$), and (3) alternating 120°

rotations repeating every other layer, generating monoclinic symmetry ($2M$). Minerals related by such structural geometries are special kinds of polymorphs called **polytypes.**

One structural consequence of polytypism is the formation of twins having a composition plane parallel to the sheets; such twins occur when the stacking sequence has no inherent regularity. This disorder is indicated by the subscript d; for example, $1M_d$ indicates a disordered one-layer monoclinic polytype.

CLAY FAMILY

The term *clay* is variously applied to mean a particle size, a group of physical properties, or a related group of minerals. The clay minerals usually have a very fine grain size and in bulk usually exhibit plasticity.

The structure of all clay minerals (see figure 21.26) is based on a sheet of linked silicon-oxygen tetrahedra. The octahedrally coordinated cation is predominately Al^{3+} but may be Mg^{2+} or Fe^{2+}. Both $(OH)^-$ and H_2O are always present, and alkali metals or alkaline earths are absent except as adsorbed ions.

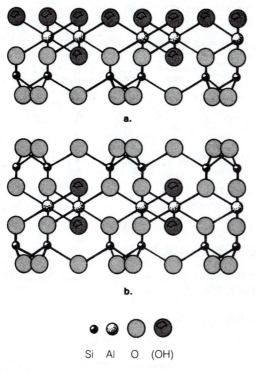

Figure 21.26
Comparison of the bonding for (a) one-layer clay minerals and (b) two-layer clay minerals.

The clay minerals can be divided in the following manner:

Single-Sheet Clays	Double-Sheet Clays
Kaolinite Group	Smectite Group
Kaolinite	Montmorillonite
Nacrite	Beidellite
Dickite	Hectorite
Halloysite-endellite	Nontronite
Series	Sauconite
	Saponite
	Illite Group
	Members not well established; includes clays that are similar and transitional to muscovite (e.g., brammalite) or hydromuscovite (illite)
	Palygorskite Group

The clay minerals are typically found as very fine-grained hydrous aggregates, often as alteration products from the decomposition of aluminous silicates.

The economic importance of clay minerals and their structural complexity and microscopic grain size have given rise to a special subdiscipline, **clay mineralogy.** The structural complexity of clays is due to several factors: clay minerals may be based on either one or two silica sheets per layer (i.e., open-faced or regular sandwiches), polytypic shifts are common between the layers, and these shifts may occur in both an orderly or disorderly way. Additionally, different kinds of clay sheets or related minerals may be interleaved into mixed-layer structures.

Intersheet bonding in clay minerals is typically accomplished by the attachment of water molecules to the (001) sheet surfaces. Water is a polar molecule, and its positive (hydrogen) end is attracted to the negatively charged sheet surface where interaction of the water molecules and the surface causes the water layer to assume an icelike structure. Additional water layers, perhaps five to ten, can be added before the structure's water finally loses its rigidity. Water added beyond this amount acts as a lubricant. For engineering purposes, the mechanical properties of clay mineral aggregates arising from different amounts of adsorbed water are described in terms of the **Atterberg limits:**

State	Limit
Liquid	
	Liquid limit
Plastic	
	Plastic limit
Solid	

A simple test of the liquid limit is to add enough water to cause a small sample to liquify briefly when jarred. A test for the plastic limit is to add enough water to allow the sample to be rolled into a tiny cylinder in the palm of the hand.

Addition of intersheet water to an air-dried clay mineral aggregate causes it to expand, often by large amounts: 5%–20% for kaolinite, 90%–150% for a montmorillonite with adsorbed calcium ions in the water layer, and up to 1,500% for montmorillonite with adsorbed sodium ions. Obviously, the mechanical properties of clay are highly sensitive to mineral species, water content, and the nature of the adsorbed ions. This property of swelling may be employed for sealing permeable rocks. It may also represent a serious hazard for buried utility lines or for building foundations exposed to water infiltration.

Clays, either in a natural or refined state, are widely used for both ceramic and nonceramic purposes. Ceramic uses include the manufacture of brick, tile, pottery, china, and related materials. Nonceramic uses typically take advantage of the very fine grain, absorptive capacity, chemical stability, and low unit cost of clays. Such uses include sizing or filling of porous substances, extending of medicines, insecticides or cleansers, and filtering and decolorizing oils.

KAOLINITE
Al₂(Si₂O₅)(OH)₄

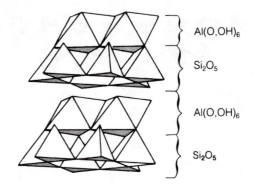

Figure 21.27
Structure of kaolinite.

(kaolin)

kā ʹə-lĭ-nīt, from Chinese *Kao-ling,* the name of the hill in Jiangxi Province which was the earliest source of samples sent to Europe.

Crystallography, Structure, and Habit Triclinic. $\bar{1}$. *a* 5.155, *b* 8.959, *c* 7.407, α 91.7°, β 104.9°, γ 89.9°. *P$\bar{1}$*. Pseudohexagonal tabular crystals.

Structure (see figure 21.27) is essentially an Si₂O₅ sheet combined with an Al(OH)₃ (gibbsite) layer in the manner of an open-faced sandwich. Coupling of the layers involves considerable distortion; the resulting strain limits the crystal size. Two formula units per cell.

Usually found as earthy masses and also in folia; granular.

Physical Properties Perfect {001} cleavage, yielding flexible but inelastic folia. Hardness 2–2.5. Specific gravity 2.6. Luster pearly for crystals and dull and earthy for massive specimens. Transparent to translucent. Color white but often stained grayish, brownish, bluish, or reddish. Optical: α = 1.553–1.561, β = 1.539–1.569, γ = 1.560–1.570; negative; 2V = 25–50°. Usually unctuous and plastic or, if dry, adheres to the tongue. Earthy smell when dampened. Infusible.

Distinctive Properties and Tests Habit, feel, and smell. Insoluble in acids. Yields water in the closed tube.

Association and Occurrence A very common mineral and widespread in occurrence. It is secondary in origin, usually forming through the decomposition of aluminous silicates with which it is associated. It is found (1) in rocks undergoing weathering, (2) in extensive beds, (3) as a constituent of soils, and (4) associated with hydrothermal veins.

Alteration From aluminous silicates.

Confused with Chalk and other clay minerals.

Variants The Al:Si ratio may vary, usually favoring Si (anauxite is a high-silica kaolin). Endellite is a hydrated analog with interlayer water. Absorbed cations, especially Na⁺, K⁺, and Ca²⁺. May contain ferric iron or chromium.

Related Minerals *Polymorphs:* Common stacking polymorphs (polytypes) are one-layer triclinic (1*T*) and one-layer monoclinic with disorder along the *b*-axis (1*M$_d$*); 6*M*, nacrite; 2*M*, dickite; and 1*M$_d$* with disorder along the *a*- and *b*-axes, halloysite.

SMECTITE GROUP

smĕk′tīt, from Greek *smektikos,* to cleanse, possibly an allusion to its soaplike feel.

MONTMORILLONITE
X(Al,Mg)₂(Si₄O₁₀)(OH)₂ · nH₂O

mŏnt-môr-ĭlʹə-nīt, from its discovery locality at Montmorillon, northwest of Limoges, France.

BEIDELLITE
XAl₂[(Al,Si)₄O₁₀](OH)₂ · nH₂O
X = exchangeable cation, mainly Na, Ca

bī-dĕlʹīt, from its discovery in a mine at Beidell, Colorado.

Crystallography, Structure, and Habit Monoclinic. Montmorillonite: *a* 5.17, *b* 8.94, *c* 15.2 to 9.6, β ≈ 90°. Beidellite: *a* 5.23, *b* 9.06, *c* 15.8 to 9.2, β ≈ 90°.

Structure based on that of pyrophyllite, with a gibbsite layer, Al(OH)₃, sandwiched between inward-pointing sheets of SiO₄ tetrahedra (see figure 21.28). See also mica.

Earthy masses.

Physical Properties Perfect {001} cleavage into flexible, inelastic folia. Very soft. Specific gravity 2.0–2.7. White or gray in color or stained bluish, reddish, or greenish. Optical: Highly variable because of large range in chemistry. Unctuous feel and earthy smell when wet. Swells and disaggregates when dried and rewet.

Distinctive Properties and Tests Unctuous feel and slaking. Infusible. Yields abundant water in the closed tube. Insoluble in acids.

Association and Occurrence A secondary mineral formed by alteration of aluminous silicates.

Alteration From aluminous silicates.

Confused with Chalk and kaolin.

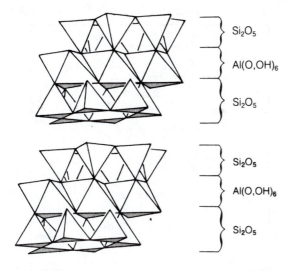

Figure 21.28
Structure of montmorillonite.

Variants *Form varieties:* bentonite, montmorillonitic rock derived from the alteration of volcanic ash or tuff. *Chemical varieties:* aluminian (sobotkite), magnesian (saponite), and zincian (sauconite). A wide variety of adsorbed ions may be present.

Related Minerals *Solid solutions:* montmorillonite, beidellite, and nontronite, $X(Fe,Mg)_2(Si_4O_{10})(OH)_2$. *Isotypes:* hectorite, $X(Mg,Li)_3(Si_4O_{10})(F,OH)_2$; stevensite, $Mg_3(Si_4O_{10})(OH)_2$; and swinefordite, $(Li,Ca,Na)(Al,Li,Mg)_4[(Si,Al)_8O_{20}](OH,F)_4$. *Other similar minerals:* talc and pyrophyllite.

TALC GROUP

The isotypic minerals pyrophyllite, $Al_2(Si_4O_{10})(OH)_2$, and talc, $Mg_3(Si_4O_{10})(OH)_2$, are the only important species in this group. Structurally, these minerals are built up from paired Si-O sheets bound together by Al^{3+} or Mg^{2+} in octahedral coordination with oxygen ions. The manner in which these structures are built is the same as for the other sheet structures.

Pyrophyllite and talc are very soft (hardness 1) and have a platy or micaceous habit, a greasy feel, and a perfect basal cleavage that supplies flexible but inelastic folia. Pyrophyllite and talc occur as lenses in dolomitic marble or other metamorphic rocks and as the alteration product of ultramafic rocks.

Massive impure talc (soapstone or steatite) finds use as heat and reagent-resistant bench tops, sinks, and stoves. It is readily carvable and is commonly used as a sculpture medium by the Inuit. Very pure and fine-grained massive talc can be machined and retains its configuration when calcined into an electrical nonconductor. This material, called "lava," is much used in the electrical and scientific communities. In finely ground form, talc and pyrophyllite are used as paint pigments and extenders, in fine ceramics, and, of course, in toilet powders.

PYROPHYLLITE
$Al_2(Si_4O_{10})(OH)_2$

(pencil stone)
pī-rŏ'fĭ-līt, from Greek *pyro*, fire, and *phyllon*, leaf, an allusion to its exfoliation when strongly heated.

Crystallography, Structure, and Habit Monoclinic. $2/m$. a 5.14, b 8.90, c 18.55, β 99.9°. $C2/c$. Crystals not known.

Isostructural with talc. For general structural arrangement, see also muscovite. Four formula units per cell.

Foliated, massive, spherulitic aggregates, or radiate.

Physical Properties Perfect {001} cleavage yielding flexible, inelastic folia. Hackly fracture. Sectile. Hardness 1–2. Specific gravity 2.8. Pearly to dull luster. Translucent. Color white, greenish, or yellowish. Optical: $\alpha = 1.566$, $\beta = 1.589$, and $\gamma = 1.601$; negative; $2V = 60°$. Greasy feel. Infusible.

Distinctive Properties and Tests Greasy feel, cleavage, and association. May exfoliate on heating. Partially broken down with H_2SO_4. Yields water in the closed tube.

Association and Occurrence Pyrophyllite is a relatively rare mineral found in metamorphic rocks and frequently associated with kyanite.

Confused with Talc.

Variants *Chemical varieties:* ferripyrophyllite, $Fe_2(Si_4O_{10})(OH)_2$. May contain a little Mg, Ca, Na, or K.

Related Minerals *Isotypes:* talc; minnesotaite, $(Fe,Mg)(Si_4O_{10})(OH)_2$; and willemseite, $(Ni,Mg)(Si_4O_{10})(OH)_2$.

TALC
$Mg_3(Si_4O_{10})(OH)_2$

(soapstone, steatite)
tălk, indirectly from Arabic, *talq*, pure (?).

Crystallography, Structure, and Habit Monoclinic. $2/m$. a 5.287, b 9.158, c 18.95, β 99.5°. $C2/c$. Rare, pseudohexagonal crystals.

Talc structure is generally similar to muscovite. Four formula units per cell.

Compact or foliated massive; sometimes in globular or radiating groups.

Physical Properties Perfect {001} cleavage yielding flexible, inelastic folia. Hackly fracture. Sectile. Hardness 1–1.5. Specific gravity 2.7. Luster vitreous; usually greasy to pearly. Translucent. Color apple-green to white or brown. Colorless streak. Optical: $\alpha = 1.54$, $\beta = 1.59$, and $\gamma = 1.59$; negative; $2V = 5-30°$. Greasy feel. Fusible with difficulty (5).

Distinctive Properties and Tests Greasy feel and cleavage. Insoluble in acids. Yields water in the closed tube with intense heating.

Association and Occurrence Found in igneous and metamorphic rocks as an alteration product of magnesian silicates. Associated with serpentine, chlorite, dolomite, magnetite, magnesite, apatite, tourmaline, pyrite, and actinolite.

Alteration From magnesian silicates, such as pyroxene, olivine, or amphibole.

Confused with Pyrophyllite, chlorite, and serpentine.

Variants *Form varieties:* soapstone, steatite, or French chalk—impure, massive talc often forming large rock masses. May contain small amounts of Ti, Ni, Fe, and Mn and variable quantities of (OH).

Related Minerals *Isotypes:* minnesotaite, $(Fe,Mg)_3$ $(Si_4O_{10})(OH)_2$, and willemseite, $(Ni,Mg)_3(Si_4O_{10})(OH)_2$. *Polymorphs:* stevensite (monoclinic).

CHLORITE GROUP
$Y_6(Z_4O_{10})(OH)_8$
where Y = octahedral cations, usually Mg, Fe, Al, Ni, Mn, Cr and Z = tetrahedral cations, Si, Al.

klō'rīt, from Greek *chloros,* green; named by mineralogist and chemist A. G. Werner in 1798.

CLINOCHLORE
$(Mg_5Al)(Si_3Al)O_{10}(OH)_8$

klī'nō-klōr, from Greek *klino,* inclined, for the obliquity between its optic axes, and *chloros,* green.

CHAMOSITE
$(Fe_5^{2+}Al)(Si_3Al)O_{10}(OH)_8$

shă'mə-sīt, from its discovery locality near Chamoson, Switzerland.

NIMITE
$(Ni_5Al)(Si_3Al)O_{10}(OH)_8$

nĭ'mīt, after the National Institute of Metallurgy, South Africa.

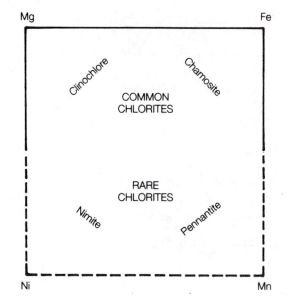

Figure 21.29
A classification of chlorites. (After Bayliss, 1975.)

PENNANTITE
$(Mn,Al)_6(Si,Al)_4O_{10}(OH)_8$

pĕn'ən-tīt, after Thomas Pennant (1726–1798), Welsh mineralogist and zoologist.

Chlorites vary widely in composition, especially in the ratio of Mg to Fe and in the variable occupancy of octahedral and tetrahedral sites by Al. Minerals of the chlorite group are classified on the basis of the principal octahedral cation (see figure 21.29). Recognized end members are clinochlore and chamosite. Albeit rare, two other species are recognized, nimite and pennantite.

The members of the chlorite group have very similar chemical, crystallographic, and physical properties. It is difficult to distinguish between the various species without recourse to quantitative chemical analysis, petrographic examination, or x-ray studies.

Crystallography, Structure, and Habit Monoclinic. $2/m$. a 5.3, b 9.2, c 14.3, β 97°. $C2/m$. Rare pseudohexagonal tablets.

The chlorites are structurally based on a sheet of linked (SiO_4) tetrahedra in which some replacement of Si^{4+} by Al^{3+} has taken place. Charge deficiencies of the sheets are neutralized by such cations as Al^{3+}, Fe^{3+}, Fe^{2+}, and Mg^{2+}, and more rarely by nickel, chromium, or manganese. Alkali metals and alkaline earths are absent. The structure of chlorite may be idealized as alternating sheets of talc and brucite (see figure 21.30). Random stacking is common. One formula unit per cell.

Foliated masses, scaly aggregates, and disseminated flakes.

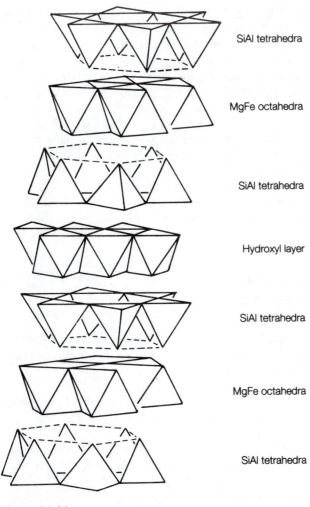

SiAl tetrahedra

MgFe octahedra

SiAl tetrahedra

Hydroxyl layer

SiAl tetrahedra

MgFe octahedra

SiAl tetrahedra

Figure 21.30
Exploded view of the chlorite structure; alternating layers of corner-sharing tetrahedra and edge-sharing octahedra.

Physical Properties Perfect {001} cleavage, yielding flexible, inelastic folia. Hardness 2–2.5. Specific gravity 2.7. Vitreous to pearly or dull luster. Transparent to translucent. Color dark green or, rarely, yellow, white, or rose. Optical: $\alpha = 1.56$–1.67, $\beta = 1.57$–1.68, and $\gamma = 1.57$–1.68. Chlorites rich in Mg and Al are positive; chlorites rich in Fe and Si are negative; $2V = 0$–60°. Fusible with difficulty (5–5.5). Sometimes has a slightly soapy feel.

Distinctive Properties and Tests Color, habit, and cleavage. Yields water in the closed tube. Insoluble in HCl but decomposes in boiling H_2SO_4 to yield a milky solution.

Association and Occurrence Chlorite is very widespread. It is often secondary and is found (1) as the alteration product of ferromagnesian silicates, commonly as pseudomorphs or earthy cavity fillings. Many altered igneous rocks owe their green color to the presence of secondary chlorite; (2) in sediments and sedimentary rocks; and (3) in schists and serpentinites with muscovite, biotite, garnet, diopside, magnesite, magnetite, and apatite.

Alteration From ferromagnesian silicates such as pyroxene, amphibole, biotite, or garnet.

Confused with Talc.

Related Minerals *Polymorphs:* orthochamosite (orthorhombic). Stacking polymorphs (polytypes) are common. *Other similar species:* cookeite, $LiAl_4(Si_3AlO_{10})(OH)_8$; baileychlor, $(Zn, Fe^{2+}, Al, Mg)_6(Si,Al)_4O_{10}(OH)_8$; gonyerite, $(Mn, Mg)_5Fe^{3+}(Si_3Fe^{3+}O_{10})(OH)_8$; odinite, $(Fe^{3+}, Mg, Al, Fe^{2+})_5(Si,Al)_4O_{10}(OH)_8$; sudoite, $Mg_2(Al, Fe^{3+})_3(Si_3AlO_{10})(OH)_8$.

BRITTLE MICA GROUP

The brittle micas are intermediate between the chlorites and the elastic micas in their structure and properties. Brittle micas are closely related structurally to the chlorite group on the one hand and to the elastic mica group on the other. Their principal chemical distinction from the elastic micas is in containing calcium as a principal cation and from the chlorites in their lower water content. Brittle micas are minor constituents of metamorphic rocks and have no commercial value.

MARGARITE
$CaAl_2(Al_2Si_2O_{10})(OH)_2$

mär′gə-rīt, from Greek *margarites,* pearl, alluding to its pearly lustre.

Crystallography, Structure, and Habit Monoclinic. $2/m$. a 5.13, b 8.92, c 19.50, β 95.0°. $C\,2/c$. Crystals very rare. Twins with composition plane (001) and axis [310].

Modified mica structure with intrasheet bonding by Al^{3+} and 50% proxy of Al^{3+} for Si^{4+} in tetrahedral coordination. Typically as a $1M$ polytype. Four formula units per cell.

Usually in foliated aggregates.

Physical Properties Perfect {001} cleavage, yielding brittle folia. Hardness 3.5–5. Specific gravity 3.0. Vitreous to pearly luster. Translucent. Color pink, white, or gray. Sometimes pale yellow or green. Optical: $\alpha = 1.630$–1.638, $\beta = 1.642$–1.648, $\gamma = 1.644$–1.650; negative; $2V = 40$–67°. Fusible (4–4.5).

Distinctive Properties and Tests Brittle cleavage folia, color, and association. Slowly and incompletely decomposed by boiling HCl.

Association and Occurrence Usually found associated with corundum and, rarely, as a component of chlorite schists.

Alteration　From corundum.

Confused with　Lepidolite.

Variants　*Chemical varieties:* sodian and beryllian. May also contain Ba, Sr, K, Mn, Fe^{3+} or Mg. Frequently contains excess $(OH)^-$.

Related Minerals Solid solutions: bityite, $Ca(Al,Li)_2[(Al,Be)_2Si_2(O,OH)_{10}](OH)_2$, and ephesite, $NaLiAl_2(Al_2Si_2O_{10})(OH)_2$. *Isotypes:* clintonite, $Ca(Mg,Al)_3(Al_3Si)O_{10}(OH)_2$; anandite, $(Ba,K)(Fe,Mg)_3(Si,Al,Fe)_4O_{10}(O,OH)_2$. *Other similar species:* stilpnomelane, $K(Fe,Al)_{10}(Si_{12}O_{30})(OH)_{12}$.

ELASTIC MICA GROUP

Elastic micas are an extensive group of platy minerals with a perfect basal cleavage that yields elastic folia. They are widespread and fairly abundant and can be found in all rock types as either essential or common accessory minerals. Typical occurrences are as follows:

Muscovite: felsic igneous rocks, pegmatites, schists, gneisses, and detrital sediments

Phlogopite: ultramafic igneous rocks and metamorphosed calcareous rocks

Biotite: felsic and mafic igneous rocks, schists, and gneisses

Lepidolite: pegmatites and high-temperature veins

Mineralogically, the principal members of the group can be divided into three solid-solution series: the muscovite series, the biotite series, and the lepidolite series. The structure of all the elastic micas is based on two opposed silica sheets in which one-fourth of the silicon ions have been replaced by aluminum ions. The intrasheet bonding is accomplished by octahedrally coordinated Al^{3+} ions in the muscovite series, by Mg^{2+} and Fe^{2+} ions in the biotite series, and by Li^+ and Al^{3+} ions in the lepidolite series. The net negative charge on the double sheet occasioned by the proxy of Al^{3+} for Si^{4+} in the tetrahedral sites is neutralized by K^+ located between the double sheets.

Muscovite and phlogopite exhibit a unique combination of properties, notably their ability to cleave into thin, tough, strong, and flexible sheets with a high dielectric constant, low heat conductivity, and high-temperature resistance that make them essential to the electric industry. In finely ground form, they are used as fillers, rubber dusting powder, lubricants, and luster-producing additives in wallpaper. Of particular importance is their ability to be mixed with adhesives and molded into electrically nonconductive forms.

Broadly speaking, there are four forms in which mica enters international trade: as block mica, representing unsplit books; as split mica (splittings, thins, sheets, and films), its most familiar form; as ground mica, either wet or dry; and as scrap, or flake mica, which is waste.

The production of block and split mica is highly labor-intensive. The principal producing countries are India, Russia, Madagascar, and Norway, where large books of mica are found in pegmatite bodies. These deposits are mined by small-scale methods, and the ore is cobbed, cleaned, trimmed, split, and graded by hand. The United States dominates the production of ground mica, principally from mines in North Carolina but also from nine other states.

Biotite is not commercially exploited because of its electrical conductivity from the presence of iron. Lepidolite is a minor ore of lithium and is sometimes used in the manufacture of glass and ceramics.

MUSCOVITE
$KAl_2(AlSi_3O_{10})(OH,F)_2$

(white mica, potash mica, muscovy glass)
mŭs′kə-vīt, after *Muscovy,* an ancient Russian province. Muscovite was formerly used in Russia for windowpanes and known as muscovy glass.

Crystallography, Structure, and Habit　Monoclinic. $2/m$. a 5.203, b 8.995, c 20.030, β 94.5°. $C2/c$. Distinct crystals are rare; usually in the form of tabular rhombs or pseudohexagonal prisms (see figure 21.31). Contact twins united on {001} with twin axis [310] are fairly common.

Muscovite structure is shown in figure 21.32. Four formula units per cell.

Tabular or foliated (micaceous) habits predominate. Discrete flakes, foliated masses; plumose, stellate, or globular.

Physical Properties　Perfect {001} cleavage, yielding elastic folia. Ragged fracture. Tough. Hardness 2.5–3. Specific gravity 2.75–3.0. Vitreous to silky luster. Transparent to translucent. Colorless, but often tinted gray, brown, pale green, violet, yellow, olive-green, or rose. Optical: $\alpha = 1.552–1.578$, $\beta = 1.582–1.615$, and $\gamma = 1.587–1.617$; negative; $2V = 30–45°$. Whitens and fuses (5) when heated.

Distinctive Properties and Tests　Cleavage, habit, and color. Insoluble in acids except HF. Yields a small amount of water in the closed tube.

Association and Occurrence　Muscovite is the most common of the micas. It is a primary constituent of potash and alumina-rich igneous rocks, such as granites and pegmatites, and is an essential mineral in many schists and gneisses. Muscovite is also found as detrital flakes in sedimentary rocks. Usually associated with such minerals as potash feldspar, quartz, albite, beryl, tourmaline, or garnet.

Figure 21.31
Muscovite showing pseudohexagonal form. Governados Valadares, Minas Gerais, Brazil. The specimen is 20 × 11 cm. (Courtesy Royal Ontario Museum, Toronto.)

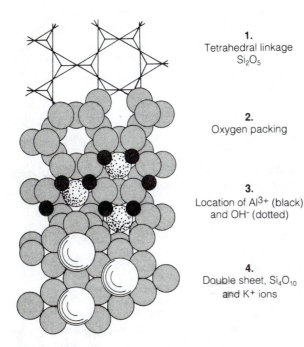

1.
Tetrahedral linkage
Si_2O_5

2.
Oxygen packing

3.
Location of Al^{3+} (black)
and OH^- (dotted)

4.
Double sheet, Si_4O_{10}
and K^+ ions

Figure 21.32
Sequence of layers in muscovite.

Alteration From topaz, kyanite, and feldspars.

Confused with Other micas.

Variants Form varieties: sericite, a fine scaly muscovite; ruby mica, red. *Chemical varieties:* sodian; chromian (fuchsite, mariposite), green; vanadian (roscoelite), green-brown; ferrian; zincian (hendricksite); silician (phengite); fluorian; barian (chernykhite), green. May also contain Rb, Cs, Ca, Li, and Ti.

Related Minerals Solid solutions: paragonite $NaAl_2(AlSi_3O_{10})(OH)_2$. *Polymorphs: $1M_d$, $1M$, $2M_1$,*

$2M_2$, and $3T$ muscovite (sheet-stacking polymorphs). *Isotypes:* biotite, phlogopite, and lepidolite.

PHLOGOPITE
$KMg_3(AlSi_3O_{10})(OH,F)_2$

flŏg'ə-pīt, from Greek *phlogopos,* fiery, an allusion to its coppery-brown color.

Crystallography, Structure, and Habit Monoclinic. $2/m$. a 5.326, b 9.210, c 10.311, β 100.2°. $C2/m$. Rare tapering or pseudohexagonal platy crystals. Twinning on (001).

Mica structure. See biotite and muscovite. Two formula units per cell.

Found in disseminated scales and plates, often in large crystals.

Physical Properties Perfect basal cleavage, yielding flexible folia. Ragged fracture. Tough. Hardness 2.5–3. Specific gravity 2.8. Pearly to submetallic luster. Transparent to translucent. Color reddish brown or yellowish brown. Optical: $\alpha = 1.530–1.590$, $\beta = 1.557–1.640$, and $\gamma = 1.558–1.630$; negative; $2V = 0–15°$. Asterism often displayed when viewed in transmitted light. Fusible (4.5–5).

Distinctive Properties and Tests Cleavage, color, and occurrence. Decomposes in boiling H_2SO_4 and yields a milky solution. Yields small amounts of water in the closed tube.

Association and Occurrence Commonly found in marble associated with pyroxenes, amphiboles, and serpentine. Rarely a constituent of igneous rocks.

Confused with Other micas.

Variants Chemical varieties: fluorian, ferroan, manganoan, and sodian. May also contain Rb, Cs, and Ba.

Related Minerals Solid solutions: biotite and taeniolite, $KLiMg_2(Si_4O_{10})F_2$. *Isotypes:* other micas. *Polymorphs: $1M$* and $3T$ phlogopite (sheet-stacking polymorphs).

BIOTITE
$K(Mg,Fe)_3(AlSi_3O_{10})(OH,F)_2$

bī'ə-tīt, after Jean Baptiste Biot (1774–1862), French physicist and chemist who first called attention to the optical differences in the micas.

Crystallography, Structure, and Habit Monoclinic. $2/m$. a 5.31, b 9.23, c 20.36, β 99.3°. $C2/m$. Rare tabular pseudohexagonal or pseudorhombohedral crystals. Contact twins joined on {001} with twin axis [310].

Isostructural with muscovite. In biotite all the intrasheet octahedral sites are filled with Mg^{2+} or Fe^{2+}, whereas in muscovite some of these sites (normally occupied by Al^{3+}) are vacant. One formula unit per cell.

Commonly in disseminated scales or irregular foliated masses.

Physical Properties Perfect {001} cleavage, yielding flexible folia. Ragged fracture. Tough. Hardness 2.5–3. Specific gravity 2.7–3.1. Vitreous luster. Transparent to opaque. Color black, greenish black, or brownish black. Optical: $\alpha = 1.570$–1.625, $\beta = 1.610$–1.696, and $\gamma = 1.610$–1.696; negative; $2V = 0$–$25°$. Whitens and fuses (5) when heated. Becomes weakly magnetic after heating in a reducing flame.

Distinctive Properties and Tests Cleavage, habit, color, and postheating magnetism. Completely decomposes in boiling H_2SO_4, yielding a milky solution. Yields water in the closed tube.

Association and Occurrence Biotite is an important constituent of many igneous rocks, especially those rich in potassium and aluminum. It is also a common mineral in metamorphic rocks. It is associated with muscovite, alkali feldspar, amphiboles.

Alteration From augite, hornblende, scapolite, garnet, and other ferromagnesian minerals. To chlorite, epidote, quartz, iron oxides, and vermiculite.

Confused with Other micas, especially phlogopite and chlorite. Chloritoid.

Variants *Chemical varieties:* ferrian (lepidomelane, celadonite) black; manganoan (montdorite and manganophyllite), bronze to copper-red; sodian-magnesian (preiswerkite); titanian; calcian, brown to colorless; ferroan (annite); ferroaluminian (siderophyllite); barian (anandite and kinoshitalite); lithian (masutomilite), purple to pink; and fluorian. Biotite may also contain small amounts of Na, Rb, and Cs.

Related Minerals *Solid solutions:* wide solid-solution range, with most compositions lying within the field defined by phlogopite; annite, $KFe_3(AlSi_3O_{10})(OH)_2$; and siderophyllite, $K(Fe,Al)_3[(Si,Al)_4O_{10}](OH)_2$. *Isotypes:* other micas. *Polymorphs:* $1M$ (most common), $2M$, and $3T$ biotite (sheet-stacking polymorphs). *Other similar species:* vermiculite, an altered biotite or chlorite depleted in potassium and hydrated $(Mg,Fe^{2+},Al)_3(Al,Si)_4O_{10}$ $(OH)_2 \cdot 4H_2O$; glauconite.

LEPIDOLITE SERIES
$KX_3[(Si,Al)_4O_{10}](OH,F)_2$
where X = Li, Al, Fe, Mg, Mn

(lithium mica)
li-pĭd'l-īt, from Greek *lepid,* scale, because of its usual occurrence as scaly aggregates.

Crystallography, Structure, and Habit Monoclinic. $2/m$. a 9.2, b 5.3, c 20.0, β 98.0°. $C2/m$. Very rare short prismatic crystals. Contact twins with (001) as composition plane.

Mica structure (see muscovite). Two formula units per cell.

Coarse- to fine-grained scaly or granular aggregates.

Physical Properties Perfect {001} cleavage, yielding elastic folia. Ragged fracture. Tough. Hardness 2.5–4. Specific gravity 2.9. Pearly luster. Translucent. Color rose-red to lilac and, less commonly, yellowish or grayish to white. Optical: $\alpha = 1.525$–1.548, $\beta = 1.548$–1.585, and $\gamma = 1.551$–1.587; negative; $2V = 30$–$50°$. Fuses easily (2) with intumescence and colors the flame red.

Distinctive Properties and Tests Color and occurrence. Some water in the closed tube. Attacked by acids after fusion.

Association and Occurrence Found in granitic pegmatites with tourmaline, spodumene, muscovite, cassiterite, albite, quartz, and topaz.

Confused with Rose muscovite and margarite.

Variants *Chemical varieties:* ferroan. Lepidolite may contain Rb, Cs, Zn, Nb, and Mg.

Related Minerals *Solid solutions:* The lepidolite series is a complex group of solid solutions extending from Li-Al members to Fe- and Mg-rich species and to muscovite (see figure 21.33). *Polymorphs:* $1M$, $2M_2$, $2O$, and $3T$ lepidolite sheet-stacking polymorphs are fairly common, and $2M_1$ and $6H$ are known. *Isotypes:* other micas.

MISCELLANEOUS PHYLLOSILICATES

Serpentine is usually found as a greasy-looking green to yellow mineral associated with mafic rocks where it develops by alteration from Mg-rich silicates. It is probably best known as the asbestiform variety, chrysotile, whose hair-fine fibers are used as insulation, woven into heat-resistant clothing, or compressed into shingles and brake linings. Chrysotile asbestos (see figure 21.34) has been recognized as a hazardous material. When the tiny fibers

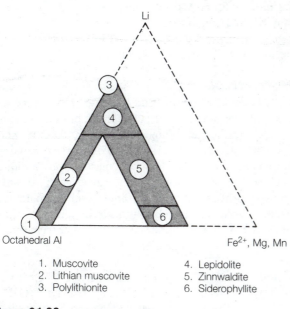

Octahedral Al Fe^{2+}, Mg, Mn

1. Muscovite   4. Lepidolite
2. Lithian muscovite 5. Zinnwaldite
3. Polylithionite   6. Siderophyllite

Figure 21.33
Lepidolite and related minerals.

are inhaled, they damage the lungs and cause emphysema. Currently, considerable effort is being expended in identifying and alleviating the dangerous conditions brought about by the use of exposed asbestos insulation in buildings.

Asbestos is properly used to describe a mineral habit and is not a species name. Minerals showing an asbestiform habit are serpentine and some members of the amphibole group (tremolite, riebeckite [crocidolite], and grunerite [amosite]).

The principal North American production of chrysotile asbestos is from the eastern townships of southeastern Quebec, although the zone of ultramafic bodies that are the basis for production extend the length of the Appalachian belt from Alabama to Newfoundland. Other important producing areas are in southern Africa (Zimbabwe and South Africa) and in the Russian Urals.

Glauconite is a green phyllosilicate forming at low temperatures and characteristically found in certain sandstones. Its presence colors the sedimentary rocks that are called **greensands.**

Prehnite is of uncertain classification. It is most commonly found in cavities of mafic volcanic rocks, often associated with zeolites. It also occurs in some low-grade metamorphic rocks.

SERPENTINE GROUP
$(Mg,Fe)_3(Si_2O_5)(OH)_4$

sûr'pən-tēn, from French *serpentin,* sinuous.

Figure 21.34
Asbestiform chrysotile. Note the kink band due to postcrystallization deformation. (Courtesy J. S. Hudnall Mineralogy Museum, University of Kentucky. Photograph T. A. Moore.)

CHRYSOTILE

krĭs'ə-tīl, from Greek *chrysos,* golden, and *tilos,* fiber.

ANTIGORITE

ăn-tĭ'gə-rīt, first found in the Valle Antigorio of northern Italy.

LIZARDITE

lĭz'ər-dīt, from its occurrence on the Lizard Peninsula in Cornwall, England.

Crystallography, Structure, and Habit Monoclinic. $2/m$. Chrysotile: a 5.34, b 9.25, c 14.65, β 93.3°. Antigorite: a 5.31, b 9.22, c 7.27, β 91.6°. Lizardite: a 5.31, b 9.20, c 7.31, β 90°. $C2/m$. Crystals unknown.

Structure similar to kaolinite, but the slight difference between octahedral edge length and apical oxygen

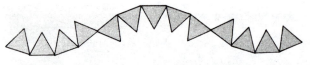

Antigorite — alternation of facing direction

a.

Lizardite — distortion of tetrahedra

b.

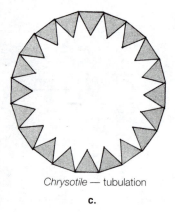

Chrysotile — tubulation

c.

Figure 21.35
Diagrammatic representation of the polymorphic forms of serpentine.

spacing causes the sheets to curl. In antigorite this tendency is offset by an alternation of the facing direction of the sheets (see figure 21.35a); in lizardite the tetrahedra are distorted (figure 21.35b); and in chrysotile the sheets curl into tubes (figure 21.35c). Four formula units per cell.

Usually massive, platy (antigorite, lizardite), or fibrous (chrysotile).

Physical Properties Perfect {001} cleavage but rarely seen. Conchoidal to splintery fracture. Hardness 2.5–3.5. Specific gravity 2.2–2.4. Greasy or waxy luster. Translucent. Color various shades of green, yellow-green, and yellow. Optical: $\alpha \approx \beta = 1.53$–$1.56$, and $\gamma = 1.545$–1.584; negative; $2V = 20$–$60°$. Smooth to greasy feel. Infusible.

Distinctive Properties and Tests Color, luster, and feel. Decomposed by HCl, yielding silica but not a jelly. Yields water in the closed tube.

Association and Occurrence Serpentine is a common and widespread secondary mineral, often forming large rock masses. It is associated with magnesite, chromite, spinel, and garnet. Chrysotile is the principal fibrous mineral of commerce under the name asbestos.

Alteration From Mg-rich silicates, especially olivine, pyroxene, and amphibole.

Confused with Epidote and chlorite.

Variants The serpentines can be divided into three species: chrysotile, antigorite, and lizardite. These minerals are usually considered to be polymorphs, but small chemical differences in their compositions suggest that they may represent distinct chemical species. *Form varieties:* antigorite, common serpentine (compact masses); verd antique, a rock composed of antigorite and carbonates; asbestos, fibrous serpentine (chrysotile); and picrolite, columnar. *Chemical varieties:* nickelian (garnierite and nepouite), manganoan (kellyite), aluminian (amesite), ferrian (cronstedtite), titanian, and zincian (fraipontite).

Related Minerals *Isotypes:* berthierine, $(Fe,Mg)_{2-3}[(Si,Al)_2O_5](OH)_4$; greenalite, $(Fe,Fe)_3(Si_2O_5)(OH)_4$. *Polymorphs:* clino-, ortho-, and parachrysotile are stacking polymorphs of chrysotile.

GLAUCONITE
$(K,Na)(Fe,Mg,Al)_2[(Al,Si)_4O_{10}](OH)_2$

glô′kə-nīt, from Greek *glaukos,* bluish green.

Crystallography, Structure, and Habit Monoclinic. $2/m.$ a 5.25, b 9.09, c 20.07, β 95.0°. No crystals.

Dioctahedral mica structure but with more K in the interlayer positions and high Fe^{3+} in the octahedral sheet.

Resembles earthy chlorite; usually found in small pellets or scaly grains.

Physical Properties {001} cleavage. Hardness 2. Specific gravity 2.3–2.8. Color dull green. Optical: $\alpha = 1.585$–1.616, $\beta \approx \gamma = 1.600$–$1.644$; negative; $2V = 0$–$20°$.

Distinctive Properties and Tests Color, habit, and occurrence. Evolves water in the closed tube.

Association and Occurrence Found typically as pellets or granules in sedimentary rocks (greensands) deposited near a continental shore; may be a major component of such rocks. Associated with chalk, fragmental calcite, and quartz.

Alteration Perhaps from ferromagnesian silicates, such as pyroxene, amphibole, and biotite; to goethite.

Confused with Chlorite.

Variants May be mixed-layer structure with illite and smectite. *Chemical varieties:* sodian.

Related Minerals *Similar species:* greenalite, a mineral resembling glauconite but without potassium.

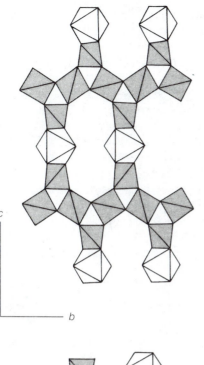

(Al,Si)O₄   AlO₆

Figure 21.36
Structure of prehnite.

Figure 21.37
Barrel-shaped crystals of prehnite. Asbestos, Quebec. (Courtesy Royal Ontario Museum, Toronto. Photograph by Arthur Williams.)

PREHNITE
$Ca_2Al(AlSi_3O_{10})(OH)_2$

prĕ'nīt, after its discovery in South Africa in 1774 by the Dutch Colonel Hendrick van Prehn.

Crystallography, Structure, and Habit Orthorhombic. 2*mm*. *a* 4.61, *b* 5.47, *c* 18.48. *P2cm* or *Pncm* depending on Si-Al ordering. Rare tabular crystals. Sometimes fine lamellar twinning.

Complex layers of corner-sharing (Al,Si)-O tetrahedra paralleling the *b*-axis are cross-linked by edge-sharing Al-O octahedra. The linking forms are staggered elongate loops parallel to the *c*-axis (see figure 21.36). Two formula units per cell.

Barrel-shaped, colloform, or stalactitic, with a crystalline surface. Often in rounded aggregates of tabular crystals (see figure 21.37).

Physical Properties Distinct {001} and poor {110} cleavage. Uneven fracture. Brittle. Hardness 6–6.5. Specific gravity 2.9. Vitreous luster. Translucent. Color light green to yellow, gray or white. Optical: $\alpha = 1.615$, $\beta = 1.624$, and $\gamma = 1.644$; positive; $2V = 68°$. Fuses (2.5) with intumescence to a white enamel.

Distinctive Properties and Tests Habit, color, and association. Slowly reacts with HCl, but is readily dissolved and yields gelatinous silica on evaporation after simple fusion.

Association and Occurrence Prehnite is a secondary mineral found as crusts and cavity fillings in silica-poor igneous rocks, usually lavas. Associated with zeolites, pectolite, calcite, and native copper. Also found in contact-metamorphosed impure limestones.

Alteration To zeolites or chlorite.

Confused with Smithsonite, hemimorphite, beryl, and chalcedony.

Variants Chemical varieties: ferrian.

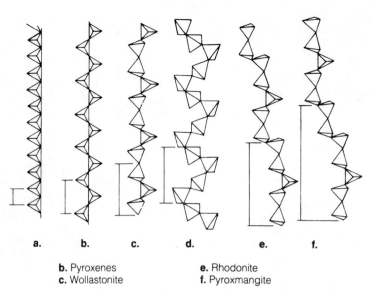

b. Pyroxenes **e.** Rhodonite
c. Wollastonite **f.** Pyroxmangite

Figure 21.38
Possible arrangements of single tetrahedral chains with unit translations indicated. Mineral examples identified. (After Liebau, 1959.)

Table 21.4	Cell Parameters of Some Pyroxenes and Amphiboles					
Mineral	Type*	Crystal Glass	Cell Constants			
			a	b	c	β
Enstatite	opx	2/m 2/m 2/m	18.2	8.8	5.2	
Anthophyllite	oam	2/m 2/m 2/m	18.6	18.0	5.3	
Diopside	cpx	2/m	9.7	8.9	5.3	105°
Tremolite	cam	2/m	9.8	18.0	5.3	105°

*opx = orthopyroxene, cpx = clinopyroxene; oam = orthoamphibole, cam = clinoamphibole

INOSILICATES (CHAINS) AND CYCLOSILICATES (RINGS)

The sharing of two oxygen ions per Si-O tetrahedron leads to the formation of single chains of tetrahedra that may be infinite in length or formed into a closed loop. In such structures, illustrated in figure 21.3 and table 21.1, the ratio of Si to O is 1:3, giving an overall composition of $(SiO_3)^{2-}$. Single chains of infinite length are most commonly of the type b illustrated in figure 21.38. These are found in the **pyroxene** group. Types c, e and f occur in minerals known as **pyroxenoids,** but types a and d are known only in nonmineral compounds.

Isolated closed loops of single-chain tetrahedra are found in the **cyclosilicates.** These minerals may be made up of 3-, 4-, 6-, or 8-membered rings. Beryl is an example of a 6-membered ring structure.

Double chains or bands of tetrahedra are formed when two single chains of the pyroxene type share oxygens between alternate tetrahedra (see figure 21.3 and table 21.1). This results in a Si:O ratio of 4:11, for an overall composition of $(Si_4O_{11})^{6-}$. The double chain is the structural unit of the **amphiboles.**

In both pyroxenes and amphiboles the chains are aligned parallel to the c-axis and bound together by ions such as Na^+, Ca^{2+}, Fe^{2+}, Mg^{2+}, and Al^{3+}, principally in 6-fold coordination. Sequential staggering of successive chains parallel to their length leads to a dominance of monoclinic symmetries in the amphiboles and pyroxenes. Orthorhombic types generated by an alternating staggering of chains are known in both groups, however.

The similarities in the structural arrangements of pyroxenes and amphiboles are highlighted by comparing their symmetries and cell dimensions (see table 21.4). If the b cell edge is chosen parallel to the chain width, only the doubling of this dimension in the amphiboles distinguishes them.

Because of their close similarity in structure and chemistry, the amphiboles and pyroxenes have very similar physical properties, and members of the two groups are often confused in hand specimens. Some of the more useful distinguishing features are listed in table 21.5.

Table 21.5	Properties of Pyroxenes and Amphiboles	
Property	Pyroxene	Amphibole
Structure	Single (SiO$_3$) chains	Double (Si$_4$O$_{11}$) chains
Composition	Anhydrous (may alter to hydrous minerals)	Contains hydroxyl groups and yields water in the closed tube
Crystal habit	Short, complex, 4- or 8-sided prisms (pseudotetragonal)	Long, simple, 6-sided crystals (pseudohexagonal)
Cleavage	Two directions at $\approx 90°$	Two directions at $\approx 60°$ and $120°$
Parting	{001} often prominent	Not common
Mineral habit	Lamellar or granular	Columnar or fibrous masses
Occurrence	In more mafic igneous rocks	In more felsic igneous rocks; common in metamorphosed rocks

Double Chains

The skeleton of an Si-O double chain (see figure 21.39) is found in the structures of members of only one large mineral group, the amphiboles. These are widespread as principal and accessory minerals in igneous and metamorphic rocks. Hornblende and riebeckite are common igneous species, whereas the other amphiboles are typically metamorphic in origin.

In the structure of an amphibole, two facing double chains have their apical oxygens octahedrally coordinated by such small cations as Mg^{2+} or Fe^{2+}, resulting in a strong, serrated rod. The rod is about five oxygen diameters wide and four high and is infinitely extended parallel to the c crystallographic axis. These rods are cross-linked by larger cations such as Na$^+$ or Ca^{2+} that coordinate unsatisfied oxygen ions on the edges of the chains and neutralize the residual charge of basal oxygens due to Al^{3+} proxy for Si^{4+}. (OH)$^-$ groups are positioned in the center of the ring of apical oxygens as in the sheet structures.

Coordination requirements for the small intrachain cations necessitate a shift in equivalent positions of the chains that make up the rod (see figure 21.40). This staggering is in the direction of the rod length (c crystallographic axis). If the stagger is always in the same direction, the structure will be referable to a monoclinic cell (clinoamphibole); alternation in an orderly way produces orthorhombic symmetry. This shift is analogous to that found in mica polytypes but is restricted in direction.

The bond density between adjacent rods is less than within the rods, and therefore mechanical rupture will occur between the rods (see the cleavage of pyroxene in figure 9.3). The steplike character of amphibole cleavage planes cannot be directly observed, and the microscopic effect is that of two smooth cleavage planes intersecting at about 56° and 124°, with the line of intersection parallel to the rod. The rodlike skeleton of amphiboles also leads to the prismatic, columnar, bladed, or fibrous habits typically exhibited by this group of minerals.

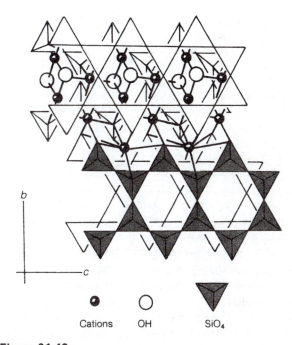

1.
Tetrahedral linkage, Si$_4$O$_{11}$

2.
Oxygen packing

3.
Location of Mg^{2+} and Fe^{2+} (small stippled) and (OH)$^-$ (large shaded) ions

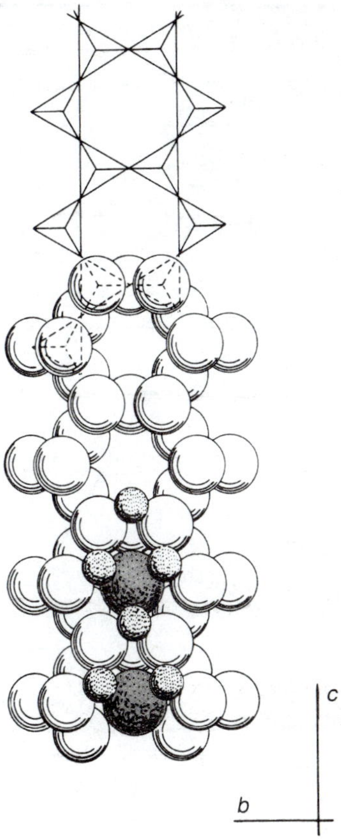

Figure 21.39
Sequence of layers in a double-chain silicate.

Figure 21.40
Articulation in double-chain structures. Interchain bonds shown in top portion.

Cations   OH   SiO$_4$

| Table 21.6 | Site Occupancies in Amphiboles |

Formula Site	Cation Site	Cation Coordination	Representative Elements
W	A	10 or 12	Na^+, K^+
X	$M4$	6 or 8	Na^+, Li^+, Ca^{2+}, Mg^{2+}, Fe^{2+}, Mn^{2+}
Y	$M1, M2, M3$	6	Mg^{2+}, Fe^{2+}, Zn^{2+}, Fe^{3+}, Al^{3+}, Cr^{3+}, Ti^{4+}
Z	T	4	Si^{4+}, Al^{3+}
(OH)			$(OH)^-$, O^{2-}, F^-, Cl^-

| Table 21.7 | Classification of Amphiboles Based on X-Site Occupancy |

Criteria	Group
$(Ca+Na)_x < 1.34$	Iron-magnesium-manganese amphiboles
$(Ca+Na)_x \geq 1.34$ and $Na_x < 0.67$	Calcic amphiboles
$(Ca+Na)_x \geq 1.34$ and $0.67 \leq Na_x < 1.34$	Sodic-calcic amphiboles
$Na_x \geq 1.34$	Alkali amphiboles

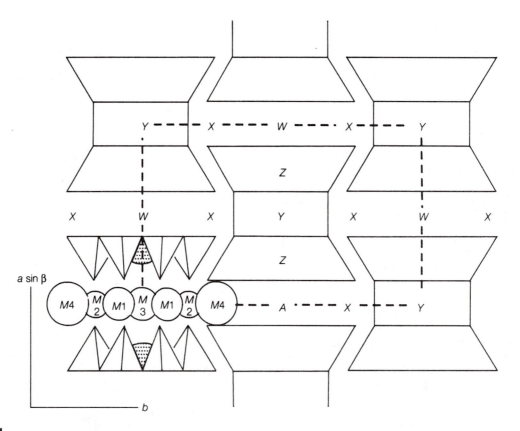

Figure 21.41

Schematic view of amphibole down [001] showing the position of (OH) (stippled) in the tetrahedral rings and the distinct cation sites W (A), X ($M4$), Y ($M1$, $M2$, $M3$), and Z. Unit cell dashed. (After Colville et al., 1966.)

The general formula for an amphibole is $W_{0-1}X_2Y_5Z_8O_{22}(OH,O,F,Cl)_2$, where W, X, Y, and Z identify cations of different coordination and structural location, as shown in table 21.6. Although solid solution is common among the amphiboles, it is not complete, and two amphiboles may occur together. For example, anthophyllite can coexist with hornblende, actinolite, or gedrite; and cummingtonite with hornblende or actinolite.

The sites for cation occupancy in the amphibole structure are shown in the schematic [001] projection of figure 21.41. The double chains are viewed end-on and appear as interleaved I beams composed of two facing Si_4O_{11} double chains held together by Y-site (6-fold) cations and cross-linked by W- and X-site cations. Note that the X-site cations ($M4$) are in 6- or 8-fold coordination with oxygen. In the Y site, all the cations are in 6-fold coordination; $M2$ with 6O, and $M1$ and $M3$ with 4O and

2(OH). The W site, sometimes designated A, is in 10- or 12-fold coordination with O and (OH).

The large number of possible combinations of elements in amphiboles leads to many distinct species, and extensive solid solution further complicates the chemistry of this mineral group. For convenience, however, the amphiboles can be subdivided according to the occupancy of the two X cation sites. An accurate classification requires a chemical analysis followed by calculation of a structural formula, as outlined in chapter 5. Amphiboles are classified into one of four principal groups on the basis of the $(Ca+Na)$ and Na contents of the X ($M4$) position (see table 21.7). Further subdivisions are made on the basis of the atomic concentrations of Fe^{2+}, Mg, Fe^{3+}, Al, and K. A review of the principal amphibole types along with their site occupancies is given in table 21.8. A complete review of amphibole nomenclature and structure is given by

Table 21.8 **Variation of End Members in the Amphibole Group**

	W	X	Y	Z
Iron-Magnesium-Manganese Amphiboles				
Orthorhombic forms				
Anthophyllite	—	$(Mg,Fe^{2+},Mn)_2$	$(Mg,Fe^{2+},Mn)_5$	Si_8
Gedrite	—	$(Mg,Fe^{2+},Mn)_2$	$(Mg,Fe^{2+},Mn)_{5-y}\,Al_y$	Al_2Si_6
Holmquistite	—	Li_2	$(Mg,Fe^{2+})_3(Al,Fe^{3+})_2$	Si_8
Monoclinic forms				
Cummingtonite Series	—	$(Mg,Fe^{2+},Mn)_2$	$(Mg,Fe^{2+},Mn)_5$	Si_8
Calcic Amphiboles				
Tremolite	—	Ca_2	Mg_5	Si_8
Ferro-actinolite	—	Ca_2	Fe_5^{2+}	Si_8
Edenite	Na	Ca_2	Mg_5	$AlSi_7$
Ferro-edenite	Na	Ca_2	Fe_5^{2+}	$AlSi_7$
Pargasite	Na	Ca_2	Mg_4Al	Al_2Si_6
Ferro-pargasite	Na	Ca_2	$Fe_4^{2+}Al$	Al_2Si_6
Hastingsite	Na	Ca_2	$Fe_4^{2+}Fe^{3+}$	Al_2Si_6
Magnesio-hastingsite	Na	Ca_2	Mg_4Fe^{3+}	Al_2Si_6
Alumino-tschermakite	—	Ca_2	Mg_3Al_2	Al_2Si_6
Ferro-alumino-tschermakite	—	Ca_2	$Fe_3^{2+}Al_2$	Al_2Si_6
Ferri-tschermakite	—	Ca_2	$Mg_3Fe_2^{3+}$	Al_2Si_6
Ferri-ferro-tschermakite	—	Ca_2	$Fe_3^{2+}Fe_2^{3+}$	Al_2Si_6
Alumino-magnesio-hornblende	—	Ca_2	Mg_4Al	$AlSi_7$
Alumino-ferro-hornblende	—	Ca_2	$Fe_4^{2+}Al$	$AlSi_7$
Kaersutite	Na	Ca_2	Mg_4Ti	Al_2Si_6
Ferro-kaersutite	Na	Ca_2	$Fe_4^{2+}Ti$	Al_2Si_6
Sodic-Calcic Amphiboles				
Richterite	Na	$NaCa$	Mg_5	Si_8
Ferro-richterite	Na	$NaCa$	Fe_5^{2+}	Si_8
Ferri-winchite	—	$NaCa$	Mg_4Fe^{3+}	Si_8
Alumino-winchite	—	$NaCa$	Mg_4Al	Si_8
Ferro-alumino-winchite	—	$NaCa$	$Fe_4^{2+}Al$	Si_8
Ferro-ferri-winchite	—	$NaCa$	$Fe_4^{2+}Fe^{3+}$	Si_8
Alumino-barrosite	—	$NaCa$	Mg_3Al_2	$AlSi_7$
Ferro-alumino-barrosite	—	$NaCa$	$Fe_3^{2+}Al_2$	$AlSi_7$
Ferri-barrosite	—	$NaCa$	$Mg_3Fe_2^{3+}$	$AlSi_7$
Ferro-ferri-barrosite	—	$NaCa$	$Fe_3^{2+}Fe_2^{3+}$	$AlSi_7$
Magnesio-ferri-katophorite	Na	$NaCa$	Mg_4Fe^{3+}	$AlSi_7$
Magnesio-alumino-katophorite	Na	$NaCa$	Mg_4Al	$AlSi_7$
Ferri-katophorite	Na	$NaCa$	$Fe_4^{2+}Fe^{3+}$	$AlSi_7$
Alumino-katophorite	Na	$NaCa$	$Fe_4^{2+}Al$	$AlSi_7$
Ferri-taramite	Na	$NaCa$	$Fe_3^{2+}Fe_2^{3+}$	Al_2Si_6
Magnesio-ferri-taramite	Na	$NaCa$	$Mg_3Fe_2^{3+}$	Al_2Si_6
Alumino-taramite	Na	$NaCa$	$Fe_3^{2+}Al_2$	Al_2Si_6
Magnesio-alumino-taramite	Na	$NaCa$	Mg_3Al_2	Al_2Si_6
Alkali Amphiboles				
Glaucophane	—	Na_2	Mg_3Al_2	Si_8
Ferro-glaucophane	—	Na_2	$Fe_3^{2+}Al_2$	Si_8
Magnesio-riebeckite	—	Na_2	$Mg_3Fe_2^{+3}$	Si_8
Riebeckite	—	Na_2	$Fe_3^{2+}Fe_2^{3+}$	Si_8
Eckermannite	Na	Na_2	Mg_4Al	Si_8
Ferro-eckermannite	Na	Na_2	$Fe_4^{2+}Al$	Si_8
Magnesio-arfvedsonite	Na	Na_2	Mg_4Fe^{3+}	Si_8
Arfvedsonite	Na	Na_2	$Fe_4^{2+}Fe^{3+}$	Si_8
Kozulite	Na	Na_2	$Mn_4(Fe^{3+},Al)$	Si_8

Hawthorne (1981). Calcic amphiboles are by far the most common and abundant, alkali amphiboles are widespread but not abundant, and Fe-Mg amphiboles are restricted in occurrence. This section gives complete descriptions of only the most common amphiboles.

The accurate characterization of an amphibole species requires chemical analysis because of the overlapping nature of the physical and optical properties within the group. Thus, instead of assigning a precise name, the general term *amphibole* should be used with an appropriate adjectival modifier, as in *tremolitic amphibole* or *riebeckitic amphibole*. *Hornblende* should be used to describe all calcic amphiboles. For asbestiform varieties, again a general composition modifier can be used, as in actinolitic asbestos.

The principal commercial uses of amphiboles are restricted to their asbestiform varieties, which enter world trade as minor competitors to chrysotile. Amosite is the commercial name of an asbestiform variety of iron-rich gedrite, and crocidolite is a blue asbestiform variety of the soda amphibole riebeckite. Both are mined in South Africa. There are also asbestiform varieties of tremolite and actinolite, the latter being the original material given the name asbestos. Nephrite, a tough and compact variety of actinolite, usually in mixtures with the pyroxene jadeite, is the gem jade.

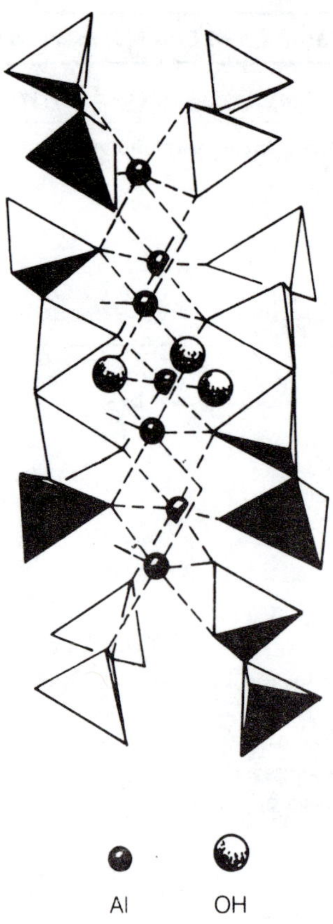

ANTHOPHYLLITE SERIES
$X_2 Y_5(Z_8 O_{22})(OH)_2$

	X	Y	Z
Anthophyllite	Mg,Fe^{2+},Mn	Mg,Fe^{2+},Mn	Si
Gedrite	Mg,Fe^{2+},Mn	Mg,Fe,Al	Si,Al

ăn-thŏ′fĭl-īt, from Latin *anthophyllum,* clove leaf, a reference to its clove-brown color.

gĕd′rīt, from its discovery locality near Gedres, in the French Pyrenees.

Crystallography, Structure, and Habit   Orthorhombic. $2/m\ 2/m\ 2/m$. *a* 18.61, *b* 18.01, *c* 5.24. *Pnma.* Rare prismatic crystals. No twinning.

Amphibole structure (see figure 21.42). Four formula units per cell.

Bladed, fibrous, or columnar with diamond-shaped cross section.

Physical Properties   Perfect {110} cleavage at 54.5° and 125.5°. Subconchoidal fracture. Brittle. Hardness 5.5–6. Specific gravity 2.8–3.2. Vitreous luster. Translucent. Color shades of brown to green. Streak uncolored or grayish. Optical: $\alpha = 1.60$–1.68, $\beta = 1.61$–1.68, and $\gamma = 1.62$–1.70; positive or negative; $2V \approx 10°$. Refractive indices increase with Fe content. Fusible (5) to a black magnetic enamel.

Al   OH

Figure 21.42
Cross-linkage of double chains in anthophyllite.

Distinctive Properties and Tests   Habit and clove-brown color. Yields water in the closed tube.

Association and Occurrence   Found in low-grade schists.

Alteration   From olivine or other Mg-rich minerals by metamorphism; to talc or serpentine.

Confused with   Other amphiboles.

Variants   Form varieties: asbestiform (amosite). Chemical varieties: sodian, ferrian, and fluorian.

Related Minerals   Solid solutions: magnesio-anthophyllite, (Mg,Fe)$_7$(Si$_8$O$_{22}$)(OH)$_2$; ferro-anthophyllite, (Fe,Mg)$_7$(Si$_8$O$_{22}$)(OH)$_2$; magnesiogedrite, (Mg,Fe)$_5$ (Al$_2$Si$_6$O$_{22}$)(OH)$_2$; and ferrogedrite, (Fe,Mg)$_5$ (Al$_2$Si$_6$O$_{22}$) (OH)$_2$. *Polymorphs:* cummingtonite (monoclinic) and grunerite (monoclinic). *Other similar species:* holmquistite, Li$_2$(Mg,Fe)$_3$Al$_2$(Si$_8$O$_{22}$)(OH)$_2$.

CUMMINGTONITE SERIES
Cummingtonite
(Mg,Fe)$_7$(Si$_8$O$_{22}$)(OH)$_2$, 30–70 mol % Mg

kŭm'ĭng-tə-nīt, from Cummington, Massachusetts, its type locality.

Grunerite
(Fe,Mg)$_7$(Si$_8$O$_{22}$)(OH)$_2$, 0–30 mol % Mg

grōōn'ə-rīt, named for Louis Emmanuel Gruner of St. Etienne, France, who first analyzed the iron-rich member of the series.

Crystallography, Structure, and Habit Monoclinic. $2/m$. a 9.6, b 18.3, c 5.3, β 101.8°. $C2/m$. Fibrous, bladed, or columnar crystals. Contact or repeated twins common on {100}.

Amphibole structure; facing double tetrahedral chains cross-linked by Mg and Fe. Two formula units per cell.

Aggregates of fibrous crystals, often radiating.

Physical Properties Perfect {110} cleavage at 56° and 124°. Fibrous varieties are flexible. Hardness 5–6. Specific gravity 3.2–3.6. Silky luster. Translucent. Color dark green or brown. Optical: $\alpha = 1.64$–1.68, $\beta = 1.65$–1.71, and $\gamma = 1.66$–1.73; positive; $2V = 60$–90°. Refractive indices increase with iron content. Fusible (5) to a black magnetic glass.

Distinctive Properties and Tests Cleavage, habit, color, and association. Insoluble in acids.

Association and Occurrence. Cummingtonite occurs with anthophyllite or hornblende, plagioclase, cordierite, and garnet in regionally or contact-metamorphosed gneissose or schistose rocks. Grunerite is found with such iron-rich minerals as magnetite, hematite, fayalite, hedenbergite, and garnet in metamorphosed, iron-rich siliceous sedimentary rocks.

Alteration To hornblende, talc, or serpentine.

Confused with Other amphiboles.

Variants *Form varieties:* asbestiform (amosite). May contain minor amounts of Fe^{3+}, Ti, Zn, and Ca.

Related Minerals *Solid solutions:* forms a series with magnesio-cummingtonite, Mg$_7$Si$_8$O$_{22}$(OH)$_2$; grunerite, Fe$_7$Si$_8$O$_{22}$(OH)$_2$; tirodite, Mn$_2$Mg$_5$Si$_8$O$_{22}$(OH)$_2$; dannemorite, Mn$_2$Fe$_5$Si$_8$O$_{22}$(OH)$_2$. *Isotypes:* other amphiboles. *Polymorphs:* anthophyllite series, of which it is the monoclinic analog. *Other similar species:* clinoholmquistite, Li$_2$(Mg,Fe^{2+},Mn)$_3$(Fe^{3+},Al)$_2$Si$_8$O$_{22}$(OH)$_2$.

ACTINOLITE SERIES
X$_2$Y$_5$(Si$_8$O$_{22}$)(OH)$_2$

	X	Y
Tremolite	Ca	Mg
Actinolite	Ca	Mg,Fe

ăk-tin'ə-līt, from Greek *aktis,* ray, an allusion to its frequent occurrence as radiating needles.
trĕm'ə-līt, discovered in Val Tremola, near St. Gotthard, Switzerland.

Crystallography, Structure, and Habit Monoclinic. $2/m$. a 4.840, b 18.052, c 5.275, β 104.7°. $C2/m$. Prismatic crystals, often flattened or columnar. Common contact and lamellar twinning on {100}.

Amphibole structure. Two formula units per cells.

Fibrous, felted, and asbestiform aggregates; compact or columnar masses.

Physical Properties Perfect {110} cleavage intersecting at 56° and 124°. Parting on {100}. Subconchoidal fracture. Brittle. Hardness 5–6. Specific gravity 3.0–3.3. Vitreous luster. Transparent to translucent. Color white or gray (tremolite) to dark green (actinolite). Optical: $\alpha = 1.66$–1.67, $\beta = 1.62$–1.68, and $\gamma = 1.63$—1.69; negative; $2V = 70$–80°. Refractive indices increase with Fe content. Fusible (3–4).

Distinctive Properties and Tests Habit and color. Yields a little water in the closed tube.

Association and Occurrence Tremolite is typically found as a contact metamorphic mineral in impure limestones and dolomites and also in talc schists. Associated with garnet, diopside, pyrite, and calcite. Actinolite is found in green-colored schists and greenstones.

Alteration Tremolite: to talc. Actinolite: from pyroxene; to chlorite, epidote, serpentine, and calcite.

Confused with Epidote, vesuvianite, and wollastonite.

Variants *Form varieties:* fibrous (asbestos) or matted (mountain leather); nephrite (jade), tough and compact, in part jadeite. *Chemical varieties:* manganoan, aluminian, fluorian, and chromian.

Related Minerals *Solid solutions:* ferro-actinolite, Ca$_2$Fe$_5^{2+}$Si$_8$O$_{22}$(OH)$_2$, and tremolite, Ca$_2$Mg$_5$Si$_8$O$_{22}$(OH)$_2$. *Isotypes:* hornblende series.

HORNBLENDE SERIES
Ferro-hornblende
Ca$_2$(Fe,Mg)$_4$Al(AlSi$_7$O$_{22}$)(OH)$_2$

hôrn'blend, from German *horn,* horn, and *blenden,* to blind or deceive, an allusion to its luster.

Magnesio-hornblende
Ca$_2$(Mg,Fe)$_4$Al(AlSi$_7$O$_{22}$)(OH)$_2$

Crystallography, Structure, and Habit Monoclinic. 2/*m.* *a* 9.87, *b* 18.42, *c* 5.37, β 105.7°. *C2/m.* Prismatic crystals with a pseudohexagonal outline, usually terminated with a low dome. Contact twins on {100}.

Amphibole structure. Two formula units per cell.

Columnar, bladed, and fibrous; also granular massive.

Physical Properties Perfect prismatic {110} cleavage at 56° and 124°. Parting on {001} and {100}. Subconchoidal to uneven fracture. Brittle. Hardness 5–6. Specific gravity 2.9–3.3. Vitreous luster. Translucent on thin edges. Color black to dark green. Optical: α = 1.62–1.70, β = 1.62–1.71, and γ = 1.63–1.73; negative; 2V = 30–80°. Hornblende is fusible with difficulty (4); soda-rich varieties fuse more easily (2) with intumescence and yield a black, magnetic globule.

Distinctive Properties and Tests Crystal habit, cleavage, and color. Yields a small amount of water in the closed tube.

Association and Occurrence Hornblende is widely distributed in igneous rocks over their entire compositional range. It is the most characteristic mafic mineral in intermediate plutonic rocks. The composition of hornblende usually reflects that of its igneous host—Mg-rich, Fe-poor in mafic rocks and Fe- and Al-rich in granites, syenites, and pegmatites. It is commonly associated with quartz, feldspar, pyroxene, and chlorite. Hornblende is a common component of metamorphosed basaltic rocks from greenschist to granulite facies.

Alteration From pyroxene. To chlorite, epidote, calcite, siderite, quartz, and biotite.

Confused with Augite.

Variants *Form varieties:* uralite, a pseudomorph of hornblende after pyroxene. *Chemical varieties:* oxyhornblende, typical in lavas, is common hornblende in which substitution of O^{2-} for (OH)$^-$ has taken place with concomitant charge balance by the oxidation of Fe^{2+} to Fe^{3+}. The substituting O^{2-} is derived from the breakdown of hydroxyl radicals by strong heating. Hornblende commonly contains titanium and fluorine.

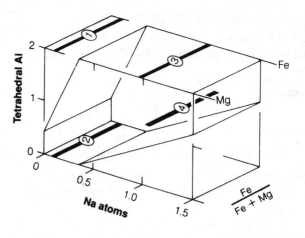
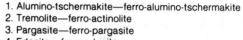

1. Alumino-tschermakite—ferro-alumino-tschermakite
2. Tremolite—ferro-actinolite
3. Pargasite—ferro-pargasite
4. Edenite—ferro-edenite

Figure 21.43
Solid solution in the hornblende series.

Related Minerals *Solid solutions:* The very complex solid solutions that are found in hornblendes can be considered intermediate to three compositional series:

Edenite-ferro-edenite,
NaCa$_2$(Mg,Fe)$_5$(AlSi$_7$O$_{22}$)(OH)$_2$
Pargasite-ferro-pargasite,
NaCa$_2$(Mg,Fe)$_4$Al(Al$_2$Si$_6$O$_{22}$)(OH)$_2$
Alumino-tschermakite-ferro-alumino-tschermakite,
Ca$_2$(Mg,Fe)$_3$Al$_2$(Al$_2$Si$_6$O$_{22}$)(OH)$_2$

These, together with the tremolite-ferro-actinolite series, approximately bound the hornblende field. The chemical relations are shown in figure 21.43. *Isotypes:* tremolite series. *Other similar species:* other amphiboles.

Single Chains

Nearly all the common silicate minerals with this structure are members of the pyroxene family. Also included, however, are a number of pyroxenoids that have generally analogous structures but are based on single chains with a different motif.

PYROXENE GROUP

The structure of a pyroxene (see figure 21.44) is based on rods consisting of two facing single chains paralleling the *c*-axis that are bound together by octahedrally coordinated ions, such as Mg^{2+} or Fe^{2+}. These rods are stacked and cross-linked in a manner analogous to those in the amphiboles, but because of the different cross-sectional shape of the rod, the intercleavage angles are different. The angle is about 87° and 93° in pyroxenes and 56° and 124° in amphiboles (see figure 9.3).

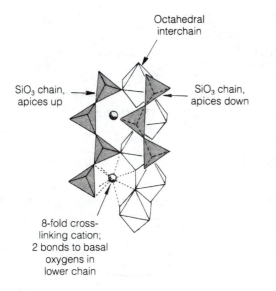

Figure 21.44
Components of the pyroxene structure.

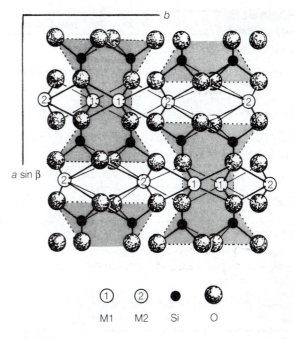

Figure 21.45
The structure of clinopyroxene looking down [001]. Cation positions *M*1 and *M*2 are shown along with schematic "I beams" (shaded). (After Cameron and Papike, 1981.)

①	②	●	◐
M1	M2	Si	O

In pyroxenes, sequential or alternating stagger of successive chains parallel to their length leads, as it does in amphiboles, to monoclinic and orthorhombic symmetries.

The general formula for the pyroxenes may be given as XYZ_2O_6 with single chains of tetrahedra extending parallel to the c-axis. Each tetrahedron shares two oxygens with adjacent tetrahedra giving a Z:O ratio of 1:3 and an overall composition of $(ZO_3)^{2-}$. The repeat distance along the chain, or the length of two tetrahedra, is 5.2Å. The chains are cross-linked by X cations in distorted octahedral or 8-fold polyhedral coordination (M2 site) and by Y cations in regular octahedral coordination (M1 sites). A generalized illustration of the pyroxene structure showing cation sites is given in figure 21.45.

The Z site is dominated by Si^{4+} but often contains Al^{3+}. The M1 site generally contains trivalent ions such as Al^{3+} and Fe^{3+}, and the M2 site contains divalent or monovalent ions. The order of ideal site occupancy in pyroxene based on crystal-chemical parameters such as ionic size, charge, and crystal field stabilization is given in table 21.9. The actual site occupancy is determined by a number of factors including chemical availability, pressure, and temperature. Coupled substitution to maintain charge balance is common, for example, Na^+–Al^{3+} for Ca^{2+}–Mg^{2+}. Excellent reviews of the structures and crystal chemistry of the pyroxenes are given by Cameron and Papike (1981) and Morimoto (1989).

The wide variety of possible cation site occupancies and extensive solid solution afford a large number of possible pyroxene compositions. On a chemical basis, the group is divided into five subdivisions comprising a total of twenty recognized species (see table 21.10), of which the Mg-Fe pyroxenes and some of the Ca pyroxenes are the most common. The great range of compositions exhibited by the pyroxenes can be illustrated on a triangular

Table 21.9 Site-Occupancy of Cations in the Pyroxene Structure. (Arrows indicate the ideal order of filling.)

Tetrahedral Z-Site $\Sigma = 2.000$	Octahedral M1 Site (Y) $\Sigma = 1.000$	Octahedral M2 Site (X) $\Sigma = 1.000$
$Si^{4+} \geq 1.000$		
$Al^{3+} \longrightarrow$	Al^{3+}	
$Fe^{3+} \longrightarrow$	Fe^{3+}	
	Ti^{4+}	
	Cr^{3+}	
	V^{3+}	
	Zr^{4+}	
	Sc^{3+}	
	Zn^{2+}	
	$Mg^{2+} \longrightarrow$	Mg^{2+}
	$Fe^{2+} \longrightarrow$	Fe^{2+}
	$Mn^{2+} \longrightarrow$	Mn^{2+}

diagram (see figure 21.46) with apices of enstatite, $Mg_2(Si_2O_6)$, ferrosilite, $Fe_2(Si_2O_6)$, and wollastonite, $Ca_2(Si_2O_6)$. The last mineral, although possessing a single chain structure, is a pyroxenoid and will be discussed separately. Na pyroxenes are relatively rare in a pure form, but they form continuous solid solutions with the Ca-Mg-Fe species. This is especially true of the aluminous and ferric end members, respectively jadeite and aegirine. The compositional relations of the more common clinopyroxenes are shown diagrammatically as a trigonal prism in figure 21.47.

Table 21.10   Accepted Species in the Pyroxene Group

Name	End-Member Composition	Solid-Solution Composition	System
Mg-Fe Pyroxenes			
Enstatite	$Mg_2Si_2O_6$		
Ferrosilite	$Fe_2Si_2O_6$	$(Mg,Fe)_2Si_2O_6$	Orthorhombic
Clinoenstatite			
Clinoferrosilite		$(Mg,Fe)_2Si_2O_6$	Monoclinic
Pigeonite		$(Mg,Fe,Ca)_2Si_2O_6$	Monoclinic
Mn-Mg Pyroxenes			
Donpeacorite		$(Mn,Mg)MgSi_2O_6$	Orthorhombic
Kanoite		$(Mn,Mg)MgSi_2O_6$	Monoclinic
Ca Pyroxenes			
Diopside	$CaMgSi_2O_6$		
Hedenbergite	$CaFe^{2+}Si_2O_6$	$Ca(Mg,Fe)Si_2O_6$	Monoclinic
Augite		$(Ca,Mg,Fe^{2+})_2Si_2O_6$	Monoclinic
Johannsenite	$CaMnSi_2O_6$		Monoclinic
Petedunnite	$CaZnSi_2O_6$		Monoclinic
Esseneite	$CaFe^{3+}AlSi_2O_6$		Monoclinic
Ca-Na Pyroxenes			
Omphacite		$(Ca,Na)(Fe^{2+},Mg,Al)Si_2O_6$	Monoclinic
Aegirine-augite		$(Ca,Na)(Fe^{2+},Mg,Fe^{3+})Si_2O_6$	Monoclinic
Na Pyroxenes			
Jadeite	$NaAlSi_2O_6$		
Aegirine	$NaFe^{3+}Si_2O_6$	$Na(Al,Fe^{3+})Si_2O_6$	Monoclinic
Kosmochlore	$NaCrSi_2O_6$		Monoclinic
Jervisite	$NaScSi_2O_6$		Monoclinic
Li Pyroxenes			
Spodumene	$LiAlSi_2O_6$		Monoclinic

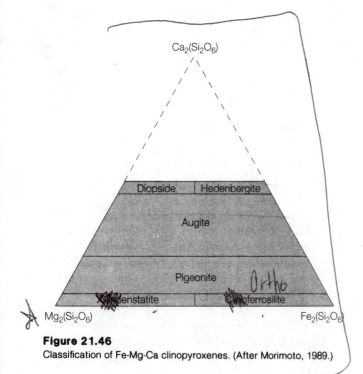

Figure 21.46

Classification of Fe-Mg-Ca clinopyroxenes. (After Morimoto, 1989.)

The actual naming of pyroxenes, as with the amphiboles, demands a chemical analysis and a calculated structural formula (see chapter 5). For field identification or laboratory study of hand specimens, the general terms orthopyroxene or clinopyroxene are useful; these can be modified with compositional adjectives. One can also use general pyroxene names such as hypersthene, augite, or omphacite if warranted by physical properties and association.

Except for spodumene and to a lesser extent jadeite, there are no significant commercial uses of minerals in the pyroxene group. Jadeite is used as a gem and carving medium. Spodumene is recovered as an ore of lithium that is used in a wide range of industrial applications, as an additive to greases, in storage batteries, in pharmaceuticals, in pyrotechnics, in purifying helium, in welding aluminum, and in making ammonia.

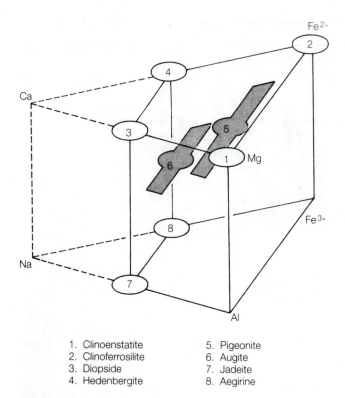

1. Clinoenstatite
2. Clinoferrosilite
3. Diopside
4. Hedenbergite
5. Pigeonite
6. Augite
7. Jadeite
8. Aegirine

Figure 21.47
Chemical variation in the clinopyroxenes.

ENSTATITE SERIES
Enstatite
Mg₂(Si₂O₆)

(hypersthene)
ĕn'stə-tīt, from Greek *enstates,* adversary; so named by
G. A. Kengott in 1855 because the mineral is nearly
infusible.

Ferrosilite
Fe₂(Si₂O₆)

fĕ-rō-sĭl'ĭt, named for its composition of iron (Latin
ferrum) plus silicon.

Crystallography, Structure, and Habit Orthorhombic.
$2/m\ 2/m\ 2/m$. Enstatite: *a* 18.22, *b* 8.81, *c* 5.18. Ferro-
silite: *a* 18.42, *b* 9.08, *c* 5.24. *Pbca.* Rare prismatic or tab-
ular crystals. Contact and lamellar twinning on {100}.
  Pyroxene structure (see figures 21.44 and 21.45).
Eight formula units per cell.
  Usually massive, fibrous, or lamellar.

Physical Properties Good {210} cleavage; two directions intersecting at 87° and 93°. Good {010} and {100}
parting. Uneven fracture. Brittle. Hardness 5–6. Specific
gravity 3.2–4.0. Vitreous to pearly luster. Translucent.
Color grayish, yellowish, or greenish white to olive-green
and brown. Sometimes a metalloidal luster or schiller because of oriented opaque inclusions. Optical: α = 1.655–

1.755, β = 1.658–1.785, and γ = 1.663–1.775; enstatite
is positive, $2V$ = 55–90°; iron-rich varieties are negative.
$2V$ = 90–50°. Refractive indices increase with Fe content. Enstatite is almost infusible; ferrosilite fuses to a
magnetic black enamel.

Distinctive Properties and Tests Color and cleavage.
Enstatite is insoluble in HCl and ferrosilite is
decomposed.

Association and Occurrence Enstatite is a common constituent of calcium-poor mafic and ultramafic igneous
rocks. Ferrosilite is found in mafic igneous rocks, especially lavas, associated with calcic plagioclase. Ferrosilite
is also found in high-grade metamorphic rocks. Both species are found in stony meteorites.

Alteration To chlorite, hornblende, serpentine (bastite
with schiller, uralite if fibrous).

Confused with Other pyroxenes and anthophyllite.

Variants *Form varieties:* bronzy luster. *Chemical varieties:* ferroan enstatite (bronzite), magnesian ferrosilite
(hypersthene). May also contain Ca, Mn, Ni, Cr, Al,
or Ti.

Related Minerals *Polymorphs:* clinoenstatite (monoclinic) and clinoferrosilite (monoclinic). *Isotypes:* other
pyroxenes. *Other similar species:* donpeacorite
(Mn,Mg)₂Si₂O₆.

DIOPSIDE SERIES
Diopside
CaMg(Si₂O₆)

dī-ŏp'sĭd, from Greek *dis,* twice, and *opsis,* appearance,
referring to the fact that two views can be taken of its
prismatic form.

Hedenbergite
CaFe(Si₂O₆)

hē'dĕn-bər-gīt, after M. A. Ludwig Hedenberg, the
Swedish chemist who analyzed and described the mineral.

Crystallography, Structure, and Habit Monoclinic,
$2/m$. Diopside: *a* 9.743, *b* 8.923, *c* 5.251, β 105°. Hedenbergite: *a* 9.854, *b* 9.024, *c* 5.263, β 104.2°. *C2/c.* Prismatic crystals (see figure 21.48), often stubby with a
square, rectangular, or 8-sided cross section. Common
contact and rare polysynthetic twins on {100} or {001}.
  Pyroxene structure. Four formula units per cell.
  Granular, lamellar, or columnar masses. Discrete
grains or crystals.

Physical Properties   Imperfect {110} cleavage; two directions intersecting at 87° and 93°. Parting on {100} often prominent (diallage). Uneven to conchoidal fracture. Brittle. Hardness 5.5–6.5. Specific gravity 3.2–3.6. Vitreous luster. Transparent to translucent. Color white to pale green (diopside), and green (hedenbergite). Optical: $\alpha = 1.66$–1.73, $\beta = 1.67$–1.74, and $\gamma = 1.70$–1.76; positive; $2V = 50$–$60°$. Refractive indices increase with Fe content. Fusible (4) to a green glass.

Distinctive Properties and Tests   Crystal form, cleavage, parting, and color. Insoluble in acids except HF.

Association and Occurrence   Diopside is found (1) in contact-metamorphosed limestones with tremolite, scapolite, vesuvianite, garnet, and titanite; (2) in mafic and ultramafic igneous rocks and their altered equivalents; and (3) in gneiss and schist. Hedenbergite is a constituent of mafic and ultramafic igneous rocks.

Alteration   Diopside alters to serpentine, talc, chlorite, limonite, and hedenbergite.

Confused with   Amphiboles, scapolite, and spodumene.

Variants   *Form varieties:* diallage, lamellar or micaceous resulting from ilmenite or magnetite dust on {100} parting; malacolite, translucent white to pale green; sahlite, dingy green. *Chemical varieties:* sodian-aluminian (omphacite, the pyroxene of garnet-bearing eclogites); manganoan; manganoan-zincian; manganoan-sodian; chromian; vanadian; titanian; and ferrian-aluminian.

Related Minerals   *Solid solutions:* johannsenite, $CaMn(Si_2O_6)$; petedunnite, $CaZn(Si_2O_6)$; esseneite, $CaFe^{3+}(AlSiO_6)$. *Isotypes:* aegirine, $NaFe(Si_2O_6)$; jadeite, $NaAl(Si_2O_6)$; jervisite, $NaSc(Si_2O_6)$; and kosmochlor, $NaCr(Si_2O_6)$. *Other similar species:* pigeonite, $(Mg,Fe,Ca)_2Si_2O_6$, a calcian member of the Fe-Mg pyroxenes.

AUGITE
$XY(Z_2O_6)$

X	Y	Z
Ca	Mg	Si
Mg	Fe^{2+}	Al
Fe^{2+}	Fe^{3+}	
Na	Al	
	Ti	
	Cr	

ô′jīt, from Greek *augites,* brightness, an allusion to the shining appearance of the mineral's cleavage surfaces.

Augite is difficult to define as a species because of its intermediate chemical position in a very complex solid solution system (see table 21.10 and figure 21.47). It is,

Figure 21.48
Prismatic crystals of diopside. Bancroft, Ontario. The longest crystal is 15 cm. (Courtesy University of Windsor. Photograph by D. L. MacDonald.)

however, the most common rock-forming pyroxene in contrast to its related end members which are comparatively rare.

Crystallography, Structure, and Habit   Monoclinic. $2/m$. a 9.8, b 9.0, c 5.25, β 105°. $C2/c$. Stubby octagonal crystals with a square to octagonal cross section. Simple or repeated twinning on {100} is common. Polysynthetic twinning on {001} is responsible for the often present basal parting.

Pyroxene structure (see figures 21.44 and 21.45).
Usually in poorly formed stubby crystals.

Physical Properties   Distinct cleavage on {110} and {1$\bar{1}$0} intersecting at 87° and 93°. Parting on {100} and {001}. Brittle. Hardness 5–6. Specific gravity 3.2–3.4. Vitreous luster. Translucent on very thin edges. Color black to dark green. Optical: $\alpha = 1.671$–1.735, $\beta = 1.672$–1.741, $\gamma = 1.703$–1.761; positive, $2V = 25$–$60°$. Fusible at 4–4.5. Insoluble in HCl.

Distinctive Properties and Tests   Association. Stubby black crystals with a square or octagonal cross section. Good but interrupted cleavages intersecting at close to 90°. Often a well-marked basal parting.

Association and Occurrence   The common pyroxene of mafic igneous rocks, especially gabbro; fairly common in intermediate and even felsic plutonic rocks; as phenocrysts in mafic volcanic rocks. Commonly in association with calcic to intermediate plagioclase.

Alteration   To uralite, a fibrous aggregate of secondary minerals, principally amphiboles; to amphibole and biotite as surrounding reaction rims; to chlorite.

Confused with   Hornblende (see table 21.5).

Variants   *Chemical varieties:* significant solid solution with Na pyroxene end members, especially jadeite and aegirine; titanian; aluminian-ferrian. May contain Li, Mn, Sr, and rare earth elements. *Form varieties:* diallage and malacolite, with well-developed parting on {100} or {001} respectively, often with films of ilmenite on the parting surfaces.

Related Minerals   Pigeonite, a Ca-bearing Fe-Mg pyroxene typically found in the groundmass of mafic volcanic rocks. Other pyroxenes and pyroxenoids.

SPODUMENE
LiAl(Si$_2$O$_6$)

(triphane)

späj'ə-mēn, from Greek *spodoumenos,* to burn to ashes, in recognition of the grayish-white mass that is formed when the mineral is ignited.

Crystallography, Structure, and Habit   Monoclinic. 2/m. a 9.451, b 8.387, c 5.208, β 110.1°. C2/m. Prismatic crystals; often flattened and deeply striated parallel to the c-axis. Crystals may be very large (several meters in length). Polysynthetic twins on (100) common.

Isostructural with the pyroxenes but does not form solid solutions with them because of the smaller size and charge of the Li$^+$ ion. Four formula units per cell.

Cleavable masses and crystals.

Physical Properties   Perfect {110} cleavages intersecting at 87° and 93°. Good {010} parting. Uneven to splintery or subconchoidal fracture. Brittle. Hardness 6.5–7. Specific gravity 3.0–3.2. Vitreous luster. Translucent to transparent. Color white or tinted green, gray, yellow-green, purple, or pink. Changes color when heated or irradiated. Fe, Cr, and Mn are the principal chromophores, apparently replacing Li. Streak white. Optical: α = 1.653, β = 1.659–1.668, and γ = 1.676–1.679; positive; 2V = 60–70°. Fusible (3) with branching and yields a clear or white glassy globule.

Distinctive Properties and Tests   Cleavage and parting, hardness, color, manner of fusion, and association. Insoluble in acids.

Association and Occurrence   Found in granitic pegmatites associated with tourmaline, beryl, garnet, lepidolite, alkali feldspars, micas, and quartz.

Alteration   To eucryptite, Li(AlSiO$_4$) and albite, or to muscovite and albite.

Confused with   Feldspar, scapolite, amblygonite, and tremolite.

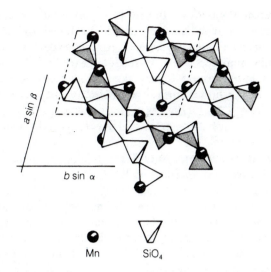

Figure 21.49
Structure of rhodonite. Unit cell dashed.

Variants   *Form varieties:* hiddenite (chromian, emerald green), kunzite (manganoan, lilac or pink), and triphane (colorless or yellow). *Chemical varieties:* sodian. Spodumene commonly contains small amounts of Fe, Ca, Mn, Mg, or Ga.

Related Minerals   *Polymorphs:* β-spodumene above 690° C. *Other similar species:* pyroxene group.

PYROXENOIDS

In addition to the pyroxene family proper, a number of other silicate species incorporate single chains in their structure (see figure 21.38). In these pyroxenoids the chain has a more complex repeat that is a function of cation size. Wollastonite has a three-tetrahedra repeat, rhodonite five, and pyroxmangite, (Mn,Fe)(SiO$_3$), seven.

Wollastonite is used in the manufacture of high-quality tile, rhodonite is sometimes used as a decorative stone, and chrysocolla is a minor ore of copper.

RHODONITE
Mn(SiO$_3$)

(manganese spar)

rō'də-nīt, from Greek *rhodon,* rose, alluding to its color.

Crystallography, Structure, and Habit   Triclinic. $\bar{1}$. a 7.682, b 11.818, c 6.707, α 92.4°, β 93.95°, γ 105.7°. P $\bar{1}$. Large and rough tabular crystals. Rare lamellar twins on {010}.

Structure based on kinked chains of SiO$_4$ tetrahedra with a repeating unit of 5 (see figure 21.49). Closely similar to wollastonite. Ten formula units per cell.

Cleavable masses; compact, discrete grains.

Physical Properties Perfect {010} and {100} cleavages intersecting at 88° and 92°. Distinct {001} cleavage. Conchoidal to uneven fracture. Tough. Hardness 5.5–6.5. Specific gravity 3.4–3.7. Vitreous to pearly luster. Translucent to transparent. Color pink or red, sometimes masked by black or brown surface oxidation. Streak white. Optical: $\alpha = 1.70$–1.729, $\beta = 1.725$–1.734, and $\gamma = 1.733$–1.741; positive; $2V = 60$–80°. Fuses (2.5) and blackens.

Distinctive Properties and Tests Color and hardness. Black alteration. Insoluble in HCl.

Association and Occurrence Found with iron and zinc ores associated with other manganese minerals (pyrolusite, tephroite, rhodochrosite, zincite, and willemite), calcite, and quartz in hydrothermal ore deposits and contact metamorphic rocks.

Alteration To rhodochrosite and pyrolusite.

Confused with Rhodochrosite.

Variants *Chemical varieties:* zincian (fowlerite), calcian (bustamite), ferroan, and magnesian.

Related Minerals *Solid solutions:* babingtonite, $Ca_2(Fe,Mn)[Si_5(O,OH)_{15}]$, and inesite, $Ca_2Mn_7[Si_{10}(O,OH)_{30}] \cdot 5H_2O$. At high temperatures, to bustamite, $(Mn,Ca,Fe)SiO_3$, and wollastonite. *Other similar species:* pyroxene family and pyroxmangite, $(Mn,Fe)(SiO_3)$.

CHRYSOCOLLA
$(Cu,Al)(SiO_3) \cdot nH_2O$

(bisbeeite)
kris′ə-kō′la, from Greek *chrysos,* golden, and *kolla,* glue, an allusion to its ancient use as a flux for soldering gold.

Crystallography, Structure, and Habit Possibly amorphous.

Structure unknown; included with inosilicates on the basis of Si:O.

Encrustations and fillings of enamel-like or earthy texture; colloform, never as crystals.

Physical Properties Conchoidal fracture. Sectile to brittle. Hardness 2–4. Specific gravity 2–3. Vitreous, shining, or earthy luster. Translucent to opaque. Color green to green-blue or brown to black when impure. Infusible.

Distinctive Properties and Tests Habit and color. Decomposed by HCl without gelatinization. Blackens and yields water in the closed tube.

Association and Occurrence Chrysocolla is a secondary copper mineral usually associated with other secondary copper minerals such as malachite, cuprite, and native copper.

Alteration From primary copper minerals, such as chalcopyrite.

Confused with Opal, turquoise, malachite, chalcanthite, and garnierite.

Variants Chrysocolla is typically impure, and the formula given is only an approximation.

Related Minerals *Similar species:* shattuckite, $Cu_5(SiO_3)_4(OH)_2$, blue; and dioptase, $Cu(SiO_3)(OH)_2$, emerald green.

WOLLASTONITE-1T
$Ca(SiO_3)$

(tabular spar)
wŏl′ə-stə-nīt, after William Hyde Wollaston (1766–1828), English chemist and physicist, discoverer of palladium (1804) and rhodium (1805), inventor of the reflecting goniometer (1809) and the camera lucida (1812).

Crystallography, Structure, and Habit Triclinic and pseudomonoclinic. $\bar{1}$. a 7.94, b 7.32, c 7.07, α 90.0°, β 95.4°, γ 103.4°. $P\bar{1}$. Short prismatic or tabular crystals. Multiple twinning common with axis [010] and composition plane {100}.

SiO_3 chains with a three-tetradron repeat and composed of tetrahedral pairs linked by a differently oriented single-tetrahedron run parallel to the b-axis (see figure 21.50). Ca ions are surrounded by seven oxygen ions, and the Ca-O polyhedra share faces to form a sheet parallel to (010). Six formula units per cell.

Cleavable masses to fibrous aggregates; also compact.

Physical Properties Perfect {100} and good {001} cleavages intersecting at 84° and 96°. Uneven fracture. Brittle. Hardness 4.5–5. Specific gravity 2.8. Vitreous to silky luster. Translucent. White tending toward grayish, yellowish, reddish, or brownish. Streak white. Optical: $\alpha = 1.618$, $\beta = 1.628$, and $\gamma = 1.631$; negative; $2V = 40°$. Fusible (4) to a subglassy globule. Often phosphorescent.

Distinctive Properties and Tests Cleavage, habit, and occurrence. Decomposed by HCl with separation of silica.

Association and Occurrence Found in contact-metamorphosed limestones with grossularite, diopside, tremolite, calcic plagioclase, vesuvianite, epidote, and calcite. Rarely as a component of alkaline igneous rocks.

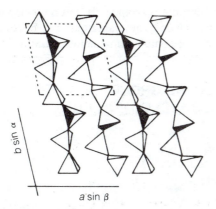

Figure 21.50
Structure of wollastonite. Unit cell dashed.

Figure 21.51
Sheaflike bundles of acicular crystals of pectolite. Asbestos, Quebec. Specimen is 5 × 2.5 × 1.5 cm. (Courtesy Royal Ontario Museum, Toronto.)

Alteration　Replaced by calcite.

Confused with　Tremolite, pectolite.

Variants　May contain Fe, Sr, Mg, or Mn.

Related Minerals　*Solid solutions:* possibly to pectolite. Extensive solid solution at high temperatures to bustamite, $(Mn,Ca,Fe)_3(Si_3O_9)$, and to rhodonite. *Polymorphs:* wollastonite-2M (monoclinic); and wollastonite-7T (triclinic). *Other similar species:* alamosite, $Pb(SiO_3)$; rhodonite; tobermorite, $Ca_5[Si_6(O,OH)_{18}] \cdot 4H_2O$; and bustamite, $(Mn,Ca)_3(Si_3O_9)$.

PECTOLITE
$NaCa_2H(Si_3O_9)$

pĕk′tə-līt, from Greek *pektos,* well put together, referring to its compact structure.

Crystallography, Structure, and Habit　Triclinic. $\bar{1}$. a 7.99, b 7.04, c 7.02, α 90.05°, β 95.3°, γ 102.5°. $P\bar{1}$. Acicular crystals (see figure 21.51).
　Structurally similar to wollastonite. Kinked chains of Si-O tetrahedra parallel the b-axis and are cross-linked by Na and Ca ions in 6-fold coordination. Hydrogen bonds some oxygen in adjacent tetrahedra. Two formula units per cell.
　Fibrous radiated masses.

Physical Properties　Perfect {100} and {001} cleavages, yielding needlelike fragments. Uneven fracture. Brittle. Hardness 4.5–5. Specific gravity 2.8. Vitreous to silky luster. Opaque. Color and streak white. Optical: $\alpha = 1.600$, $\beta = 1.605$, $\gamma = 1.635$; positive; $2V = 50°$. Fuses (2.5–3) readily to a glass. Sometimes phosphoresces when crushed.

Distinctive Properties and Tests　Habit, cleavage, opacity, and occurrence. Yields water in the closed tube. Decomposed by HCl with the separation of silica but without gelatinization.

Association and Occurrence　Found as crusts and fillings in mafic lavas. Associated with zeolites, prehnite, and calcite. Rarely as a primary mineral in alkaline igneous rocks or in calcium-rich metamorphic rocks.

Confused with　Wollastonite and fibrous zeolites.

Variants　*Chemical varieties:* magnesian and manganoan (schizolite). May contain a little Fe, K, or Al.

Related Minerals　*Solid solutions:* serandite, $Na(Mn,Ca)_2[Si_3O_8(OH)]$.

CYCLOSILICATES (Rings)

Endless chains or rings of Si-O tetrahedra exist in tectosilicates and phyllosilicates, for example, in scapolite and kaolinite, where they are articulated by oxygen-sharing into networks or sheets. Rings, however, are also found as isolated complex radicals in a few silicate minerals.
　Energetically, the most stable ring contains six shared tetrahedra, $(SiO_3)_6^{12-}$, although structures based on 3-, 4-, and 8-membered rings are also known (see figure 21.52). Typical structures are composed of stacks of these hexagonal rings and impart a hexagonal or trigonal columnar crystalline habit to the mineral. The open channels in this structure accommodate nonformulary components, such as water and alkali metals.

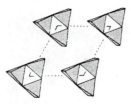

View along *c*-axis of arrangement of 3-membered tetrahedral rings in benitoite, $BaTi(Si_3O_9)$

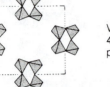

View along *c*-axis of arrangement of 4-membered tetrahedral rings in papagoite, $Ca_2Cu_2Al_2(Si_4O_{12})(OH)_6$

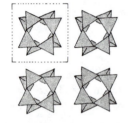

View along *c*-axis of arrangement of 8-membered tetrahedral rings in ekanite, $ThK(Ca, Na)_2(Si_8O_{20})$

Figure 21.52
Kinds and arrangements of rings of tetrahedra. Unit cells dashed.

Some minerals are difficult to classify as cyclosilicates because cross-linkage between the ring stacks may be by ions in tetrahedral coordination. In cordierite, for example, 6-membered $(AlSi_5O_{18})$ rings are cross-linked by tetrahedrally coordinated Al, and the mineral formula may be written $MgAl_3(AlSi_5O_{18})$. If, however, we focus attention only on the articulation of tetrahedra, the structure is a network whose formula is $Mg(Al_4Si_5O_{18})$. Cordierite is here classified as a tectosilicate because of the participation of Al in both rings and cross-linkages.

The situation with respect to beryl is similar because Be is in tetrahedral coordination. The formula may be written as a cyclosilicate, $Be_3Al_2(Si_6O_{18})$, or as a tectosilicate, $Al_2(Be_3Si_6O_{18})$. Because Be is not included as an ion in the Si-O tetrahedral articulation elsewhere in this book, however, beryl is here classified as a ring structure.

Beryl, and to a lesser extent tourmaline, is the source of a number of precious and semiprecious gems. Additionally, beryl is the ore of beryllium, which has a number of special uses. Beryllium alloyed with copper or, less commonly, with cobalt, nickel, or aluminum, forms a non-sparking compound with high strength and excellent fatigue resistance appropriate for use in tools, springs, control parts, and instruments that must be operated in explosive atmospheres. Beryllium compounds are also used in x-ray tubes, as phosphors in fluorescent lamps, in neon tubes, and to form refractory crucibles.

Many beryllium compounds, but not beryl itself, are toxic. When inhaled, they cause cumulative lung damage leading to severe emphysema and eventual death.

TOURMALINE GROUP
$WX_3Y_6(BO_3)_3(Si_6O_{18})(OH,F)_4$

	W	*X*	*Y*
Schorl	Na	Fe^{2+}	Al
Dravite	Na	Mg	Al
Elbaite	Na	Li,Al	Al

tōŏr'mə-lēn, from Singhalese *toramalli*, carnelian, describing brown gemstones found in Ceylon early in the eighteenth century.

shôrl, from Swedish *skör*, brittle.
drā'vīt, after Drave, Austria.
ĕl'bə-īt, after the island of Elba in the Mediterranean.

Crystallography, Structure, and Habit Hexagonal. 3*m*. Schorl: *a* 16.032, *c* 7.149. Dravite: *a* 15.942, *c* 7.224. Elbaite: *a* 15.842, *c* 7.009. *R*3*m*. Usually slender, columnar crystals with prism faces showing strong vertical striations. The cross section is either hexagonal or has the form of a spherical triangle (see figure 21.53).

The tourmaline structure is illustrated in figure 21.54. Si_6O_{18} tetrahedral rings, MgO_6 octahedra, and BO_3 triangles in rodlike aggregates parallel the *c*-axis and are cross-linked by Al^{3+}. Alkali metals are centered in the rings of the tetrahedra. Three formula units per cell.

Parallel or radiating crystal groups; massive and compact.

Physical Properties No cleavage. Subconchoidal to uneven fracture. Brittle, sometimes friable. Hardness 7–7.5. Specific gravity 3.0–3.2. Vitreous to resinous luster. Transparent to opaque. Color various—black, blue, green, red, colorless. Colors may be zoned. Streak uncolored. Optical: $\omega = 1.635–1.675$ and $\epsilon = 1.610–1.650$; negative. Pyroelectric. Electrified by friction. Piezoelectric. Fusibility varies with composition; iron-rich varieties are fused with difficulty, magnesium-rich varieties at 3, and lithium-rich varieties are infusible.

Distinctive Properties and Tests Crystal habit, hardness, lack of cleavage, and association. Insoluble in acids.

Association and Occurrence Tourmaline is one of the more common minerals formed by the action of hot, high-pressure vapors. It is an accessory mineral in pegmatites, at granitic contacts, and, less commonly, in gneisses and schists. Associated with potash feldspar, quartz, micas (including lepidolite), beryl, apatite, fluorite, topaz, and cassiterite. Common accessory mineral in detrital sediments.

a.

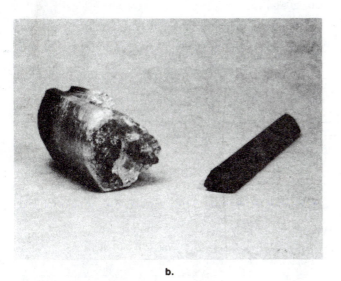

b.

Figure 21.53

(a) Hexagonal prism of schorl showing distinct striations along [001].
(b) Trigonal prisms of elbaite (left) and schorl (right). (Courtesy University of Windsor. Photograph by D. L. MacDonald.)

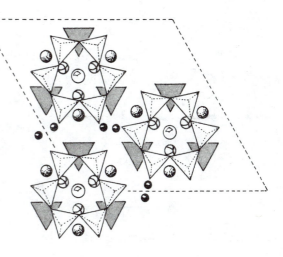

Mg   Al   OH   Na   (BO₃)   (SiO₄)

Figure 21.54
Tourmaline structure. View along *c*-axis; unit cell dashed.

Alteration   To muscovite, biotite, and chlorite.

Confused with   Hornblende, staurolite.

Variants   *Form varieties:* uvite (brown), achroite (colorless), verdelite (green), rubellite (red), and dravite (brown). *Chemical varieties:* complete composition range between dravite and schorl and between schorl and elbaite. Other members of the tourmaline group are as follows:

	W	*X*	*Y*
Buergerite	Na	Fe^{3+}	Al
Chromdravite	Na	Mg	Cr,Fe^{3+}
Ferridravite	Na,K	Mg,Fe^{2+}	Fe^{3+}
Feruvite	Ca	Fe^{2+},Mg	Al,Mg
Liddicoatite	Ca	Li,Al	Al
Olenite	Na	Al	Al
Uvite	Ca,Na	Mg,Fe^{2+}	Al,Mg

Tsilaisite is a manganoan elbaite.

BERYL
Be₃Al₂(Si₆O₁₈)

bĕr′əl, from Greek *beryllos,* a blue-green gem.

Crystallography, Structure, and Habit   Hexagonal. 6/*m* 2/*m* 2/*m*. *a* 9.215, *c* 9.192, increasing with alkali content. P6/*mmc.* Simple hexagonal prisms, often large. Vertically striated faces. Pinacoidal termination.

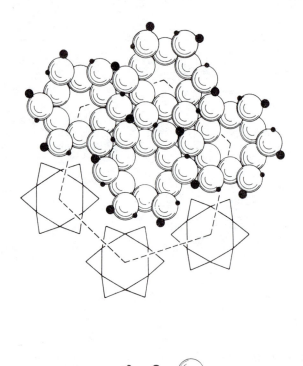

Be   Al   O

Figure 21.55
Structure of beryl. View along *c*-axis.

Beryl structure (see figure 21.55). Hexagonal rods made up of 6-membered tetrahedral rings and cross-linked by Be and Al. Two formula units per cell.

Found as distinct crystals; columnar aggregates; and massive.

Physical Properties No cleavage. Conchoidal to uneven fracture. Brittle. Hardness 7.5–8. Specific gravity 2.6–2.9, increasing with alkali content. Vitreous to resinous luster. Transparent to translucent. Color blue-green, green, blue, yellow, and white. The color may be modified by heating; greenish yellow to clear green (275° C), green to blue (300–400° C). The color of emerald is unaffected by heating. Optical: $\omega = 1.566–1.608$ and $\epsilon = 1.562–1.600$; negative. Infusible to fusible with difficulty (5.5); whitens on heating. Slowly dissolves in HF.

Distinctive Properties and Tests Crystal form, lack of cleavage, color, hardness, and association.

Association and Occurrence Found in granitic rocks, especially pegmatites. Also in mica schist. Association with feldspar, mica, tourmaline, spodumene, and garnet.

Alteration To mica and kaolin; also to rare Be-minerals.

Confused with Quartz, apatite, and green tourmaline.

Variants *Form varieties:* common beryl, green to yellow-green pigmented by Fe^{3+}; emerald, bright green because of traces of Cr; aquamarine, pale greenish blue; morganite or rose beryl; heliodor or golden beryl; goshenite, colorless. *Chemical varieties:* In some beryl, Li may replace Al in octahedral positions while Al replaces Be in tetrahedral sites. Large alkali ions in structural channels provide the additional positive charge needed for electrical balance. Sodian, pale green to white; sodian-lithian, greenish white to colorless; and lithian-cesian, colorless to pink. May also contain K, Rb, Cs, Na, Ca, Sc, and noble gases, mainly He. Water always present in channels.

Related Minerals *Isostructures:* bazzite, $Be_3(Sc,Al)_2 (SiO_3)_6$. *Other similar species:* cordierite.

SOROSILICATES (Pairs)

Silica tetrahedra sharing a single oxygen ion to form paired tetrahedra make up a relatively unstable type of silicate articulation. Both tetrahedral units, which share no oxygen ions, and tetrahedral chains, in which two oxygen ions per tetrahedron are shared, provide a more stable linkage for a silicate structure. The formation of minerals whose structure incorporates tetrahedral pairs, therefore, requires rather special conditions—either chemical restrictions or high temperatures. The melilite series is an example of minerals for which both of these conditions are met. Melilites form only in silica-deficient environments under high-temperature conditions.

Included with the sorosilicates are the mineral vesuvianite and the epidote series. These exceptional silicate minerals incorporate two kinds of tetrahedral silicate groups in their structure; that is, they include both pairs and individual tetrahedral units.

Hemimorphite represents an interesting problem in mineral classification. In its usual formularization, $Zn_4(Si_2O_7)(OH)_2 \cdot H_2O$, zinc appears as a normal cation coordinating the $(Si_2O_7)^{6-}$ complex ions. However, the structure of this mineral shows zinc to coordinate four oxygen ions into a regular $(ZnO_4)^{6-}$ tetrahedral group. Each of these groups, in turn, shares oxygen ions with adjacent $(ZnO_4)^{6-}$ or $(SiO_4)^{4-}$ tetrahedra. In sum, the structure of hemimorphite is a network of ZO_4 tetrahedra, where Z is either Zn or Si. Hemimorphite is here classified as a tetrahedral pair structure, but future studies of the complete range of structures in which some other ion occupies the silicon site will probably require reclassification of this and other minerals. Aluminum proxy for silicon is generally recognized as not affecting the structural type, but substitution of other ions in the same way is not accorded the same treatment.

MELILITE GROUP
$X_2Y(Z_2O_7)$

	X	Y	Z
Gehlenite	Ca	Al	Si,Al
Åkermanite	Ca	Mg	Si

měl′ə-līt, from Greek *meli,* honey, an allusion to its color.

gāl′ə-nīt, named by J. F. Fuchs in 1815 for his friend and colleague R. F. Gehlen.

ō′kər-mə-nīt, after A. R. Åkerman, Swedish metallurgist.

Crystallography, Structure, and Habit Tetragonal. $\bar{4}2m$. Gehlenite: a 7.690, c 5.068. Åkermanite: a 7.844, c 5.010, $P\bar{4}2_1m$. Short square or octagonal prisms. Twinning on {100} and {001}.

Structure (see figure 21.56) consists of $(Si,Al)_2O_7$ pairs and AlO_6 or MgO_6 octahedra in sheetlike array parallel to (001) with sheets held together by Ca-O linkages. Two formula units per cell.

Found as tubular crystals and embedded grains.

Physical Properties Indistinct {001} and poor {110} cleavage. Conchoidal to uneven fracture. Brittle. Hardness 5. Specific gravity 3.0. Vitreous to resinous luster. Color white, yellowish, or often tinted yellow, green, red, or brown. Optical: $\omega = 1.632$–1.669 and $\epsilon = 1.640$–1.658; gehlenite, negative; åkermanite, positive. Fusible (3) to a yellow or green glass.

Distinctive Properties and Tests Crystal form.

Association and Occurrence Melilites are found in mafic extrusive rocks poor in silica and alkalies but rich in lime and alumina. They are associated with nepheline, leucite, augite, and hornblende. Gehlenite is found in undersaturated basalts and is formed in limestone by contact metamorphism. Åkermanite is found in thermally metamorphosed siliceous limestones and dolostones with other calc-silicates. The melilites are also typical minerals of carbonatite complexes.

Alteration To garnet, vesuvianite, diopside, calcite, and prehnite. Also to cebollite, $Ca_4Al_2(Si_4O_{12})(OH)_2$, and juanite, $Ca_{10}Mg_4[(Si,Al)_{11}O_{30}] \cdot 4H_2O$.

Confused with Leucite.

Variants *Chemical varieties:* ferroan. May also contain small amounts of Na, Zn, and Mn.

Related Minerals *Solid solutions:* soda-melilite, $CaNaAl(Si_2O_7)$. *Isotypes:* hardystonite, $Ca_2Zn(Si_2O_7)$. *Other similar species:* leucophanite, $(Na,Ca)_2Be[Si_2(O,OH,F)_7]$, and meliphanite, $(Ca,Na)_2Be[(Si,Al)_2(O,OH,F)_7]$.

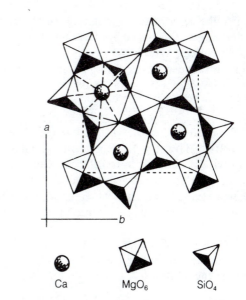

Ca MgO_6 SiO_4

Figure 21.56
Structure of melilite. Unit cell dashed.

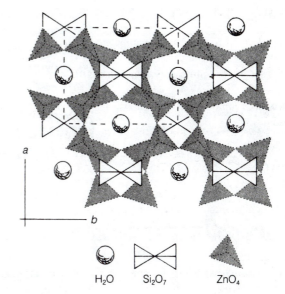

H_2O Si_2O_7 ZnO_4

Figure 21.57
Structure of hemimorphite. Unit cell dashed.

HEMIMORPHITE
$Zn_4(Si_2O_7)(OH)_2 \cdot H_2O$

(calamine, galmei)

hěm-ĭ-môr′fīt, from Greek *hemi,* half, and *morphos,* form, allusion to its symmetry. The lack of a center or mirror perpendicular to the *c*-axis causes different terminations to appear on the prism.

Crystallography, Structure, and Habit Orthorhombic. $2mm$. a 8.370, b 10.719, c 5.120. *Imm*2. Tabular or prismatic crystals with vertical striations.

(Si_2O_7) and $[Zn_2(O,OH)_7]$ pairs are articulated into a framework of shared tetrahedra (see figure 21.57). The structure may be described as alternating Zn-O and Si-O tetrahedral sheets cross-linked by O-Si-O-bridging. Two formula units per cell.

Usually in rounded forms; divergent crystal groups, stalactitic, or colloform; also massive or earthy, encrusting.

Physical Properties Perfect {110} cleavage. Uneven fracture. Brittle. Hardness 4.5–5. Specific gravity 3.5. Vitreous luster. Transparent to translucent. Color white with tints of blue or green; also yellowish to brown. Streak white. Decrepitates and whitens in the closed tube, yielding water. Fusible with difficulty (5–6). Strongly pyroelectric.

Distinctive Properties and Tests Habit, color, and fusibility. Decomposes in boiling HCl and gelatinizes on evaporation of the solution.

Association and Occurrence A secondary mineral found in the oxidized portion of zinc deposits in carbonate sedimentary rocks, where it is associated with smithsonite, sphalerite, cerussite, anglesite, and galena.

Alteration From other zinc minerals.

Confused with Smithsonite and prehnite.

Variants Small amounts of Al, Fe, and Pb may be present.

Related Minerals *Isotypes:* cuspidine, $Ca_4 (Si_2O_7)(F,OH)_2$.

VESUVIANITE
$Ca_{10}(Mg,Fe)_2Al_4(Si_2O_7)_2(SiO_4)_5(OH,F)_4$

(idocrase)

vi-soo′vē-ə-nīt, after Mt. Vesuvius, Italy, where it is found in ejected blocks.

Crystallography, Structure, and Habit Tetragonal. $4/m\ 2/m\ 2/m$. *a* 15.55, *c* 11.81. *P4nnc*. Prisms or pyramids, often vertically striated.

Structurally similar to grossular garnet but with some silica in paired tetrahedra. Four formula units per cell.

Striated columnar aggregates; massive, granular, or fibrous.

Physical Properties Very poor cleavage on {110} or {001}. Subconchoidal to uneven fracture. Brittle. Hardness 6.5. Specific gravity 3.4. Vitreous to resinous luster. Translucent. Color brown to green; rarely red or blue. Optical: $\omega = 1.703$–1.746 and $\epsilon = 1.700$–1.746; negative. Fusible (3) with intumescence to a greenish brown glass.

Distinctive Properties and Tests Habit, color, and association. Gelatinizes with HCl after simple fusion.

Association and Occurrence Found in contact-metamorphosed limestones associated with garnet (grossular), phlogopite, diopside, wollastonite, epidote, and tourmaline and in serpentines, chlorite schists, and gneisses.

Confused with Epidote, tourmaline, garnet.

Variants *Chemical varieties:* Composition highly variable. Mn, Na, and K replace Ca. Ti, Al, and, more rarely, Cu substitute for (Fe,Mg). Also may contain B, Be, Cr, Li, Zn, and rare earths.

EPIDOTE SERIES
$X_2Y_3(SiO_4)(Si_2O_7)(O,OH)$

	X	Y
Clinozoisite	Ca	Al
Epidote	Ca	Al,Fe
Piemontite	Ca	Al,Mn,Fe
Allanite-Ce	Ce,Ca,Y	Al,Fe

ē′pi-dōt, from Greek *epididonai,* to give in addition, an allusion to the enlargement of the base of some crystals.
klī′nō-zō′ə-sīt, named for Siegmund Zois, Baron von Edelstein (1747–1819), an Austrian scholar who financed mineral-collecting expeditions.
pē′mŏn-tīt, named for the Piemonte region in northwestern Italy.
ăl′ə-nīt, named for Thomas Allan (1777–1833), the Scottish mineralogist who first noticed the mineral.

Crystallography, Structure, and Habit Monoclinic. $2/m$. Clinozoisite: *a* 8.887, *b* 5.581, *c* 10.14, β 115.9°. Epidote: *a* 8.89, *b* 5.63, *c* 10.19, β 115.4°. Piemontite: *a* 8.95, *b* 5.70, *c* 9.41, β 115.7°. Allanite-Ce: *a* 8.98, *b* 5.75, *c* 10.23, β 115°. $P2_1/m$. Prismatic to acicular crystals, striated. Common lamellar twins on {100}.

Structure consists of chains of edge-sharing AlO_6 and $AlO_4(OH)_2$ octahedra parallel to *b* cross-linked by SiO_4 and Si_2O_7 groups (see figure 21.58). Two formula units per cell.

Granular, massive, and fibrous; crystals striated parallel to *b*.

Physical Properties All have perfect basal cleavage. See table 21.11 for other properties.

Distinctive Properties and Tests Color and habit. Will gelatinize with HCl when previously ignited.

Association and Occurrence Epidote minerals are formed during metamorphism of calcium-rich rocks. Common and widespread, they are often associated with iron oxides, quartz, feldspar, actinolite, chlorite, and hornblende. Epidotes are also produced by hydrothermal alterations of plagioclase (saussuritization) and are often

Table 21.11   **Epidote Series**

	Clinozoisite	Epidote	Piemontite	Allanite-Ce
Color	pale green to brown	pale to dark green	purple to red	brown to black
Luster	vitreous to resinous	vitreous to resinous	vitreous to resinous	pitchy to submetallic
Hardness	6–6.5	6–7	6–7	5.5–6
Specific gravity	3.2	3.2–3.5	3.4	3.5–4.6
HCl	insoluble	partly soluble	partly soluble	gelatinizes
α	1.670–1.715	1.715–1.751	1.730–1.794	1.690–1.791
β	1.674–1.725	1.725–1.784	1.750–1.807	1.700–1.815
γ	1.690–1.734	1.730–1.797	1.760–1.832	1.706–1.828
Sign	positive	negative	positive	negative
$2V$	14–90°	64–90°	64–99°	0–123°

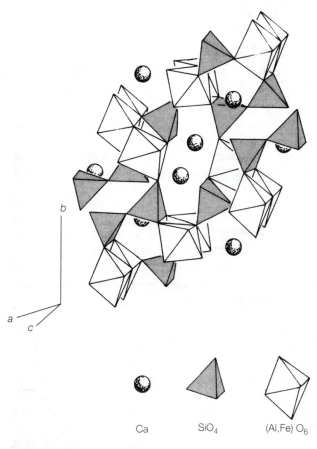

Ca   SiO₄   (Al,Fe) O₆

Figure 21.58
Structure of epidote.

found along joints and in vugs. Allanite-Ce is an accessory mineral in many granitic rocks. It also occurs in some limestone skarns.

Alteration   From garnet, hornblende, plagioclase, augite, biotite, scapolite, and olivine.

Variants   *Form varieties:* tanzanite, blue gemstone; thulite, rosy. *Chemical varieties:* chromian; magnesian, plumbian, and strontian (hancockite, brownish red); vanadian (mukhinite), manganoan, stannian, and yttrian (allanite-Y).

Related Minerals   *Polymorphs:* zoisite, $Ca_2Al_3(SiO_4)$ $(Si_2O_7)(O,OH)$(orthorhombic). *Similar species:* dollaseite-Ce, $CaCeMg_2AlSi_3O_{11}(F,OH)_2$, and ilvaite, $CaFe_2^{2+}Fe^{3+}(Si_2O_7)O(OH)$.

Nesosilicates (Units)

Minerals employing individual $(SiO_4)^{4-}$ tetrahedra as their structural basis embrace several cohesive mineral groups and many miscellaneous species.

The olivine group, which shows complete solid solution between Mg, Fe, and Mn end members, includes a Ni-bearing species. All members of the olivine group crystallize in the orthorhombic system. Structurally, the cations lie between opposed faces of individual tetrahedra, coordinating three oxygens in each tetrahedron.

The garnet group includes two series, each of which shows fairly complete solid solution within the series. There is, however, only limited solid solution between the series. The general formula for garnet is $X_3Y_2(SiO_4)_3$. X is always Ca^{2+} in the andradite series, whereas Y may be Cr^{3+}, Al^{3+}, or Fe^{3+}. In the almandine series, Y is always Al^{3+}, and X may be Fe^{2+}, Mg^{2+}, Mn^{2+}, or Ca^{2+}.

All garnets crystallize in the isometric system and have the dodecahedron and the trapezohedron as common forms. Garnets are usually some shade of red, although grossular is often cinnamon-brown, and uvarovite is emerald-green as a consequence of its chromium content. The physical characteristics of garnets are so similar, and the intraseries substitutions so extensive, that chemical analysis or measurement of specific gravity or refractive index must be used to distinguish them.

Garnets are found in both igneous and metamorphic rocks. Almandine and spessartine are common minerals in schists and gneisses. The garnet found in metamorphosed limestone is usually grossular. Pyrope is a common accessory mineral in ultramafic rocks such as peridotite and is also found in the derived serpentines. Uvarovite also has a periodotitic association, where it frequently occurs with chromite. Spessartine is a common constituent of siliceous lavas and complex pegmatites. Garnet has been used as an abrasive, and good quality almandine garnets are used as gems.

Minerals of the humite group have the general formula $n\text{Mg}_2(\text{SiO}_4) \cdot \text{Mg(OH,F)}_2$, where n may be 1, 2, 3, or 4. The structure of the various members of this group is essentially one or more $\text{Mg}_2(\text{SiO}_4)$ layers separated by an Mg(OH,F)_2 layer. The packing is such that when n is odd the mineral is orthorhombic and when it is even the mineral is monoclinic.

Zircon is a ubiquitous accessory mineral in igneous rocks, and because of its high resistance to weathering, it is often found in the heavy detrital fraction of sediments and their metamorphic equivalents. Transparent varieties serve as gemstones, either with their natural yellowish or reddish colors or made blue by heat treatment. The principal use of zircon is as the ore for zirconium metal, most being produced from beach sands.

Topaz is formed from high-temperature fluorine-bearing vapors such as those that cause contact alteration near igneous bodies. It is a typical mineral of the greisen in which cassiterite is found. The principal use of topaz is as a minor gemstone.

OLIVINE GROUP
$Y_2(\text{SiO}_4)$

	Y
Forsterite	Mg
Fayalite	Fe
Tephroite	Mn

(chrysolite, peridot)
ŏl'ə-vēn, from Latin *oliva,* olive, an allusion to its olive-green color.

Crystallography, Structure, and Habit Orthorhombic. $2/m\,2/m\,2/m$. Forsterite: a 4.758, b 10.214, c 5.984. Fayalite: a 4.817, b 10.477, c 6.105. Tephroite; a 4.871, b 10.636, c 6.232. *Pbnm.* Flattened prisms. Rare contact twins.

Olivine structure (see figure 21.59). Independent SiO_4 tetrahedra linked by divalent ions in 6-fold coordination. Oxygen in approximately hexagonal packing parallel to (100) and tetrahedral apices alternate in $+a$ and $-a$ directions. Four formula units per cell.

Imbedded grains or granular masses.

Physical Properties No distinct cleavage. Conchoidal fracture. Brittle. Hardness 6.5–7. Specific gravity 3.3 (forsterite) to 4.4 (fayalite). Vitreous luster. Transparent to translucent. Color olive-green, sometimes brownish or reddish. May turn yellow-brown or red on oxidation of iron. Optical: $\alpha = 1.635$–1.827, $\beta = 1.651$–1.869, and $\gamma = 1.670$–1.879; positive to negative; $2V \approx 90°$. Infusible unless iron-rich. Whitens when heated.

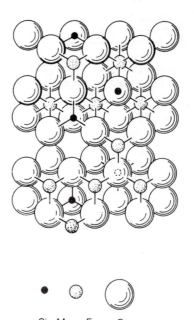

Si   Mg or Fe   O

Figure 21.59
Structure of olivine. View along c-axis.

Distinctive Properties and Tests Color, lack of cleavage, and association. Slowly dissolves in boiling HCl and gelatinizes on evaporation of the solution.

Association and Occurrence A necessary constituent of certain mafic and ultramafic rocks (the sole constituent of dunite). Usually associated with orthopyroxenes and clinopyroxenes, calcic plagioclase, chromite, spinel, and garnet. Also found in metamorphosed limestones associated with dolomite and magnesite.

Alteration Readily to serpentine, magnesite, brucite, chlorite, talc, and iron oxides.

Confused with Quartz and epidote.

Variants *Form varieties:* precious olivine (peridot), pale yellow-green and transparent. *Chemical varieties:* titanian (possible proxy for silicon); nickelian forsterite (liebenbergite); and manganoan zincian fayalite (roepperite). Sn, Zn, and Ca may be present in minute amounts.

Related Minerals *Solid solutions:* complete among forsterite, fayalite, and tephroite end members (see figure 5.3). *Polymorphs:* wadsleyite (orthorhombic), ringwoodite (isometric). *Isostructures:* chrysoberyl, $\text{BeAl}_2(\text{O}_4)$. *Other similar species:* larnite, $\beta\text{-Ca}_2(\text{SiO}_4)$; merwinite, $\text{Ca}_3\text{Mg(SiO}_4)_2$; sinhalite, $\text{MgAl(BO}_4)$; larsenite, $\text{PbZn(ZiO}_4)$; triphylite-lithiophilite series, Li(Fe,Mn)PO_4; monticellite, $\text{CaMg(SiO}_4)$; glaucochroite, $\text{CaMn(SiO}_4)$; and kirschsteinite, $\text{CaFe(SiO}_4)$.

GARNET GROUP
$X_3Y_2(SiO_4)$

gär'nĭt, from Latin *granatum,* pomegranate, because of the resemblance of garnet crystals to the seeds of this fruit in color and shape.

Garnets are particularly characteristic minerals of metamorphic rocks, and their appearance and composition can be important in determining metamorphic grade. Less commonly they are found as accessory minerals in certain types of igneous rock and as grains in detrital sediments.

The common garnets can be divided into two series:

Ugrandites: *uva*rovite, *gros*sular, *and*radite
Pyralspites: *py*rope, *al*mandine, *sp*essartine

Their names originated as described below.

Uvarovite, yōō-vär'ə-vīt, named for Count S. S. Uvarov (1785–1855), president of the St. Petersburg Academy.

Grossular, grŏs'yə-lər, from Latin *grossularia,* gooseberry, referring to the resemblance of pale-green specimens to gooseberries, *R. grossularia.*

Andradite, ăn'drä-dīt, named for J. B. d'Andrada e Silva (1763–1838), Brazilian mineralogist who first described the mineral.

Pyrope, pī'rōp, from Greek *pyropos,* firey, because of its characteristic color.

Almandine, ăl'mən-dēn, from Latin *alabandina,* after Alabanda, an ancient country in Asia Minor from whence came the "alabandic carbuncles" mentioned by Pliny in A.D. 77.

Spessartine, spĕs'ər-tēn, from Spessart, a hilly district in central Germany.

Each series shows fairly continuous compositional variation. Limited solid solution may also exist between the series (see figure 21.60). Compositions corresponding to particular end members are rare, and species are designated according to the dominant composition.

Other substitutions involving particular elements are fairly common in some of the species. Phosphate radicals may replace some SiO_4, and Y may replace some Mn in spessartine. Andradite is sometimes titanian (melanite, schorlomite), zirconian (kimzeyite), or vanadian (goldmanite).

As might be expected, many of the physical properties of garnets vary regularly with their chemistry. Their color is typically a tint or shade of red but may range to shades of brown or to black with increase in Ca or Ti, or to shades of green with increase in Cr. Linear relationships among composition, length of the cell edge, and refractive index have been demonstrated in the almandine-pyrope series, and a similar relationship be-

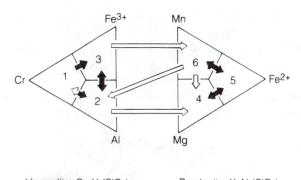

Ugrandite, $Ca_3Y_2(SiO_4)_3$

1. Uvarovite
2. Grossular
3. Andradite

Pyralspite, $X_3Al_2(SiO_4)_3$

4. Pyrope
5. Almandine
6. Spessartine

Figure 21.60
Principal (solid arrows) and less common (open arrows) solid solutions in the garnet group.

Figure 21.61
Almandine garnet showing trapezohedral form. (Courtesy D. K. Jones. Photograph by T. A. Moore.)

tween composition and specific gravity has been shown in both the almandine-pyrope and almandine-spessartine series.

Crystallography, Structure, and Habit Isometric. $4/m$ $\bar{3}$ $2/m$. Ia3d. Well-formed dodecahedra or trapezohedra (see figure 21.61).

Structure composed of alternating SiO_4 tetrahedra and YO_6 octahedra that share corners to form a continuous three-dimensional network (see figure 21.62). XO_8 triangular dodecahedra fill the interstices of the net and share two edges with adjacent tetrahedra, two with neighboring octahedra, and four with other dodecahedra. Eight formula units per cell.

Usually in well-formed crystals; also embedded grains; massive.

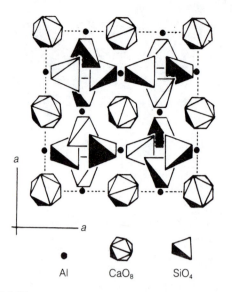

| Al | CaO$_8$ | SiO$_4$ |

Figure 21.62
Structure of grossular. Unit cell dashed.

Physical Properties No cleavage. Fair {110} parting. Subconchoidal to uneven fracture. Brittle. Hardness 6.5–7.5. Specific gravity 3.5–4.3, varying with composition. Vitreous to subadamantine luster. Transparent to translucent. Color usually red; also brown, yellow, white, green, black.

Species	Formula	Usual Color	Specific Gravity	a
Uvarovite	Ca$_3$Cr$_2$(SiO$_4$)$_3$	Emerald-green	3.8	11.999
Grossular	Ca$_3$Al$_2$(SiO$_4$)$_3$	Cinnamon-brown	3.56	11.851
Andradite	Ca$_3$Fe$_2$(SiO$_4$)$_3$	Various	3.84	12.048
Pyrope	Mg$_3$Al$_2$(SiO$_4$)$_3$	Red	3.54	11.459
Almandine	Fe$_3$Al$_2$(SiO$_4$)$_3$	Deep red	4.33	11.526
Spessartine	Mn$_3$Al$_2$(SiO$_4$)$_3$	Pink	4.19	11.621

Optical: The refractive indices are given for pure end members and will vary in natural specimens depending on the composition.

	n
Uvarovite	1.85–1.87
Grossular	1.73–1.75
Andradite	1.87–1.89
Pyrope	1.71–1.72
Almandine	1.79–1.83
Spessartine	1.79–1.81

Fusible (3–3.5) to a light brown or black glass; uvarovite fuses at 6.

Distinctive Properties and Tests Crystal form, color, and association. All species, except uvarovite, dissolve in boiling HCl after fusion and gelatinize on evaporation of the solution.

Association and Occurrence Pyrope is an accessory constituent of mafic and ultramafic igneous rocks where it is associated with olivine, pyroxene, spinel, and sometimes diamond. Almandine-pyrope is the garnet of eclogite where it is associated with the pyroxene omphacite.

Almandine is the typical garnet of medium- to high-grade regionally metamorphosed schists and gneisses. It forms from chlorite and biotite as temperature rises and thus serves as an index of metamorphic grade. Almandine may be associated with micas, staurolite, cordierite, tourmaline, andalusite, kyanite, or sillimanite. Almandine is also found in hornfelses of contact metamorphic zones where it is accompanied by cordierite and spinel. It is also occasionally observed as an accessory mineral in granites.

Spessartine is found in skarns with other manganiferous minerals such as rhodonite and tephroite, in low-grade metamorphic rocks, and in granitic, rhyolitic, and pegmatitic igneous rocks. In the igneous environment it is associated with alkali feldspar, quartz, tourmaline, muscovite, and topaz.

Grossular and andradite are formed in contact-metamorphosed, impure limestones. Typical associated minerals are vesuvianite, wollastonite, scapolite, and diopside. Andradite is also found in chlorite schists. Andradite is an accessory mineral in igneous rocks; light-colored varieties (topazolite) occur in silica-poor alkalic rocks such as syenite, and dark-colored varieties (schorlomite) in serpentinites derived from ultramafic rocks.

Uvarovite is found in gneiss, serpentinite, and metamorphosed limestone. Its most common occurrence is in seams and veins in peridotite as a gangue of chromite ore.

Alteration To chlorite, serpentine, talc, and iron oxides.

Confused with Vesuvianite and apatite.

Variants *Chemical varieties:* (1) uvarovite: complete solid solutions with grossular and andradite; vanadian (goldmanite); (2) grossular: ferroan and chromian; (3) andradite: alumian; manganoan (calderite); titanian (schorlomite), black; and zirconian; (4) pyrope: chromian (knorringite); (5) almandine: magnesian and manganoan; (6) spessartine: phosphatian, magnesian, ferroan, calcian, yttrian, and ferrian.

Related Minerals *Solid solutions:* grossular-hibschite, Ca$_3$Al$_2$(SiO$_4$)$_{3-x}$(OH)$_{4x}$, and andradite-hydroandradite, Ca$_3$Fe$_2$(SiO$_4$)$_{3-x}$(OH)$_{4x}$. *Isotypes:* kimzeyite, Ca$_3$(Zr,Ti, Mg,Fe,Nb)$_2$[(Al,Fe,Si)$_3$O$_{12}$]; and majorite, Mg$_3$(Fe,Al, Si)$_2$(SiO$_4$)$_3$. *Isostructures:* berzeliite, Ca$_2$NaMg$_2$(AsO$_4$)$_3$; and cryolithionite, Na$_3$Al$_2$Li$_3$F$_{12}$. *Other related species:* henritermierite, Ca$_3$(Mn,Al)$_2$ (SiO$_4$)$_2$(OH)$_4$.

ZIRCON
$Zr(SiO_4)$

zûr′kŏn, from Persian *azargun*, gold-colored.

Crystallography, Structure, and Habit Tetragonal. $4/m\,2/m\,2/m$. a 6.604, c 5.979. $I4_1/amd$. Simple square prisms with a pyramidal termination. Rare simple twins on {111}.

Isostructural with rutile. Chains of alternating, edge-sharing SiO_4 tetrahedra and ZrO_8 triangular dodecahedra extend parallel to the c-axis (see figure 21.63). The chains are joined laterally by edge-sharing ZrO_8 groups. Four formula units per cell.

Found as distinct crystals, usually tiny, and embedded grains.

Physical Properties Rare imperfect {110} and poor {111} cleavage. Uneven to conchoidal fracture. Brittle. Hardness 7.5 (normal) and 5 (metamict). Specific gravity 4.7 (normal) and 3.6–4.2 (metamict). Subadamantine to vitreous or dull luster. Transparent to opaque. Color usually shades of brown; also colorless, gray, green, lilac, or red. Optical: $\omega = 1.92$–1.96 and $\epsilon = 1.97$–2.01; positive. Infusible. Colored varieties lose color on heating. Often weakly radioactive.

Distinctive Properties and Tests Crystals, luster, and color. Usually insoluble but sometimes slowly attacked by hot concentrated H_2SO_4.

Association and Occurrence Zircon is a ubiquitous accessory mineral in igneous rocks, especially syenite, granite, or diorite. It is also found in schists, in iron formations, and in beach sands and placers.

Alteration Resistant to change.

Confused with Diamond, topaz, and spinel.

Variants *Form varieties:* metamict, structurally damaged by recoil of fission products from included radioelements; hyacinth, orange, reddish, or brownish and transparent; malacon, altered zircon; also colorless or smoky. *Chemical varieties:* Hf, Th, U, Fe, Ca, and rare earths (cyrtolite) may be present in small amounts. Y and P usually present but may be as exsolved xenotime, $Y(PO_4)$.

Related Minerals *Isotypes:* thorite, $(Th,U)(SiO_4)$; xenotime-(Y), $Y(PO_4)$; huttonite, $ThSiO_4$; tombarthite-(Y), $Y_4(Si,H_4)_4O_{12-x}(OH)_{4+2x}$; and hafnon, $Hf(SiO_4)$. *Other similar species:* baddeleyite, ZrO_2.

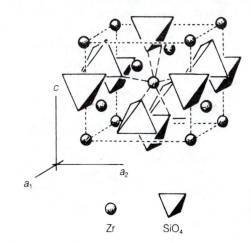

Figure 21.63
Structure of zircon. Unit cell dashed.

HUMITE GROUP
Norbergite
$Mg_3(SiO_4)(F,OH)_2$

nôr′bər-gīt, from the type locality at Norberg, Sweden.

Chondrodite
$(Mg,Fe)_5(SiO_4)_2(F,OH)_2$

kŏn′drō-dīt, from Greek *khondros*, grain, for its frequent granular appearance.

Humite
$(Mg,Fe)_7(SiO_4)_3(F,OH)_2$

hyōō′mīt, after Sir Abraham Hume (1749–1839), English mineral collector.

Clinohumite
$(Mg,Fe)_9(SiO_4)_4(F,OH)_2$

klī-nə-hyōō′mīt, from Greek *klinein*, to slope, and *humite*.

Crystallography, Structure, and Habit Norbergite and humite are orthorhombic, $P2_1/bnm$. Chondrodite and clinohumite are monoclinic, $P2_1/b$.

Highly modified orthorhombic or pseudo-orthorhombic crystals. Simple or multiple twinning on {001} is common in chondrodite and clinohumite.

Species	Class	a	b	c	β
Norbergite	$2/m\,2/m\,2/m$	8,727	10.271	4.709	
Chondrodite	$2/m$	7.89	4.743	10.29	109.0°
Humite	$2/m\,2/m\,2/m$	10.243	20.72	4.735	
Clinohumite	$2/m$	13.68	4.75	10.27	100.8°

Structure is an alternation of "forsterite" $Mg_2(SiO_4)$ and "brucite-sellaite" $Mg(F,OH)_2$ layers parallel to (001). Layers are one tetrahedron thick in norbergite, two in chondrodite, three in humite, and four in clinohumite. The chondrite and norbergite structures are illustrated in figure 21.64. Four formula units per cell for norbergite and humite and two for chondrodite and clinohumite.

Embedded formless grains, crystals, and massive.

Physical Properties Absent to poor {001} cleavage. Subconchoidal to uneven fracture. Brittle. Hardness 6–6.4. Specific gravity 3.1–3.4. Vitreous to resinous luster. Translucent. Color yellow, brown, or red; also white. Optical: all biaxial positive:

Species	α	β	γ	2V
Norbergite	1.56	1.57	1.59	44–50°
Chondrodite	1.60	1.62	1.63	60–90°
Humite	1.60	1.62	1.58	70–80°
Clinohumite	1.63	1.64	1.60	73–77°

Infusible; some varieties blacken and then turn white on heating.

Distinctive Properties and Tests Color and association. Yields water in the closed tube. Dissolves in boiling HCl and gelatinizes on evaporation of the solution.

Association and Occurrence Found typically in dolomitic marbles associated with phlogopite, spinel, pyroxene, olivine, wollastonite, grossular, magnetite, pyrrhotite, and graphite. Also found in skarns.

Alteration To serpentine and brucite, and iron oxides.

Confused with Olivine and garnet.

Variants Chemical varieties: titanian, ferroan, and manganoan (manganhumite).

Related Minerals Isotypes: alleghanyite, $Mn_5(SiO_4)_2(OH)_2$; leucophoenicite, $Mn_7(SiO_4)_3(OH)_2$; and sonolite, $Mn_9(SiO_4)_4(OH)_2$.

TOPAZ
$Al_2(SiO_4)(F,OH)_2$

tō'păz, from Greek *topazos,* original meaning lost; probably originally applied to an unidentified gemstone.

Crystallography, Structure, and Habit Orthorhombic. $2/m\ 2/m\ 2/m$. a 8.394, b 8.792, increasing with (OH) content, c 4.649. $P2_1/bnm$. Prisms, often highly modified, with vertically striated faces.

Structure (see figure 21.65) composed of chains made up of two edge-sharing $Al(OH,F)_6$ octahedra alternating

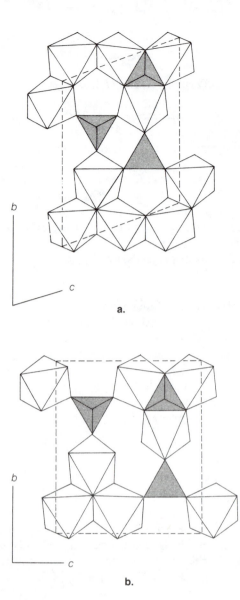

a.

b.

Figure 21.64
a-axis projections of the structures of chondrodite (a) and norbergite (b). Unit cells dashed. Note similarity of articulation of Mg(O,OH,F) octahedra and SiO_4 tetrahedra (shaded). (After Smyth and Bish, 1988.)

with SiO_4 tetrahedra that run parallel to the c-axis. Chains are cross-linked at the octahedral pairs by SiO_4 tetrahedra. Four formula units per cell.

Found as stubby prisms with a square, diamond-shaped, or 8-sided cross section.

Physical Properties Perfect {001} cleavage. Subconchoidal to uneven fracture. Brittle. Hardness 8. Specific gravity 3.5. Vitreous luster. Transparent to translucent. Colorless, light yellow, white or tinted greenish, bluish, or reddish. Optical: $\alpha = 1.606–1.630$, $\beta = 1.610–1.631$, and $\gamma = 1.616–1.638$; positive; $2V = 50–70°$. Infusible. Yellow varieties turn pink on heating.

Distinctive Properties and Tests Hardness, crystals, cleavage, and color. Insoluble in acids.

Association and Occurrence Topaz forms in the late hydrothermal or pneumatolitic stages of igneous activity. It is found as an accessory mineral in highly siliceous rocks, such as granite, rhyolite, or granitic pegmatite; it is also found in veins and cavities in schists near granite contacts. Associated with lithium minerals, fluorite, cassiterite, and tourmaline, and also with apatite, beryl, quartz, mica, and feldspar.

Alteration To muscovite plus kaolin; also to fluorite.

Confused with Quartz.

Variants *Form varieties:* brazilian ruby, rose; oriental topaz. *Chemical varieties:* hydroxylian; traces of Fe^{3+}, Cr^{3+}, and V^{3+} may be present.

Related Minerals *Isotypes:* euclase, $BeAl(SiO_4)(OH)$.

CHLORITOID
$(Fe^{2+}, Mg, Mn)_2(Al, Fe^{3+})Al_3O_2(SiO_4)_2(OH)_2$

klôr′ə-toid, from its resemblance to chlorite.

Crystallography, Structure, and Habit Monoclinic. $2/m$. a 9.48, b 5.48, c 18.18, β 101.74°. $C\,2/c$. Crystals are rare. Twinning on $\{001\}$ is common.

Structure (see figure 21.66) of closely packed brucite-type layers and corundum-type layers with compositions $(Fe^{2+}, Mg, Mn)_4Al_2O_4(OH)_8$ and Al_6O_{16}, respectively. The layers, linked by *individual* SiO_4 tetrahedra and hydrogen bonds, alternate perpendicular to (001). Four formula units per cell.

Often found as coarsely foliated masses or thin scales.

Physical Properties Perfect $\{001\}$ and fair $\{110\}$ cleavages. Parting on $\{010\}$. Hardness 6.5. Specific gravity 3.5–3.8. Pearly luster. Translucent. Color dark green to black. Optical: $\alpha = 1.705-1.730$, $\beta = 1.708-1.734$, $\gamma = 1.712-1.740$ (lower values refer to Mg-rich varieties); positive; $2V = 45-70°$.

Distinctive Properties and Tests Cleavage, color, and occurrence. Optical properties distinguish chloritoid from chlorite, biotite, and stilpnomelane. Soluble in H_2SO_4.

Association and Occurrence Often found in iron- and aluminum-rich pelitic schists metamorphosed to low or medium grade. In such rocks chloritoid is typically associated with muscovite, chlorite, biotite, garnet, staurolite, and hematite. Chloritoid breaks down to form staurolite at higher temperatures. In high-pressure metamorphic rocks, chloritoid may occur with glaucophane.

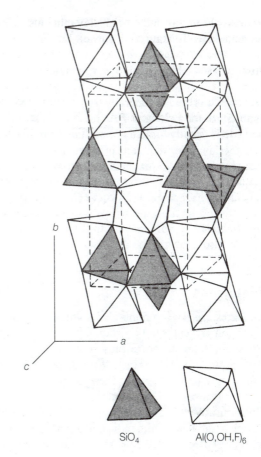

SiO_4 $Al(O,OH,F)_6$

Figure 21.65
Structure of topaz. Unit cell dashed. (After Smyth and Bish, 1988.)

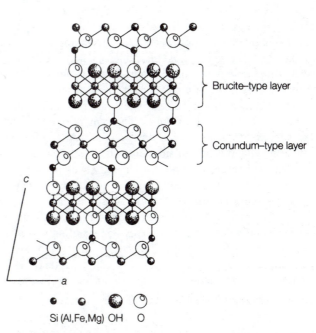

Brucite–type layer

Corundum–type layer

Si   (Al,Fe,Mg)   OH   O

Figure 21.66
The structure of chloritoid projected onto (010). (Data from Brindley and Harrison, 1952.)

Alteration From Fe- and Al-rich clay minerals, chlorite, or stilpnomelane. To staurolite, garnet, biotite.

Confused with Chlorite, biotite, stilpnomelane.

Variants Chloritoid is generally iron-rich but (Mn,Mg) may substitute for Fe^{2+}, and Fe^{3+} for Al. Substitution of Mg and Mn is usually limited to less than 40 atom percent but magnesian and manganoan varieties are known. *Chemical varieties:* magnesiochloritoid, $Mg_2Al_4O_2$ $(SiO_4)_2(OH)_4$, triclinic; ottrelite, $(Mn,Fe^{2+},Mg)_2Al_4O_2$ $(SiO_4)_2(OH)_4$; carboirite, $Fe_2^{2+}Al_4O_2(GeO_4)_2(OH)_4$.

Related Minerals *Polymorphs:* triclinic, $C\bar{1}$. A trigonal polytype is also known.

Subsaturates

Subsaturates are silicate minerals in which some portion of the oxygen present is not coordinated by silicon into Si-O tetrahedra. The structure thus contains both anionic $(SiO_4)^{4-}$ groups and O^{2-} anions. With increasing proportions of simple oxygen anions, these minerals provide a link between silicates and oxides.

Subsaturates are divided into an aluminum group containing andalusite, sillimanite, and kyanite; a titanium group represented by titanite; and some miscellaneous species.

Andalusite, sillimanite, and kyanite are polymorphs of $AlAl(SiO_4)O$ with rather similar structures. Phase relationships are shown in figure 6.6b. In all three minerals chains of Al-O octahedra parallel the *c*-axis. Each octahedron shares two oxygens with the octahedron above and two oxygens with the octahedron below, thus giving the Al-O chains the composition AlO_4. The chains are cross-linked by the remaining Al and Si ions. The variations in structure among the polymorphs result from different cross-linking and chain positioning. In all cases, silicon is tetrahedrally coordinated by oxygen ions, resulting in discrete $(SiO_4)^{4-}$ structural units. The remaining Al ion is presumed to have 6-fold coordination in kyanite, 5-fold in andalusite, and 4-fold in sillimanite.

Subsaturates are typically formed at high temperatures and are occasionally major constituents of metamorphic rocks derived from aluminous sediments.

Andalusite and kyanite are used in the manufacture of high-grade porcelain capable of withstanding extreme temperatures. Spark plugs, laboratory ware, and thermocouple tubing are some typical products. Titanite is a minor source of titanium. Andalusite and staurolite have a minor use as gemstones.

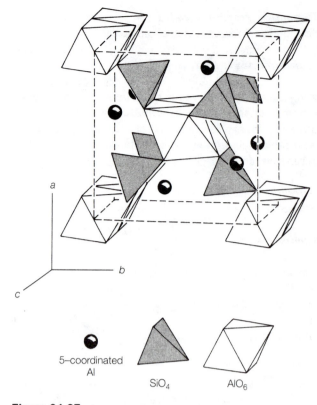

Figure 21.67
Structure of andalusite. Unit cell dashed. (After Smyth and Bish, 1988.)

ANDALUSITE
AlAl(SiO₄)O

(macle)
an'də-lōō-sīt, discovered in Andalusia, Spain.

Crystallography, Structure, and Habit Orthorhombic. $2/m\,2/m\,2/m$. *a* 7.796, *b* 7.893, *c* 5.558. $P2_1/nnm$. Square prisms. Rare twinning on {101}.

Andalusite structure (see figure 21.67). Chains of AlO_6 edge-sharing polyhedra parallel the *c*-axis and are cross-linked by SiO_4 tetrahedra and 5-coordinated Al-O groups. Four formula units per cell.

Rough rounded or elongate crystals; massive and granular.

Physical Properties Distinct {110} cleavage intersecting at about 90°. Uneven fracture. Brittle. Hardness 7.5. Specific gravity 3.2. Vitreous to dull luster. Transparent to opaque. Color white, rose-red, violet, pearl-gray, and olive-green. Optical: $\alpha = 1.63–1.64$, $\beta = 1.63–1.64$, and $\gamma = 1.64–1.65$; positive; $2V = 74–86°$. Infusible.

Distinctive Properties and Tests Crystal form, hardness, and association. Insoluble in acids. Maltese cross on (001) sections of the variety chiastolite. Nearly square cross section of crystals.

Association and Occurrence Commonly found in metamorphosed aluminous rocks near intrusive igneous bodies, often as chiastolite in metamorphosed shale. Associated with sillimanite, kyanite, cordierite, garnet, micas, quartz, corundum, and tourmaline.

Alteration To muscovite (sericite), kyanite, and kaolinite.

Confused with Scapolite.

Variants *Form varieties:* chiastolite; carbonaceous impurities pushed aside during growth form a maltese cross on a (001) section. *Chemical varieties:* manganian (kanonaite, viridine), $(MnAl)Al(SiO_4)O$; ferrian; chromian; titanian.

Related Minerals *Polymorphs:* kyanite and sillimanite.

SILLIMANITE
AlAl(SiO₄)O
$AlAl(SiO_4)O$

(fibrolite)
si'lə-mə-nīt, after Benjamin Silliman (1779–1864), a professor at Yale University.

Crystallography, Structure, and Habit Orthorhombic. $2/m\ 2/m\ 2/m$. *a* 7.484, *b* 7.673, *c* 5.771, $P2_1/nnm$. Slender to acicular prisms with striated faces.

Structure consists of chains of edge-sharing AlO_6 octahedra parallel to the *c*-axis (see figure 21.68). Cross-linkage of chains is accomplished by a chain of alternating SiO_4 and AlO_4 tetrahedra. (On the basis of tetrahedral articulation, sillimanite may be classed as a tectosilicate.) Four formula units per cell.

Usually in parallel fibrous crystal groups; also columnar.

Physical Properties Very good {010} cleavage. Uneven fracture. Brittle. Hardness 6–7. Specific gravity 3.2. Vitreous to adamantine luster. Transparent to translucent. Color brown, gray-green, or white. Optical: $\alpha = 1.66$, $\beta = 1.66$, and $\gamma = 1.68$; positive; $2V = 30°$. Infusible.

Distinctive Properties and Tests Habit, cleavage, and association. Insoluble in acids.

Association and Occurrence Found in high-grade metamorphic aluminous rocks, where it is associated with

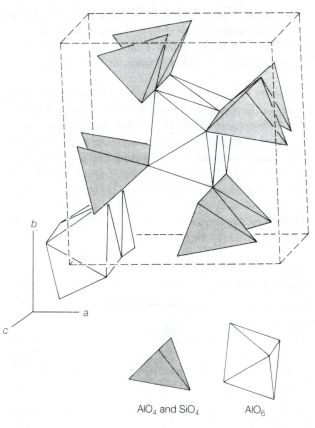

AlO₄ and SiO₄   AlO₆

Figure 21.68
Structure of sillimanite. Unit cell dashed. (After Smyth and Bish, 1988.)

quartz, micas, orthoclase, hypersthene, garnet, corundum, cordierite, spinel, and andalusite.

Alteration To muscovite and kaolin.

Confused with Anthophyllite.

Variants May contain a little iron.

Related Minerals *Polymorphs:* kyanite and andalusite.

KYANITE
AlAl(SiO₄)O
$AlAl(SiO_4)O$

(cyanite, disthene)
kī'ə-nīt, from Greek *kyanos,* blue in reference to its common color.

Crystallography, Structure, and Habit Triclinic. $\bar{1}$. *a* 7.123, *b* 7.848, *c* 5.564, α 89.9°, β 101.25°, γ 106.0°. $P\bar{1}$. Bladed crystals, rarely terminated. Lamellar twinning on {100}.

Chains of edge-sharing Al-O octahedra parallel the *c*-axis and are cross-linked by Si-O tetrahedra and Al-O octahedra (see figure 21.69). Four formula units per cell.

Aggregates of bladed crystals, usually subparallel.

Physical Properties Perfect {100} and good {010} cleavage. Sometimes a {001} parting. Uneven fracture. Brittle. Hardness 4.5 parallel to the *c*-axis, 6.5 normal to the *c*-axis. Specific gravity 3.6. Vitreous to pearly luster. Translucent to transparent. Color blue and white; also gray, green, or black. Color may be patchy. Optical: $\alpha = 1.71$, $\beta = 1.72$, and $\gamma = 1.73$; negative; $2V = 82°$. Infusible.

Distinctive Properties and Tests Habit, color, and differential hardness. Insoluble in acids.

Association and Occurrence Found in schist and gneiss associated with garnet, muscovite, staurolite, rutile, and corundum.

Alteration To kaolin, muscovite, and pyrophyllite.

Variants May contain a little Fe, or Mn, occasionally Cr.

Related Minerals *Polymorphs:* andalusite and sillimanite.

TITANITE
CaTi(SiO$_4$)O

(sphene)

tīt′ə-nīt, named for its titanium content.

Crystallography, Structure, and Habit Monoclinic. $2/m$. a 6.56, b 8.72, c 7.44, β 119.85°. $C 2/c$. Varied habit; often wedge-shaped and flattened or stout prismatic. Contact and cruciform penetration twins common with twin plane {100}. Occasional lamellar twinning on {221}.

Structure (see figure 21.70) contains independent SiO$_4$ and TiO$_6$ polyhedra that share corners to form kinked strings. Ca is in 7-fold coordination. Four formula units per cell.

Usually as distinct crystals; also lamellar, compact, or massive.

Physical Properties Good {110} cleavage. {221} parting sometimes present. Conchoidal fracture. Brittle. Hardness 5–5.5. Specific gravity 3.5. Subadamantine to resinous luster. Transparent to opaque. Colorless, brown, gray, green, red, or black. Optical: $\alpha = 1.86$, $\beta = 1.93$, and $\gamma = 2.10$; positive; $2V = 26–50°$. Some specimens when heated change color to yellow and fuse (3–4) with intumescence to a yellow, brown, or black glass.

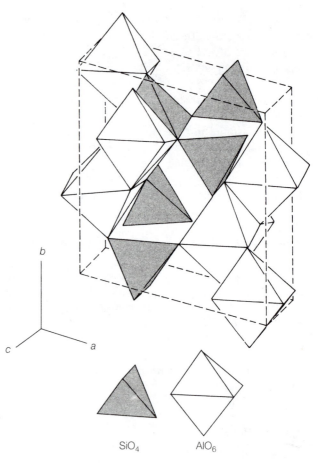

SiO$_4$ AlO$_6$

Figure 21.69
Structure of kyanite. Unit cell dashed. (After Smyth and Bish, 1988.)

Distinctive Properties and Tests Sphenoidal crystals, luster, and color. Imperfectly soluble in HCl but decomposes in H$_2$SO$_4$.

Association and Occurrence Titanite is a widespread accessory mineral, especially in coarse-grained igneous rocks of intermediate composition. It is also common in metamorphic rocks. Associated with pyroxene, amphibole, epidote, chlorite, scapolite, zircon, apatite, magnetite, and ilmenite.

Alteration To anatase, TiO$_2$, rutile, or leucoxene.

Confused with Staurolite, sphalerite, and zircon.

Variants *Form varieties:* greenovite, red or pinkish. *Chemical varieties:* yttrian (keilhauite). Numerous substitutions. For Ca: Th, Sr, Na, Ba, Mn, and rare earths. For Ti: Al, Fe, Mg, Sn, Nb, Ta, V, and Cr. For O: OH, F.

Related Minerals *Isostructures:* tilasite, CaMg(AsO$_4$)F; malayaite, CaSn(SiO$_4$)O; and fersmanite, (Ca,Na)$_4$(Ti,Nb)$_2$(Si$_2$O$_{11}$)(F,OH)$_2$.

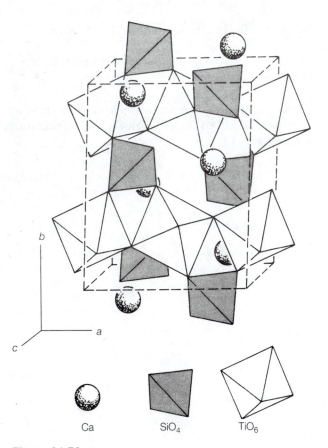

Figure 21.71
Staurolite with (231) twin plane. (Courtesy J. S. Hudnall Mineralogy Museum, University of Kentucky. Photograph by T. A. Moore.)

Ca   SiO₄   TiO₆

Figure 21.70
Structure of titanite. Unit cell dashed. (After Smyth and Bish, 1988.)

STAUROLITE
(Fe,Mg,Zn)$_2$Al$_9$[(Si,Al)$_4$O$_{16}$]O$_6$(OH)$_2$

stôr′ə-līt, from Greek *stauros,* cross, in recognition of its common occurrence as cruciform penetration twins.

Crystallography, Structure, and Habit Monoclinic; pseudo-orthorhombic. $2/m$. a 7.90, b 16.65, c 5.63, $\beta \approx 90°$. $C2/m$. Prismatic crystals; often flattened and having a rough surface. Penetration twins (see figure 21.71) are very common and may be at 60° or 90°.

Structure (see figure 21.72) is similar to that of kyanite with intercalated layers of Fe(OH)$_2$. Two formula units per cell.

Usually found in distinct crystals.

Physical Properties Interrupted {010} cleavage. Subconchoidal fracture. Brittle. Hardness 7–7.5 decreasing to 3–4 in altered specimens. Specific gravity 3.7. Resinous to dull luster. Translucent to opaque. Color dark reddish brown to brown-black. Streak uncolored to grayish. Optical: $\alpha = 1.74–1.75$, $\beta = 1.74–1.76$, and $\gamma = 1.75–1.76$; positive; $2V = 80–90°$. Infusible.

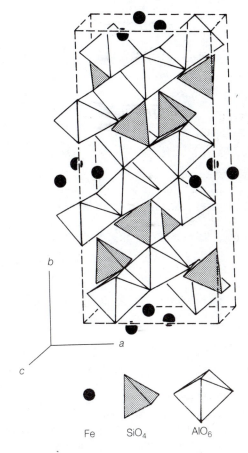

Fe   SiO₄   AlO₆

Figure 21.72
Structure of staurolite. Unit cell dashed. (After Smyth and Bish, 1988.)

Distinctive Properties and Tests Crystal form and twinning, hardness, color, and association. Yields water in the closed tube. Slowly attacked by H_2SO_4; insoluble in cold HF.

Association and Occurrence Found in metamorphic rocks (schist) associated with garnet, kyanite, micas, and tourmaline. Sometimes in parallel growth with kyanite.

Alteration To chlorite, kaolinite, and iron oxides.

Confused with Andalusite, pyroxene, amphibole, tourmaline, and titanite.

Variants *Chemical varieties:* normally magnesian; also manganoan, cobaltoan (lusakite, blue), zincian, titanian, and nickelian.

Appendix A

Table of Atomic Parameters

Element	Symbol	Atomic Number	Atomic Weight[1], $^{12}C = 12.000$	Usual Valence	Usual Coordination	Ionic Radius (Å)[2]	Covalent Radius (Å)	First Ionization Potential, eV
Actinium	Ac	89	227.03	3+	—	—		6.9
Aluminum	Al	13	26.98	3+	4	0.47	1.18	5.98
				3+	6	0.61		
Antimony	Sb	51	121.75	3+	4	0.85	1.40	8.64
Argon	Ar	18	39.95	0	1	—		15.76
Arsenic	As	33	74.92	5+	4	0.42	1.20	9.81
				3+	6	0.58		
Barium	Ba	56	137.33	2+	8	1.50	1.98	5.21
				2+	12	1.68		
Beryllium	Be	4	9.01	4+	4	0.35	0.90	9.32
Bismuth	Bi	83	208.98	3+	6	1.10	1.46	7.29
				3+	8	1.19		
Boron	B	5	10.81	3+	3	0.10	0.82	8.30
				3+	4	0.20		
Bromine	Br	35	79.90	1−	6	1.88	1.14	11.84
Cadmium	Cd	48	112.41	2+	6	1.03	1.48	8.99
				2+	8	1.15		
Calcium	Ca	20	40.08	2+	6	1.08	1.74	6.11
				2+	8	1.20		
				2+	12	1.43		
Carbon	C	6	12.01	4+	3	0.16	0.77	11.26
Cerium	Ce	58	140.12	3+	6	1.09	1.65	5.60
				3+	8	1.22		
Cesium	Cs	55	132.91	1+	8	1.82	2.35	3.89
				1+	12	1.96		
Chlorine	Cl	17	35.45	1−	6	1.72	0.99	13.01
Chromium	Cr	24	51.99	3+	6	0.70	1.18	6.76
Cobalt	Co	27	58.93	2+	6	0.73–0.83[3]	1.16	7.86
Copper	Cu	29	63.55	1+	6	0.96	1.17	7.72
				2+	6	0.81		
Dysprosium	Dy	66	162.50	3+	6	0.99	1.59	6.80
				3+	8	1.11		
Erbium	Er	68	167.26	3+	6	0.97	1.57	6.08
				2+	8	1.08		

Element	Symbol	Atomic Number	Atomic Weight[1], $^{12}C = 12.000$	Usual Valence	Usual Coordination	Ionic Radius (Å)[2]	Covalent Radius (Å)	First Ionization Potential, eV
Europium	Eu	63	151.97	2+	6	1.25	1.85	5.67
				2+	8	1.33		
Fluorine	F	9	18.99	1−	6	1.25	0.72	17.42
Gadolinium	Gd	64	157.25	3+	6	1.02	1.61	6.16
				3+	8	1.14		
Gallium	Ga	31	69.72	3+	4	0.55	1.26	5.97
Germanium	Ge	32	72.61	4+	4	0.48	1.22	7.88
				2+	6	0.62		
Gold	Au	79	196.97	3+	4	0.78	1.34	9.22
Hafnium	Hf	72	178.49	4+	6	0.79	1.44	~7
				4+	8	0.91		
Helium	He	2	4.00	0	1	—		24.48
Holmium	Ho	67	164.93	3+	6	0.98	1.58	—
Hydrogen	H	1	1.008	1+	—	—	0.32	13.60
					8	1.10		
Indium	In	49	114.82	3+	6	0.88	1.44	5.79
				3+	8	1.00		
Iodine	I	53	126.90	1−	6	2.13	1.33	10.45
				1−	8	1.97		
Iridium	Ir	77	192.22	4+	6	0.71	1.44	~9
Iron	Fe	26	55.85	2+	6	0.69–0.86[3]	1.17	7.87
				3+	4	0.57		
				3+	6	0.63–0.73[3]		
Krypton	Kr	36	83.80	0	1	—	1.12	14.00
Lanthanum	La	57	138.91	3+	6	1.13	1.69	5.61
				3+	8	1.26		
Lead	Pb	82	207.21	2+	6	1.26	1.47	7.42
				2+	8	1.37		
				2+	12	1.57		
Lithium	Li	3	6.94	1+	4	0.68	1.23	5.39
Lutetium	Lu	71	174.97	3+	6	0.94	1.56	~7
				3+	8	1.05		
Magnesium	Mg	12	24.31	2+	6	0.80	1.36	7.64
				2+	8	0.97		
Manganese	Mn	25	54.94	2+	6	0.75–0.91[3]	1.17	7.43
				3+	6	0.66–0.73[3]		
				4+	6	0.62		
Mercury	Hg	80	200.59	2+	6	1.10	1.90	10.43
				2+	8	1.22		
Molybdenum	Mo	42	95.94	4+	6	0.73	1.30	7.10
				6+	4	0.50		
				6+	6	0.68		
Neodymium	Nd	60	144.24	3+	6	1.06	1.64	5.51
				3+	8	1.20		
Neon	Ne	10	20.18	0	1	—		21.56
Nickel	Ni	28	58.69	2+	6	0.77	1.15	7.63
Niobium	Nb	41	92.91	5+	4	0.40	1.34	6.88
				5+	6	0.72		
Nitrogen	N	7	14.01	5+	—	0.13	0.75	14.53
Osmium	Os	76	190.20	4+	6	0.71	1.26	8.50
Oxygen	O	8	15.999	2−	4	1.30	0.73	13.61
				2−	6	1.32		
				2−	8	1.34		

Element	Symbol	Atomic Number	Atomic Weight[1], $^{12}C = 12.000$	Usual Valence	Usual Coordination	Ionic Radius (Å)[2]	Covalent Radius (Å)	First Ionization Potential, eV
Palladium	Pd	46	106.42	4+	4	0.72	1.28	8.33
Phosphorus	P	15	30.97	5+	4	0.25	1.06	10.48
Platinum	Pt	78	195.08	2+	4	0.68	1.30	9.0
				4+	6	0.71		
Polonium	Po	84	209	4+	8	1.16	1.46	8.43
Potassium	K	19	39.10	1+	8	1.59	2.03	4.34
				1+	12	1.68		
Praseodymium	Pr	59	140.91	3+	6	1.08	1.65	5.46
				3+	8	1.22		
Promethium	Pm	61	147	3+	8	1.06	1.63	5.74
Protactinium	Pa	91	231.04	4+	8	1.09		
Radium	Ra	88	226.03	2+	8	1.56		5.28
				2+	12	1.72		
Radon	Rn	86	222	2+	8	1.56		10.75
				2+	12	1.72		
Rhenium	Re	75	186.21	4+	6	0.71	1.28	7.87
Rhodium	Rh	45	102.91	3+	6	0.75	1.25	7.46
				4+	6	0.71		
Rubidium	Rb	37	85.47	1+	8	1.68	2.16	4.18
				1+	12	1.81		
Ruthenium	Ru	44	101.07	3+	6	0.76	1.25	7.36
				4+	6	0.70		
Samarium	Sm	62	150.36	3+	6	1.04	1.62	5.6
				3+	6	1.17		
Scandium	Sc	21	44.96	3+	6	0.83	1.44	6.54
				3+	8	0.95		
Selenium	Se	34	78.96	2−	6	1.88	1.16	9.75
				2−	8	1.90		
Silicon	Si	14	28.09	4+	4	0.34	1.11	8.15
Silver	Ag	47	107.87	1+	4	1.10	1.34	7.57
				1+	6	1.23		
				1+	8	1.38		
Sodium	Na	11	22.99	1+	6	1.10	1.54	5.14
				1+	8	1.24		
Strontium	Sr	38	87.62	2+	6	1.21	1.91	5.69
				2+	8	1.33		
				2+	12	1.48		
Sulfur	S	16	32.07	2−	4	1.56	1.06	10.36
				2−	6	1.72		
				2−	8	1.78		
Tantalum	Ta	73	180.95	5+	6	0.72	1.34	7.88
				5+	8	0.77		
Technetium	Tc	43	98.91	4+	6	0.72	1.27	7.28
Tellurium	Te	52	127.60	2−	—	0.85	1.36	9.01
				4+	3	0.60		
Terbium	Tb	65	158.93	3+	6	1.00	1.59	5.98
				3+	8	1.12		
Thallium	Tl	81	204.38	1+	6	1.58	1.48	6.11
				1+	8	1.68		
				1+	12	1.84		
Thorium	Th	90	232.04	4+	6	1.08	1.65	6.95
				4+	8	1.12		

Element	Symbol	Atomic Number	Atomic Weight[1], $^{12}C = 12.000$	Usual Valence	Usual Coordination	Ionic Radius (Å)[2]	Covalent Radius (Å)	First Ionization Potential, eV
Thulium	Tm	69	168.93	3+	6	0.96	1.56	5.81
				3+	8	1.07		
Tin	Sn	50	118.71	2+	8	1.30	1.41	7.34
Titanium	Ti	22	47.88	2+	6	0.94	1.32	6.82
				3+	6	0.75		
				4+	6	0.69		
Tungsten	W	74	183.85	4+	6	0.73	1.30	7.98
				6+	4	0.50		
				6+	6	0.68		
Uranium	U	92	238.03	4+	8	1.08	1.42	6.08
				6+	4	0.56		
				6+	6	0.81		
Vanadium	V	23	50.94	5+	4	0.44	1.22	6.74
				5+	6	0.62		
Xenon	Xe	54	131.29	0	1	—		12.13
Ytterbium	Yb	70	173.04	3+	6	0.95		6.2
				3+	8	1.06		
Yttrium	Y	39	88.91	3+	6	0.98	1.62	6.38
				3+	8	1.10		
Zinc	Zn	30	65.39	2+	4	0.68	1.25	9.39
				2+	6	0.83		
Zirconium	Zr	40	91.22	4+	6	0.80	1.45	6.84
				4+	8	0.92		

[1]From International Union of Pure and Applied Chemistry 1986. Atomic weights of the elements 1985. *Pure and Appl. Chem.* 58:1677–1692.
[2]From Whittaker, E. J. W., and Muntus, R. (1970) Ionic radii for use in geochemistry. *Geochim. Cosmochim. Acta* 34:945–56.
[3]Low and high spin.

Determinative Tables

The inherent difficulties of mineral identification can be alleviated by grouping together minerals that have closely similar physical characteristics. This greatly limits the number of minerals among which a choice must be made. Within each small group of minerals, final identification can be made by performing a simple chemical test, identifying some distinctive property, noting the mineral occurrence, or considering associated minerals.

For a number of reasons, it is difficult both to construct and to use determinative tables on a completely logical and objective basis. Mineral determination is, to a large degree, accomplished by subjective summation of all the mineral characteristics and is ordinarily not a step-by-step process. Physical properties observed in hand specimens show relatively large variations, and unless the mineral under study is an average specimen, its classification may be impossible without recourse to chemical, optical, x-ray, or other testing. Small grain size, incipient alteration, pseudomorphism, staining, and many other phenomena may make the determination of critical properties difficult or impossible. To judge luster, cleavage, and color, small mineral grains should be separated from their matrix and examined against a white paper background with a good lens or binocular microscope. Hardness can be tested on fine-grained material in matrix by using a needle.

Two listings and a set of determinative tables are given on the following pages to help you identify the minerals described in this book. A number of other minerals included in the tables are found under the entries "Related Minerals" in part 3 of the text.

Table B.1 lists minerals that may be found in some of the more common geological environments. In general, the minerals in any column may be found together, but associations should be checked by referring to the mineral descriptions. The order of the columns follows the order in which the mineral classes were presented in part 3 of the text, from native elements to silicates.

Table B.2 groups minerals that display some relatively striking physical properties. Again, the order in the columns is that of the mineral classes.

Table B.3 is a determinative table in which the minerals are grouped into sections according to luster, cleavage, and color. The table is divided into two principal parts: minerals with metallic luster and minerals with nonmetallic luster. Each of these parts is further divided into sections based on cleavage and color, in accordance with the following outline:

Part 1: Minerals with a Metallic Luster

(opaque on thinnest edge and having a dark-colored streak)

    Section A: No Cleavage
    Section B: Distinct Cleavage

Part 2: Minerals with a Nonmetallic Luster

(transparent on thinnest edge and having a colorless or light-colored streak)

Section A: No Cleavage

Section B: Distinct Cleavage. Cleavage fragments are tabular or micaceous.

Section C: Distinct Cleavage. Cleavage fragments are polyhedral.

Section D: Distinct Cleavage. Colorless or lightly tinted minerals; cleavage fragments are not micaceous, tabular, or polyhedral.

Section E: Distinct Cleavage. Black, brown, or red minerals; cleavage fragments are not micaceous, tabular, or polyhedral.

Section F: Distinct Cleavage. Orange, yellow, green, blue, or violet minerals; cleavage fragments are not micaceous, tabular, or polyhedral.

The minerals in each section are arranged in the order of their increasing hardness, which is shown graphically in the left-hand column. Tick marks at 3, 5.5, and 7 on the hardness scale represent, respectively, the hardnesses of a copper penny, a knife blade or window glass, and quartz. Names of minerals described in detail in part three of the text are italicized. Other minerals listed in the table are mentioned as variants or related minerals in the mineral descriptions.

The mineral formulas will provide clues to critical chemical spot tests that can be used to distinguish otherwise similar species. Specific gravity may be a useful criterion, especially if its value has been determined or the difference in specific gravities of two possible minerals is large. Color and especially streak can be essential clues in the identification of a mineral species, but color must be used with caution. Features of particular value in determination are given in the remarks column, and the page number provides a cross reference to the mineral description, which should be checked before final identification is made.

Confusing species are listed in more than one place in the table because the critical parameters are difficult to assess or because physical properties are highly variable. For example, sphalerite will be found under both metallic and nonmetallic luster; gypsum in sections B, C, and D of part 2, and orthoclase under both uncolored and red cleavable minerals.

To see how the tables can be used, here are a few examples.

1. *Observations:* The mineral is nonmetallic, blue-green, shows cleavage, is softer than 3, and is water soluble.

Determination: The mineral will be found in part 2, section F. The only minerals fitting the observations are melanterite and chalcanthite. A test for copper or iron will readily distinguish these two species.

2. *Observations:* The mineral has a metallic luster, no cleavage, hardness between 3 and 5.5, yellowish color, and a black streak.

Determination: The mineral will be found between pyrolusite and magnetite, inclusive, in part 1, section A. The mineral is gold, chalcopyrite, or pyrrhotite based on color, and chalcopyrite or pyrrhotite based on streak. These two minerals can be readily distinguished by a test for copper (chalcopyrite, if positive) or magnetism (pyrrhotite, if positive).

3. *Observations:* The mineral is nonmetallic, has a micaceous cleavage yielding flexible folia, can be scratched by a copper penny, and is sectile. Further, the mineral does not have an unctuous or greasy feel or an earthy smell when wet.

Determination: The mineral will be found between smectite and barite, inclusive, in part 2, section B. Color and streak somewhat limit the choices, and feel and smell further reduce the possibilities. The mineral must be gypsum, vivianite, sericite, brucite, muscovite, lepidolite, or barite. The occurrence of the mineral in a vein in serpentine rock, together with the evolution of copious water when heated, low specific gravity, and sectility, identifies the mineral as brucite.

4. *Observations:* The mineral is metallic and shows distinct cleavage. The hardness is about 3, and the color and streak are black.

Determination: The mineral will be found in part 1, section B. The black color and streak coupled with the hardness make jamesonite, bournonite, and enargite the only possibilities. The distinctive habits of these minerals may distinguish them, or combinations of tests for Cu, Pb, and Fe can be used.

5. *Observations:* The mineral is nonmetallic, yields rhombic cleavage fragments, has a hardness of about 3.5, and is pink in color.

Determination: The mineral will be found in part 2, section C. Only one pink mineral, rhodochrosite, is listed, but a confirmatory test with cold dilute HCl is negative because strong effervescence is observed. The mineral must therefore be a carbonate, and thus only calcite satisfies the observations.

Igneous Rocks			Metamorphic Rocks		
Silica-Rich	**Silica-Poor**	**Mafic Volcanics**	**Crystalline Schists and Gneisses**	**Calcareous Metamorphic Rocks**	**Contact Metamorphic Rocks**
Molybdenite	Platinum	Copper	Graphite	Graphite	Sphalerite
		Diamond			Chalcopyrite
Fluorite	Chalcopyrite		Pyrrhotite	Arsenopyrite	Arsenopyrite
	Pyrrhotite		Arsenopyrite		Molybdenite
Cassiterite	Nickeline	Anhydrite		Calcite	
Magnetite			Rutile	Dolomite	Corundum
	Corundum		Corundum		Hematite
Monazite	Ilmenite	Chalcedony	Spinel	Scapolite	Cassiterite
Apatite	Rutile	Opal		Phlogopite	Spinel
	Spinel	Leucite	Diaspore	Olivine	Magnetite
Quartz	Magnetite	Analcime		Tremolite	
Tridymite	Chromite	Plagioclase	Rhodocrosite	Garnet	Gibbsite
Microcline		Stilbite		Humite	
Orthoclase	Apatite	Heulandite	Monazite	Epidote	Scheelite
Sanidine	Zoisite	Chabazite	Apatite		
Plagioclase		Thomsonite	Pyromorphite		Tremolite
Muscovite	Plagioclase	Natrolite			
Biotite	Nepheline	Apophyllite	Quartz		Wollastonite
Garnet	Sodalite	Prehnite	Orthoclase		Tourmaline
Allanite	Leucite	Pectolite	Plagioclase		Cordierite
Hornblende	Cancrinite	Melilite	Chlorite		Melilite
Zircon	Chrysotile	Datolite	Chloritoid		Vesuvianite
Topaz	Enstatite		Margarite		Axinite
Titanite	Diopside		Muscovite		Staurolite
	Augite		Biotite		Lazurite
	Olivine		Anthophyllite		
	Garnet		Hornblende		
	Titanite		Diopside		
			Vesuvianite		
			Garnet		
			Zircon		
			Staurolite		

Sedimentary Rocks		Veins		Oxidized Zones	Volcanic Sublimates
Sulfur	Apatite	Gold	Barite	Copper	Sulfur
	Collophane	Silver	Celestine		
Bornite	Carnotite	Arsenic	Anhydrite	Chalcocite	Realgar
Chalcopyrite			Jarosite	Bornite	Orpiment
Marcasite	Quartz	All sulfides	Alunite	Covellite	
	Microcline	Sulfosalts		Pyrargyrite	Halite
Halite	Analcime		Apatite		Fluorite
Sylvite	Kaolinite	Fluorite	Amblygonite	Chlorargyrite	
Fluorite	Smectite				Anhydrite
Carnallite	Muscovite	Hematite	Wolframite	Cuprite	Gypsum
	Glauconite	Ilmenite	Scheelite	Hematite	Alunite
Hematite		Rutile		Cassiterite	
Uraninite		Cassiterite	Quartz	Romanechite	
		Uraninite	Microcline		
Gibbsite		Magnetite	Rhodonite	Smithsonite	
Diaspore			Topaz	Aragonite	
Bauxite		Brucite			
		Manganite		Malachite	
Magnesite				Azurite	
Siderite		Calcite			
Aragonite		Magnesite		Anglesite	
Strontianite		Siderite		Gypsum	
		Rhodochrosite		Melanterite	
Borax		Witherite		Alunite	
Colemanite		Strontianite		Epsomite	
		Dolomite		Pyromorphite	
Barite		Cerussite		Vanadinite	
Celestine					
Anhydrite				Wulfenite	
Melanterite				Chrysocolla	
Polyhalite				Hemimorphite	

Table B.2 | Minerals Showing Some Distinctive Physical Properties

Sectile Minerals	Magnetic Minerals	Radioactive Minerals	Luminescent or Fluorescent Minerals	Piezoelectric, Pyroelectric, Thermoelectric, and Triboelectric Minerals	Minerals with Fusibility ≤ 1
Gold	Platinum	Uraninite	Diamond	Diamond	Sulfur
Silver				Graphite	
Platinum	Pyrrhotite	Monazite	Sphalerite		Realgar
Graphite		Carnotite		Arsenopyrite	Orpiment
Sulfur	Ilmenite[1]	Autunite	Fluorite	Sphalerite	Stibnite
	Hematite[1]	Torbernite	Carnallite		Bismuthinite
Acanthite	Magnetite			Alunite	Skutterudite
Chalcocite[2]	Franklinite	Zircon	Corundum	Jarosite	
Cinnabar	Chromite	Titanite		Pyromorphite	Pyrargyrite
Realgar			Calcite		Proustite
Orpiment	Wolframite		Aragonite	Quartz	Enargite
Stibnite[2]			Witherite	Thomsonite	Bournonite
Bismuthinite[2]			Strontianite	Tourmaline	
Molybdenite			Cerussite	Hemimorphite	Carnallite
			Dolomite		Chlorargyrite
Chlorargyrite			Magnesite		
					Ice
Brucite			Brucite		
			Celestine		
Wulfenite[2]			Anglesite		
Soda niter[2]					
			Apatite		
Pyrophyllite					
Talc			Scheelite		
Chrysocolla[2]					
Garnierite[2]			Quartz		
			Scapolite		
			Wollastonite		
			Pectolite		
			Willemite		

Minerals Whose Color May Change on Heating	Water-Soluble Minerals	Minerals with Specific Gravity > 3.5 but < 5.0		Minerals with Specific Gravity > 5.0	
Rutile	Sylvite	Diamond	Witherite	Gold	Cuprite
Corundum	Halite		Strontianite	Silver	Hematite
Spinel	Carnallite	Sphalerite	Malachite	Platinum	Pyrolusite
	Soda niter	Greenockite	Azurite	Arsenic	Cassiterite
Apatite		Chalcopyrite			Uraninite
Lazulite	Borax	Pyrrhotite	Barite	Acanthite	Magnetite
Turquoise	Kernite	Covellite	Celestine	Chalcocite	Columbite-Tantalite Series
Vivianite	Mirabilite	Realgar	Carnotite	Bornite	
Crocoite		Orpiment		Galena	Cerussite
	Melanterite	Stibnite	Wolframite	Nickeline	
Sodalite	Epsomite	Marcasite		Cinnabar	Anglesite
Cancrinite	Polyhalite	Molybdenite	Enstatite	Pyrite	Monazite
Muscovite	Chalcanthite	Pentlandite	Acmite	Cobaltite	Pyromorphite
Biotite		Tetrahedrite	Diopside	Arsenopyrite	Vanadinite
Rhodonite		Enargite	Jadeite	Skutterudite	Carnotite
Beryl			Rhodonite	Bismuthinite	
Olivine		Corundum	Hemimorphite		Scheelite
Zircon		Ilmenite	Garnet	Pyrargyrite	Wulfenite
Humite		Spinel	Zircon	Proustite	Crocoite
Topaz		Chrysoberyl	Willemite	Tetrahedrite	
Titanite		Chromite	Epidote	Bournonite	
			Allanite	Jamesonite	
		Goethite	Topaz		
		Limonite	Humite	Chlorargyrite	
		Romanechite	Titanite		
		Siderite			
		Rhodochrosite			
		Smithsonite			

[1] Included magnetite
[2] Subsectile

Hardness 1 2 3 4 5 6 7 8 9 10	Hardness Range	Mineral Name	Crystal System	Formula
	2–2.5	*Acanthite*	Mon	Ag_2S
	2–6.5	*Pyrolusite*	Tet	MnO_2
	2.5–3	*Gold*	Isom	Au
	2.5–3	*Silver*	Isom	Ag
	2.5–3	*Copper*	Isom	Cu
	2.5–3	*Chalcocite*	Mon	Cu_2S
	2.5–3	*Bournonite*	Orth	$PbCuSbS_3$
	3	*Bornite*	Tet	Cu_5FeS_4
	3–4.5	*Tetrahedrite Series*	Isom	$(Cu,Fe)_{12}(Sb,As)_4S_{13}$
	3.5–4	*Pentlandite*	Isom	$(Ni,Fe)_9S_8$
	3.5–4	*Chalcopyrite*	Tet	$CuFeS_2$
	3.5–4.5	*Pyrrhotite*	Mon + hex	$Fe_{1-x}S$
	4–4.5	*Platinum*	Isom	Pt
	5–5.5	*Nickeline*	Hex	$NiAs$
	5–6	*Romanechite*	Mon	$BaMnMn_8O_{16}(OH)_4$
	5–6.5	*Hematite*	Hex	Fe_2O_3
	5–6	*Ilmenite Series*	Hex	$(Fe,Mg,Mn)TiO_3$
	5–6	*Uraninite*	Isom	UO_2
	5.5	*Chromite*	Isom	$FeCr_2O_4$
	5.5–6	*Allanite*	Mon	$X_2^{2+}X_3^{3+}(SiO_4)$ $(Si_2O_7)(O,OH)$
	5.5–6.5	*Magnetite*	Isom	$FeFe_2O_4$
	5.5–6.5	Franklinite	Isom	$(Fe,Zn,Mn)Fe_2O_4$
	6–6.5	*Pyrite*	Isom	FeS_2
	6–6.5	*Marcasite*	Orth	FeS_2

Quartz

Knife blade

Copper penny

Specific Gravity	Color	Streak	Remarks	Page
7.3	Dark lead-gray	Black and shining	Sectile, darkens on exposure to light	235
5.0	Steel-gray	Black	Pulverulent forms may soil fingers, crystals hard	262
19.3	Yellow	Yellow	Sectile, does not tarnish	227
10.5	Silver-white	Silver-white	Sectile, black tarnish	227
8.9	Copper-red	Copper-red	Sectile, brown tarnish	228
5.7	Dark lead-gray	Dark lead-gray	Subsectile, massive	235
5.8	Gray to black	Gray to black	Wheel-like twins, easily fused	250
5.1	Copper-red	Gray-black	"Peacock" tarnish	236
4.4–5.4	Gray to black	Black to red	Very thin splinters are red in transmitted light	249
4.6–5	Light bronze-yellow	Light bronze-brown	Brittle, with pyrrhotite in mafic rocks	239
4.2	Brass-yellow	Greenish black	Common sulfide	237
4.6	Bronze-yellow	Black	Weakly magnetic	239
14–19	Gray	Gray	Sectile, may be weakly magnetic, in ultramafic rocks	228
7.3–7.7	Pale copper-red	Brownish black	Green alteration "bloom"	238
4.7	Black to gray	Black and shining	Colloform or earthy	268
5.3	Red-brown to black	Brick-red	Common mineral	260
4.7	Black	Black	Common mineral, may be weakly magnetic	261
10.8 or less	Black	Brownish black	Radioactive, usually massive or colloform	264
4.6	Black	Brown	May be weakly magnetic, in mafic rocks	267
3.5–4.6	Brown to black	Grayish brown	Accessory in granitic rocks, usually radioactive	354
5.2	Black	Black	Common mineral, strongly magnetic	266
5.1–5.2	Black	Brown	With willemite and zincite, weakly magnetic	265
5.0	Pale brass-yellow	Green-black to brown-black	Common mineral, crystals as cubes or pyritohedra	243
4.9	White to pale yellow	Gray-black	Tarnishes to brass-yellow, coxcomb crystals	241

Hardness (1 2 3 4 5 6 7 8 9 10)	Hardness Range	Mineral Name	Crystal System	Formula
	1–1.5	*Molybdenite*	Hex	MoS_2
	1–2	*Graphite*	Hex	C
	1.5–2	*Covellite*	Hex	CuS
	2	*Stibnite*	Orth	Sb_2S_3
	2–2.5	*Acanthite*	Mon	Ag_2S
	2–2.5	*Cinnabar*	Hex	HgS
	2.5	*Galena*	Isom	PbS
	2.5	*Pyrargyrite*	Hex	Ag_3SbS_3
	2.5	*Jamesonite*	Mon	$Pb_4FeSb_6S_{14}$
	2.5–3	*Bournonite*	Orth	$PbCuSbS_3$
	3	*Bornite*	Tet	Cu_5FeS_4
	3–3.5	*Enargite*	Orth	Cu_3AsS_4
	3–3.5	Millerite	Hex	NiS
	3.5	*Arsenic*	Hex	As
	3.5–4	*Sphalerite*	Isom	ZnS
	3.5–4	*Pentlandite*	Isom	$(Ni,Fe)_9S_8$
	3.5–4	*Cuprite*	Isom	Cu_2O
	3.5–4.5	*Pyrrhotite*	Mon+hex	$Fe_{1-x}S$
	4	Manganite	Mon	MnO(OH)
	4–4.5	*Wolframite Series*	Mon	$(Fe,Mn)(WO_4)$
	5–5.5	*Goethite*	Orth	FeO(OH)
	5–6	*Ilmenite Series*	Hex	$(Fe,Mg,Mn)TiO_3$
	5.5	*Cobaltite Group*	Orth	CoAsS
	5.5–6	*Arsenopyrite*	Mon	FeAsS
	5.5–6	*Skutterudite Series*	Isom	$(Co,Ni,Fe)As_{3-x}$
	5.5–6	*Allanite–ce*	*Mon*	$X_2^{2+}X_3^{3+}(SiO_4)(Si_2O_3)(O,OH)$
	5.5–6.5	*Magnetite*	Isom	$FeFe_2O_4$
	5.5–6.5	Franklinite	Isom	$(Fe,Zn,Mn)Fe_2O_4$
	6–6.5	*Rutile*	Tet	TiO_2
	6–6.5	Braunite	Tet	$Mn^{2+}Mn_6^{3+}SiO_{12}$
	6–6.5	*Columbite-Tantalite Series*	Orth	$(Fe,Mn)(Nb,Ta)_2O_6$
	6–6.5	*Pyrolusite*	Tet	MnO_2
	6–7	*Cassiterite*	Tet	SnO_2

Quartz

Knife blade

Copper penny

Specific Gravity	Color	Streak	Remarks	Page
4.7	Gray	Gray to black	Foliated, distinguished from graphite by green streak on glazed porcelain	247
2.1–2.3	Steel-gray to black	Black	Foliated, greasy feel	231
4.7	Indigo-blue	Black to gray	Platy aggregates	240
4.6	Gray	Gray	Columnar, black tarnish	242
7.3	Dark lead-gray	Black and shining	Sectile	235
8.1	Red	Red	Subsectile, often earthy	240
7.6	Lead-gray	Lead-gray	Cubic crystals and cleavage	236
5.8	Deep red	Red	Translucent	248
5.6	Gray-black	Gray-black	Plumose, iridescent tarnish	251
5.8	Gray to black	Gray to black	Wheel-like twins	250
5.1	Copper-red	Gray-black	"Peacock" tarnish	236
4.4	Black	Black	Bladed aggregates	250
5.3–5.6	Brass-yellow	Greenish black	Hair-like crystals	239
5.7	Tin-white	Gray	Dark gray tarnish, concentric layers	229
3.9–4.2	Black to brown	Brown to yellow	Common sulfide, resinous luster	236
4.6–5.0	Light bronze-yellow	Yellow-brown	Octahedral parting, with pyrrhotite and chalcopyrite in mafic rocks	239
6.1	Dark red	Brown-red	Highly lustrous octahedra	258
4.6	Bronze-yellow	Black	Weakly magnetic	239
4.2–4.4	Black	Brown	Prismatic crystals, with other Mn oxides	271
7.1–7.5	Black to brown-black	Black to brown-black	Sometimes weakly magnetic	301
3.3–4.3	Brown	Brownish yellow	Common mineral, colloform	271
4.5–5.0	Black to red	Black to reddish to yellow	Sometimes weakly magnetic, massive	261
6.3	Silver-white	Gray-black	Pink alteration "bloom"	245
6.1	Silver-white	Black	Prismatic crystals, common	246
6.1–6.9	Tin-white to silver-gray	Black	Cubic or octahedral crystals	247
3.5–4.6	Brown to black	Grayish brown	Accessory in granitic rocks, usually radioactive	354
5.2	Black	Black	Common mineral, strongly magnetic	266
5.1–5.2	Black	Brown	With willemite and zincite, weakly magnetic	265
4.2	Red to black	Pale brown	Prismatic crystals	261
4.72–4.8	Brown-black to steel-gray	Dark gray	Perfect cleavage, infusible	
5.2–7.9	Black to brown-black	Dark red to black	In granitic pegmatites	267
5.0	Steel-gray	Black	Pulverulent forms may stain fingers crystals hard	262
6.8–7.1	Brown or black	White to brown	In greisen, highly lustrous	263

Hardness 1 2 3 4 5 6 7 8 9 10	Hardness Range	Mineral Name	Crystal System	Formula
	1–3	Bauxite	—	Mixture of Al hydroxides
	1–1.5	Limonite	—	$Fe_2O_3 \cdot nH_2O$
	1.5	*Ice*	Hex	H_2O
	2–2.5	*Cinnabar*	Hex	HgS
	2–3	*Chlorargyrite*	Isom	$AgCl$
	2–3	Garnierite	—	$(Ni,Mg)_3(Si_2O_5)(OH)_4$
	2–4	*Chrysocolla*	—	$(Cu,Al)(SiO_3) \cdot nH_2O$
	2–5	Collophane	Mon	$Ca_3(PO_4)_3 \cdot H_2O$
	2.5	Cryolite	Mon	Na_3AlF_6
	2.5	Carnallite	Orth	$KMgCl_3 \cdot 6H_2O$
	2.5–3.5	*Serpentine Group*	Mon	$(Mg,Fe)_3(Si_2O_5)(OH)_4$
	3	*Vanadinite*	Hex	$Pb_5(VO_4)_3Cl$
	3–4	*Pyromorphite Series*	Hex	$Pb_5[(PO_4),(AsO_4),(VO_4)]Cl$
	3.5–4	*Alunite*	Hex	$KAl(SO_4)_3(OH)_6$
	5	*Apatite Series*	Hex	$Ca_5(PO_4)_3(F,Cl,OH)$
	5–5.5	*Analcime*	Isom	$Na(AlSi_2O_6) \cdot H_2O$
	5–5.5	Datolite	Mon	$CaBSiO_4(OH)$
	5–5.5	Lazurite	Isom	$(Na,Ca)_{7-8}$ $[(Al,Si)_{12}(O,S)_{24}]$ $[(SO_4),Cl,(OH)_2]$
	5–5.5	Lazulite	Mon	$MgAl_2(PO_4)_2(OH)_2$
	5–6	*Uraninite*	Isom	UO_2
	5–6	*Turquoise*	Tric	$CuAl_6(PO_4)_4(OH)_8 \cdot 5H_2O$
	5–6	*Opal*	—	$SiO_2 \cdot nH_2O$
	5.5–6	*Leucite*	Tet	$K(AlSi_2O_6)$
	5.5–6	*Allanite–ce*	*Mon*	$X_2^{2+}X_3^{3+}(SiO_4)(Si_2O_7)(O,OH)$
	5–6.5	*Hematite*	Hex	Fe_2O_3
	6–6.5	*Humite Group*	Orth+mon	$(Mg,Fe)_x(SiO_4)_y(F,OH)_2$
	6–7	*Cassiterite*	Tet	SnO_2
	6.5	*Vesuvianite*	Tet	$Ca_{10}(Mg,Fe)_2Al_4(Si_2O_7)_2$ $(SiO_4)_5(OH,F)_4$
	6.5–7	*Olivine Group*	Orth	$(Mg,Fe,Mn)_2(SiO_4)$
	6.5–7.5	*Garnet Group*	Isom	$X_3^{2+}Y_2^{3+}(SiO_4)_3$
	7	Chalcedony	Hex	SiO_2
	7	*Quartz*	Hex	SiO_2
	7	Cristobalite	Tet	SiO_2
	7	Danburite	Orth	$CaB_2(Si_2O_8)$
	7–7.5	*Tourmaline Series*	Hex	$WX_3Y_6(BO_3)_3(Si_6O_{18})(OH,F)_4$
	7.5–8	Spinel	Isom	$MgAl_2O_4$
	7.5–8	Beryl	Hex	$Be_3Al_2(Si_6O_{18})$
	9	*Corundum*	Hex	Al_2O_3

Quartz

Knife blade

Copper penny

Specific Gravity	Color	Streak	Remarks	Page
2.0–2.6	White, stained gray, yellow, or red	White	Often pisolitic, Al-ore	272
3.6–4.0	Brown to black	Yellow-brown	Alteration of iron-bearing minerals	269
0.9	Colorless to white	White	Melts at 0° C	259
8.1	Dark red	Red	Subsectile, earthy	240
5.5	Gray to green	White	Texture of horn, color deepens on exposure to light	254
2.5	Green to white	White	Unctuous, earthy	334
2.3	Green to brown	White	Encrustations and fillings, enamel-like	348
2.6–2.9	White, gray, or brown	White	Colloform variety of apatite	298
2.97	White, brownish, grayish	White	Easily fusible	
1.6	White to reddish	White	In salt beds, bitter taste	254
2.2–2.4	Green	White	Smooth or greasy feel, common alteration product	333
6.7–7.1	Red to yellow	White to yellow	Hollow prisms	299
6.5–7.2	Green, brown, or yellow	White	Hexagonal prisms, often barrel-shaped	299
2.6–2.9	White to gray or pink	White	Alteration of felsic volcanics	293
3.1–3.4	Green or red	White	Hexagonal prisms	297
2.3	White	White	Trapezohedra	323
3.0	White, greenish	White	Vitreous, often crystalline	
2.4–2.5	Blue	White	In limestones near granite contacts	317
3.0–3.1	Blue	White	Loses color when heated	295
10.8 or less	Black	Brownish black	Radioactive	264
2.6–2.8	Blue to green	White or greenish	Turns brown when heated	298
1.9–2.2	Various	White	Milky or opalescent, never in crystals	310
2.6	White or gray	White	Pseudotrapezohedra, in silica-poor igneous rocks	315
3.5–4.6	Brown to black	Grayish brown	Accessory in granitic rocks, usually radioactive	354
5.3	Red-brown to black	Brick-red	Common mineral	260
3.1–3.4	Yellow, brown, red, or white	White	In dolomitic marbles	359
6.8–7.1	Brown or black	White to brown	Highly lustrous, in greisen	263
3.4	Brown to green	White	Columnar aggregates, in metamorphic rocks	354
3.3–4.4	Green to brown	White	Rock-forming mineral, in mafic rocks	356
3.5–4.3	Red, brown, or green	White	Dodecahedra or trapezohedra	357
2.6	Various	White	Often banded	380
2.6	Colorless or white	White	Common rock-forming mineral, glassy	308
2.3	Colorless to white	White	In siliceous volcanic rocks	310
3.0	Colorless to yellow	White	Orthorhombic crystals	315
3.0–3.2	Black, green, or red	White	Spherical triangle cross section	350
3.6	Various	White	Octahedra	265
2.6–2.9	Blue to green	White	Hexagonal prisms in pegmatite	351
4.0	Brown, gray, pink, or blue	White	Good parting, in metamorphic rocks	259

Part 2: Nonmetallic Luster, Section B: Distinct Cleavage
(Cleavage fragments are tabular or micaceous.)

Hardness 1 2 3 4 5 6 7 8 9 10	Hardness Range	Mineral Name	Crystal System	Formula
▪	1	*Smectite Group*	Mon	$X(Al,Mg)_2(Si_4O_{10})(OH)_2 \cdot nH_2O$
▬	1–1.5	Talc	Mon	$Mg_3(Si_4O_{10})(OH)_2$
▬	1–2	*Pyrophyllite*	Mon	$Al_2(Si_4O_{10})(OH)_2$
▬	1–2	*Graphite*	Hex	C
▬	1–2	*Carnotite*	Mon	$K_2(UO_2)_2(VO_4)_2 \cdot 3H_2O$
▪	1.5	Vermiculite	Mon	$Mg_3(Si_4O_{10})(OH)_2 \cdot nH_2O$
▬	1.5–2	*Orpiment*	Mon	As_2S_3
▬	1.5–2	Vivianite	Mon	$Fe_3(PO_4)_3 \cdot 8H_2O$
▪	2	*Gypsum*	Mon	$Ca(SO_4) \cdot 2H_2O$
▬	2–2.5	Kaolinite	Tric	$Al_2(Si_2O_5)(OH)_4$
▬	2–2.5	*Chlorite Group*	Mon	$Y_n(Z_4O_{10})(OH)_8$
▬	2–2.5	Autunite	Tet	$Ca(UO_2)_2(PO_4)_2 \cdot nH_2O$
▪	2.5	*Brucite*	Hex	$Mg(OH)_2$
▬	2.5–3	*Muscovite*	Mon	$KAl_2(AlSi_3O_{10})(OH,F)_2$
▬	2.5–3	*Biotite*	Mon	$K(Mg,Fe)_3(AlSi_3O_{10})(OH)_2$
▬	2.5–3	*Phlogopite*	Mon	$KMg_3(AlSi_3O_{10})(OH,F)_2$
▬	2.5–3.5	*Gibbsite*	Mon	$Al(OH)_3$
▬	2.5–3.5	Jarosite	Hex	$KFe_3(SO_4)_2(OH)_6$
▬	2.5–4	*Lepidolite Series*	Mon	$KX_3(Al,Si)_4O_{10}(OH,F)_2$
▪	3	*Wulfenite*	Tet	$Pb(MoO_4)$
▬	3–3.5	Barite	Orth	$Ba(SO_4)$
▪	3.5	*Polyhalite*	Tric	$K_2Ca_2Mg(SO_4)_4 \cdot 2H_2O$
▪	3.5	*Arsenic*	Hex	As
▬	3.5–4	Heulandite	Mon	$(Ca,Na)_2(Al_2Si_7O_{18}) \cdot 6H_2O$
▬	3.5–5	*Margarite*	Mon	$CaAl_2(Al_2Si_2O_{10})(OH)_2$
▬	4–4.5	Zincite	Hex	ZnO
▬	4.5–5	Apophyllite	Tet	$KCa_4(Si_4O_{10})_2F$
▪	6–6.5	*Albite*	Tric	$Na(AlSi_3O_8)$
▬	6.5	*Chloritoid*	Mon	$(Fe^{2+}MgMn)_2(Al, Fe^{3+})$ $Al_3O_2(SiO_4)_2(OH)_2$

| Quartz
| Knife blade
| Copper penny

Specific Gravity	Color	Streak	Remarks	Page
2.0–2.7	White or gray	White	Unctuous feel, earthy smell when wet	326
2.7	Green to white	White	Greasy feel, plastic cleavage folia	327
2.8	White	White	Greasy feel, plastic cleavage folia	342
2.1–2.3	Steel-gray to black	Black	Greasy feel	231
4–5	Yellow	Pale yellow	Radioactive, in sandstones	299
2.4	Yellow to brown	White	Expands when heated	332
3.5	Yellow	Yellow	Sectile, melts at 300° C	242
2.6–2.7	Colorless, blue, or green	White to bluish	Color darkens with oxidation	295
2.3	White, gray, or brown	White	Common mineral, plastic	290
2.6	White, often stained	White	Unctuous when wet, earthy	326
2.7	Dark green	White	Common alteration product, flexible cleavage folia	328
3.1	Yellow	Yellow	Tabular crystals, radioactive	300
2.4	White or pale tints	White	In veins in Mg-rich metamorphic rocks	270
2.8–3.0	Colorless or tints	White	Common rock-forming mineral, elastic cleavage folia	330
2.7–3.1	Black or brown-black	White	Common rock-forming mineral, elastic cleavage folia	331
2.8	Brown	White	Rare in igneous rocks, elastic cleavage folia	332
2.4	White, often stained	White	Earthy smell when wet, in bauxite deposits and talc schists	272
2.9–3.3	Yellow to brown	Pale yellow	Crusts and coatings	292
2.9	Pink, violet, or white	White	Elastic cleavage folia, in pegmatites	332
6.5–7.0	Orange to yellow	White	Square tablets	302
4.5	White	White	Often tinted yellow, red, green, or blue; tabular crystals	287
2.78	Colorless, white, tinted	White	Massive, also fibrous or foliated, bitter taste	254
5.7	Tin-white	Gray	Dark gray tarnish, concentric layers	229
2.2	White to reddish	White	A zeolite	320
3.0	Pink to white	White	Brittle cleavage folia, in chlorite schists	329
5.4–5.7	Red to orange-yellow	Orange-yellow	With franklinite and willemite	257
2.3–2.4	White to gray	White	In mafic volcanics	
2.6	White	White	A plagioclase	314
3.5–3.8	Dark green, black	White	Resembles chlorite	361

Part 2: Nonmetallic Luster, Section C: Distinct Cleavage
(Cleavage fragments are polyhedral.)

Hardness 1 2 3 4 5 6 7 8 9 10	Hardness Range	Mineral Name	Crystal System	Formula
■	1–1.5	Nitratine	Hex	$Na(NO_3)$
■	2	*Gypsum*	Mon	$Ca(SO_4) \cdot 2H_2O$
■	2	*Halite*	Isom	$NaCl$
■	2	Sylvite	Isom	KCl
■	2.5–3	Thenardite	Orth	Na_2SO_4
■	3–3.5	*Calcite*	Hex	$Ca(CO_3)$
■	3–3.5	*Celestine*	Orth	$Sr(SO_4)$
■	3.5	*Anhydrite*	Orth	$Ca(SO_4)$
■	3.5–4	*Sphalerite*	Isom	ZnS
■	3.5–4	*Rhodochrosite*	Hex	$Mn(CO_3)$
■	3.5–4	*Dolomite*	Hex	$CaMg(CO_3)_2$
■	3.5–4	*Cuprite*	Isom	Cu_2O
■	4–4.5	*Siderite*	Hex	$Fe(CO_3)$
■	4	*Magnesite*	Hex	$Mg(CO_3)$
■	4	*Fluorite*	Isom	CaF_2
■	4–4.5	Smithsonite	Hex	$Zn(CO_3)$
■	4–5	*Chabazite*	Hex	$Ca(Al_2Si_4O_{12}) \cdot 6H_2O$
■	4.5–5	*Pectolite*	Tric	$NaCa_2H(Si_3O_9)$
■	5–6	*Cancrinite*	Hex	$(Na,Ca)_4(AlSiO_4)_3(CO_3)$
■	5.5–6	*Sodalite*	Isom	$Na,Ca(AlSiO_4)_3Cl$
■	10	*Diamond*	Isom	C

Quartz
Knife blade
Copper penny

Specific Gravity	Color	Streak	Remarks	Page
2.2–2.3	White to brown-gray, or yellow	White	Rhombs, cool taste	276
2.3	White, gray, or brown	White	Flattened rhombohedra, fibrous, massive	290
2.2	Colorless or white	White	Cubes, salty taste	253
2.0	White to bluish or reddish	White	Cubes, bitter salty taste, often tinted	255
2.7	Colorless, white	White	Salt lake deposits, also around fumaroles	
2.7	Colorless or white	White	Rhombs, often colored, very common	275
4.0	Colorless to pale blue	White	Oblongs, in sedimentary rocks	288
3.0	Colorless to blue-white	White	Oblongs, often tinted	289
3.9–4.2	Yellow, brown, black	White, Yellow	Dodecahedra, resinous luster	236
3.3–3.7	Pink	White	Rhombs	277
2.9	White	White	Rhombs, common sedimentary rock-forming mineral	280
6.1	Red	Brown-red	Octahedra, highly lustrous	258
3.9	Brown	White	Rhombs	277
3.0	Colorless or white	White	Rhombs or white enamel as an alteration of Mg-rich rock	276
3.2	Violet, green, white, or blue	White	Octahedral cleavage, cubic crystals, often fluorescent	255
4.4	White, green	White	Rhombs, usually colloform, variously colored	278
2.1	Flesh-red to white	White	Rhombs, in cavities in mafic lavas	320
2.8	Colorless or white	White	Radiating groups of needles with a silky luster	349
2.5	Yellow	White	Rare prismatic crystals, in silica-poor igneous rocks	317
2.1–2.4	Blue	White	Nodular, in silica-poor igneous rocks, variously colored	316
3.5	Colorless, bluish, pink, green, yellow, or black	None	Octahedra	230

Part 2: Nonmetallic Luster, Section D: Distinct Cleavage
(Colorless or lightly tinted minerals; cleavage fragments are not micaceous, tabular, or polyhedral.)

Hardness (1 2 3 4 5 6 7 8 9 10)	Hardness Range	Mineral Name	Crystal System	Formula
	1.5–2	Mirabilite	Mon	$Na_2(SO_4) \cdot 10H_2O$
	2	Gypsum	Mon	$Ca(SO_4) \cdot 2H_2O$
	2–2.5	Borax	Mon	$Na_2(B_4O_7) \cdot 10H_2O$
	2–2.5	Epsomite	Orth	$Mg(SO_4) \cdot 7H_2O$
	2.5–3	Anglesite	Orth	$Pb(SO_4)$
	2.5–3	Kernite	Mon	$Na_2(B_4O_7) \cdot 4H_2O$
	3–3.5	Witherite	Orth	$Ba(CO_3)$
	3–3.5	Cerussite	Orth	$Pb(CO_3)$
	3–3.5	Celestine	Orth	$Sr(SO_4)$
	3–3.5	Barite	Orth	$Ba(SO_4)$
	3.5–4	Strontianite	Orth	$Sr(CO_3)$
	3.5–4	Aragonite	Orth	$Ca(CO_3)$
	3.5–4	Stilbite	Mon	$NaCa_2(Al_5Si_{13}O_{36}) \cdot 14H_2O$
	3.5–4	Wavellite	Orth	$Al_3(PO_4)_2(OH)_3 \cdot 5H_2O$
	4	Alunite	Hex	$KAl_3(SO_4)_2(OH)_6$
	4–4.5	Colemanite	Mon	$Ca_2(B_6O_{11}) \cdot 5H_2O$
	4.5–5	Scheelite	Tet	$Ca(WO_4)$
	4.5–5	Wollastonite	Tric	$Ca(SiO_3)$
	4.5–5	Hemimorphite	Orth	$Zn_4(Si_2O_7)(OH)_2 \cdot nH_2O$
	4.5,6.5	Kyanite	Tric	$AlAl(SiO_4)O$
	5	Pectolite	Tric	$Ca_2NaH(Si_3O_9)$
	5	Melilite Group	Tet	$X_2Y(Z_2O_7)$
	5–5.5	Natrolite	Orth	$Na_2(Al_2Si_3O_{10}) \cdot 2H_2O$
	5–6	Nepheline	Hex	$(KNa)(AlSiO_4)$
	5–6	Tremolite Series	Mon	$X_2Y_5(Si_8O_{22})(OH)_2$
	5–6.5	Scapolite Group	Tet	$(Na,Ca)_4[(Al,Si)_4O_8]_3(Cl,CO_3)$

Quartz

Knife blade

Copper penny

Specific Gravity	Color	Streak	Remarks	Page
1.5	White	White	Cool then bitter taste	291
2.3	Colorless to white, gray, brown	White	Common mineral, plastic	290
1.7	Colorless to white	White	Sweetish taste, in evaporites, tinted gray, blue, or green	284
1.7	Colorless to white	White	Bitter taste, tinted pink or green	292
6.4	Colorless to white	White	Alteration of galena, tinted gray, yellow, brown, or green	289
1.9	Colorless to white	White	Long splintery cleavage fragments	285
4.3	Colorless to white	White	Tinted yellow, brown, or green	279
6.6	Colorless to white	White	Reticulated aggregates; gray, blue, or green tints; very brittle	280
4.0	Colorless to pale blue	White	In sedimentary rocks, tinted red or green	288
4.5	Colorless to white	White	Tinted yellow, brown, red, gray, green, or blue; tabular crystals, heavy	287
3.8	Colorless to gray	White	Columnar or fibrous; tinted yellow, green, or red	280
2.9	Colorless to white	White	Columnar or fibrous	278
2.2	White to brownish	White	Sheaflike aggregates, in cavities in basalt	319
2.3	White to yellow, green, gray, or brown	White	Radiated globular aggregates	
2.6–2.8	White to gray or pink	White	Alteration of felsic volcanics	293
2.4	Colorless to white	White	In stratified lake deposits, tinted yellow or gray	285
6.1	Colorless to white and gray	White	Blue fluorescence; tinted yellow, green, or red	301
2.8	White	White	In contact metamorphic limestones; tinted gray, yellow, or brown	348
3.5	Green-white	White	Tinted blue or brown	353
3.6	Blue-white	White	Bladed crystals in metamorphic rocks	363
2.8	White	White	Fibrous radiating masses in mafic lavas	349
3.0	White	White	Tinted yellow, green, red, or brown	353
2.2–2.4	Colorless to white	White	Radiating acicular crystals; tinted green, yellow, or red	321
2.6	Colorless to white	White	In silica-poor igneous rocks; tinted yellow, green, gray, or red	316
3.0–3.3	Gray to white	White	Prismatic crystals, in contact metamorphic zones	341
2.5–2.8	White or gray	White	Square prisms with a woody surface; tinted green, blue, or red	321

Table B.3 **Part 2: Nonmetallic Luster, Section D (*continued*)**

Hardness (1 2 3 4 5 6 7 8 9 10)	Hardness Range	Mineral Name	Crystal System	Formula
	5.5–6	*Amblygonite Group*	Tric	$LiAl(PO_4)(OH,F)$
	5.5–6	*Leucite*	Tet	$K(AlSi_2O_6)$
	5.5–6.5	*Diopside Series*	Mon	$Ca(Mg,Fe)(Si_2O_6)$
	6	Sanidine	Mon	$(K,Na)(AlSi_3O_8)$
	6	Adularia	Mon	$(K,Na)(AlSi_3O_8)$
	6–6.5	*Orthoclase Series*	Mon	$(K,Na)(AlSi_3O_8)$
	6–6.5	*Microcline Series*	Tric	$(K,Na)(AlSi_3O_8)$
	6–6.5	*Clinozoisite*	Mon	$Ca_2Al_3(SiO_4)(Si_2O_7)(O,OH)$
	6–6.5	*Plagioclase Series*	Tric	$(Ca,Na)[(AlSi)_4O_8]$
	6–6.5	*Prehnite*	Orth	$Ca_2Al(AlSi_3O_{10})(OH)_2$
	6–6.5	*Humite Group*	Orth + mon	$(Mg,Fe)_x(SiO_4)_y(F,OH)_2$
	6–7	*Sillimanite*	Orth	$AlAl(SiO_4)O$
	6.5–7	*Diaspore*	Orth	$AlO(OH)$
	6.5–7	*Spodumene*	Mon	$LiAl(Si_2O_6)$
	7.5	*Zircon*	Tet	$Zr(SiO_4)$
	7.5	*Andalusite*	Orth	$AlAl(SiO_4)O$
	8	*Topaz*	Orth	$Al_2(SiO_4)(F,OH)_2$

Quartz

Knife blade

Copper penny

Specific Gravity	Color	Streak	Remarks	Page
3.1	White	White	In pegmatites or greisen; tinted yellow, pink, green, blue, or gray	296
2.6	White or gray	White	Pseudotrapezohedral crystals, in silica-poor volcanic rocks	315
3.2–3.6	White to green	White	Stubby prismatic crystals	345
2.6	Glassy	White	In salic extrusive rocks	312
2.6	Colorless	White	Pseudo-orthorhombic crystals, often opalescent	314
2.6	White to flesh-pink	White	Common rock-forming mineral	313
2.6	White to cream to reddish, yellowish, or greenish	White	Common rock-forming mineral, distinguished from orthoclase by fine twinning striations	314
3.2	Pale green to brown	White	Vertically striated prisms	354
2.6–2.7	White or gray	White	Common rock-forming mineral, polysynthetic twinning	314
2.9	Light green to white	White	Colloform with a crystalline surface	335
3.1–3.4	Yellow, brown, or white	White	In dolomitic marbles	359
3.2	Brown, green, or white	White	Fibrous groups in metamorphic rocks	363
3.2–3.5	Colorless or white	White	Platy crystals, very brittle	270
3.0–3.2	White	White	In pegmatites; tinted green, gray, yellow, purple, or pink	347
4.7	Brown to colorless	White	Square prisms; tinted gray, green, or red	359
3.2	White, rose, or green	White	Maltese cross on (001) section, in metamorphic rocks	362
3.5	Colorless to white	White	Highly modified prisms; tinted green, blue, or red	388

Table B.3 Part 2: Nonmetallic Luster, Section E: Distinct Cleavage
(Black, brown, or red minerals; cleavage fragments are not micaceous, tabular, or polyhedral.)

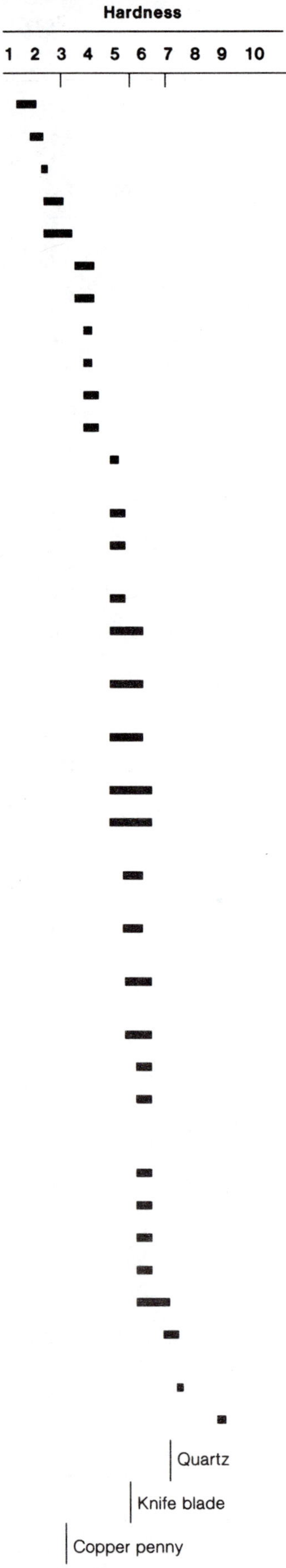

Hardness (1 2 3 4 5 6 7 8 9 10)	Hardness Range	Mineral Name	Crystal System	Formula
	1.5–2	*Realgar*	Mon	AsS
	2–2.5	Proustite	Hex	Ag_3AsS_3
	2.5	*Pyrargyrite*	Hex	Ag_3SbS_3
	2.5–3	*Crocoite*	Mon	$Pb(CrO_4)$
	2.5–3.5	*Jarosite*	Hex	$KFe_3(SO_4)_2(OH)_6$
	3.5–4	*Sphalerite*	Isom	ZnS
	3.5–4	*Cuprite*	Isom	Cu_2O
	4	Manganite	Mon	$MnO(OH)$
	4	Riebeckite	Mon	$Na_2Fe_3Fe_2(Si_8O_{22})(OH)_2$
	4–4.5	*Wolframite*	Mon	$(Mn,Fe)(WO_4)$
	4–4.5	Zincite	Hex	ZnO
	5	*Apatite Series*	Hex	$Ca_5(PO_4)_3(F,Cl,OH)$
	5–5.5	Goethite	Orth	$\alpha\text{–}FeO(OH)$
	5–5.5	Monazite	Mon	$(Ce,La,Nd,Th)(PO_4)$
	5–5.5	Titanite	Mon	$CaTi(SiO_4)O$
	5–6	*Ilmenite Series*	Hex	$(Fe,Mg,Mn)TiO_3$
	5–6	*Hornblende Series*	Mon	$Ca_2(Fe,Mg)_4$ $Al(AlSi_7O_{22})(OH)_2$
	5–6	*Enstatite Series*	Orth	$(Mg,Fe)_2(Si_2O_6)$
	5–6.5	*Hematite*	Hex	Fe_2O_3
	5–6.5	*Scapolite Group*	Tet	$(Na,Ca)_4[(Al,Si)_4O_8]_3$ (Cl,CO_3)
	5.5–6	*Allanite–ce*	*Mon*	$X_2^{2+}Y_3^{3+}(SiO_4)$ $(Si_2O_7)(O,OH)$
	5.5–6	*Anthophyllite Series*	Orth	$(Mg,Fe,Al)_7$ $[(Al,Si)_8O_{22}](OH)_2$
	5.5–6.5	*Augite*	Mon	$Ca(Mg,Fe,Al)[(Al,Si)_2O_6]$
	5.5–6.5	*Rhodonite*	Tric	$Mn(SiO_3)$
	6–6.5	*Orthoclase Series*	Mon	$(K,Na)(AlSi_3O_8)$
	6–6.5	*Microcline Series*	Tric	$(K,Na)(AlSi_3O_8)$
	6–6.5	*Rutile*	Tet	TiO_2
	6–6.5	*Columbite-Tantalite Series*	Orth	$(Fe,Mn)(Nb,Ta)_2O_6$
	6–6.5	*Humite Group*	Orth + mon	$(Mg,Fe)_x(SiO_4)_y(F,OH)_2$
	6–6.5	Aegirine	Mon	$NaFe(Si_2O_6)$
	6–7	*Cassiterite*	Tet	SnO_2
	7–7.5	*Staurolite*	Mon	$(Fe,Mg,Zn)_2Al_9$ $[(Si,Al)_4O_{16}]O_6(OH)_2$
	7.5	*Zircon*	Tet	$Zr(SiO_4)$
	9	*Corundum*	Hex	$\alpha\text{–}Al_2O_3$

Quartz

Knife blade

Copper penny

Specific Gravity	Color	Streak	Remarks	Page
3.5	Red to orange-yellow	Orange to yellow	Melts at 310° C	241
5.6	Red	Red	With pyrargyrite	249
5.8	Deep red	Red	Translucent	248
5.9–6.1	Red	Orange	Secondary mineral in lead deposits	300
2.9–3.3	Brown to yellow	Pale yellow	Crusts and coatings	292
3.9–4.2	Black, brown, or yellow	Brown, yellow, or white	Resinous luster	236
6.1	Dark red	Brown-red	Octahedra, highly lustrous	258
4.3	Black	Brown	Prismatic crystals, with other Mn oxides	271
3.4	Dark blue to black	White	In Na,Fe-rich igneous rocks	339
7.1–7.5	Black to brown-black	Brown-black	Sometimes weakly magnetic	301
5.4–5.7	Red to orange-yellow	Orange-yellow	With franklinite and willemite	257
3.1–3.4	Dark red-brown, blue, green	White	Hexagonal prisms	297
3.3–4.3	Brown	Brown	Common mineral, colloform	271
5.1	Brown	White	Usually radioactive, accessory in salic igneous rocks	296
3.5	Brown, red, or black	White	Wedge-shaped crystals	364
4.5–5.0	Black, brown, red	Black, brown, or yellow	Common mineral, sometimes weakly magnetic	261
2.9–3.3	Black	White	Common rock-forming mineral	342
3.1–3.5	Brown	White	Common rock-forming mineral in mafic rocks	345
5.3	Black to red-brown	Brick-red	Common mineral, good parting	260
2.5–2.8	Red, gray	White	Square prisms with woody surface	321
3.5–4.6	Brown to black	Gray-brown	Accessory in granitic rocks, usually radioactive	354
2.8–3.2	Brown	White or gray	Lamellar or fibrous, in crystalline schists	340
3.2–3.6	Black to dark green	White	Common rock-forming mineral, cleavage at right angles	346
3.4–3.7	Red or pink	White	Black surface stains	347
2.6	Red or pink	White	Common rock-forming mineral	313
2.6	Red to pink	White	Common rock-forming mineral, distinguished from orthoclase by fine twinning striations	314
4.2	Black to red	Pale brown	Prismatic crystals, highly lustrous	261
5.2–8.2	Black to brown-black	Dark red to black	In granitic pegmatites	267
3.1–3.4	Brown, red	White	Formless grains in dolomitic marble	359
3.5	Brown to green	Yellowish gray	In Na,Fe-rich igneous rocks	346
7.0	Brown or black	White to brown	Highly lustrous, in greisen	263
3.7	Brown to black	White	Distinct crystals in metamorphic rocks	365
4.7	Brown	White	Square prisms	359
4.0	Brown	White	In metamorphic rocks	259

Part 2: Nonmetallic Luster, Section F: Distinct Cleavage
(Orange, yellow, green, blue, and violet minerals; cleavage fragments are not micaceous, tabular, or polyhedral.)

Hardness (1 2 3 4 5 6 7 8 9 10)	Hardness Range	Mineral Name	Crystal System	Formula
	1.5–2	*Orpiment*	Mon	As_2S_3
	1.5–2	*Realgar*	Mon	AsS
	1.5–2.5	*Sulfur*	Orth	S
	2	*Melanterite*	Mon	$Fe(SO_4)\cdot7H_2O$
	2	*Glauconite*	Mon	$(K,Na)(Fe,Mg,Al)_2[(Al,Si)_4O_{10}](OH)_2$
	2.5	Chalcanthite	Tric	$Cu(SO_4)\cdot5H_2O$
	2.5–3.5	*Jarosite*	Hex	$KFe_3(SO_4)_2(OH)_6$
	2.5–3.5	*Serpentine Group*	Mon	$Mg_3(Si_2O_5)(OH)_4$
	2.5–5.5	Chrysotile Group	Mon	$(Mg,Fe)_3(Si_2O_5)(OH)_4$
	3	*Wulfenite*	Tet	$Pb(MoO_4)$
	3.5–4	*Malachite*	Mon	$Cu_2(CO_3)(OH)_2$
	3.5–4	*Azurite*	Mon	$Cu_3(CO_3)_2(OH)_2$
	4	*Fluorite*	Isom	CaF_2
	4	Riebeckite	Mon	$Na_2Fe_3Fe_2(Si_8O_{22})(OH)_2$
	4.5,6.5	*Kyanite*	Tric	$AlAl(SiO_4)O$
	5	*Apatite Group*	Hex	$Ca_5(PO_4)_3(F,Cl,OH)$
	5	*Melilite Group*	Tet	$Ca_2(Mg,Al)[(Al,Si)_2O_7]$
	5–5.5	*Monazite*	Mon	$(Ce,La,Nd,Th)(PO_4)$
	5–5.5	Lazurite	Isom	$(Na,Ca)_{7-8}[(Al,Si)_{12}(O,S)_{24}][(SO_4),Cl,(OH)_2]$
	5–5.5	Lazulite	Mon	$MgAl_2(PO_4)_2(OH)_2$
	5–6	*Cancrinite*	Hex	$Na_3Ca(SiAlO_4)_3(CO_3)$
	5–6	*Actinolite*	Mon	$Ca_2(Mg,Fe)_5(Si_8O_{22})(OH)_2$
	5–6	*Hornblende Series*	Mon	$Ca_2(Fe,Mg)_4Al(AlSi_7O_{22})(OH)_2$
	5–6	*Enstatite Series*	Orth	$(Mg,Fe)_2(Si_2O_6)$
	5–6	*Turquoise*	Tric	$CuAl_6(PO_4)_4(OH)_8\cdot5H_2O$
	5–6.5	*Scapolite Group*	Tet	$(Na,Ca)_4[(Al,Si)_4O_8]_3(Cl,CO_3)$
	5.5	Willemite	Hex	$Zn_2(SiO_4)$
	5.5–6	*Anthophyllite Series*	Orth	$Y_7(Z_8O_{22})(OH)_2$
	5.5–6	Sodalite	Isom	$Na_4(AlSiO_4)_3Cl$
	5.5–6.5	*Diopside Series*	Mon	$Ca(Mg,Fe)_2(Si_2O_6)$

Quartz

Knife blade

Copper penny

Specific Gravity	Color	Streak	Remarks	Page
3.5	Yellow	Yellow	Sectile, melts at 300° C	242
3.5	Orange-yellow to red	Orange-yellow to red	Sectile, melts at 310° C	241
2.1	Yellow	White	Melts and burns at 113° C	230
1.9	Blue or green	White	Sweetish taste	291
2.3–2.8	Dull green	White	Pellets in sedimentary rocks	334
2.1–2.3	Blue	White	Metallic taste	292
2.9–3.3	Yellow to brown	Pale yellow	Crusts and coatings	292
2.2–2.4	Green	White	Smooth to greasy feel	333
2.2	Green	White	Asbestiform	333
6.5–7.0	Orange to yellow	White	Square tablets	302
4.1	Green	Green	Colloform	281
3.8	Blue	Blue	Encrusting	282
3.2	Green or violet	White	Cubic crystals with octahedral cleavage, frequently fluorescent	255
3.4	Dark blue to black	White	In Na,Fe-rich igneous rocks	339
3.6	Blue-white	White	Bladed crystals in metamorphic rocks	363
3.1–3.4	Dark green	White	Hexagonal crystals	297
3.0	Yellow or green to white	White	In silica-poor igneous rocks	353
5.1	Yellow-brown	White	Usually radioactive	296
2.4–2.5	Blue	White	In limestones near granite contacts	317
3.0–3.1	Blue	White	Loses color when heated	295
2.5	Yellow, green, or blue	White	In silica-poor igneous rocks	317
3.0–3.3	Green	White	In contact metamorphic rocks	341
2.9–3.3	Dark green	White	Common rock-forming mineral	342
3.1–3.5	Greenish white	White	Common rock-forming mineral in mafic rocks	345
2.6–2.8	Blue to green	White or greenish	Turns brown when heated	298
2.5–2.8	Green or blue	White	Square prisms with a woody surface	321
3.9–4.2	Green to brown or white	White	With other zinc minerals, fluorescent	
2.8–3.2	Green	White to grayish	In crystalline schists	340
2.1–2.4	Blue or green	White	In silica-poor igneous rocks	316
3.2–3.6	Green to white	White	Common mineral in mafic rocks and marbles	345

Hardness 1 2 3 4 5 6 7 8 9 10	Hardness Range	Mineral Name	Crystal System	Formula
■ (≈6)	6–6.5	*Prehnite*	Orth	$Ca_2Al(AlSi_3O_{10})(OH)_2$
■ (≈6)	6–6.5	*Humite Group*	Orth + mon	$(Mg,Fe)_x(SiO_4)_y(F,OH)_2$
■ (≈6)	6–6.5	Glaucophane	Mon	$Na_2Mg_3Al_2(Si_8O_{22})(OH)_2$
■ (≈6)	6–6.5	*Microcline Series*	Tric	$(K,Na)(AlSi_3O_8)$
▬ (6–7)	6–7	*Sillimanite*	Orth	$AlAl(SiO_4)O$
▬ (6–7)	6–7	*Epidote Group*	Mon	$X_2Y_3(SiO_4)(Si_2O_7)(O,OH)$
■ (6.5–7)	6.5–7	Jadeite	Mon	$NaAl(Si_2O_6)$
■ (7–7.5)	7–7.5	*Cordierite*	Orth	$(Mg,Fe)_2(Al_4Si_5O_{18})$
▪ (8.5)	8.5	Chrysoberyl	Orth	Al_2BeO_4
■ (9)	9	*Corundum*	Hex	$\alpha\text{–}Al_2O_3$

Quartz

Knife blade

Copper penny

Specific Gravity	Color	Streak	Remarks	Page
2.9	Light green	White	Colloform with a crystalline surface	335
3.1–3.4	Yellow to white	White	Formless grains in dolomitic marble	359
3.0–3.2	Blue	Grayish blue	In metamorphic rocks	339
2.6	White to reddish	White	In siliceous igneous rocks and pegmatites	314
3.2	Gray-green to white	White	Fibrous groups in metamorphic rocks	363
3.2–3.5	Green	White	Common mineral	394
3.3–3.5	Green	White	Compact and tough	346
2.6	Blue to green	White	In metamorphic rocks	322
3.7–3.8	Green	White	Tabular crystals in pegmatite or crystalline schist	356
4.0	Blue	White	Good parting, in metamorphic rocks	259

References

Abbe, E. 1873. *Arch. Mikros. Anat.* 9: 413.

Anderson, O. F. 1915. The system anorthite-forsterite-silica. *Am. J. Sci.* (4th Ser.) 29: 407–54.

Bambauer, H. O.; Corlett, M.; Eberhard, E.; and Viswanathan, K. 1967. Diagrams for the determination of plagioclase using x-ray powder methods. *Schweiz. Mineral. Petrog. Mitt.* 42: 333–49.

Barth, T. F. W.; Correns, C. W.; and Eskola, P. 1939. *Die Enstehung der Gesteine.* Berlin: Springer. 422 pp.

Bayliss, P. 1975. Nomenclature of trioctahedral chlorities. *Can. Mineral.* 13: 178–180.

Bayliss, P. 1986. Subdivision of the pyrite group, and a chemical and x-ray diffraction investigation of ullmannite. *Can. Mineral.* 24: 27–33.

Bowen, N. L. 1913. The melting phenomena of the plagioclase feldspars. *Am. J. Sci.* (4th Ser.) 34: 577–99.

Bowen, N. L. 1915. The crystallization of haplobasalt, haplodiorite, and related magma. *Am. J. Sci.* (4th Ser.) 40: 161–85.

Boyd, F. R., and England, J. L. 1960. The quartz-coesite transition. *J. Geophys. Res.* 65: 749–56.

Brindley, G. W., and Harrison, F. W. 1952. The structure of chloritoid. *Acta Cryst.* 5: 698–99.

Buerger, M. J. 1961. Polymorphism and phase transformations. *Fortsch. Mineral.* 39: 9–24.

Buerger, M. J. 1963. *Elementary Crystallography.* Revised printing. New York: Wiley. 528 pp.

Burns, R. G., 1970. *Mineralogical Applications of Crystal Field Theory.* Cambridge: Cambridge University Press. 224 pp.

Cameron, M., and Papike, J. J. 1981. Structural and chemical variations in pyroxenes. *Am. Mineral.* 66: 1–50.

Choudhry, A. G. 1981. *The Petrology and Geochemistry of the Gamitagama Lake Igneous Complex, Wawa, North Central Ontario.* M. Sc. Thesis, University of Windsor.

Colville, P. A.; Ernst, W. G.; and Gilbert, M. C. 1966. Relationships between cell parameters and chemical compositions of monoclinic amphiboles. *Am. Mineral.* 51: 1727–54.

de Broglie, L. 1924. Thèse de Doctorat. 1925 Ann. de Physique 3: 22. 1968. Investigations on Quantum Theory, Ludvig, G. ed. *Selected Readings Physics-Wave Mechanics.* Elmsford, NY: Pergamon. 230 pp.

Deer, W. A.; Howie, R. A.; and Zussman, J. 1963. *Rock-forming Minerals.* Vol. 2. *Chain Silicates.* London: Longmans. 379 pp.

Dennen, W. H., and Moore, B. R. 1986. *Geology and Engineering.* Dubuque, IA: Wm. C. Brown. 378 pp.

Ford, W. E. 1932. *Dana's Textbook of Mineralogy.* 4th ed. New York: Wiley. 851 pp.

Garrels, R. M., and Christ, C. L. 1965. *Solutions, Minerals, and Equilibria.* San Francisco: Freeman. 450 pp.

Goldsmith, J. R.; Graf, D. L.; and Joensuu, O. I. 1955. The occurrence of magnesian calcites in nature. *Geochim. Cosmochim. Acta* 7: 212–30.

Harker, R. T., and Tuttle, O. F. 1956. Experimental data on the P_{CO_2}-T curve for the reaction calcite + quartz = wollastonite + carbon dioxide. *Am. J. Sci.* 254: 239–56.

Harris, D. C., and Cabri, L. J. 1991. Nomenclature of platinum-group elements: review and revision. *Can. Mineral.* 29: 231–37.

Hartree, D. 1948. The calculation of atomic structures. *Reports on Progress in Physics* 11: 113.

Hassan, I., and Grundy, H. D. 1991. The crystal structure of basic cancrinite. *Can. Mineral.* 29: 377–83.

Hawthorne, F. 1981. Crystal chemistry of the amphiboles. *Reviews in Mineralogy.* Min. Soc. Amer. 9A. 1–102.

Holdaway, M. J. 1971. Stability of andalusite and the aluminum silicate phase diagram. *Am. J. Sci.* 271: 97–131.

International Union of Pure and Applied Chemistry 1986. Atomic Weights of the Elements 1985. *Pure and Appl. Chem.* 58: 1677–92.

Jenkins, R., and Devries, J. L. 1969. *Practical X-ray Spectrometry.* 2nd ed. New York: Springer-Verlag. 182 pp.

Johnson, N. E.; Craig, J. R.; and Rimstidt, J. D. 1988. Crystal chemistry of tetrahedrite. *Am. Mineral.* 73: 389–97.

Kostov, I. 1968. *Mineralogy.* Trans. ed. P. G. Embrey and J. Phemister, eds. Edinburgh: Oliver & Boyd. 587 pp.

Leake, B. E. 1978. Nomenclature of amphiboles. *Am. Mineral.* 63: 1023–52.

Liebau, F. 1959. Uber die kristallstruktur des pyroxmangits (Mn, Fe, Ca, Mg)SiO_3. *Acta Crystal.* 12: 177–81.

Mandarino, J. A. 1977. Old mineralogic techniques. *Can. Mineral.* 15: 1–2.

Morimoto, N. 1989. Nomenclature of pyroxenes. *Can. Mineral.* 27: 143–56.

Mosely, H. G. J. 1913. The high frequency of the elements. *Phil. Mag.* 26: 1024.

Osborne, E. F. 1942. The system $CaSiO_3$-diopside-anorthite. *Am. J. Sci.* 240: 751–88.

Osborne, E. F., and Tait, D. B. 1952. The system diopside-forsterite-anorthite. *Am. J. Sci.* Bowen Volume: 413–34.

Pauling, L. 1960. *The Nature of the Chemical Bond.* 3rd ed. Ithaca: Cornell University Press. 644 pp.

Povarennykh, A. S. 1972. *Crystal Chemical Classification of Minerals.* New York: Plenum. 766 pp.

Putnis, A., and McConnell, J. D. C. 1980. *Principles of Mineral Behaviour.* New York: Elsevier.

Schairer, J. F., and Bowen, N. L. 1947. The anorthite-leucite-silica system. *Comm. Geol. Finland Bull.* 140: 67–87.

Smith, J. V. 1974. *Feldspar Minerals 1. Crystal Structure and Physical Properties.* Berlin: Springer-Verlag. 627 pp.

Smyth, J. R., and Bish, D. L. 1988. *Crystal Structures and Cation Sites of the Rock-forming Minerals.* London: Allen & Unwin. 388 pp.

Sosman, R. B. 1965. *The Phases of Silica.* New Brunswick: Rutgers University Press. 388 pp.

Streckeisen, A. 1974. Classification and nomenclature of igneous rocks. *Geol. Rundsch.* 63: 773–86.

Taylor, W. H. 1933. Structure of sanidine and other feldspars. *Zeit. Kristal.* 85: 425–42.

Weinschenk, E., and Clark, R. W. 1912. *Petrographic Methods.* New York: McGraw-Hill. 396 pp.

Whittaker, E. J. W., and Muntus, R. 1970. Ionic radii for use in geochemistry. *Geochim. Cosmochim. Acta* 34: 945–56.

Yoder, H. S., and Sahama, Th. G. 1957. Olivine x-ray determinative curve. *Am. Mineral.* 42: 465–81.

Credits

Chapter 1

Figure 1.1: Courtesy of International Business Machines Corporation, Research Division, Yorktown Heights, New York.

Chapter 2

Figures 2.8, 2.9, 2.18, 2.19, 2.20, 2.22, 2.46, and 2.48: From Buerger, M. J. 1956. *Elementary Crystallography* (Revised Printing, 1963). John Wiley & Sons, Inc., New York. Redrawn by permission.

Chapter 4

Table 4.5: From Linus Pauling: *The Nature of the Chemical Bond,* 3rd Edition. Copyright © 1960 by Cornell University. Used by permission of the publisher, Cornell University Press.

Chapter 5

Table 5.6: From Burns, R. G. 1970. *Mineralogical Applications of Crystal Field Theory.* Copyright ©, Cambridge University Press. Reprinted by permission.

Figure 5.19: From *The Phases of Silica* by Robert B. Sosman. Copyright © 1965 by Rutgers, The State University.

Figure 5.35: From Ford, W. E. 1932. *Dana's Textbook of Mineralogy.* 4th ed., 22nd printing. Figs. 387 and 388, p. 161. Copyright © 1966 John Wiley & Sons, Inc. Reprinted by permission.

Figure 5.36: From Ford, W. E. 1932. *Dana's Textbook of Mineralogy.* 4th ed., 22nd printing. Fig. 408, p. 166. Copyright © 1966 John Wiley & Sons, Inc. Reprinted by permission.

Figure 5.38: From Ford, W. E. 1932. *Dana's Textbook of Mineralogy.* 4th ed., 22nd printing. Fig. 422, p. 168; Fig. 444, p. 170; Fig. 391, p. 163; Fig. 445, p. 170; Fig. 438, p. 170; Fig. 430, p. 169; Fig. 398, p. 164; Fig. 458, p. 172. Copyright © 1966 John Wiley & Sons, Inc. Reprinted by permission.

Chapter 6

Figure 6.4: From Harker, R. I., and Tuttle, O. F. 1956. Experimental data on the P_{CO_2}-T curve for the reaction calcite + quartz = wollastonite + carbon dioxide. *American Journal of Science,* v. 254, pp. 239–256. Copyright © 1956 American Journal of Science. Redrawn by permission.

Figure 6.6: (a) From Boyd, F. R., and England, J. L. 1960. The quartz-coesite transition. *Journal of Geophysical Research,* v. 65, pp. 749–756. Copyright © 1960 by the American Geophysical Union. (b) From Holdaway, M. J. 1971. Stability of andalusite and the aluminum silicate phase diagram. *American Journal of Science,* v. 271, pp. 97–131. Copyright © 1971 American Journal of Science. Redrawn by permission.

Figure 6.7: From Osborn, E. F. 1942. The system $CaSiO_3$-diopside-anorthite. *American Journal of Science,* v. 240, pp. 751–788. Copyright © 1942 American Journal of Science. Redrawn by permission.

Figure 6.9: From Schairer, J. F., and Bowen, N. L. 1947. The anorthite-leucite-silica system. *Comm. Geol. Finland Bull.,* no. 140, pp. 67–87. Redrawn by permission.

Figure 6.10: From Bowen, N. L. 1913. The melting phenomena of the plagioclase feldspars. *American Journal of Science,* v. 35, pp. 577–599. Copyright © 1913 American Journal of Science. Redrawn by permission.

Figure 6.13: From Bowen, N. L. 1915. The crystallization of haplobasalt, haplodiorite, and related magma. *American Journal of Science,* 4th series, v. 29, pp. 161–185. Copyright © 1915 American Journal of Science. Redrawn by permission.

Figure 6.14: From Anderson, O. 1915. The system anorthite-forsterite-silica. *American Journal of Science,* 4th series, v. 29, pp. 407–454. Copyright © 1915 American Journal of Science. Redrawn by permission.

Figures 6.16, 6.17, and 6.18: From Garrels, R. M., and Christ, C. L. *Solutions, Minerals, and Equilibria.* Fig. 7.27a, p. 239; Fig. 7.27b, p. 240; and Fig. 7.28b, p. 243. Copyright © 1965. Reprinted by permission of authors.

Chapter 7

Figure 7.22: From Kostov, I. 1968. *Mineralogy.* Translated edition, eds. P. G. Embrey and J. Phemister. Edinburgh: Oliver & Boyd. By permission of Jus Autor Copyright Agency, Sofia, Bulgaria.

Chapter 8

Figures 8.6 and 8.7: From Streckeisen, A. 1974. Classification and nomenclature of igneous rocks. *Geol. Rundschau,* v. 63, pp. 773–786. Redrawn by permission.

Chapter 11

Figure 11.21: Dennen, W. H., and Moore, B. R. 1986. *Geology and Engineering.* © Wm. C. Brown Communications, Inc., Dubuque, Iowa. All rights reserved. Reprinted by permission.

Chapter 12

Figure 12.1: Data from Jenkins, R., and DeVries, J. L. 1969. *Practical X-ray Spectrometry,* 2nd. ed. Philips Technical Library, Springer-Verlag, New York. Copyright 1969. Redrawn by permission of authors.

Figure 12.23: Redrawn with permission from *Geochimica et Cosmochimica Acta,* v. 7. Goldsmith, J. R., Graf, D. L., and Joensuu, O. I. The occurrence of magnesian calcites in nature. pp. 212–230. Copyright 1955 Pergamon Press Ltd., Headington Hill Hall, Oxford OX3 0BW, U.K.

Figure 12.24: From Yoder, H. S., and Sahama, Th. G. 1957. Olivine X-ray determinative curve. *American Mineralogist,* v. 42, pp. 475–491. Modified from data in Table 3. Copyright by the Mineralogical Society of America.

Figures 12.26 and 12.27: Bambauer, H. V., Corlett, M., Eberhard, E., and Viswanathan, K. 1967. Diagrams for the determination of plagioclase using x-ray powder methods. *Schweiz. Mineral. Petrog. Mitt.,* v. 47, 333–349. Redrawn by permission.

Chapter 13

Figure 13.5: From Harris, D. C., and Cabri, L. J. 1991. Nomenclature of platinum-group elements: review and revision. *Canadian Mineralogist,* v. 29, pp. 231–237. Redrawn by permission of authors.

Chapter 14

Figure 14.11: Povarennykh, A. S. 1972. *Crystal Chemical Classification of Minerals.* Plenum: New York. Fig. 90b, p. 260. Redrawn by permission.

Chapter 17

Figure 17.8: From Smyth, J. R., and Bish, D. L., 1988. *Crystal Structures and Cation Sites of the Rock-Forming Minerals.* Allen & Unwin: London. Fig. 9.3, p. 273. Redrawn by permission.

Figure 17.11: Povarennykh, A. S. 1972. *Crystal Chemical Classification of Minerals.* Plenum: New York. Fig. 267b, p. 615. Redrawn by permission.

Chapter 20

Figure 20.3: Povarennykh, A. S. 1972. *Crystal Chemical Classification of Minerals.* Plenum: New York. Fig. 241b, p. 534. Redrawn by permission.

Chapter 21

Figure 21.5: Povarennykh, A. S. 1972. *Crystal Chemical Classification of Minerals.* Plenum: New York. Fig. 106c, p. 291. Redrawn by permission.

Figures 21.7 and 21.8: From Smith, J. V. 1974. *Feldspar Minerals.* v. 1, *Crystal Structure and Physical Properties.* Copyright 1974 Springer-Verlag. Redrawn by permission of author.

Figure 21.9: Taylor, W. H. 1933. The structure of sanidine and other feldspars. *Zeit. Kristallographie,* v. 85, p. 438. Redrawn by permission.

Figure 21.15: From Hassan, I., and Grundy, H. D. 1991. The crystal structure of basic cancrinite. *Canadian Mineralogist,* v. 29, pp. 377–383. Redrawn by permission of authors.

Figure 21.21: Povarennykh, A. S. 1972. *Crystal Chemical Classification of Minerals.* Plenum: New York. Fig. 143, p. 348. Redrawn by permission.

Figure 21.38: Liebau, F. 1959. Uber die kristallstruktur des pyroxmangits (Mn,Fe,Ca,Mg)SiO$_3$. *Acta Crystallographica,* v. 12. Fig. 5, p. 180. Redrawn by permission.

Figure 21.41: From Colville, P. A., Ernst, W. G., and Gilbert, M. C. 1966. Relationships between cell parameters and chemical compositions of monoclinic amphiboles. *American Mineralogist,* v. 51, pp. 1727–1754. Modified from Fig. 9, p. 1739, Copyright by the Mineralogical Society of America.

Figure 21.45: From Cameron, M., and Papike, J. J. 1981. Structural and chemical variations in pyroxenes. *American Mineralogist,* v. 66, pp. 1–50. Modified from Fig. 5, p. 7. Copyright by the Mineralogical Society of America.

Figure 21.46: From Morimoto, N. 1989. Nomenclature of pyroxenes. *Canadian Mineralogist,* v. 27, pp. 143–156. Redrawn by permission of author.

Figure 21.58: Povarennykh, A. S. 1972. *Crystal Chemical Classification of Minerals.* Plenum: New York. Fig. 183, p. 405. Redrawn by permission.

Figures 21.64a and b, 21.65, 21.67, 21.68, 21.69, 21.70, and 21.72: Smyth, J. R., and Bish, D. L. 1988. *Crystal Structures and Cation Sites of the Rock-Forming Minerals.* Allen & Unwin: London. Figs. 4.6a, 4.6b, 4.6c, 4.6d, 4.7a and b, 4.8, and 4.9, pp. 100, 101, 102, 103, 107, 108, 115, and 118. Redrawn by permission.

Mineral Index

Accepted mineral species described in Part 3 are given in **boldface** along with the page where the description can be found. Other accepted names mentioned as variants or related species are given in ordinary type. *Italicized* names include form and chemical variants or traditional group names that are recognized as mineral species. Some commonly used synonyms are also given in italics with reference to the accepted species names.

A

Abukumalite. See Britholite-(Y)
Acanthite, 152, 233, **235,** 248, 376, 378
Achroite, 351
Acmite. See Aegirine
Actinolite, 145, 172, 304, 338, **341,** 392
Adelite, 295
Adularia, 314, 388
Aegirine, 344, 345, 346, 347, 390
Aeschynite, 257
Afghanite, 318
Agate, 310
Aguilarite, 236
Aikinite, 251
Ainalite, 268
Akaganeite, 261, 271
Åkermanite, 306, **353**
Alabandite, 112, 113, 236
Alabaster, 287, 291
Alamosite, 349
Albite, 94, 107, 109, 145, 172, 217, 309, 311, 312, 313, **314,** 315, 316, 382
Alkali feldspar, 78, 87, 129, 137, 304, 311, 312
Allanite-(Ce), **354,** 355, 376, 378, 380, 390
Allanite-(Y), 355
Alleghanyite, 360
Allemontite, 229
Alluaudite, 295
Almandine, 172, 355, 357, 358
Almandine spinel, 266
Alstonite, 279, 281

Altaite, 236, 254
Alum, 287
Alumino-barroisite, 339
Alumino-ferro-hornblende, 339
Alumino-katophorite, 339
Alumino-magnesio-hornblende, 339
Alumino-taramite, 339
Alumino-tschermakite, 339, 342
Alumino-winchite, 339
Alunite, 154, 287, **293,** 380, 386
Amakinite, 270
Amalgam, 224, 228
Amazonite, 314
Amblygonite, 295, **296,** 388
Amesite, 334
Amethyst, 7, 168, 310
Amicite, 319
Ammoniojarosite, 293
Amosite, 333, 340, 341
Amphibole, 123, 136, 161, 215, 304, **336, 337**
Analcime, 175, 304, 308, 316, 319, 321, **323,** 380
Anandite, 330, 332
Anatase, 262, 364
Andalusite, 105, 144, 161, 304, 308, **362,** 363, 354, 388
Andesine, 217, 311, 313, 315
Andorite, 248
Andradite, 175, 357, 358
Anglesite, 287, 288, **289,** 386
Anhydrite, 141, 142, 161, 253, 254, 287, **289,** 384
Ankerite, 274, 281
Annabergite, 238, 248
Annite, 332
Anorthite, 106, 107, 109, 110, 172, 217, 309, 311, 312, 313, **314,** 315
Anorthoclase, 311, 314
Anthophyllite, 145, 304, 308, 336, 338, 339, **340,** 341, 390, 392
Antigorite, 333
Antimonite. See Stibnite
Antimony, 229
Antozonite, 255
Apatite, 137, 143, 152, 163, 190, 295, **297,** 380, 390, 392
Apophylite, 382
Apthitalite, 290

Aquamarine, 352
Aragonite, 85, 94, 102, 142, 147, 164, 165, 274, **278**
Ardealite, 291
Arfvedsonite, 339
Argentite, 233, 235
Argentojarosite, 293
Argentopentlandite, 240
Argyrodite, 248
Arquerite, 228
Arsenic, 161, **229,** 231, 378, 382
Arsenoferrite, 245
Arsenolamprite, 229
Arsenolite, 164, 229, 242, 257
Arsenopyrite, 153, 233, 234, **246,** 378
Asbestos. See Chrysotile
Ashcroftine-(Y), 319
Atacamite, 253
Attapulgite. See Palygorskite
Augite, 123, 139, 162, 304, 308, 344, 345, **346,** 390
Aurichalcite, 281
Aurocupride, 227
Aurostibite, 245
Autunite, 295, 300, 382
Avogadrite, 288, 289
Azurite, 16, 155, 274, **282,** 392

B

Babingtonite, 348
Baddeleyite, 262, 264, 359
Baileychlor, 329
Balas ruby, 266
Ballas, 231
Banalsite, 314
Barite, 143, 152, 154, 161, 253, **287,** 289, 382, 386
Barrerite, 319
Barytes. See Barite
Barytocalcite, 281
Bassanite, 291
Bastite, 345
Bauxite, 270, 272, 380
Bayerite, 272
Bazzite, 352
Beaverite, 293
Beidellite, 325, **326,** 327
Belovite, 298

Subject Index

Molybdates, 7, 300
　　test for, 295
Molybdenum porphyry, 153
Monochromater, crystal, 205
Monoclinic system 39, 178
Monohedron, 45
Morphology, **43**
Mosely, H. G. J., 194
Mössbauer, 192
Mössbauer spectrometry, 189, **192**
Motif, **13**
Motion of atoms, 3
　　rotation, 3
　　translation, 3
　　vibration, 3
Mudstone, 141, 142
Multiple cell, **13,** 19
Multiple structure, 70

N

NAA, 190
Native elements, 7, 70, 71, 224
Native metals, 224
Nesosilicates, 304, **305, 355**
Network structures, 304, **305, 307**
Neutron activation analysis, 188, **190**
Neutrons, 3, 53, 175
　　thermal, 190
Nickeline structure, 233, **238,** 239
Nicol, W., 8
NMR, 191
Noble gases, **58**
Nonconsistent mirror, **14**
Nonequivalent axes, **14**
Nonmetalliferous deposits, **148**
Nonmetals, 224
Normal distribution, 200
Nuclear magnetic resonance, 189, **191**
Nuclear progression, 53
Nucleation, **117**
　　of crystals, 117
　　heterogeneous, 117
　　homogeneous, 117
　　rate of, 118, 133
　　secondary, 118
Nucleus
　　atomic, 191
　　crystal, 117
Nuclide, 190

O

Obsidian, 118
Occurrence of minerals, 220
Ocher, 269
Octet rule, **225**
Oil immersion method, **176**
Oil shale, 142
One-component systems
　　$Al_2(SiO_4)O$, 105
　　H_2O, 104
　　SiO_2, 105
Oölitic habit, 128
Opalescence, **169**
Open form, 44
Open system, **96**
Optic
　　angle, **178**
　　axis, **178**
　　sign, 183
Optical emission spectrography, 189, **198**
Optical indicatrix, **177**

Orbital, 54, 82
Order-disorder, **86,** 191, 217, 312
　　dynamic, **87**
　　static, **87**
Ordinary ray, **180**
Ore, **148**
Ore deposits, 148
　　bauxite, 140, 150, 154, **270,** 271, 272
　　enriched, 155
　　epithermal, **154**
　　hydrothermal, 152
　　hypothermal, **153**
　　magmatic, 150
　　manganese, 143
　　mesothermal, **153**
　　Mississippi Valley-type, 154, 253, 255
　　nickel, 150, 154, 235
　　phosphorus, 141, 143, 295
　　placer, 154, 227, 263
　　porphyry copper, 153, 234
　　porphyry molybdenum, 153, 234
　　sedimentary, 154
　　telethermal, **154**
Orthopyroxene, 344
Orthorhombic system, 39, 178
Oxidation, **59**
Oxides, 7, 70, 139, 216, 257

P

Packing
　　body-centered cubic, 66, **224**
　　cubic close packing, 66, **224,** 225, 253
　　hexagonal close packing, 66, **224,** 225
　　of ions, 165
Packing index, 165
Pair structures, 304, 352
Parallelohedron, 45
Paramagnetism, 56, **170**
Parametral plane, **23**
Parting, **160**
Pauli exclusion principle, **55,** 61
Pauling, L., 62
Peacock ore, 236
Pedion, 45
Pegmatite, **138,** 150
Pelitic rocks, **144**
Penetration twin, **92**
Peraluminous rocks, **144**
Peridotite, 145
Periodic table, 56, 57
　　actinide series, **58**
　　alkali metals, **58**
　　groups, **56**
　　halogens, **58**
　　noble gases, **58**
　　periods, **56**
　　rare earths, **58**
　　transition series, **58**
Peristerite gap, 217, 311
Peritectic, 107, 134
Perthite, **79,** 129, 311, 314
Petrographic microscope, 183, 187, 188
PGE, 229
pH, 110, 111
Phase, **96**
Phase diagrams
　　albite-anorthite, 107, 311
　　albite-orthoclase, 311
　　$Al_2(SiO_4)O$, 105
　　anorthite-diopside-silica, 109
　　$CaCO_3 + SiO_2 = CaSiO_3 + CO_2$, 103
　　calcite-aragonite, 102

　　$Cu-H_2O-O_2$, 112
　　$Cu-H_2O-S-CO_2$, 113
　　diopside-albite-anorthite, 109
　　diopside-anorthite, 106
　　graphite-diamond, 101
　　H_2O, 104
　　leucite-silica, 106
　　$Mn-H_2O-S-CO_2$, 113
　　silica-forsterite-anorthite, 110
　　SiO_2, 105
Phase rule
　　Gibbs, **104**
　　mineralogic, **104**
Phenocrysts, **136**
Phosphates, 6, 7, 295
　　test for, 295
Phosphorescence, **169**
Phosphorite, 141, **143,** 295
Photoelectron, 194
Photon, 194, 198
Phyllite, **144**
Phyllosilicates, 304, **305,** 323
Physical properties, 220, 371, 375
Piezoelectricity, 170
Pinacoid, 45
Pinite, 322
Pisolitic habit, 128
Pitchblende, 264
Placer deposit, 154, 227, 263
Planck's constant, 169, 194
Plane indices, **23**
Plaster of paris, 287, 291
Platinum group elements, 229
Play of colors, **168**
Pleochroism, **169,** 183
Pliny, 7, 276, 357
Pluton, **137**
Plutonic rock, **137**
Point counting, **138**
Point groups, **32**
　　crystallographic, **34**
Points, 231
Polarizability, **68**
Polarization
　　of ions, 60
　　of light, **179**
Polarizer, microscope, 184
Polarizing microscope, 8, 183
Polarizing power, **68**
Polarography, 188, **193**
Polymer, 177
Polymorphism, **84**
　　displacive, **85,** 89
　　reconstructive, **84**
　　thermodynamics of, 100, 101, 102
　　transformations, 100, 144
Polymorphs, high and low, **85**
Polytypism, **86,** 120, **324**
Portland cement, 274, 275
Powder
　　camera, 208
　　diffraction file, 209, 211
　　diffractometry, 208
　　method, 207
　　photography, 208
Precipitates, sedimentary, **140,** 142
Precision, **199**
Primitive cell, **13**
Prism, 45
Probability volume, 54
　　relation to covalent bonding, 65
Prograde metamorphism, **143,** 146
Proper operation, **17**
Protolith, **144**

phosphates, 295
sulfates, 287
tungstates, 295
vanadates, 295
Tetragonal system, 39, 177
Tetrahedron, 47
Texture, **134**
TGA, 187
Theophrastus, 7
Theory of rational indices, **8**
Thermal properties, 170
conduction, 171
expansion, **171**
transmissibility, **171**
Thermal vibration, 68, 86
Thermocouple, 189
Thermodynamics, **96**
first law, **98**
second law, **99**
Thermodynamic systems
closed, **96**
heterogeneous, **96**
homogeneous, **96**
isochemical, **100**
isolated, **96**
one-component, 101, 104
open, **96**
three-component, 108
two-component, 104, 106
Thermoelectricity, 170
Thermogravimetric analysis, **187,** 188
Thermoluminescence, **169,** 170
Thin section, **183,** 185
Three-component diagrams, 75, 76
Three-component systems, 108
anorthite-diopside-SiO₂, 109
diopside-albite-anorthite, 109
silica-forsterite-anorthite, 110
Tie line, **107**
Tiling a plane, 12
Transformation twins, 89
Transition elements, **58,** 77
Translation, **11**
Transmission electron microscope (TEM), 8,
197
Trapezohedron, 46
Triangular diagram, 75, 76
Triboluminescence, **169,** 170
Triclinic system, 39, 178
Trioctahedral structure, **323**
Triple point, 104
Tungstates, 7, 300
test for, 295

Twin, **84**
axis, **92**
composition plane, **89**
plane, **92**
Twinning, 81, 88, 312, 313
kinds of, 88
mechanics of, 89
symmetry of, 92
Twins
Albite, 313
Baveno, 313
Brazil, 308
Carlsbad, 313
contact, **92**
cyclic, **93**
Dauphine, 308
glide, **90**
growth, **89**
Iron cross, 243
lamellar, **93**
Manebach, 313
penetration, **92**
Pericline, 313
polysynthetic, **93**
repeated, **93**
Spinel law, 93
tartan, 314
transformation, 89
Two-circle goniometer, 42
Two-component diagram, 74, 75, 106
Two-component systems, 106
albite-anorthite, 107
diopside-anorthite, 106
leucite-silica, 106
2V, **178,** 183
Type symbol, **24**

U

Ugrandite group, 304, 357
Ultramafic rock, **137**
Umber, 269
Uniaxial indicatrix, **178,** 182
Unit cell, **4,** 13, 124
Unit structures, 304, **305, 355**
Unmixing, **79**
Uranyl radical, 296
Uses of minerals. *See* Mineral group
descriptions
uvw, 22, 25

V

Vacancy, 75
Valence, **58,** 74, 221
Vanadates, 295
test for, 295
Van der Waals bond, 226, 270
Variance, **104**
Variants, 220
Vesicles, **132**
Vibration direction, 182
Vicinal faces, **129**
Viscosity of magmas, 132
Volatile species, 132
Volcaniclastic sediments, **140**
Volcanic rock, **137**
von Laüe, M., 8, 205

W

Wad, **262,** 268
Wallrock alteration, 138
Water
coordinated, 71, 172, 221, 269
structural, 71, 172, 221, 269
Water softening, 319
Wavelength, 54, 167, 204
Wave mechanics, **54**
Weathering, **138,** 154
chemical, 138
mechanical, 138
Werner, A. G., 8
Westphal balance, 165
Wet chemistry, 188
Whewell, W., 21
Wollaston, W. H., 8, 41
Work, 98
Wulff, G. V., 27
Wulff net 27, 28

X

X radiation, 194, 204
X-ray diffraction, 187, 188
X-ray fluorescence spectrometry, 189, **194**
XRD, 187, 188
XRF, 194

Z

Zeolite group, 304, 318
Zone, 25, 45
Zone axis, **25,** 45
Zoning, compositional, 134, 135